Biochemistry of Bacterial Growth

EDITED BY

JOEL MANDELSTAM

Microbiology Unit
Department of Biochemistry
University of Oxford

KENNETH McQUILLEN

Sub-Department of Chemical Microbiology
Department of Biochemistry
University of Cambridge

IAN DAWES

Department of Microbiology
University of Edinburgh

THIRD EDITION

BLACKWELL SCIENTIFIC PUBLICATIONS
OXFORD LONDON EDINBURGH
BOSTON MELBOURNE

© 1968, 1973, 1982 by Blackwell Scientific Publications
Editorial offices:
Osney Mead, Oxford, OX2 0EL
8 John Street, London, WC1N 2ES
9 Forrest Road, Edinburgh, EH1 2QH
52 Beacon Street, Boston, Massachusetts 02108, USA
99 Barry Street, Carlton, Victoria 3053, Australia

First published 1968
Second edition 1973
Reprinted 1976
Third edition 1982

Set by Santype Ltd
Salisbury and printed and
bound in Great Britain by
Clark Constable Ltd., Edinburgh

Distributed in the USA and Canada by
Halsted Press, a Division of John Wiley & Sons Inc,
605 Third Avenue, New York, NY 10016, USA

British Library
Cataloguing in Publication Data

Biochemistry of bacterial growth.—3rd ed.
  1. Bacterial growth   2. Bacteria—Physiology
  3. Biological chemistry
  I. Mandelstam, Joel   II. McQuillen, Kenneth
  III. Dawes, Ian William
  589.9'03'1   QR84
  ISBN 0-632-00323-5
  ISBN 0-632-00596-3 Pbk

Contributors

S. BAUMBERG, *Department of Genetics, University of Leeds*

E. A. DAWES, *Department of Biochemistry, University of Hull*

IAN W. DAWES, *Department of Microbiology, School of Agriculture, University of Edinburgh*

D. KERRIDGE, *Sub-Department of Chemical Microbiology, Department of Biochemistry, University of Cambridge*

P. J. LARGE, *Department of Biochemistry, University of Hull*

KENNETH McQUILLEN, *Sub-Department of Chemical Microbiology, Department of Biochemistry, University of Cambridge*

J. MANDELSTAM, *Microbiology Unit, Department of Biochemistry, University of Oxford*

R. H. PRITCHARD, *Department of Genetics, University of Leicester*

P. E. REYNOLDS, *Sub-Department of Chemical Microbiology, Department of Biochemistry, University of Cambridge*

JOHN G. SCAIFE, *Department of Molecular Biology, University of Edinburgh*

D. W. TEMPEST, *Laboratorium voor Microbiologie, Universiteit van Amsterdam*

MICHAEL J. WARING, *Department of Pharmacology, University of Cambridge*

Contents

Preface to Third Edition

Because it is conventional to reprint prefaces from all previous editions there is little point in reiterating the aims set out in the First Edition (since they remain the same) or in attempting to rebut the opinions of the few reviewers who misunderstood them. We do, however, again express our thanks to colleagues who have made many helpful suggestions and have provided valuable material for illustrations.

Since the second edition was published, knowledge of the biochemistry of bacterial growth has increased much as bacteria themselves increase in a favourable environment. This new edition reflects these changes, in particular, with new presentations of the subjects of Growth, Genetics, and Differentiation. More detailed treatment is given to the division of individual cells and to the replication of chromosomes, while growth of cell populations in continuous cultures has received much more attention. The chapter on Genetics has been re-written to take account of major advances in the biochemistry of DNA and of the application of techniques for genetic manipulation both *in vivo* and *in vitro*. Previously, the treatment of differentiation was restricted to sporulation and germination. The chapter now deals with a much wider range of cellular structures and developmental processes, and the emphasis is on what can be explained in molecular terms. The form of the remaining chapters has been retained but the content has been up-dated where necessary.

In both of the earlier editions a very brief summary was used to introduce a general development of the whole subject in about fifty pages and this was followed by detailed treatment of the same material expanded about tenfold. This presentation seems to have been found useful both by teachers and by students. We have retained it.

JOEL MANDELSTAM
KENNETH McQUILLEN
IAN W. DAWES

Preface to First Edition

This book is not intended to be a comprehensive textbook on the biochemistry of bacteria. It has been written in the belief that the advances in biochemistry in the last ten years provide a basis for a fairly comprehensive description of bacterial life in biochemical terms, and that such a view of the bacterial cell can with advantage be presented to beginners. We also believe that the most recent concepts are as readily intelligible as the older and more basic ideas. For this reason we have, for example, thought it just as easy, and much more interesting, for the student to learn the modern view of replication of the bacterial chromosome before he learns the structural formulae of the nucleotides. Similarly we have presented the 'coding problem' in protein synthesis before introducing the chemical structures of the twenty common amino acids. As far as possible this method of approach has been followed throughout.

We have also attempted to build up from the start a coherent picture. Too often in the teaching of biochemistry the student is taken through one detailed aspect of the subject after another. Only at the end does he have all the information which will allow him to construct some sort of integrated picture. By this time his mind may be so clogged with details that the process of fitting them together is needlessly difficult. Our method of presentation will, we hope, avoid this danger and it has resulted in a book written in three parts. The introduction is a summary of the book in a few pages and it is based upon a very general account of what a bacterial cell does during growth. This is followed in Part I by a somewhat more detailed description of the same material; it presupposes very little knowledge apart from some basic chemistry. If we have been successful in our exposition, the student should, at this stage, have a clear picture—still in very general terms—of a bacterial cell as an integrated biochemical system. The detailed biochemistry will be found in Part II, the third and largest part of the book. We realize that the subject may seem to be so oversimplified by this treatment that the impression is given that everything in bacterial life is now explicable in biochemical terms apart from a few minor gaps. We have tried to avoid this by stressing, particularly in the conclusion, those phenomena which are as yet not reducible to biochemistry, and which are likely to be the growing points of the subject.

Finally, we have avoided the historical approach which, while it may be the most scholarly, is for the reader the dullest. It can, furthermore, reasonably be argued that it usually fails in its object. The significance of early discoveries and controversies is best appreciated by those who already know the subject fairly well, and not by beginners: it is for the latter that the book has been written.

In our attempt to co-ordinate the chapters of this book we have inflicted our views and prejudices upon the contributors to a considerable extent. This was particularly true during the preparation of Part I which has now been written and re-written so many times that it is impossible to attribute individual authorship to the sections. We are grateful to the authors for their tolerance and patience. We are also most appreciative of the willing help and co-operation of everyone at Blackwell Scientific Publications who was concerned with this book, in particular Mr Per Saugman, Mr John Robson and Miss Yvonne Prince.

JOEL MANDELSTAM
KENNETH MCQUILLEN

Introduction

Introduction
Abstract of the Book

This chapter is a highly condensed introduction to the biochemical events that underlie bacterial growth. It is, at the same time, intended to be a summary description of the contents of the rest of the book.

For a model system it will be convenient to choose an unspecified bacterium that can grow in a medium with glucose as the carbon source. Its nitrogen requirements are satisfied by ammonium ions and its sulphur requirements by sulphate. Magnesium and phosphate are essential and it needs trace amounts of other metals (e.g. iron). It can be considered as a 'generalized bacterial cell' and we shall attribute to it a mixture of the properties found in several different kinds of bacteria. It should be regarded as an abstraction in much the same way as the 'average man'. Real bacteria will be considered in the main section of the book and some of the ways in which they differ from the model will become apparent.

The organism is represented schematically in Figure 1. It is rod-shaped and has a rigid outer wall that maintains and supports the membrane that it encloses. The wall is made of a polymer substance, the peptidoglycan, and the membrane contains proteins and lipids. These coats surround the cytoplasm, which consists mainly of polymers: deoxyribonucleic acid (DNA); ribonucleic acid (RNA); proteins and polysaccharides. In terms of dry weight the polymers account for about 90% of the cell (see Figure 2). The remaining 10% of the cell is made up of a large variety of small molecules: amino acids, nucleotides, growth factors, fats. Although these constitute so small a fraction of the cell mass, they are metabolically of great importance.

Not only do the macromolecules make up the bulk of the bacterial cell, they also give it the characteristics that distinguish it from all other types of bacteria. The small molecules, on the other hand, are common to all types of bacteria and, indeed, to other forms of life.

When the organism is in a suitable environment or growth medium, more of all these materials is produced and in due course the cell divides into two daughter cells indistinguishable from one another and from their parent. The subject of this book is a description of the way in which the simple organic and inorganic constituents of the medium are transformed into new cell material with its enormous diversity of molecular species.

Classification of biochemical reactions

The number of chemical reactions involved in growth is unknown but is probably of the order of a thousand. Of these a few may occur spontaneously but the vast majority have to be catalysed by specific

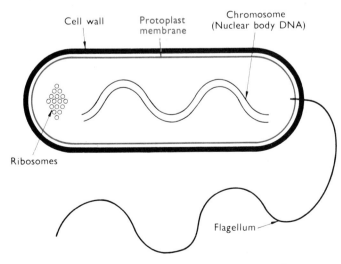

Cell wall Protoplast membrane Chromosome (Nuclear body DNA)

Ribosomes

Flagellum

Figure 1. Diagrammatic representation of a bacterial cell.

3

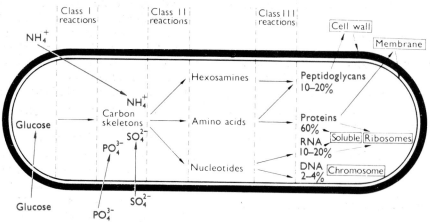

Figure 2. Flow diagram of synthesis of bacterial constituents.

proteins, the *enzymes*. Each of these catalyses a specific reaction such as the addition or removal of water, or hydrogen, or 1-C residues, or amino groups, etc. (see Appendix B, p. 405). For our purposes enzyme reactions can be grouped into three classes.

Class I (degradative reactions)
There is first a complex of enzymes which degrades glucose to smaller aliphatic carbon compounds. This class will be called degradative enzymes. The net process is exergonic, i.e. produces energy. It also results in the supply of carbon skeletons for synthetic reactions.

Class II (biosynthetic reactions)
From these carbon skeletons a further series of enzymes catalyses the formation of the small molecules which are the basic components of the macromolecules. Many of these intermediates (amino acids, nucleotides, hexosamines) contain nitrogen which is derived from the NH_4^+. Some contain sulphur which comes from SO_4^{2-}. At the same time some small molecules (vitamins, co-factors) are synthesized but are not incorporated into macromolecules; rather, they are needed for the proper functioning of the enzymes. The enzymes producing all these substances will be called biosynthetic, Class II. As a group they largely require energy and are therefore endergonic. The energy is produced by the Class I reactions.

Class III (biosynthetic reactions)
A further series of enzymes then converts the basic small molecules into macromolecules. When enough

of these have been synthesized the cell divides. Since the distinctive character of the cell is determined by its macromolecules, much of this book will be concerned with the mechanisms by which these complicated structures are reproduced so exactly.

The genetic information for copying the cell is carried in its DNA which is the 'blueprint' for the whole cell, that is, *all* the information determining what the biochemical machinery shall be, and how it will be put together, is encoded in the DNA. When the cell divides each of the daughter cells must, apart from anything else, receive a complete copy of the 'blueprint'. It is essential that the DNA molecules should be copied correctly at every division because, as in any highly organized system, a random error will almost certainly be damaging. It is only *very* rarely that such an error will be advantageous. We have thus two separate problems to consider. Firstly, how an exact copy of the DNA is made and then, when this has happened, how the information in it is translated into the other types of molecules.

The DNA molecule is a very long polymer made up of four kinds of nucleotide joined through their phosphates. All four contain the sugar *deoxy*ribose and are represented by dA, dG, dC, and dT because of the four different bases (see Figure 3, p. 12 for their chemical structures). The properties of these nucleotides are such that dA and dT have an affinity for each other and so have dG and dC. Thus if we have a chain as follows:

$$—dA—dA—dG—dT—dC—dG—$$

then free nucleotides will be assembled into a

sequence in accordance with their pairing properties, thus:

—dA—dA—dG—dT—dC—dG— Original chain

—dT—dT—dC—dA—dG—dC— Complementary sequence

Cells possess an enzyme (DNA polymerase) that links these nucleotides covalently to give a polymer that will be exactly complementary to the original template. This complementary chain can itself be copied, again in accordance with the pairing properties of the nucleotides, thus giving a strand identical with the original chain:

—dT—dT—dC—dA—dG—dC— Complementary chain

—dA—dA—dG—dT—dC—dG— Identical copy of original chain

The complementary chain can be regarded as the biochemical equivalent of a photographic negative. This explains in an over-simplified way the principle of DNA replication. In fact, the DNA exists as a double-stranded structure that is unravelled during the copying process (see p. 26 and Chapter 5).

So far, then, we have accounted for the formation of a DNA molecule containing all the necessary information for the hundreds of enzymes which will catalyse the three classes of reactions we have described. Now, these reactions are responsible for all the materials that the cell contains and for all the biochemical reactions it can carry out. Our problem is thus reduced essentially to that of understanding how the information in the DNA is translated into that of the enzyme proteins. The information for any particular species of protein is carried in a stretch of DNA which may contain more than 1000 nucleotides and which is known as the *structural gene* for that protein. The enzymes and other proteins consist of chains containing twenty kinds of amino acid. The number of amino acid residues and their order are different for each kind of protein. The problem was to find the way in which the four types of nucleotide in a stretch of DNA specify the 20 types of amino acid in a protein. This was formally analogous to finding out how the two-letter system (dots and dashes) of Morse code could be translated into the ordinary alphabet of 26 letters and it was therefore referred to as the coding problem.

The translation of DNA into protein occurs in a number of steps. First, the informational content of the structural gene is transferred to a strip of RNA known as the *messenger-RNA*. RNA is also a polymer consisting of four kinds of nucleotide, but in this nucleic acid they all contain ribose instead of

deoxyribose and they have U instead of T. They are represented by A, G, C and U. However, pairing can occur between these nucleotides and those of DNA. In the presence of the appropriate enzyme (RNA polymerase) and a DNA template, the four ribonucleotides are polymerized into a complementary copy of the DNA strand:

dA—dA—dG—dT—dC—dG— DNA template

U—U—C—A—G—C— Complementary RNA

Thus DNA fulfills *two* separate template functions. The first, mediated by DNA polymerase, is to serve as a template for its own replication: the second is to act as a template for the production of the complementary messenger-RNA.

The messenger-RNA acts as a template on which amino acid residues are assembled in correct sequence before being linked together. The solution of the coding problem is that the code is triplet, i.e. that a sequence of three nucleotides codes for each amino acid:

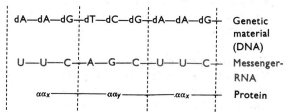

Here the DNA triplet, dA—dA—dG is transcribed into the complementary messenger-RNA triplet, U—U—C which is translated as an amino acid ($\alpha\alpha_x$), etc. Each succeeding triplet causes the insertion of the next appropriate amino acid until the protein is complete. The whole chain might easily contain 300 amino acid residues. The assembly of proteins is, however, more complex than this description suggests and involves the participation of other types of molecules.

The remaining macromolecules to be considered are the polysaccharides and peptidoglycans. The biosynthesis of these is generally simpler than that of the proteins. Some, like glycogen, are polymers containing only one type of sugar residue. Theoretically a single enzyme could string together a chain of such residues to form a polymer. In fact most polysaccharides are more complex, but even so only a few enzymes are required for the synthesis of any one of them. The peptidoglycans are somewhat similar, containing two kinds of amino sugar occurring in regular alternation. They also have short side chains of amino acids but their assembly is achieved by a fairly small number of enzymes.

The number and types of small molecules that a cell can make and degrade are determined by its content of enzymes.

Genetics

So far all the processes we have outlined may be considered as taking place in a single bacterium. They involve the conveyance of information in the DNA to the rest of the cell material. However, information can also be conveyed *inter*-cellularly. In bacteria this can be effected in one of three ways: (a) by a mating process; (b) by transformation—the direct uptake by one cell of free DNA liberated from another cell, by lysis or otherwise; (c) by transduction. Here an infective virus particle during its formation picks up some of the DNA of the host cell and then transfers it to the next bacterial cell it infects.

Growth and the regulation of biosynthesis

We can now summarize the events that take place when some viable cells are placed into growth medium. Some of their enzymes degrade glucose, some synthesize basic molecules and yet others assemble macromolecules including more of all the enzymes. With more of all these catalysts thus available, the same processes will continue but at an accelerated rate, giving yet more enzyme, and more cells. The rate of synthesis is thus proportional to the amount of cell material present and this leads to an exponential rate of growth that continues until something in the environment becomes limiting. When this hap-

pens some types of bacteria simply cease to grow but others form spores which are heat-resistant and can lie in a dormant state for many years. Subsequently, if the environment becomes favourable, these can germinate and begin to grow again. Sporulation and germination are among the most primitive forms of cell differentiation and are consequently of considerable interest.

Returning to the actively growing cell, let us consider its internal economy. It has to produce 20 types of amino acids for its proteins, four types of nucleotides containing deoxyribose for DNA, four more containing ribose for RNA and also a variety of co-factors and lipids. The synthesis of any one of these substances may easily involve ten or more specifically catalysed steps carried out by enzymes of Class II. In addition there is a considerable number of intermediates produced from glucose by the Class I enzymes.

For efficient growth all the basic materials and all the macromolecules derived from them have to be produced in the correct proportions. Under natural conditions, bacteria are often in competition for a limited amount of nutrient. The consequence is that a very efficient regulation mechanism has evolved. Since virtually all metabolic steps are enzymically catalysed, in considering metabolic regulation, we have really to consider regulation of enzymic function. There are two ways in which this can occur. One is by alteration of the *amount* of any particular enzyme, the other is by alteration of the *rate* at which it functions. Both types of regulation are found in the bacterial cell and their combined effect ensures that the cell is geared to get the maximum yield of protoplasm from its environment and to do so in the minimum time.

Part I

Section 1
The Bacterial Cell: Major Structures

Bacterial cells occur in a variety of shapes and sizes depending on the kind of organism and on the way in which it has been grown, but for many purposes it is possible to disregard these variations and to consider the common properties of the 'generalized bacterial cell'. Thus, although bacteria may be spherical or curved or spiral, the majority are rod-shaped and are about 1 μm wide and 2 μm in length (1 μm = 0·001 mm). A single bacterial cell may thus have a volume of 10^{-12} ml and contain $2·5 \times 10^{-13}$ g of dry matter (equivalent to a relative molecular mass, M_r, of $1·5 \times 10^{11}$). But this bacterium is not just an undifferentiated blob of 'protoplasm'. It is a highly organized structure with organelles corresponding in function to many of those found in higher organisms. The hereditary material (DNA) is embedded in the *cytoplasm* which, surrounded by the *cell membrane*, is called the *protoplast*. Outside this lies the *cell wall*.

Cell walls and membranes

The wall is fairly rigid and gives shape and protection to the cell. It amounts to about 10% of the weight of the entire cell. Always there is present in it peptidoglycan, and this seems to be what makes the wall rigid. The peptidoglycans are made of chains of amino sugars, *N*-acetylglucosamine alternating with *N*-acetylmuramic acid. Short peptides are linked to the muramic acid residues and separate chains are joined by these peptides to form the relatively thick structure needed for a wall (Figure 1).

The amino sugar, muramic acid, has not been found in any biological polymers other than the peptidoglycans of the cell walls of bacteria and the closely related blue-green algae. The peptides are also interesting in that they contain unusual amino acids. Besides L-alanine they contain D-alanine and D-glutamic acid, the so-called 'unnatural' isomers which are not present in proteins. Most species also contain diaminopimelic acid which, like muramic acid, is restricted in nature to peptidoglycans.

Other polymers which may occur in cell walls include teichoic acids, lipopolysaccharides and lipoproteins (see Chapter 1).

In species in which the wall is mainly peptidoglycan it is sometimes possible to digest it away with an enzyme called *lysozyme* which occurs in secretions such as tears and sweat and also in white of egg. Enzymic digestion of the wall releases the protoplast, but this is likely to burst unless given some osmotic protection. This is because the concentration of intra-cellular solutes exerts an osmotic pressure equivalent to 5–25 atmospheres. A solution of sucrose (10–20% w/v) will usually prevent this lysis of protoplasts. No matter what the shape of the cell, the naked protoplast on release from the cell wall assumes a spherical shape. It is bounded by a very delicate membrane called variously the *plasma membrane*, the *protoplast membrane*, or the *cytoplasmic membrane*. The structure consists predominantly of protein and lipid, as do all biological membranes, and it has a thickness of about 8 nm—this is dictated by the dimensions of the molecules of which it is composed. The membrane is the main permeability barrier of the cell, since the wall is freely penetrated by most molecules except very large ones. Some substances pass into and out of cells by passive diffusion but many are transported by highly specific systems which require energy and are located in the cell membrane. The name *permease* or sometimes *translocase* is given to such a system. As far as passive diffusion is concerned, smaller molecules and substances of high lipid solubility penetrate membranes more easily than do larger molecules and polar substances. For instance, C_4 sugars and some C_5 sugars may pass freely but other C_5 and all C_6 sugars (including glucose) may fail to penetrate by passive diffusion except very slowly. Bacterial cells are also generally impermeable to small cations and to inorganic phosphate ions. These non-penetrating substances have to be actively transported into the organisms.

Proteins and nucleic acids

Three classes of polymers are found in all bacteria. These are the proteins, and the two kinds of nucleic acids. Viruses, on the other hand, contain protein and *either* DNA *or* RNA.

Structure of proteins

The proteins perform various functions, some catalytic and some structural, and it is this class of sub-

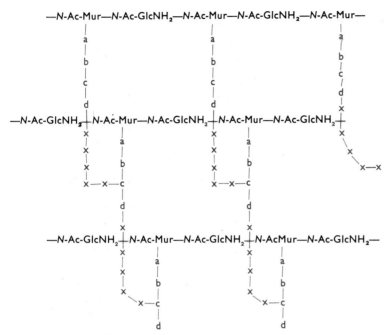

Figure 1. Structure of cell wall peptidoglycan.

Alternating amino sugars (*N*-acetyl-glucosamine and *N*-acetyl-muramic acid) form a backbone with a peptide (a-b-c-d) attached to a muramic acid residue. These peptides may be linked through further amino acids (x-x-x-)

stance which determines the identity of an organism, that is, its membership of a particular species. About half of the dry weight of a bacterium consists of protein but there may be more than 1000 different species of protein in any one cell. The size of these polymers ranges in relative molecular mass from a few thousands to millions but each has a definite, precise composition.

The constituents of proteins are the amino acids, having the general formula:

$$NH_2CH.COOH \quad \text{see Figure 2}$$
$$|$$
$$R$$

They occur with various frequencies and in various orders in the proteins. A simple protein may consist of a single polypeptide chain formed by the condensation of amino acids:

$$NH_2CH.COOH + NH_2CH.COOH + NH_2CH.COOH + \text{etc.}$$
$$\quad | \qquad\qquad | \qquad\qquad |$$
$$\quad R_a \qquad\qquad R_b \qquad\qquad R_c$$

$$\longrightarrow \ NH_2CH.CONH.CH.CONH.CH.CO— \ \text{etc.}$$
$$\qquad\qquad | \qquad\qquad | \qquad\qquad |$$
$$\qquad\qquad R_a \qquad\qquad R_b \qquad\qquad R_c$$

The sequence of amino acid residues in a protein or polypeptide is called its *primary structure*. Physical studies have shown that the chains do not exist in a straight, extended form but often coil into a spiral with $3\frac{2}{3}$ amino acid residues per turn, that is, eleven for each three turns of the helix. This α-helical structure is called the *secondary structure* and it is stabilized principally by so-called hydrogen bonds

(H-bonds) between $\diagdown NH$ and $O{=}C\diagup$ residues in the chain.

Many proteins consist not of one but of several polypeptide chains joined together, and this is frequently achieved through —S—S— bridges. The amino acid cysteine possesses an —SH group in its side-chain and residues of cysteine in two chains may couple together thus:

$$NH_2CH.CONH.CH.CONH.CH.CONH—$$
$$\quad | \qquad\qquad | \qquad\qquad |$$
$$\quad R_a \qquad\qquad R_b \qquad\qquad CH_2$$
$$\qquad\qquad\qquad\qquad\qquad\qquad |$$
$$\qquad\qquad\qquad\qquad\qquad\qquad SH$$

$$+$$

$$SH$$
$$|$$
$$CH_2$$
$$|$$
$$NH_2CH.CONH.CH.CONH.CH.CONH—$$
$$\quad | \qquad\qquad |$$
$$\quad R_p \qquad\qquad R_q$$

$$\longrightarrow$$

$$NH_2CH.CONH.CH.CONH.CH.CONH—$$
$$\quad | \qquad\qquad | \qquad\qquad |$$
$$\quad R_a \qquad\qquad R_b \qquad\qquad CH_2$$
$$\qquad\qquad\qquad\qquad\qquad\qquad |$$
$$\qquad\qquad\qquad\qquad\qquad\qquad S$$
$$\qquad\qquad\qquad\qquad\qquad\qquad |$$
$$\qquad\qquad\qquad\qquad\qquad\qquad S$$
$$\qquad\qquad\qquad\qquad\qquad\qquad |$$
$$\qquad\qquad\qquad\qquad\qquad\qquad CH_2$$
$$\qquad\qquad\qquad\qquad\qquad\qquad |$$
$$NH_2CH.CONH.CH.CONH.CH.CONH.—$$
$$\quad | \qquad\qquad |$$
$$\quad R_p \qquad\qquad R_q$$

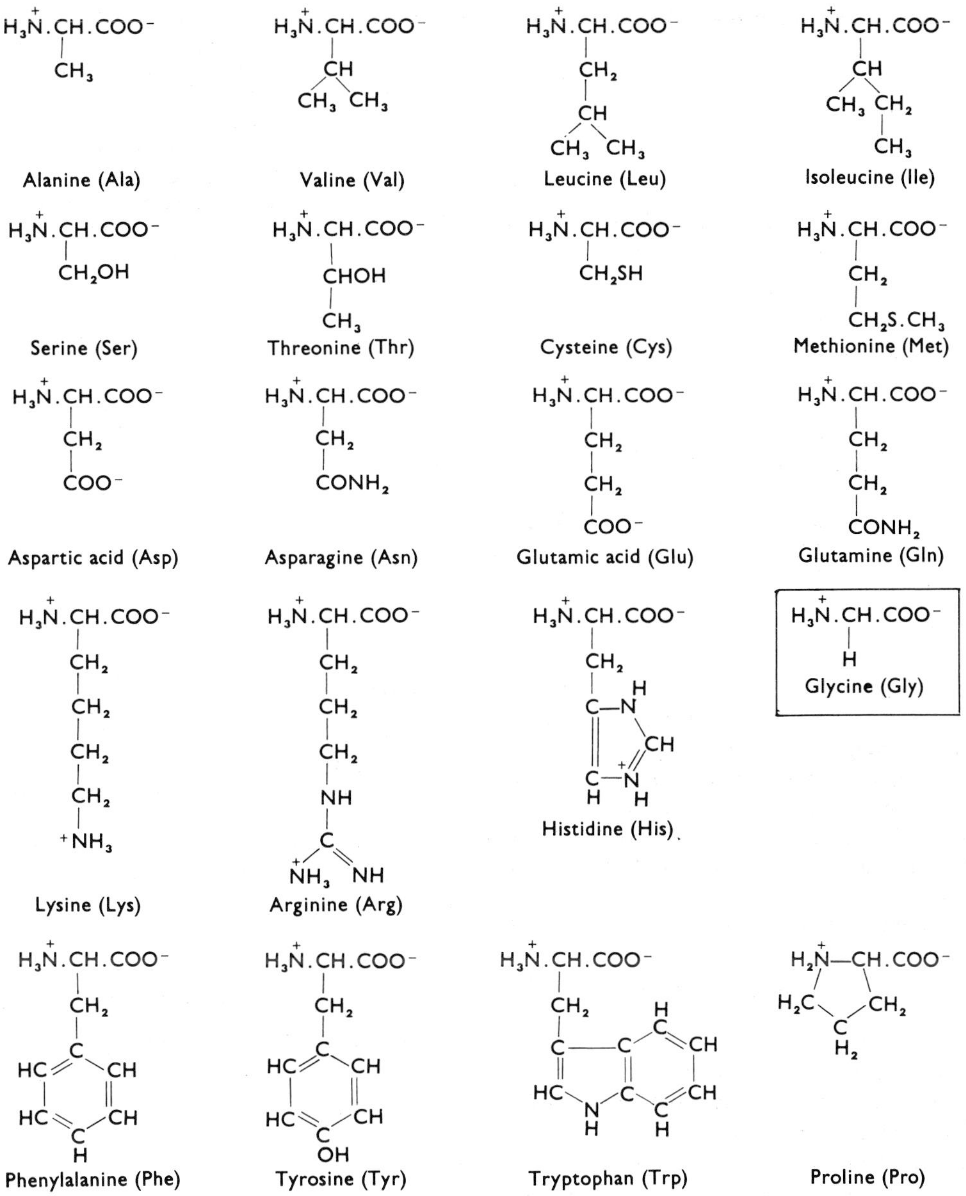

Figure 2. The 20 amino acids commonly occurring in proteins, together with their abbreviations. All the amino acids have the L-configuration when they occur in proteins except glycine which is not optically active.

Two cysteine residues in the same chain can likewise be oxidized to form a cystine residue and thus form an *intra*-chain bridge:

—NH.CH.CONH— .... —NH.CH.CONH—
     | |
     CH_2 CH_2
     | |

Other bonds can be formed between the R-groups of amino acid residues, e.g. electrostatic bonds between the $—COO^-$ of an acidic amino acid and the $—^+NH_3$ group of a basic one. These side-group interactions cause the polypeptide chain to contort into a three-dimensional *tertiary structure* characteristic of the particular sequence of amino acid residues. It will be apparent that secondary and tertiary structures are interrelated.

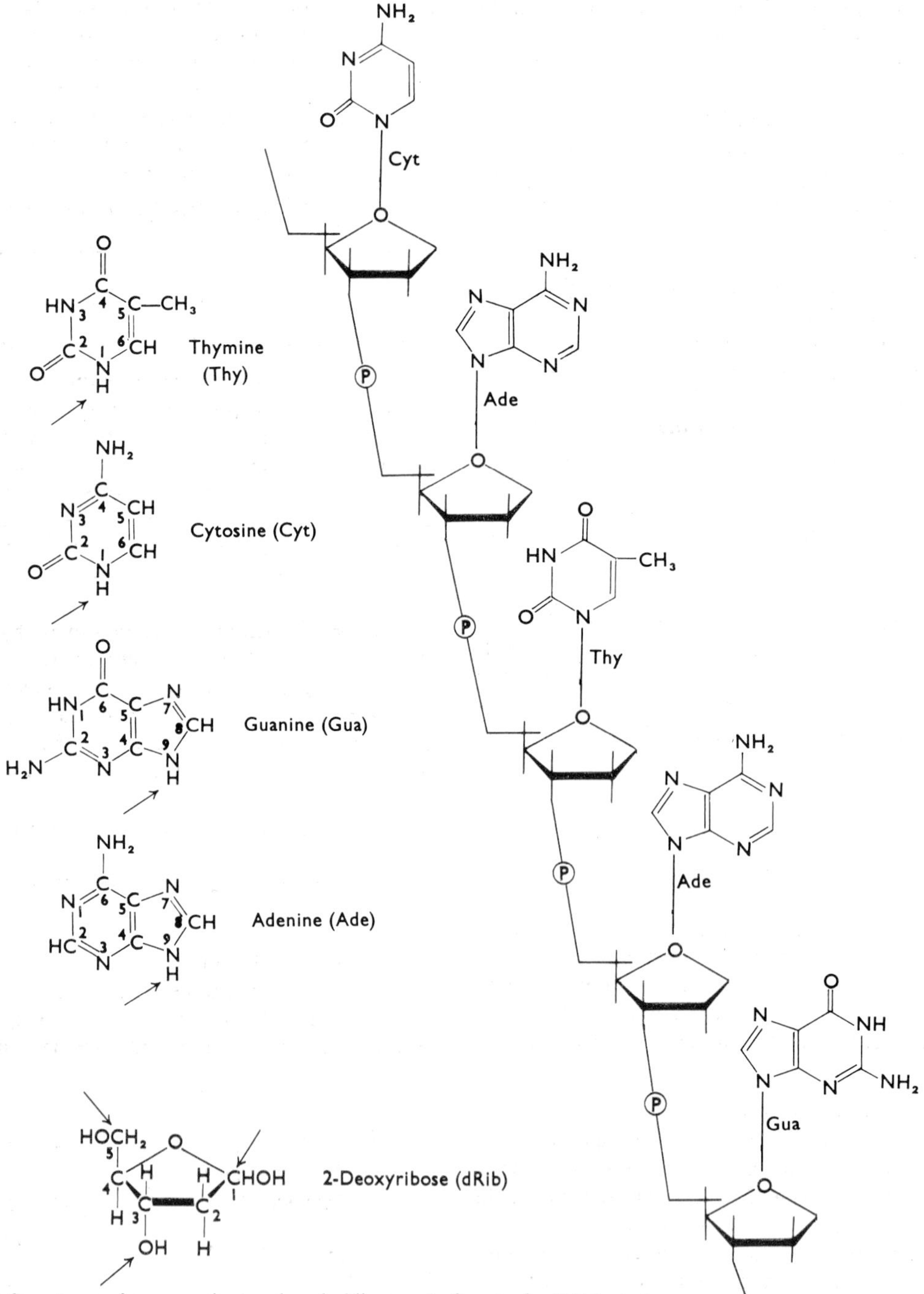

Figure 3. Structures of some purines and pyrimidines and of part of a DNA chain.

    The pyrimidines are thymine and cytosine and these are linked via N-1 to the C-1 of deoxyribose. The purines are guanine and adenine and these are linked via N-9 to C-1 of deoxyribose. The sugars themselves are joined by phosphodiester bridges between C-3 and C-5.

Sometimes a biologically functional protein consists of an aggregate of units held together by non-covalent bonds. This arrangement, which is somewhat analogous to that found in a crystal, is called the *quaternary structure* of the protein.

Thus, all proteins which have been investigated consist of specific sequences of amino acids joined together as polypeptide chains which may in turn be linked to each other. The whole molecule has a precise three-dimensional conformation which is necessary for its biological activity. Non-protein components may be associated with the polypeptide chains, as, for instance, the haem group in cytochromes. Many proteins have catalytic functions and, as has been said, over 1000 different enzymes may occur in a single bacterial cell. The *catalytic site* or *active centre* of an enzyme is the region where the substrate molecule combines, and it is usually very highly specific (e.g. being able to distinguish between glucose and galactose or between aspartic acid and glutamic acid). This is due to the topological arrangement of specific amino acid residues forming the active centre of a specific enzyme.

Other proteins have structural functions in cell walls and membranes, and the machinery for propelling motile bacteria consists wholly of protein. The motion is brought about by *flagella*, and each flagellum is composed of molecules of the protein flagellin which have a relative molecular mass of about 40 000. A solution of this protein can be caused to aggregate artificially into structures apparently identical with natural flagella. These are 2–5 μm in length, 12–30 nm in diameter, and consist of the molecules of flagellin (*c.* 5 nm diameter spheres) arranged like strings of beads with up to a dozen strands spiralling round each other.

The formation of these organs of motility seems to be spontaneous if the appropriate protein units are available at the appropriate site in the cell.

Deoxyribonucleic acid (DNA)

Nucleic acids are polymers of nucleotides, and each nucleotide consists of a nitrogenous base, a sugar, and phosphate. They are linked through the phosphates. In DNA the sugar is 2-deoxyribose (dRib) and the bases are the purines, adenine (Ade) and guanine (Gua), and the pyrimidines, cytosine (Cyt) and thymine (Thy). Thus the DNA chain can be represented:

```
—dRib—Ⓟ—dRib—Ⓟ—dRib—Ⓟ—dRib—Ⓟ—dRib—Ⓟ—
  |        |        |        |        |
 Cyt      Ade      Thy      Ade      Gua
```

The structures of the bases, the sugar, and of this polydeoxyribonucleotide are shown in Figure 3.

Usually the DNA is found to occur as a double strand with the nucleotide sequence in one strand related to that in the other. The molecule forms a double helix in which an adenine nucleotide in one chain is side-by-side with a thymine nucleotide in the other, and similarly for guanine and cytosine:

```
—dRib—Ⓟ—dRib—Ⓟ—dRib—Ⓟ—dRib—Ⓟ—dRib—Ⓟ—
  |        |        |        |        |
 Cyt      Ade      Thy      Ade      Gua
  :        :        :        :        :
 Gua      Thy      Ade      Thy      Cyt
  |        |        |        |        |
—dRib—Ⓟ—dRib—Ⓟ—dRib—Ⓟ—dRib—Ⓟ—dRib—Ⓟ—
```

Such a structure is said to have complementary base-pairing, and this is possible because H-bonds (indicated by the dashes in the diagram above) can form specifically between adenine and thymine and between guanine and cytosine, thus stabilizing the double helix. It follows that an analysis of DNA usually shows that the amounts of adenine and thymine are equal and so are those of guanine and cytosine.

Electron microscopy and autoradiography suggest that the total DNA from a bacterium is in the form of a single giant 'molecule'. It is composed of a double helix about 1000 μm in length—this from a cell only about 2 μm in length. It has a relative molecular mass (M_r) of about $2{\cdot}5 \times 10^9$ and is built up from some 8 million nucleotides. This hereditary material, representing a linear array of some thousands of genes, is stuffed into the cytoplasm but is never surrounded by a nuclear membrane as occurs in the nuclei of higher organisms.

Ribonucleic acid (RNA)

Ribonucleic acid resembles DNA in some but not in other respects. The sugar is D-ribose (Rib) and the bases are adenine, guanine and cytosine as in DNA,

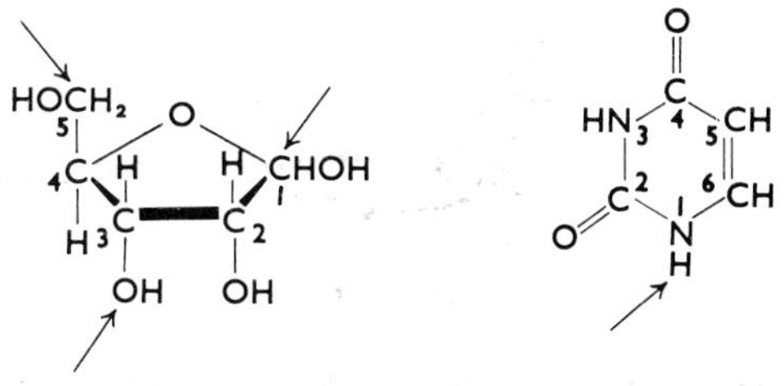

Figure 4. Structures of D-ribose and uracil.

The other purines and pyrimidine which occur in RNA are shown in Figure 3. The links between residues are the same in RNA as in DNA.

but the second pyrimidine is uracil (Ura) instead of thymine (5-methyl-uracil) which occurs in DNA (*cf.* Figures 3 and 4). Thus part of an RNA chain may be represented:

Three kinds of RNA (messenger, transfer and ribosomal; see p. 27) occur in all cells but none of these is double-stranded. The transfer-RNA's (tRNA) have M_r values of about 25 000, but the other kinds may range up to more than a million. In a cell the ribosomal-RNA is combined with protein to form ribonucleoprotein particles known as *ribosomes*. These are found in the cytoplasm and many of them may be strung like beads on a strand of messenger-RNA (mRNA) to form a polyribosome (Figure 5).

Figure 5. Diagrammatic representation of a polyribosome.

Polyribosomes consist of several ribosomes attached to the same strand of mRNA. The ribosomes are ribonucleoprotein particles consisting of a smaller and a larger part. The polyribosomes are known also as polysomes and ergosomes.

Section 2
Growth: Cells and Populations

An organism, whether it be a bacterium or not, has an identity specified by its genetic make-up and largely functioning by virtue of the enzymes it contains. For the individual and the species to survive and be perpetuated, the specification and the functional systems must be maintained and reproduced. Growth can be considered at the level of the individual cell or at that of the population, although the latter depends, of course, on the former. When one organism becomes two, everything has been duplicated—the amount of cell wall, of membrane, of ribosomes, the DNA, the RNA, the proteins, the other cytoplasmic constituents such as ions, amino acids, intermediates of metabolism, and so on. Essentially a bacterial cell increases to double its size and then divides into two. This process is repeated again and again.

Exponential growth

The time taken for the number of organisms in a culture to double can be less than one hour for bacteria. In this time one cell increases to two, two increase to four—and 1000 enzyme molecules increase to 2000. Thus, in a suitable environment a population will tend to increase exponentially for a time—1, 2, 4, 8, 16, 32, ... It is convenient to regard exponential growth as the 'normal' or 'ideal' state and then to consider changes from this state. The 'generalized bacterial cell' if put into its simple glucose/ammonium/salts medium and aerated at 37°C will grow exponentially—or rather the culture will. The ions and glucose will pass from the medium through the membrane into the cytoplasm, some by passive diffusion and some carried by specific transport systems or permeases. Within the cell will be an array of enzymes, some free in solution, some organized on particulate structures, but each coded for by a specific structural gene. Absence or alteration of the gene will result in absence or alteration of the corresponding enzyme.

The Class I enzymes (see Section 3) will degrade the carbon substrate, glucose, to other substances including carbon dioxide and compounds whose chemical energy can be used by the organism. Class II and III enzymes will then catalyse the synthesis of the smaller compounds and the macro-molecular polymers which are characteristic of the organism. Some of these substances will be organized in the form of organelles or other structures, and when everything has been duplicated, the cell will divide. If the environment remains essentially the same, the process will be repeated. A small population in a large volume of growth medium will be in this state for some time. The passage of nutrients from the environment into the cell and the passage of end-products out of it will not materially alter the composition of the medium and under these conditions the composition of the bacteria will be as given in Figure 2, p. 4.

If exponential growth continued and if the mean generation time were 60 min, the population would increase in 48 h from 1 to 281,474,976,710,656 and in 96 h would reach about 10^{29}. This is equivalent to a mass of $10^{29} \times 2 \cdot 5 \times 10^{-13}$ g or $2 \cdot 5 \times 10^{13}$ kg. Manifestly this would require an impossibly large volume of medium.

Growth curves

The *growth curve* of a culture of bacteria can be plotted as growth versus time. The abscissa may be numbers of organisms or mass or N-content or some other index of growth. It is often possible to use a turbidimeter or spectrophotometer to determine the optical density of a suspension of bacteria and by means of a calibration curve to convert this reading to a measure of growth. Figure 1 shows the curves obtained by plotting growth against time in various ways for a culture which, after a lag of 4 h, begins to grow exponentially with a mean generation time of 1 h in a medium containing 4 g of glucose per litre. It will be seen that while growth is exponential it is impractical to use a linear plot and that a more informative straight line is obtained in a semi-log plot.

Limitation of growth

In a batch culture which is not treated in this way the growth eventually stops when a constituent of the medium is all consumed. For instance, 1000 ml of the synthetic medium with 0·4% glucose will yield

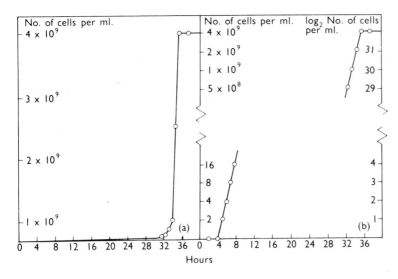

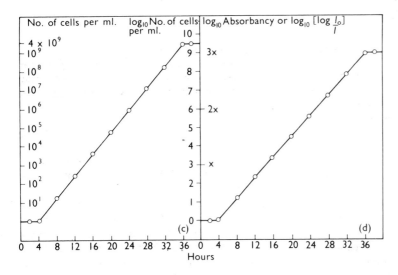

Figure 1. Growth curves.

A culture of bacteria (one organism per ml) after a lag of about 4 hours grows exponentially with a doubling time of one hour in a medium containing 4 mg of glucose per ml. This supports the development of 1 mg dry weight of cells—about 4×10^9 since each weighs 0.25×10^{-10} mg. (a) Linear plot of numbers versus time. (b) and (c) Semi-log plots (with exponential ordinate). These give a straight line but the scale has to be chosen appropriately or interrupted as in (b). Note that log can be to base 10 or to base 2. (d) Turbidity (absorbancy at some suitable wavelength, 450 nm, 600 nm, or 650 nm) can be used in place of numbers of organisms. So can dry weight, *N*-content, etc.

about 1 g dry weight of bacteria (about 4×10^{12} organisms or 4×10^9 per ml). The bacteria will have grown exponentially with a mean generation time of one hour until the glucose was exhausted. Had a lower concentration of glucose been used, the yield of organisms would have been proportionately less. With more sugar, a higher yield might have been obtained but at some concentration of glucose another component in the medium would be limiting, e.g. nitrogen or magnesium or sulphur. Moreover, it is possible that the oxygen supply might restrict growth or that the pH value of the solution might, as a result of the formation of acid products of metabolism, fall to a value at which growth no longer occurred. When growth is limited by exhaus-

tion of the carbon source, it is frequently possible to cause it to resume more or less immediately by further addition of the substrate—even after the culture has been in the *stationary phase* for many hours. This can be exploited practically since it is often useful to grow an overnight culture of bacteria under such conditions and to add more substrate the following morning. This soon results in a culture which is growing exponentially (Figure 2).

The constitution of a bacterial cell is determined to a considerable extent by the medium in which it is growing. A 'rich medium' is one which provides many substances such as amino acids, purines and pyrimidines, and bacterial growth factors (vitamins), which would otherwise have to be synthesized. This

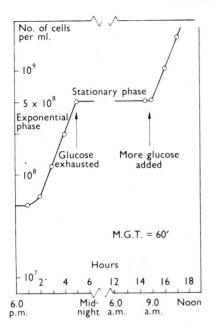

Figure 2. Growth limitation by restriction of the carbon substrate.

A culture grown with limiting glucose reaches its maximum density in a few hours and then remains in the stationary phase overnight. Addition of more glucose in the morning leads rapidly to an exponentially growing culture.

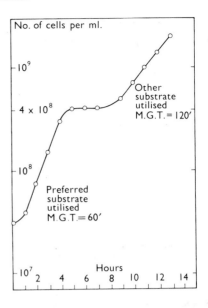

Figure 3. Biphasic growth curve.

In the presence of two carbon substrates one may be used preferentially and then, after a lag, the other may be used. The first usually gives a faster growth rate than the second.

kind of environment tends to cause cells to grow more rapidly and to be larger than does a poor medium. Merely altering the carbon source from one substance to another may provoke changes in mean generation time and size of cells. Sometimes, one substrate (e.g. glucose) is preferred to another (e.g. galactose) and organisms will grow exponentially in the presence of both, using only one until it is exhausted and then, after a short lag, will use the other but with a different generation time and yielding cells of different size and composition (Figure 3).

Continuous cultivation

It is possible to keep a culture in a state of exponential growth by diluting it continuously. The underlying principle can be understood by considering a batch culture in which, as before, growth is limited by the concentration of glucose. Suppose that the density of the culture is measured and that just as it reaches the limit imposed by the glucose restriction, half of the culture is removed and replaced by fresh medium. The culture will grow up again and the procedure can then be repeated indefinitely (Figure

4). If the amount removed each time is now reduced to a smaller fraction of the total volume, the fluctuations in cell density will be correspondingly diminished. Continuing this process we reach the point at which the flow of medium into, and out of, the vessel is continuous, the oscillations are completely damped, and the density of the culture remains fixed at a value determined by the concentration of glucose. If desired, the flow rate can be reduced to give a culture which, though growing more slowly, is still in a state of exponential growth. This is the principle of the chemostat (p. 114, Chapter 2).

Clearly, the medium must not be added so quickly that it outstrips growth capacity. If it is, the density of the culture will fall until, ultimately, all the cells have been washed out of the culture flask.

Effect of temperature on growth

Temperature also affects the rate at which cultures grow, but probably has less effect on composition than has the nature of the growth medium. Many species grow best between 30°C and 37°C but, although some tolerate much higher or lower temperatures, only a few are truly *thermophilic* or

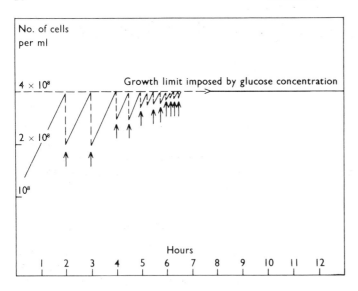

Figure 4. Maintenance of a culture in the exponential phase of growth.

A culture is grown with limiting glucose until it just reaches its maximum density. Portions are removed and replaced with fresh medium (arrows). As the volumes added and removed are made progressively smaller the culture density approaches a fixed value determined by the concentration of glucose.

psychrophilic in the sense of growing better at high or low temperatures. Most of the reactions occurring in living organisms are chemical reactions and even when catalysed by enzymes have a Q_{10} of about 2, i.e. the rate doubles for each 10°C rise in temperature. However, the catalysts are themselves heat-labile so that the optimal temperature for growth results from a balance between the increased rates of reactions and the increased rate of thermal inactivation of the enzymes.

Growth of individual organisms

So far we have been considering the growth of populations of bacteria and it is, of course, much easier to do this than to study the growth of individual cells. Indeed, it is only comparatively recently that techniques other than straightforward microscopy have been evolved which can be applied to single organisms or used to produce collections of cells all in exactly the same state. Normally, a culture of bacteria will contain organisms in every possible state ranging from those which have just divided to those which are just about to do so. Biochemical investigations using such material will, therefore, give the average results for a very heterogeneous population. In only a very few instances is it possible to assay the amount of a component such as an enzyme in a single cell, but microelectrophoresis can give information about the electrical properties of the surface layers of individuals, and autoradiography can be used to demonstrate the amount and location of radioactive atoms in individual bacteria. Because of the smallness of bacteria, the scope of these methods is limited.

Complementary to these approaches are those in which a whole population of organisms is brought into synchronous growth so that all divide simultaneously and pass through each stage of growth at the same time. It is then possible to use methods applicable to large numbers. It has, for instance, been established that the synthesis of DNA occurs during practically the whole of the cell division cycle and that RNA and protein are also made continuously. The sequence of changes on moving from one growth medium to another, and hence from one generation time to another, can also be investigated. The immediate consequence of a *shift-down* to a poorer growth medium is a reduction in the rate of synthesis of the ribonucleoprotein particles called ribosomes so that the content per cell falls. This results in a slower rate of protein synthesis and hence of growth. A *shift-up* to a richer medium leads to rapid ribosome formation, faster protein synthesis and an increased rate of growth. Changes in growth rate also affect the sizes of cells and the way in which they replicate their chromosomes. These effects are dealt with at greater length in Chapter 2.

Section 3
Class I Reactions: Supply of Carbon Skeletons

As we have seen, a bacterial cell, in order to grow, must be supplied with sources of carbon and of energy in addition to nitrogen and essential inorganic ions. Often the source of carbon serves also as the energy source, and this holds true for our 'generalized bacterial cell' which can utilize glucose for this dual purpose. The present chapter outlines the ways in which glucose is metabolized to furnish the basic components essential for synthesis of the macromolecules necessary for cellular growth. These reactions, which are referred to collectively as *intermediary metabolism*, comprise a series of enzymic degradations and syntheses carried out in a stepwise manner. In essence, the individual reactions are relatively simple, involving the addition or removal of hydrogen, water, ammonia, carbon dioxide or phosphate.

The macromolecules of the cell include proteins, nucleic acids (RNA and DNA) and the peptidoglycan of the cell wall; the essential units for their biosynthesis may be summarized as follows:

    Proteins: 20 amino acids
    RNA: 4 ribonucleotides
    DNA: 4 deoxyribonucleotides
    Peptidoglycan: hexosamines + amino acids.

These component units must be synthesized before polymerization can take place. The glucose molecule, containing six carbon atoms, is subjected to stepwise degradation by the action of various enzymes resulting in the formation of smaller molecules of five, four, three or two carbon atoms. These carbon skeletons are then utilized, either directly or after further modification, as building blocks for the synthesis of macromolecules. We may thus distinguish the *catabolic* or degradative enzymes (Class I) which break down glucose and the *anabolic* or biosynthetic (Class II) enzymes which effect the conversion of either glucose itself or its degradation products to compounds such as amino sugars (hexosamines), amino acids and nucleotides for the subsequent syntheses of peptidoglycan, proteins and nucleic acids. The overall reactions catalysed by Class I enzymes additionally yield energy to the cell and are therefore said to be *exergonic*, while Class II enzymes catalyse processes which require the expenditure of energy and are referred to as

endergonic. Class I enzymes thus furnish not only the carbon skeletons necessary for the synthesis of the basic components of macromolecules, but also the energy required for the transformations to occur under the influence of Class II enzymes.

Breakdown of glucose

The oxidation of glucose to carbon dioxide and water involves a considerable number of steps. At certain of these, energy is produced and stored in a

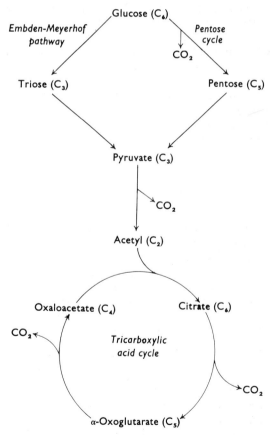

Figure 1. Class I reactions: degradation of glucose.

The breakdown of glucose to CO_2 via intermediates which are also precursors of other small molecules (see Figures 1–4, pp. 21–23). The Class I reactions also produce utilizable energy in the form of ATP.

chemical form as the compound adenosine tri-phosphate (ATP). ATP is used for cellular work of all kinds including endergonic biosynthetic reactions, mechanical work as in flagellar movement, and the osmotic work required for the maintenance of intracellular concentrations of metabolites.

Figure 1 illustrates two ways in which glucose may be degraded in bacteria; one of these results in the conversion of the C_6 sugar into two C_3 units, while the other permits the release of carbon dioxide with the formation of a pentose (C_5) sugar which, after further transformations, can also yield a C_3 compound. The first of these metabolic sequences is known as the *Embden-Meyerhof pathway*. It involves conversion of glucose to an isomeric C_6 sugar, fructose, followed by splitting to two C_3 units or trioses. The trioses are then oxidized and by further reactions yield pyruvate, so that two pyruvate molecules are derived from each molecule of glucose.

The other metabolic pathway differs in that removal of two hydrogen atoms from the glucose molecule first occurs and the resulting C_6 compound is converted to a pentose with the elimination of CO_2. The pentose can then yield pyruvate (C_3) as a result of further reactions. Thus the key C_3 compound in both these degradative sequences is pyruvic acid, a keto acid which serves as the starting point for entry to a cyclic process known as the *citric acid* or *tricarboxylic acid cycle* (Krebs cycle) which is the major aerobic mechanism for cellular oxidation and energy release. In this process pyruvic acid is first oxidized to an acetyl (C_2) unit (as will become apparent later, it is actually in combination with an organic co-factor molecule) with release of carbon dioxide, and the acetyl unit then combines with a C_4 organic acid, oxaloacetate, to form a C_6 compound, citrate. By a sequence of some nine reactions involving oxidations, the removal and addition of H_2O, and the loss of two molecules of CO_2, citrate is converted to a C_5 compound, α-oxo-glutarate, and subsequently to oxaloacetate, the C_4 acid which initiated the cycle. The net effect of one turn of the tricarboxylic acid cycle is thus the complete oxidation of a C_2 unit to CO_2, and H_2O; simultaneously, the oxidation processes make energy available to the cell.

Section 4
Class II Reactions: Biosynthesis of Small Molecules

The various degradative (Class I) enzymes we have so far discussed convert glucose to smaller aliphatic units, as summarized in Figure 1, p. 19. The complete oxidation to CO_2 and water of some C_2 units in the tricarboxylic acid cycle yields energy to the bacterial cell, some of which will be required for the conversion of C_6, C_5, C_4, C_3 and other C_2 units by Class II enzymes to the basic components necessary for the biosynthesis of macromolecules. It will be noticed that the tricarboxylic acid cycle, in addition to furnishing energy, also yields C_5 and C_4 units for biosynthesis. The relationship between these various units and the formation of amino acids, the building blocks of proteins, is illustrated below.

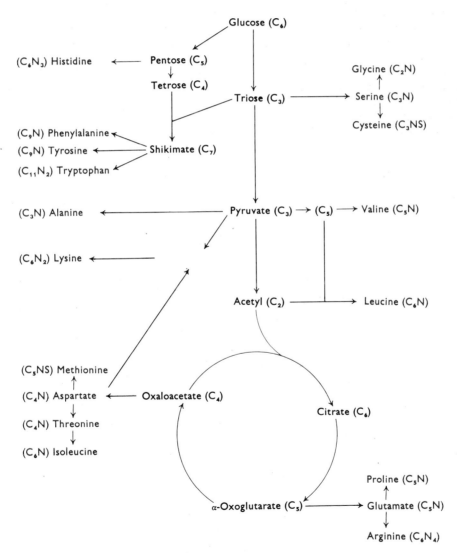

Figure 1. Class II reactions: synthesis of amino acids. The carbon is supplied by glucose, the nitrogen by NH_4^+, and the sulphur by SO_4^{2-} (diaminopimelate is represented by DAP).

Amino acids

One of the interesting features of amino acid biosynthesis is the way in which various amino acids fall into well defined 'families' sharing common pathways of synthesis. This, perhaps, is not so surprising when we consider that such similarities of chemical structure as the possession of an aromatic nucleus, or a branched carbon skeleton exist among particular amino acids. Thus the biosynthesis of the three amino acids which possess an aromatic nucleus, namely tryptophan, phenylalanine and tyrosine, involves a triose and a number of reactions common to all three, which are discussed in detail in a later chapter. Similarly, the amino acids leucine and valine, which possess a branched carbon skeleton, have pyruvate as a precursor and share a common sequence of reactions for part of their individual biosynthetic pathways.

It will be apparent that an essential feature of amino acid biosynthesis from the degradation products of glucose must be the introduction of nitrogen in the form of amino groups. This is usually achieved by the addition of ammonia to a keto acid in the process known as *amination*. The simplest examples of this are to be seen in the conversion of pyruvate to alanine and α-oxoglutarate to glutamate:

$$Pyruvate + NH_3 \longrightarrow Alanine$$

$$\alpha\text{-Oxoglutarate} + NH_3 \longrightarrow Glutamate$$

Aspartate, a very important amino acid in several biosynthetic pathways, may be formed in some bacteria by the addition of ammonia to the unsaturated dicarboxylic acid fumarate:

$$Fumarate + NH_3 \longrightarrow Aspartate$$

An amino group may also be added to a keto acid by transfer from an amino acid, a process termed *transamination*: this would, of course, produce the keto acid corresponding to the original amino acid:

$$NH_2CH.COOH + O{=}C.COOH \rightleftharpoons$$
$$\underset{R_a}{|} \qquad\qquad \underset{R_b}{|}$$

$$NH_2CH.COOH + O{=}C.COOH$$
$$\underset{R_b}{|} \qquad\qquad \underset{R_a}{|}$$

Aspartate is important in the biosynthesis of five other amino acids and Figure 1 shows that three distinct pathways are involved leading to (a) threonine and isoleucine, (b) the sulphur-containing amino acid methionine and (c) diaminopimelate and lysine.

By such biosynthetic pathways twenty different amino acids are synthesized from degradation products of glucose and are then available for assembly into proteins by Class III enzymes (Chapter 6). Certain amino acids, such as glutamate, alanine, diaminopimelate, and glycine, are required also for the synthesis of cell wall peptidoglycan where they occur in combination with amino sugars.

Hexosamines

We have already noted that the peptidoglycan of the bacterial cell wall contains various amino acids and amino sugars. The amino sugars are C_6 units, i.e. hexosamines, and their synthesis requires the introduction of an amino group into a hexose. The reaction involved is a transamination between the amide of glutamate (glutamine) and a ketohexose, e.g.

$$Ketohexose + Glutamine \longrightarrow Hexosamine + Glutamate$$

The hexosamines found in peptidoglycans include glucosamine and muramic acid and both have their amino groups acetylated as shown in Figure 2 which outlines their biosynthesis.

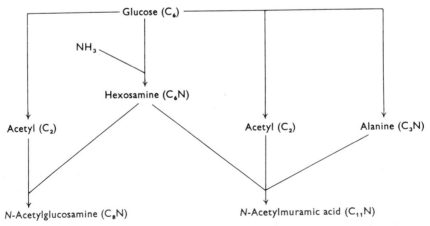

Figure 2. Class II reactions: synthesis of hexosamines.
   The acetylated hexosamines are constituents of the peptidoglycans which are essential components of bacterial cell walls.

Nucleotides

It remains to be seen how the basic components of the nucleic acids are produced. These are the purine and pyrimidine ribonucleotides and deoxyribonucleotides (see Figures 3 and 4, pp. 12 and 13 for their chemical structures). The deoxy-derivatives are formed from the ribonucleotides by reduction of the ribose to deoxyribose. The ribose itself is formed from glucose by loss of a carbon atom as CO_2. The purine ring can then be built up by stepwise addition to ribose phosphate. The carbon atoms come variously from CO_2 and the amino acids, glycine and serine; the nitrogen atoms are supplied by glycine, aspartate and glutamine. Pyrimidine carbon atoms come from CO_2 and aspartate while the nitrogen atoms are derived from aspartate and glutamine or ammonia. The pathways for the biosynthesis of these nucleotides are indicated in Figures 3 and 4.

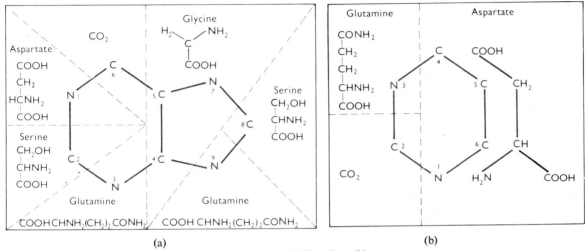

(a)   (b)

Figure 3. Origins of the atoms of purine rings (a) and pyrimidine rings (b).

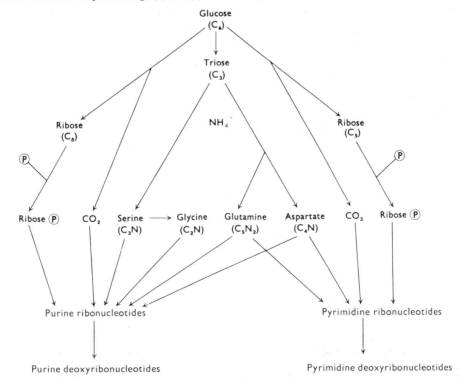

Figure 4. Class II reactions: synthesis of nucleotides.

Sections 5 and 6
Class III Reactions: Synthesis of Macromolecules

Much of a bacterial cell consists of substances of high molecular mass which are always formed by condensation of large numbers of small units. These may all be the same so that a *homopolymer* results, or they may be of several kinds as in *heteropolymers*. Examples of the latter are proteins which are made from 20 different kinds of amino acids, and nucleic acids in which four different nucleotides occur. Polysaccharides, on the other hand, may be of either class of polymer. However, an even more important distinction is between those polymers in which regular repetition of units or groups of units occurs, and those in which there is no such repetition (Figure 1).

Polysaccharides and peptidoglycans are repetitive polymers whereas proteins and nucleic acids are not. This has profound consequences on the nature of the substances and on the mechanisms by which they are made. Firstly, all the molecules of any one protein are identical to each other in composition, in sequence of units, and in molecular mass: the same is true of nucleic acids. Polysaccharides, how-

ever, may be polydisperse, i.e. a range of molecules may exist all having essentially the same repeating units but differing in size so that the term 'molecular mass' can only refer to a mean value. Secondly, the repetitive polymers can be made by successive action of a small number of enzymes whereas proteins and nucleic acids cannot—it would require an astronomically large number. For example, a polysaccharide composed of four different sugars could be put together by four different enzymes as follows:

Enz.
1: A + B $\longrightarrow$ A—B
2: A—B + C $\longrightarrow$ A—B—C
3: A—B—C + D $\longrightarrow$ A—B—C—D
4: A—B—C—D + A—B—C—D + etc. $\longrightarrow$

$$A—B—C—D$$
$$|$$
$$A—B—C—D$$
$$|$$

etc.

But in order to make a polynucleotide (nucleic acid) in which the four nucleotide units occur in a definite,

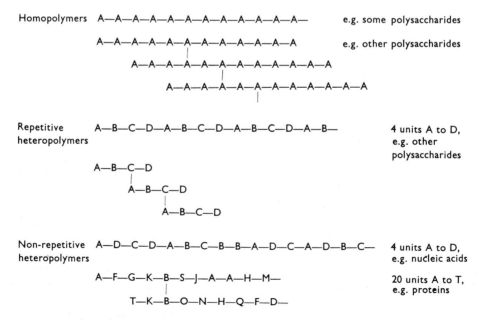

Figure 1. Types of polymer found in bacteria.

specified, but non-repetitive, sequence, a large number of enzymes would be required. For instance, the sequence shown in Figure 1 would be made as follows:

Enz.

1: A + D $\longrightarrow$ A—D
2: A—D + C $\longrightarrow$ A—D—C
3: A—D—C + D $\longrightarrow$ A—D—C—D
4: A—D—C—D + A $\longrightarrow$ A—D—C—D—A
5: A—D—C—D—A + B $\longrightarrow$ A—D—C—D—A—B
etc.

Moreover, hundreds or thousands of different nucleic acids exist in a bacterium, and there are about as many species of proteins. In these, any of the 20 amino acids may be neighbour to itself or to any of the other 19, so 400 (20^2) pairs are possible—and hence 400 enzymes would be necessary to make the specific unions. But even this would not be enough since a sequence ...K—B might be followed in some places by S and in others by O as represented in Figure 1:

A—F—G—K—B + S $\longrightarrow$ A—F—G—K—B—S

but

T—K—B + O $\longrightarrow$ T—K—B—O

Clearly the existence of the necessary enzymes for the reactions:

B + S $\longrightarrow$ B—S and B + O $\longrightarrow$ B—O

would not ensure that the right amino acid was added in each instance—there would be uncertainty unless different enzymes catalysed each of the reactions:

...G—K—B + S $\longrightarrow$ ...G—K—B—S

and

T—K—B + O $\longrightarrow$ T—K—B—O

It will be appreciated that enormously more than 20^2 enzymes would be necessary to cope with all the possibilities without ambiguity. And because each enzyme is itself a specific protein, the absurdity of this as a mechanism is apparent.

Instead, it was postulated and is now proven, that some kind of template must specify the sequence of the units in the non-repetitive heteropolymers. If the units are lined up in the correct order, there is little difficulty in envisaging a mechanism for joining them together involving only a small number of enzymes.

Whether the polymeric product is to be DNA, or RNA or protein, the template always turns out to be a nucleic acid. How this occurs will be described after examples of repetitive polymer formation are mentioned.

REPETITIVE POLYMERS

Polysaccharides

The monomeric units of polysaccharides are sugars, amino sugars and sugar acids, and the polymers have various functions and occur in various parts of bacterial cells. For instance, glycogen, a homopolymer of glucose, is often found as a storage material in the cytoplasm and some organisms such as pneumococci, surround their cell walls with a capsule made of a specific heteropolymer. In general, the mechanism of synthesis of these polysaccharides seems to be similar. The carbohydrate units are added stepwise to an existing chain or *primer* and a specific enzyme catalyses this reaction which frequently involves the transfer of the unit from a 'carrier' nucleoside diphosphate such as UDP (uridine diphosphate). Thus glycogen may be formed as follows:

Glc-Glc-Glc-Glc + UDP-Glc $\longrightarrow$

Glc-Glc-Glc-Glc-Glc + UDP etc.

The same enzyme then catalyses successive additions. Other nucleoside diphosphates may also act as carriers.

Peptidoglycans

Bacterial cell walls contain peptidoglycan (see Figure 1, p. 10) which has a repetitive backbone of *N*-acetyl amino sugars and short peptides attached to it:

—N-Ac-GlcNH$_2$—N-Ac-Mur—N-Ac-GlcNH$_2$—N-Ac-Mur—
| |
Peptide   Peptide

It is made stepwise by a series of enzymic reactions and again UDP derivatives of the carbohydrate units participate (Figure 2).

Thus even this relatively complicated peptidoglycan structure is made by a handful of enzymes catalysing

$$N\text{-Ac-GlcNH}_2 \; \textcircled{P} + \text{UTP} \longrightarrow \text{UDP-}N\text{-Ac-GlcNH}_2 + \text{PP}_i$$

$$\text{UDP-}N\text{-Ac-Mur} + \text{L-ala} \xrightarrow{\text{ATP}} \text{UDP-}N\text{-Ac-Mur-L-ala}$$

$$\text{UDP-}N\text{-Ac-Mur-L-ala} + \text{D-glu} \xrightarrow{\text{ATP}} \text{UDP-}N\text{-Ac-Mur-L-ala-D-glu}$$

$$\text{UDP-}N\text{-Ac-Mur-L-ala-D-glu} + \text{DAP} \xrightarrow{\text{ATP}} \text{UDP-}N\text{-Ac-Mur-L-ala-D-glu-DAP}$$

$$\text{UDP-}N\text{-Ac-Mur-L-ala-D-glu-DAP} + \text{D-ala-D-ala} \xrightarrow{\text{ATP}} \text{UDP-}N\text{-Ac-Mur-L-ala-D-glu-DAP-D-ala-D-ala}$$

i.e. $\text{UDP-}N\text{-Ac-Mur} + \text{amino acids} \xrightarrow[\text{ATP as energy source}]{\text{Stepwise addition}} \text{UDP-}N\text{-Ac-Mur}$
$$\qquad\qquad\qquad\qquad\qquad\qquad\qquad\qquad\qquad\qquad\qquad\qquad\qquad | $$
$$\qquad\qquad\qquad\qquad\qquad\qquad\qquad\qquad\qquad\qquad\qquad\qquad \text{Peptide}$$

$$\text{UDP-}N\text{-Ac-GlcNH}_2 + \text{UDP-}N\text{-Ac-Mur} + \text{existing peptidoglycan} \longrightarrow \text{More peptidoglycan} + \text{UDP}$$
$$\qquad\qquad\qquad\qquad\qquad | $$
$$\qquad\qquad\qquad\qquad \text{Peptide}$$

Figure 2. Biosynthesis of bacterial cell wall peptidoglycan.

The subunits are (a) *N*-acetyl glucosamine and (b) *N*-acetyl muramic acid to which amino acids have been added stepwise to form a short peptide. These subunits are then polymerized to give peptidoglycan which has a backbone of alternating amino sugars (see Figure 1, p. 10).

successive reactions and producing an unambiguous product of precise composition.

NON-REPETITIVE POLYMERS
DNA

Perhaps the reason why DNA occurs as a double helix with one strand complementary to the other (Figure 3) is because it is difficult or impossible to produce directly an exact copy of a single strand. No chemical method for this can readily be envisaged.

$$\begin{array}{ccccc} -\text{dC}- & \text{dA}- & \text{dT}- & \text{dA}- & \text{dG}- \\ | & | & | & | & | \\ -\text{dG}- & \text{dT}- & \text{dA}- & \text{dT}- & \text{dC}- \end{array}$$

Figure 3. Complementary strands of DNA.

dA, dG, dC and dT are the deoxyribonucleotides of adenine, guanine, cytosine and thymine. In DNA which is double-stranded dA is always paired with dT and dG is paired with dC.

However, the ability of adenine to pair with thymine, and of guanine and cytosine to interact similarly, forms the basis of the way in which a *double* strand can be replicated (Figure 4). Essentially, a com-

$$\begin{array}{ccccc} -\text{dG}- & \text{dT}- & \text{dA}- & \text{dT}- & \text{dC}- \quad \text{New} \\ -\text{dC}- & \text{dA}- & \text{dT}- & \text{dA}- & \text{dG}- \\ | & | & | & | & | \qquad \text{Original} \\ -\text{dG}- & \text{dT}- & \text{dA}- & \text{dT}- & \text{dC}- \\ -\text{dC}- & \text{dA}- & \text{dT}- & \text{dA}- & \text{dG}- \quad \text{New} \end{array}$$

Figure 4. Replication of double-stranded DNA.

The original double-stranded DNA is base-paired (dA with dT and dG with dC). Each original strand acts as template for the formation of its complement.

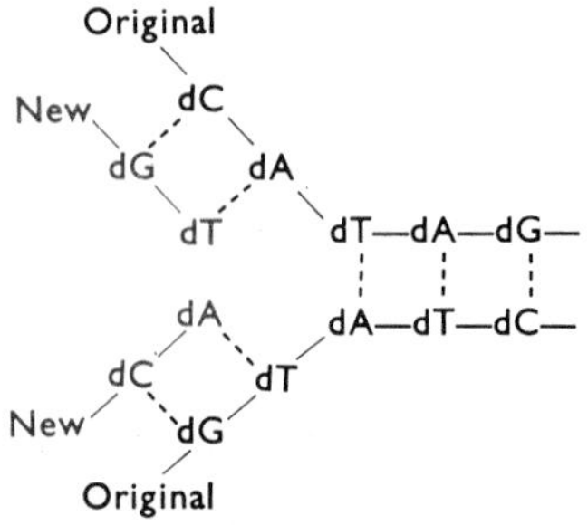

plementary copy is made of each of the original strands, and division of the resulting four strands into two pairs each consisting of one old and one new results in replication of the original pair. The synthesis begins at one end by separation of the two strands and proceeds sequentially to the other end of the original DNA chains (Figure 5).

Figure 5. Replication of DNA.

Synthesis begins at one end by separation of the two strands of the original double helix and proceeds sequentially to the other end. Each double-stranded product contains one old and one new chain.

The precursors which form the units to be polymerized are not the nucleotides themselves (e.g. Ade-dRib-$\textcircled{P}$ or dAMP) but the pyrophosphate derivatives of these, the deoxyribonucleoside triphosphates (Figure 6). Condensation occurs with the elimination of inorganic pyrophosphate and the production of polynucleotides (DNA) (Figure 7). An enzyme has been prepared and purified from bacteria which will catalyse a reaction of this kind

Ade-dRib—(P)—(P)—(P)   Cyt-dRib—(P)—(P)—(P)
dATP   dCTP

Gua-dRib—(P)—(P)—(P)   Thy-dRib—(P)—(P)—(P)
dGTP   dTTP

Figure 6. The deoxyribonucleoside triphosphates.

The combination of base and sugar is called a nucleoside; the monophosphate of this is a nucleotide—here a deoxyribonucleotide.

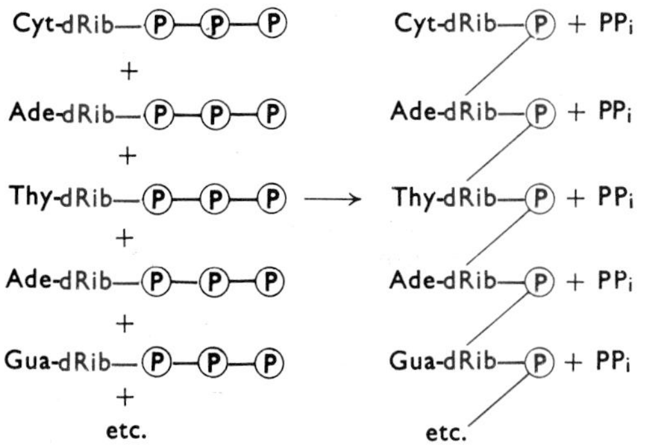

Cyt-dRib—(P)—(P)—(P)   Cyt-dRib—(P) + PP$_i$
      +

Ade-dRib—(P)—(P)—(P)   Ade-dRib—(P) + PP$_i$
      +

Thy-dRib—(P)—(P)—(P) ⟶ Thy-dRib—(P) + PP$_i$
      +

Ade-dRib—(P)—(P)—(P)   Ade-dRib—(P) + PP$_i$
      +

Gua-dRib—(P)—(P)—(P)   Gua-dRib—(P) + PP$_i$
      +
    etc.   etc.

Figure 7. Formation of DNA.

Deoxyribonucleoside triphosphates condense with elimination of inorganic pyrophosphate to form polynucleotide (DNA).

in vitro. All four deoxyribonucleoside triphosphates must be present, as must some DNA to act as a template for copying. The reaction can be formulated:

$$dATP + dGTP + dCTP + dTTP \xrightarrow[\text{DNA polymerase}]{\text{DNA template}}$$

More DNA of same kind + PP$_i$

In this way precise copies of the genes can be made.

RNA

The genetic material, DNA, has not only to be replicated exactly so that the species can be perpetuated but it must also be transcribed into functional RNA molecules—the ribosomal, amino acid transfer, and messenger RNA's mentioned earlier. In this process only one strand of the DNA is used as a template to specify the nucleotide sequence of the product but the mechanism is similar to that of DNA synthesis—the ribonucleoside triphosphates ATP, GTP, CTP and UTP (Figure 8) are needed, as is template DNA and a specific enzyme. It should be noted that it is ribonucleotides rather than

Ade-Rib—(P)—(P)—(P)   Cyt-Rib—(P)—(P)—(P)
ATP   CTP

Gua-Rib—(P)—(P)—(P)   Ura-Rib—(P)—(P)—(P)
GTP   UTP

Figure 8. The ribonucleoside triphosphates. Compare with Figure 6.

deoxyribonucleotides which are being polymerized and that one of the pyrimidines is uracil in place of thymine. However, uracil like thymine can form H-bonds to adenine so that a similar pairing mechanism operates (Figure 9).

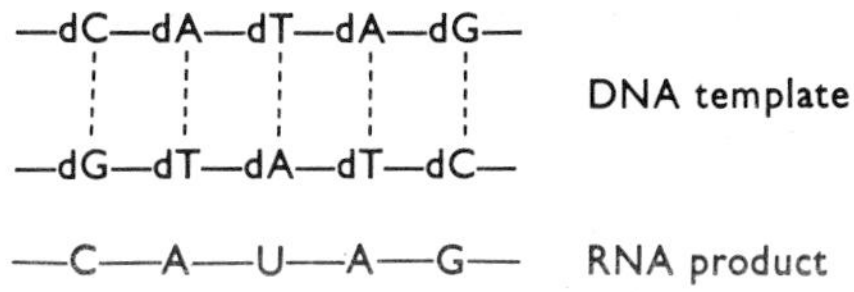

—dC—dA—dT—dA—dG—
  | | | | | DNA template
—dG—dT—dA—dT—dC—

——C——A——U——A——G——   RNA product

Figure 9. Transcription of DNA into RNA.

Only one strand of DNA is transcribed into the complementary RNA strand, the operation involving base-pairing between dA and U, dG and C, dC and G, and between dT and A.

As in DNA synthesis, the triphosphates are condensed to polynucleotides (RNA) with elimination of inorganic pyrophosphate so the reaction can be formulated:

$$ATP + GTP + CTP + UTP \xrightarrow[\text{RNA polymerase}]{\text{DNA template}}$$

RNA complementary to one strand of DNA + PP$_i$

Proteins

In the making of proteins two main problems have to be solved—how the amino acids are joined together and how they are selected in the right sequence. The formation of a peptide bond between two amino acids involves condensation and elimination of water and is a reaction which is catalysed by a peptidase:

$$H_2N.CH.COOH + H_2N.CH.COOH \rightleftharpoons$$
$$\qquad | \qquad\qquad\qquad | $$
$$\qquad R_a \qquad\qquad\qquad R_b$$

$$H_2N.CH.CONH.CH.COOH + H_2O$$
$$\qquad\quad | \qquad\qquad | $$
$$\qquad\quad R_a \qquad\qquad R_b$$

However, the equilibria of such reactions are strongly in favour of the right-to-left hydrolysis. In order to facilitate the forward reaction the amino acid has first to be 'activated' and this occurs by its reaction with ATP in a manner somewhat similar to that in which glucose is activated. In this instance, however, inorganic pyrophosphate is eliminated and an AMP-derivative of the amino acid is formed:

$$H_2N.CH.COOH + ℗-Rib \longrightarrow H_2N.CH.CO \sim ℗-Rib$$

with R (below first), ℗ Ade, ℗ and R, Ade + PP_i

$(\sim ℗$ indicates a high-energy bond to a phosphate residue.)

If the amino acid or the amino acyl residue is represented by $\alpha\alpha$, the reaction may be expressed as follows:

$$\alpha\alpha + ATP \longrightarrow \alpha\alpha \sim AMP + PP_i$$

For each of the 20 amino acids listed in Figure 2 (p. 11) there is a specific amino acid-activating enzyme which catalyses the formation of the amino acyl-AMP derivative. These could condense with the elimination of AMP to form peptides but this would not provide a mechanism for specifying the sequence in which the amino acid residues were joined together and, as has been seen, this sequence is all-important in making a specific protein.

It is abundantly clear that the ultimate control over specification in an organism resides in its genetic make-up, i.e. its DNA. It is believed that every protein which a cell is capable of making is represented by a gene, i.e. a piece of DNA. This *structural gene* specifies the amino acid sequence of the polypeptide(s) that makes up the protein.

The next problem, therefore, is how one linear polymer can specify another when the former (DNA) is composed of four species of nucleotide unit and the latter (protein) contains 20 species of amino acids. A one-for-one correspondence is clearly impossible and it is apparent that since there are only 16 (4 × 4) ways of combining any two of the four nucleotides (AA, AG, AC, AT, GA, GG, GC, etc.) the code cannot be a doublet one but must be a *triplet code* (or something more complex). There are 64 (4 × 4 × 4) possible triplet sequences which can be made (AAA, AAG, AAC, AAU, GAA, GAG, etc.) but this still does not explain *how* a trinucleotide sequence in a DNA can account for the positioning of an amino acid residue in a polypeptide.

In fact, the DNA does not participate directly in the process of protein synthesis but an active structural gene is transcribed into a messenger-RNA which does take part. There is a specific mRNA for each kind of protein which is being made, and the nucleotide sequence of the mRNA is directly related to that of the DNA template on which it was made so that it carries as much information in its sequence as does the original gene. However, there are no grounds for believing that there can be any kind of chemical 'recognition' between an amino acid and a trinucleotide sequence. The only recognitions so far established are the highly specific enzyme-substrate interactions (i.e. a specific protein recognizing a small molecule) and the polynucleotide/polynucleotide interaction based on H-bond formation between guanine and cytosine and between adenine and either thymine or uracil, i.e. base-pairing.

Together these provide a way in which a triplet code can operate. Each activated amino acid attached to AMP (e.g. $\alpha\alpha_x \sim AMP$) is transferred to a specific RNA molecule, a transfer-RNA (tRNA). This reaction is brought about by its activating enzyme which is able to recognize both the appropriate amino acid ($\alpha\alpha_x$) and the corresponding t_xRNA. The latter, by virtue of a specific trinucleotide sequence in its make-up, is able to recognize a complementary sequence in an mRNA and thus deliver the amino acid to its appropriate place so that a defined polypeptide sequence can result:

$$\alpha\alpha_x \sim AMP + t_xRNA \longrightarrow \alpha\alpha_x\text{-}t_xRNA + AMP$$

$$\alpha\alpha_x\text{-}t_xRNA + \alpha\alpha_y\text{-}t_yRNA + etc.$$

$$\xrightarrow{\text{Directed by mRNA}} \text{Specific polypeptide} + tRNA's$$

However, the mRNA's do not occur floating free in the cytoplasm of cells; they have ribosomes associated with them and before messenger-RNA was discovered or even postulated it was known that proteins were synthesized on ribosomes. These ribonucleoprotein particles have linear dimensions of 200–250 nm, have a sedimentation coefficient of 70 S in the ultracentrifuge, and consist of a smaller (30 S) and a larger (50 S) component. Each of these subunits is approximately 60% RNA and 40% protein and it is believed that they orient the mRNA and the tRNA into appropriate juxtaposition, the 30 S ribosomes having an affinity for mRNA and both for tRNA (Figure 10).

A complex is formed with the composite 70 S ribosome attached at the beginning of the message (mRNA) and t_1RNA bearing the first amino acid

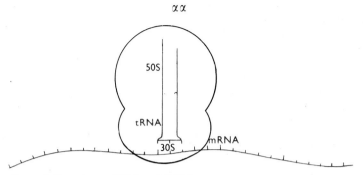

Figure 10. Interactions of ribosomes, mRNA and tRNA.

The 70 *S* ribonucleoprotein particles (ribosomes) are made up of a smaller unit (30 *S*) which has affinity for mRNA and a larger unit (50 *S*) which has affinity for tRNA. The latter has an amino acid ($\alpha\alpha$) attached to it and is thought to base-pair with a complementary triplet in the mRNA as indicated. Each amino acid transfer-RNA would have a specific complementary trinucleotide sequence.

bound both to the ribosome and to the mRNA—a triplet of nucleotides in t_1RNA being complementary to a triplet in mRNA. This first amino acid will be *N*-terminal in the polypeptide specified by this particular mRNA, the next triplet of which will select its appropriate $\alpha\alpha_2$-t_2RNA and line it up on the ribosome. The two amino acids condense with elimination of the tRNA. The process is then repeated with $\alpha\alpha_3$-t_3RNA. And so on.

The mRNA and ribosome move relative to one another so that the ribosome traverses the strand from one end to the other, 'reading' the message as it goes and with the nascent polypeptide chain increasing in length until completed. Before one ribosome has moved far along an mRNA another

may attach at the beginning and so on, so that *polyribosomes* (*polysomes, ergosomes*) are formed (Figure 11).

The amino acids have to be linked enzymically and a ribosomal protein has the function of forming the peptide bond between successive amino acids. A soluble protein is needed to release the completed product from the last tRNA and from the ribosome. The precise sequence of the amino acids has been dictated by the nucleotide sequence of the mRNA and only a relatively small number of enzymes have been required.

Subsequently several polypeptide chains may be linked together to form a native protein molecule or this may consist of a single chain.

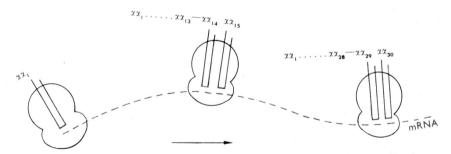

Figure 11. Polyribosomes during the course of protein synthesis.

A ribosome becomes attached to a strand of mRNA at the beginning of its message. The appropriate tRNA associates with it (Figure 10) and delivers the first amino acid ($\alpha\alpha_1$). The ribosome progresses along the mRNA with the nascent polypeptide increasing in length. Further ribosomes can become attached before the first has traversed the whole length of the mRNA. Polyribosomes are these associations of ribosomes attached to the same mRNA.

Section 7
Genetics

The primary structures of all the proteins of the cell are encoded in the genetic material, which is ordinarily DNA. The DNA controls protein structure via messenger RNA which is transcribed directly from it. The proteins, in their turn, have structural and enzymic functions which determine the form and metabolism of the cell. The genetic material thus ultimately determines the nature of the organism. The whole of a cell's genetic information, carried in the DNA, is called the *genotype*, as distinct from the *phenotype* which is the complete assembly of characteristics exhibited at a given time. The distinction is necessary for two reasons. Firstly, many potentialities provided for in the genotype may remain latent. Thus, the genotype may include the latent capacity to make an enzyme to hydrolyse lactose, but whether the enzyme is actually made will be determined by environmental factors which will therefore affect the phenotype. Secondly, in the case of diploid cells, the phenomenon of *dominance* may mask the expression of part of the genotype (see p. 35).

The entire genotype or *genome* of a bacterium is commonly carried in a single giant piece of nucleic acid which, in bacteria, usually forms a closed loop and is called a *chromosome*. The bacterial chromosome, which is a very long and fine thread, is folded to form a compact skein which can be seen (after appropriate staining) with the electron microscope as a *nuclear body*, or *chromatinic body*. Unlike cell nuclei of higher organisms it is not enclosed by a nuclear membrane.

The chromosome consists of a number of functionally distinct and non-overlapping segments, the *genes*. Each gene has a single function which can be varied or lost by mutation independently of the functions of other genes. The function in question is usually the specification of the amino acid sequence of a polypeptide chain, with one gene for each kind of polypeptide chain. Some genes, a small minority, have the function of specifying the structures of ribosomal and transfer RNA molecules.

Proof of the central position of DNA in heredity has come mainly from microbiological studies and it has been shown to be possible to transfer characters from one organism to another by three techniques of DNA transfer—*transformation, conjugation* and *transduction*. The first of these involves direct treatment of bacteria with DNA purified from another related strain with different characters, e.g. DNA from a smooth strain of pneumococcus added to cells of an appropriate rough strain may 'transform' some of them—and the ability to make the enzyme which makes the capsular polysaccharide will be inherited by the progeny of the original treated cells.

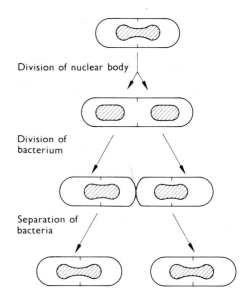

Division of nuclear body

Division of bacterium

Separation of bacteria

Figure 1. DNA replication and cell division.

First the nuclear body replicates and the chromosomes separate: then a cross wall forms between them; the cell divides and finally separates.

Conjugation in bacteria has been detected in only a few species including *Escherichia coli* but has nevertheless been of enormous importance to the development of bacterial genetics. What are equivalent to male and female forms adhere to each other and there is a slow transfer of DNA from male to female, the outcome of which is that new genetic characters can be acquired by the acceptor cell from the donor (Figure 2).

Transduction involves the use of bacterial viruses

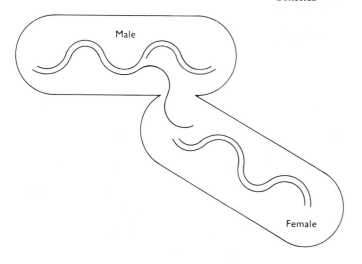

Figure 2. Bacterial conjugation. The male and female cells adhere. Replication of the chromosome occurs in the male and one old strand is injected into the female while the other is retained. There is no reciprocal transfer from the female.

or *bacteriophages*. A virus, unlike all organisms from bacteria to mammals, contains one rather than both kinds of nucleic acid. Usually this is DNA but some bacterial, plant and animal viruses contain RNA instead. The nucleic acid is the genetic material of the virus and the rest of the structure may be just a protein container and a

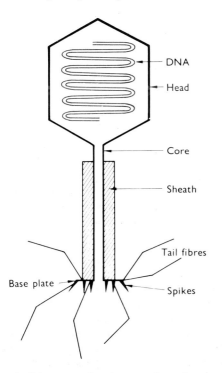

Figure 3. Diagrammatic representation of a bacteriophage.

mechanism for getting the nucleic acid into a suitable host cell. Thus a DNA-containing bacteriophage is like a miniature syringe (Figure 3). The tip of the phage tail sticks on a receptor in the wall of a sensitive bacterium and the DNA is injected into the cytoplasm. Here it acts like a group of genes and controls the formation of enzymes which in turn produce more phage DNA and *also more phage protein*. The phage DNA carries the specification for phage protein and the complete virus particles are assembled within the bacterium—often hundreds for each phage genome injected—and the bacterial cell lyses, some 30 minutes after infection, releasing the now mature infective bacteriophage particles (Figure 4).

Sometimes infection of a bacterium by a phage does not result in lysis. This can happen in a process called *lysogenization* in which the genome of the phage becomes associated with the bacterial chromosome and is replicated with it as if it were part of the host genome. This can continue for many generations and the bacteria are said to be in a lysogenic state. From time to time, in such a culture, the *temperate* phage will somehow begin to reproduce itself in a virulent fashion and produce lysis. This may happen spontaneously or it may be *induced* by ultra-violet light or by the addition of certain chemicals (Figure 5).

Now, many of the phage particles may have 'picked up' small portions of the genome of the host cell, which they carry as if these were part of their own DNA. They may then bring to the next host cell that they infect a character that it did not possess. This is called *transduction*—the transfer of

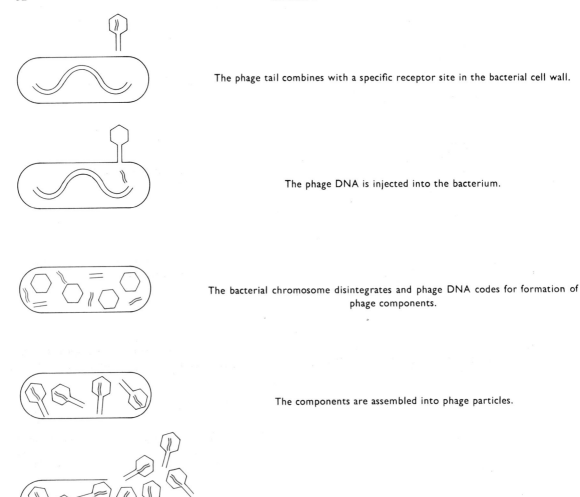

The phage tail combines with a specific receptor site in the bacterial cell wall.

The phage DNA is injected into the bacterium.

The bacterial chromosome disintegrates and phage DNA codes for formation of phage components.

The components are assembled into phage particles.

The cell lyses and releases the mature phage particles.

Figure 4. Infection of a bacterium by a lytic bacteriophage.

a character from one bacterium to another by means of bacteriophage.

We have already seen that DNA possesses two of the properties necessary for the genetic material—the ability to specify the amino acid sequence in proteins and the ability to be replicated precisely so that the specification can be passed on (Sections 5 and 6, pp. 26–29). It must also be able to undergo *mutation* and the change must be heritable. Mutations are abrupt changes—loss, addition, alteration—of genetic characters. They occur 'spontaneously' at very low frequency—perhaps about 1 in 10^6 times or less per character per cell-generation. This rate can be increased enormously by certain physical and chemical treatments—ultra-violet

and X-irradiation, nitrous acid, nitrogen mustards, etc. These treatments are all known to cause modifications to DNA structure and it is probable that 'spontaneous' mutation has a similar cause. Frequently the result of a mutation is to change one base for another in the DNA. The error is then repeated at every replication and it may manifest itself by the fact that the progeny of the mutated cell are now, because of a change in some enzyme, unable to synthesize an amino acid or a growth factor, or they can no longer grow on some particular sugar as a carbon source. The loss of a character is conventionally denoted by a negative sign. Common examples are *trp⁻*, *thr⁻*, *ade⁻*, *arg⁻*, which represent inability to synthesize tryptophan,

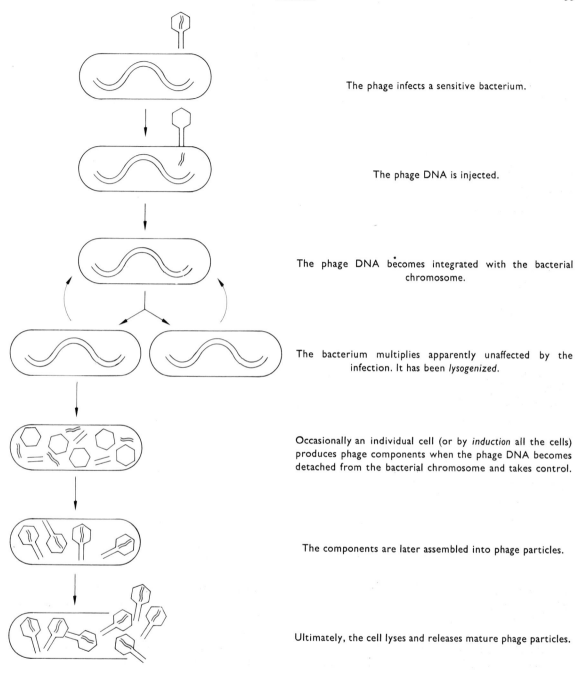

The phage infects a sensitive bacterium.

The phage DNA is injected.

The phage DNA becomes integrated with the bacterial chromosome.

The bacterium multiplies apparently unaffected by the infection. It has been *lysogenized*.

Occasionally an individual cell (or by *induction* all the cells) produces phage components when the phage DNA becomes detached from the bacterial chromosome and takes control.

The components are later assembled into phage particles.

Ultimately, the cell lyses and releases mature phage particles.

Figure 5. Infection of a bacterium by a temperate bacteriophage: lysogenization and induction. The probable integration mechanism is illustrated on p. 298.

threonine, adenine or arginine; similarly *gal⁻* and *lac⁻* represent inability to grow on galactose or lactose. The corresponding positive characters are written *trp⁺*, etc. The alternative forms of a gene, e.g. *trp⁺* and *trp⁻*, are called *alleles* and the *allele* normally found is referred to as the *wild-type* allele.

In higher organisms there is an elaborate sequence of events concerned with replication and the partitioning of the genetic material between daughter cells. The DNA in the nucleus becomes prominent in the form of visible chromosomes which thicken and then split longitudinally. The membrane surrounding the nucleus disappears and a bundle of fibres, called the *spindle*, spans the cytoplasm. The paired chromosomes line up equatorially across the spindle whose fibres are now attached to them. One of each pair of chromosomes is drawn along the spindle fibres towards each pole, the cell begins to divide, the spindle disintegrates, and the two clusters of chromosomes are surrounded by new nuclear membranes. They become diffuse, and we are back where we started. This whole process is known as *mitosis*, and it appears to be considerably more complex than is the replication in bacteria which do not possess a nuclear membrane, do not form a spindle, and have only a single circular chromosome (see Figure 8, p. 35). However, bacteria do have a great deal in common with higher organisms and it is possible to prepare genetic maps by establishing the relative positions of mutations in the genome. Our ability to do this depends on a little-understood process involving DNA. When two similar lengths of DNA come together (*in vivo*, but not *in vitro*) there is a tendency for *crossing-over* to occur and for the corresponding segments to be exchanged. The stretches of DNA must be generally, but need not be exactly, alike. The exchange process is known as *recombination*.

Immediately after conjugation, transformation or transduction the bacterium which is normally *haploid* (having only one copy of the genome) will be *diploid* or *partially diploid* because it will have an additional chromosome or portion of a chromosome. This is usually only transient and at cell division each daughter tends to receive one or other piece of DNA but not both. The pieces of *homologous* DNA become *segregated* from one another. However, recombination may have preceded this.

Let us consider a bacterium whose genome originally carried the closely linked characters thr^+ and ade^- (thr^+ meaning that it has an active functional gene for making the amino acid threonine and ade^- indicating an inactive mutated gene concerned with adenine synthesis). Hence the organism has to be supplied with adenine but can synthesize threonine. A piece of DNA from another cell with the allelic characters thr^- and ade^+ is then introduced. The recipient may receive a piece of DNA carrying thr^- or ade^+ or both. In each case a transient partial diploid is formed and after cell division (see Figure 6) four kinds of progeny are possible: thr^+ ade^- (original); thr^- ade^+; thr^- ade^-; and thr^+ ade^+ The majority of cells will be the original thr^+ ade^- variety but some may have received and exchanged the incoming piece carrying thr^- ade^+ and these now will be able to make adenine but will have lost

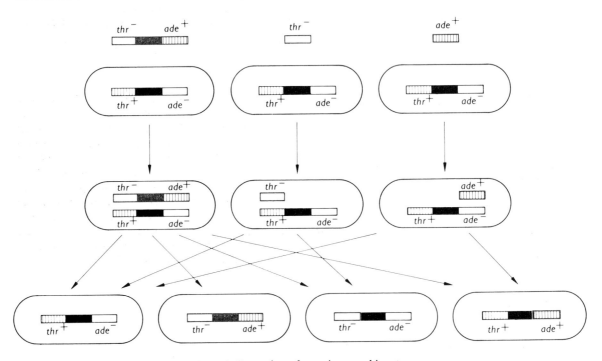

Figure 6. Formation of genetic recombinants.

the ability to make threonine. Then there will be a small proportion of the recombinants *thr⁻ ade⁻* (a double mutant requiring both threonine and adenine) and *thr⁺ ade⁺* (wild-type, able to grow without either threonine or adenine) (see Figure 6). Experimentally the last type of recombinant, having wild-type characters, is the easiest to count because only these recombinants will be able to grow when the cells are plated on a minimal medium, i.e. one containing neither threonine nor adenine. If the number of recombinant cells is compared with the number of original cells we obtain a measure of *recombination frequency*. This is an important measure because it indicates how far apart the genetic loci *thr* and *ade* are from one another. If we assume that recombination is equally likely to occur anywhere along the chromosome, it follows that the further two loci are from one another, the more likely it is that crossing-over will occur at some point between them. The recombination frequency is thus a measure of the distance between loci. If a third mutation (say one concerned with ability to utilize galactose as carbon source) is studied it can be compared in relation to the other two markers. Should the recombination frequency between *thr* and *gal* be found to be roughly equal to the sum of those between *thr* and *ade* and between *ade* and *gal*, then the sequence of the loci on the chromosome must be *thr, ade, gal* (Figure 7).

By these and other methods a large number of characters have been put into sequence, and com-

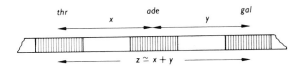

Figure 7. Linear mapping of genes by measurement of recombination frequencies.

The frequency value (*x*) for two of the genes *ade* and *thr* is first obtained (see text and Figure 6). Frequencies of recombination between *ade* and *gal* (*y*) and between *thr* and *gal* (*z*) are then measured in the same way. If *z* is roughly equal to the sum of *x + y* then the sequence of the genes is as shown.

prehensive genetic maps of *Escherichia coli* and *Salmonella typhimurium* have been prepared. Other observations, partly genetic and partly electron microscopical, show that the DNA from a bacterium is a single closed loop with no free ends (Figure 8).

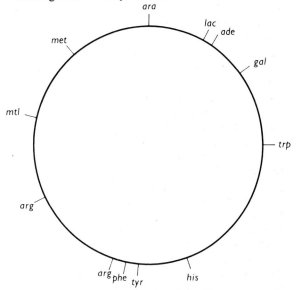

Figure 8. Genetic map of *Escherichia coli*.

The chromosome is circular and the symbols for those markers shown are as follows:

ara	arabinose	*tyr*	tyrosine
lac	lactose	*phe*	phenylalanine
ade	adenine	*arg*	arginine (2 loci shown)
gal	galactose	*mtl*	mannitol
trp	tryptophan	*met*	methionine
		his	histidine

In special circumstances the partial diploid state which we have said is normally transient may become stabilized and a cell may contain two alleles of the same gene. For instance, a *gal⁻* strain may by phage infection receive *gal⁺*. The presence of the *gal⁻* gene does not affect the functioning of the *gal⁺*, and so the cell is phenotypically *gal⁺* and can grow on galactose. The positive character is said to be *dominant* and the negative one *recessive*. The dominance of the positive character is a general rule in genetics and it is a key point in the explanation of regulation of enzyme synthesis.

Sections 8 and 9
Co-ordination and Differentiation

Regulation

A growing bacterial cell makes use—at a guess—of 1000–5000 enzymically catalysed reactions. It is essential that the rates of these should be properly coordinated, for a faulty adjustment in even one enzymic step is likely to have harmful consequences for the cell in which it occurs. Thus, suppose that we have a culture of exponentially growing cells in all of which there is a 'correct' (i.e. optimum) co-ordination of all enzymic steps. Now if in one of the cells a change occurs for some reason, which affects one Class II enzyme in such a way that an amino acid required for protein synthesis is underproduced to the extent of 20%, the supply of this amino acid will become rate-limiting for protein synthesis and hence for growth, the rate of which will consequently fall by 20%. Furthermore the enzymes and ribosome content of a cell are 'geared' to give the higher growth rate. In the new situation 20% of the cell's synthetic capacity is unused and the material and energy that went into making it initially are wasted.

If the cell had no regulatory mechanisms or if it failed to adjust the rates of its enzyme reactions, it would go on degrading glucose at the old rate and making more building blocks than it could assimilate. These might leak from the cell and be used by other cells in the same culture. A single derangement could thus have three consequences: reduction in the *increase* of protoplasm per unit time; reduction by a similar amount in the *yield* of protoplasm per unit of nutrient; utilization of part of the cell's synthetic capacity for producing materials that might be used for growth by its neighbours. It is clear that, with the passage of time, the progeny of the defective cell would form a smaller and smaller fraction of the total population. It is not surprising that in bacteria an elaborate system of control mechanisms has evolved to keep the enzymic processes in step with one another.

The rates of these reactions are controlled in two ways: in the first, the amount of an enzyme is unaltered but its activity is reduced; this is by specific *inhibition* (see below). In the second, the *amount* of enzyme is reduced. Frequently the same enzyme is subject to both types of control, i.e. it is reduced in

amount and the enzyme that is formed functions at a reduced rate. The types of regulation can usefully be considered in relation to the classes of enzymes previously enumerated.

Regulation of Class I enzymes

Our generalized cell has the enzymic capacity for degrading glucose whatever the medium in which it is grown. Many enzymes are like this and are formed in quantity in any growth medium. The cell has, however, a group of latent enzymic capacities. These are formed *only when the substrate is present* and include the enzymes that degrade other carbohydrates, e.g. maltose, galactose, lactose. The process by which the presence of a substrate evokes the synthesis of the enzyme required to degrade it is called *induction*. Also, cells have the ability, when no sugars are available, to utilize certain amino acids as sources of carbon and energy. Again the enzymes are inducible. As a rule, induced enzymes convert substrates to compounds that can be utilized in the general metabolic pathways of the organisms. Thus lactose is split to glucose and galactose; maltose is split to glucose; serine is split to give pyruvate and this in turn is 'fed' into the tricarboxylic acid cycle; tryptophan can also be degraded to give pyruvate as a product.

The mechanism of induction in the case of the proteins induced by lactose can be satisfactorily explained on the basis of the scheme proposed by Jacob and Monod. It rests largely upon the implications of the following genetic evidence. Organisms which can produce an inducible enzyme may mutate to give cells which make the enzyme *constitutively*, i.e. the enzyme is produced in large amounts whether the substrate is present or not. Furthermore, in diploids it is found that the inducible character is dominant over the constitutive. Accordingly the inducible character is represented as i^+ and the constitutive as i^-. It is proposed that the i^+ strain produces an active repressor molecule, R, which may be a protein and which specifically prevents the transcription of the relevant structural gene into messenger-RNA. If the substrate is present it interacts with the repressor and makes it inactive, thus allowing free transcription and consequent synthesis of the enzyme. The constitutive strain, i^-, makes

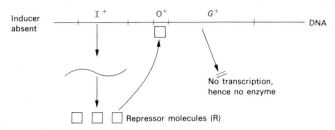

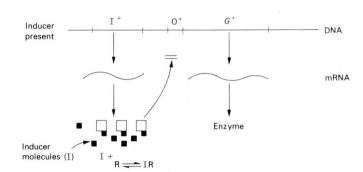

Figure 1. Jacob and Monod's model of regulation of enzyme synthesis by repression. The gene controlling inducibility (I) codes for a repressor protein (R) which interacts with the DNA and blocks transcription of the structural gene *G* for the enzyme. When inducer I (usually the substrate or a derivative) is added, it combines with R converting it to an inactive form. Transcription of *G* into mRNA and hence formation of the enzyme can now take place.

either defective molecules of R or none at all. In these cells, therefore, transcription of the structural gene into mRNA, and synthesis of the enzyme go on all the time. The model is illustrated in Figure 1 and is discussed more critically in Chapter 8.

Enzymes of Class I are also subject to a second type of control, known as catabolite repression or 'glucose effect'. That is, when the cells are growing in glucose they will not form inducible enzymes, even when the inducers for these enzymes are present in the medium. A culture grown in the presence of glucose and some other substrate such as lactose will then grow in two phases—diauxic growth—(see Figure 3, p. 17), separated by a lag. In the first phase the glucose is used exclusively as the carbon source. Then the cells, having exhausted the glucose, stop growing altogether. During the lag period they synthesize the requisite enzyme for a second phase of growth which begins at the expense of the alternative substrate. Manifestly, glucose by its presence prevents the formation of the inducible enzyme, even though the substrate is present all the time. The glucose effect is also exerted by other 'good' growth substrates— good in this context meaning that a high rate of growth is observed when these compounds are present. The mechanism of the glucose effect will be discussed in some detail later (Chapter 8).

The dual control of Class I enzymes by induction *and* the glucose effect means that (a) inducible enzymes are not formed unless the substrates are present; (b) even if they are present, the enzymes will

not be formed if the cell is already supplied with glucose or some other good source of carbon.

Regulation of Class II enzymes
Control by repression. Enymes of this class synthesize the building blocks for the macromolecules (amino acids, nucleotides, etc.) and many of them are regulated both by alteration in the amount of enzyme synthesized and by inhibition of existing enzyme. We shall first consider the control of enzyme synthesis. Suppose a series of reactions leads from a precursor S_1 to a metabolite M_1 which might be an amino acid. S_2, etc. are intermediary metabolites, and E_1, etc. are the corresponding enzymes. An example is the formation of arginine from glutamate by a series of reactions:

$$S_1 \xrightarrow{E_1} S_2 \xrightarrow{E_2} S_3 \longrightarrow \longrightarrow \longrightarrow \longrightarrow M_1$$

(S_1 is itself derived from the action of a Class I type of enzyme.)

In general it is found that if M_1 is added to the growth medium, so that the cells no longer have to synthesize it, the formation of all the enzymes E_1, E_2, etc. stops almost immediately. The process of specifically preventing the formation of an enzyme or a group of metabolically related enzymes is referred to as *repression*. The element of specificity has been emphasized because antibiotics and many other toxic substances often cause a general inhibition of enzyme formation.

Repression has been explained on the basis of the same sort of model as that used to explain induction (see Chapter 8).

Control by end-product inhibition. The allosteric effect. When M_1 is added to the growth medium it not only stops the synthesis of the enzymes E_1, E_2, etc. but *inhibits* the functioning of the first enzyme E_1. Again, the effect is highly specific and it is immediate, so that as soon as M_1 is available the cells cease to make it. This is often called feedback inhibition and it is formally analogous to a simple thermostatic regulation system.

It is important to note that the enzyme E_1 must have *two* specific recognition sites. One is the active site for its substrate, S_1, and the other, equally specific, is the site for the end-product inhibitor M_1. If S_1 and M_1 were chemically similar substances it would be reasonable to assume that they competed for the same site on the protein molecule. This would be an *isosteric* interaction. In fact, they are usually so dissimilar that it is reasonable to suppose that their interaction is *allosteric* and it is assumed that when M_1 is attached to the enzyme E_1 the protein molecule is distorted so that S_1 no longer fits properly (Figure 2).

This description of end-product inhibition is over-simplified because the same precursor, e.g. glutamate or aspartate (see Figure 1, p. 21) is often the starting point for the synthesis of several end-products. In branched pathways of this sort the regulation is more complex and will be discussed in Chapter 8.

The mechanism of end-product inhibition ensures that the materials of the environment will not be wasted in the manufacture of end-products which are already provided from the outside.

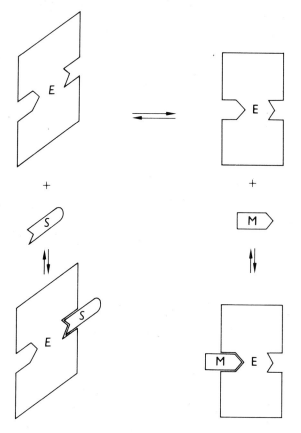

Figure 2. Schematic illustration of allosteric interaction. The enzyme molecule (E) has two conformation states which are in equilibrium with one another. It also has two recognition sites: When the enzyme is in one conformational state it can combine with M but not with S and is therefore inactive. In the other conformational state it can combine with S and it then functions as an enzymic catalyst, one state is specific for the substrate (S) and the other for a particular metabolic inhibitor (M).

Regulation of Class III enzymes
These enzymes assemble the macromolecules of the cell and it is the macromolecules, i.e. polysaccharides, proteins, DNA, etc., that will differ from those of all other types of bacteria. These substances *must* be assembled by the cells; they cannot just be picked up from the environment. Every growing cell will thus have all the enzymes of Class III and, on *a priori* grounds, it seems unlikely that these enzymes will vary greatly in amount. Their functioning is more likely to be controlled by alterations in the rate of reaction. Not much is known about regulation of these enzymes but some speculations will be mentioned in Chapter 8.

To summarize, we can say that we have a fairly good idea of the way in which the information systems built into the cell regulate the degradation and the biosynthesis of small molecules and we can see that the whole arrangement is such as to give the maximum yield of bacterial protoplasm in the minimum time. What we do not yet know is how the cell regulates the synthesis of its macromolecules, nor do we understand the way in which information, presumably stored in the DNA, is translated into structure. For instance, we do not know why the organism has its characteristic size and shape, nor do we know how certain types of enzymes come to be associated with the membrane while others are found in the ribosomes, nor do we know why the cell divides when it gets to a certain critical size.

Differentiation: sporogenesis and germination

So far we have considered the co-ordination of metabolism in actively growing bacteria. However, exponential growth is unlikely to continue for very long without some nutrient becoming growth-limiting (see Section 2). When this happens a population of bacteria can react in one of two ways. In most species, the cells will make use of endogenous reserves of carbohydrate and then begin to degrade their proteins and nucleic acids and combust the products to obtain the energy necessary to maintain the integrity of the cell. Death and lysis of the cells may ultimately ensue.

In other species deprivation of nutrients acts as a stimulus for sporulation—a sequence of events in which a spore develops within the cytoplasm and is then liberated. Briefly the process consists of the enclosing of a complete copy of the genome together with proteins and RNA within a laminated structure which is refractile. The rest of the cell disintegrates leaving the free spore which contains virtually no water, is resistant to heat and to many sorts of toxic reagents, and which can remain in a dormant state for years. When the spore is restored to a nutrient medium *germination* takes place: the refractility disappears and so do the properties of resistance to heat and to toxic chemical substances. Much of the structure of the spore breaks down, the enzymes that are needed for growth are synthesized and in due course a cell of normal shape and dimensions grows out of what is left of the spore.

The morphological changes that we have described can be regarded as a very primitive form of differentiation.

Part II

Chapter 1
The Bacterial Cell: Major Structures

The first part of this book is a brief account of the way in which bacterial cells are put together. We shall now examine the processes in more detail following the same sequence so that the succeeding chapters correspond more or less to the previous sections.

In this first chapter the anatomy of bacterial cells is considered as well as something of the chemistry of the components. The aim is to describe these constituents and to show what function they perform in the normal, healthy bacterium.

Investigations have proceeded along two main lines: firstly, the electron microscope has revealed details of surface structures and appendages and secondly, biochemical investigations of the whole cell or isolated parts of it have shown us the chemical material of which the individual structures are composed. An attempt is frequently made to relate the microscopical information obtained from the study of sections with the biochemical results obtained from studies of cell fractions and so to build up a co-ordinated picture of function and structure in the intact cell.

Not all the structures revealed by the electron microscope and by other means appear to be nec-essary for survival of the bacterium. By comparing different kinds of bacteria it is possible to arrive at the lowest common denominator in terms of those structures that are necessary for growth and division. These 'essential' structures are depicted diagrammatically in Figure 1a, together with other structures which are found in some, but not all, bacteria (Figure 1b). Thus, with the exception of a few species living in specialized environments, bacteria have a *wall*, which varies considerably in chemical composition and fine structure from one organism to another: this wall surrounds the *cytoplasm* which is bounded by a delicate *cytoplasmic* or *protoplast* or *plasma* membrane. In electron micrographs the cytoplasm has a granular appearance resulting from the presence of a large number of *ribosomes*. Within the cytoplasm the *nuclear area* (*chromatinic body*, *bacterial nucleoid*) can be seen as an electron-transparent area in thin sections. The nucleus has a very fine fibrillar network but unlike the animal or the plant cell nucleus is not separated from the cytoplasmic contents by a nuclear membrane.

In bacteria whose structure has been studied extensively, invaginations of the cytoplasmic membrane occur which are termed *mesosomes*; these are

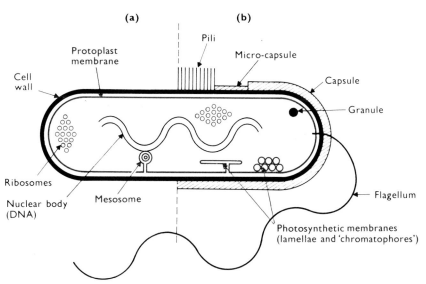

Figure 1. Cross-section of a generalized bacterial cell showing (a) essential and (b) inessential structures.

frequently associated with the nucleus and at the sites of septation and spore formation. Various *granules* (lipid, glycogen, sulphur, metaphosphate or volutin) may be present in the cytoplasm, and in photosynthetic bacteria membranous lamellae or membrane-bounded vesicles can be seen in thin sections: these are connected to the cytoplasmic membrane and are the organelles in which photosynthesis occurs.

Thin threads termed *pili* (*fimbriae*) may protrude from the cell wall: these are of several types, one of which is involved in sexual conjugation, but little is known of the function of the remaining types. Motile bacteria have at least one *flagellum*; flagella are very long thin structures composed of protein sub-units. Many bacteria have a *capsule* external to the cell wall and this structure is sometimes many times thicker than the bacterium itself. Some bacteria produce a slime which in a few instances appears to have an ordered structure. The location of these 'inessential' structures is shown in Figure 1b.

Electron microscopy

Although bacteria are easily visible when viewed by phase-contrast microscopy little information concerning the anatomical structures within them can be gained other than with the electron microscope. Three techniques are used routinely with this instrument to investigate bacterial structure: negative staining enables the topography of the whole cell or isolated parts of it to be examined, thin sectioning reveals cellular profiles, and freeze-etching permits the examination of various surfaces in the bacterium and of cross-sectional views of the cell wall, protoplast membrane and any organelles present. In addition the scanning electron microscope can be used to take pictures having a tremendous depth of field, resulting in a three-dimensional effect (Figure 2). This technique has been of value in the examination of the surface features of whole organisms and also of bacterial colonies. Samples may be examined before or after chemical fixation or after freeze-drying. The specimen on the grid is coated with a thin, 20 nm layer of metal (e.g. gold/palladium alloy) prior to examination.

Negative staining involves mixing the material to be examined with an electron-dense material such as sodium phosphotungstate and drying down a small quantity of the mixture on a grid. Regions of the cell which the stain cannot penetrate are electron-transparent and are viewed against the electron-dense background.

The preparation of thin sections of bacteria is preceded by a number of manipulative steps which

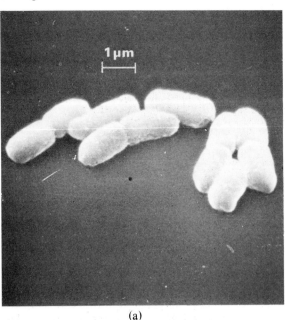

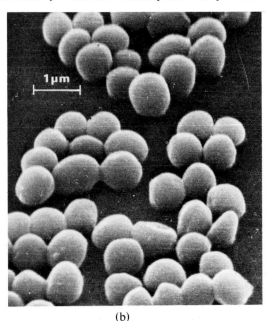

(a)   (b)

Figure 2. Normal cells of *Escherichia coli* (a) and *Staphylococcus aureus* (b) as revealed by the scanning electron microscope. Reproduced with permission of (a) F. W. O'Grady and D. Greenwood. *The Glaxo Volume* **38** (1973) 5 and (b) D. Greenwood and F. W. O'Grady. *Science* **163** (1969) 1076.

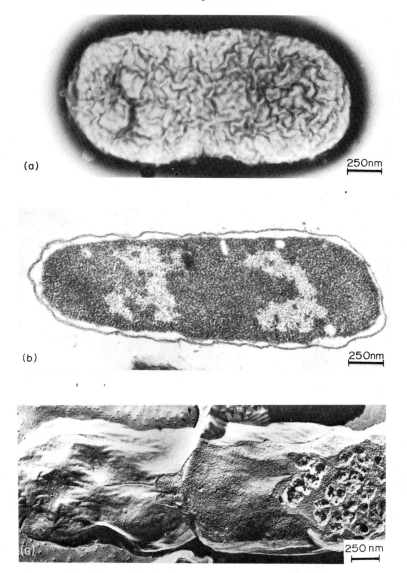

Figure 3. Comparison of *Escherichia coli* from exponentially-growing cultures after (a) negative staining, (b) fixation, dehydration, embedding and thin sectioning, (c) freeze etching (the intact cell surface is seen on the left, the cytoplasmic membrane is exposed in the centre and the cytoplasm on the right of the picture). Reproduced with permission of M. E. Bayer and C. C. Remsen. *J. Bacteriol.* **101** (1970) 304.

include drastic fixation, dehydration and embedding processes. Any one of these might introduce artifacts which may confuse those who attempt to interpret the resulting electron micrographs.

The technique of freeze-etching aims at overcoming some of the criticisms that can be applied to earlier methods and at providing high resolution pictures of the cell surface. The technique involves rapid freezing of a small sample to $-170°$ followed by

fracture of the specimen with a cold knife under high vacuum (freeze-fracturing). The cut surface is then etched by raising the temperature for a short time to $-100°$, shadowed with a heavy metal and a carbon replica is made. The specimen is dissolved away and the replica examined in the electron microscope. The preparation is not subjected at any stage to chemical procedures such as fixation and staining. Since viable bacteria have been recovered from specimens which

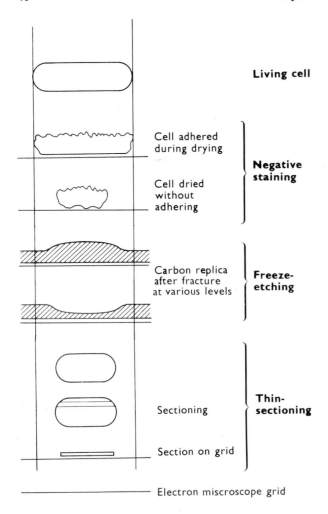

Living cell

Cell adhered
during drying

Negative
staining

Cell dried
without
adhering

Carbon replica
after fracture
at various levels

Freeze-
etching

Sectioning

Thin-
sectioning

Section on grid

Electron miscroscope grid

Figure 4. Diagrammatic representation of the influence of various preparative procedures on the dimensions of bacteria.

have undergone the freeze-etching process but which have not been fractured it is claimed that artifacts are less likely to occur than with thin sectioning or negative staining procedures.

The appearance of the bacterium *Escherichia coli* after subjection to these three electron microscopical techniques is shown in Figure 3. Negative staining reveals the convoluted appearance of the outer cell wall layer but little else, thin sectioning shows internal components and gives some idea of the thickness of the cell wall layers, while freeze-etching reveals the appearance of the surface of the protoplast membrane and cell wall in much greater detail than can be seen by negative staining. The various processes must affect the shape of the bacterium and Figure 4 gives an idea of how the cross-sectional shape may alter depending on the processes used. The removal of

water during fixation before thin sectioning also results in a drastic shrinkage of the cellular material (possibly by as much as 30%). It must therefore be appreciated that an electron micrograph may be akin to an artist's impression of the cell.

The earlier electron micrographs tended to show a limited range of bacteria which were the subjects of choice for biochemical experiments. The examination of a wider range of organisms has emphasized the diversity of structures found in bacteria—particularly the arrangement of macromolecules in the various layers of the wall. It is important not to make generalizations on the basis of what is seen in one particular electron micrograph or to assume that the structural arrangement present will be found in other bacteria of the same genus. The selection of micrographs in this chapter should illustrate this.

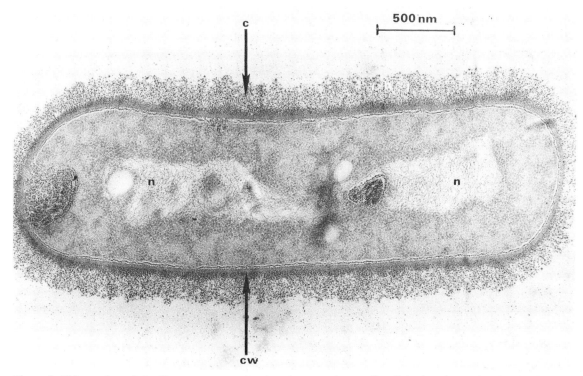

Figure 5. Thin section of *Bacillus megaterium* showing the amorphous cell wall (cw) and capsule (c). The electron-transparent nuclear material (n) is surrounded by the densely-stained ribosomes. Reproduced with permission of D. J. Ellar, D. G. Lundgren and R. A. Slepecky. *J. Bacteriol.* **94** (1967) 1189.

BACTERIAL CELL WALLS

Appearance in the electron microscope

Although there is considerable variability between genera and species, bacteria are placed in one of two main groups—referred to as Gram-positive and Gram-negative. This classification is based on the ability of some bacteria to retain the basic dye crystal violet when they are washed with ethanol: the distinction between the two groups (Gram-positive bacteria retain the stain, Gram-negative do not) is thought to be dependent on some difference in the chemistry of the surface layers.

Gram-positive bacteria

In thin sections of Gram-positive bacteria the wall appears as an amorphous structure 15–80 nm thick, apparently lacking fine structure (Figure 5). A number of layers can sometimes be distinguished on the basis of their electron-transparency, the thickness of each layer varying with the fixative and stain used but it is not clear whether this is due to a genuine difference in the chemistry of the layers or is an

artifact of fixation or staining. When the surface of different bacteria within a single genus is observed by freeze-etching many different types are found—the wall of *Bacillus megaterium* apparently lacks fine structure (Figure 6a), an observation which is in accord with pictures obtained by shadowing isolated cell walls of the same species (Figure 6b) while those of *B. anthracis* and *B. brevis* contain regular arrays of subunits on the surface of the wall overlying those non-structured inner wall layer. Arrays of subunits are also seen in some of the Gram-variable Clostridia (Figure 7) and the various layers of the wall revealed by this and other micrographs are depicted diagrammatically in Figure 8.

Gram-negative bacteria

The walls of these organisms are extremely complex both chemically and structurally. In the simplest examples thin sectioning reveals a five-layered wall (3 electron-dense and 2 electron-transparent layers) outside the three layers of the cytoplasmic membrane (Figures 9 and 10). In some organisms the innermost of the three dense wall layers is thickened so that no electron-transparent layer is evident between the L5

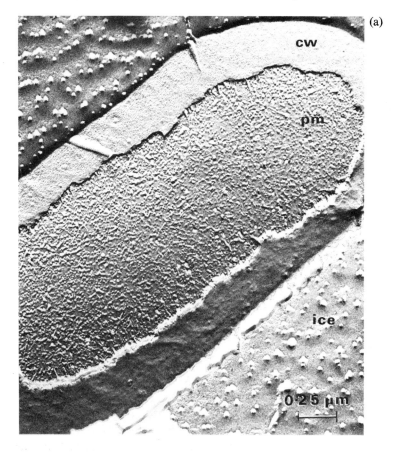

(a)

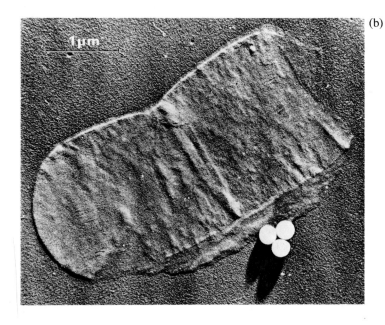

(b)

Figure 6. (a) Freeze-etched preparation of *Bacillus megaterium* showing a lack of fine structure in the cell wall (cw) which has been revealed by etching. The fracture plane has traversed the cell wall and runs along the centre of the plasma membrane (pm) exposing the outer surface of the inner half of the membrane: proteins form the raised ridges. (b) Isolated cell wall of *Bacillus megaterium*. Reproduced with permission of (6a) S. C. Holt and E. R. Leadbetter. *Bacteriol. Revs.* **33** (1969) 346 and (6b) M. R. J. Salton and R. C. Williams. *Biochim. Biophys. Acta* **14** (1955) 455.

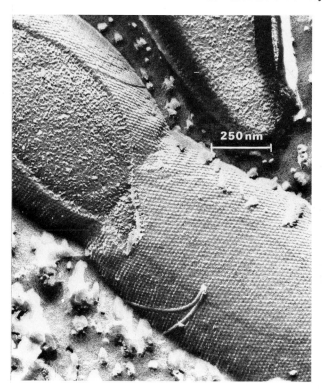

250 nm

Figure 7. Freeze-etched preparation of *Clostridium thermohydrosulfuricum* showing regular array of sub-units on the outer surface. Reproduced with permission of A. M. Glauert, M. J. Thornley, K. J. I. Thorne and U. B. Sleytr. *Microbial Ultrastructure* (1976) p. 31. Edited by R. Fuller and D. W. Lovelock, Academic Press

and L3 layers; it is this L5 layer whether thick or thin which disappears when organisms are treated with lysozyme in the presence of the metal-chelating agent, ethylenediamine tetra-acetic acid, leaving behind the outermost three layers of the wall (outer membrane). Lysozyme is an enzyme that hydrolyses bonds in the peptidoglycan, the rigid component of the wall (see page 54) so the L5 layer is probably composed of this polymer. The outer membrane can be removed by treatment of cells with hot phenol and has been shown to contain lipopolysaccharide, protein, phospholipid and lipoprotein.

It seems probable that wall constituents extend

further than the outermost electron-dense layer. In micrographs showing bacteria joined together there is a gap about 2–4 nm wide which is not found in mutants of the same strain lacking the 'O' specific chains of the lipopolysaccharide (see page 68).

Walls of many Gram-negative bacteria have an additional external layer that can be seen not only in negatively-stained preparations (Figure 11) but also in thin sections (Figure 12a) and in freeze-etched preparations (Figure 12b). The complexity of this external layer of the wall varies from the complex 'wine-glass' structure seen in Figure 12a to a simple layer only 7 nm thick.

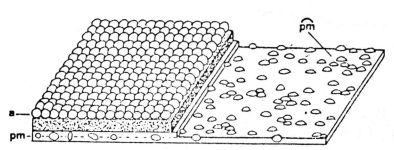

pm

a—

pm-

Figure 8. Diagrammatic representation of a cross-section through the plasma membrane and cell wall of a Gram-positive bacterium. Protein sub-units (a) are seen on the outer surface of the cell wall. p̂m is internal fracture face of the plasma membrane, pm. Reproduced with permission of A. M. Glauert, M. J. Thornley, K. J. I. Thorne and U. B. Sletyr. *Microbial Ultrastructure* (1976) p. 31. Edited by R. Fuller and D. W. Lovelock, Academic Press.

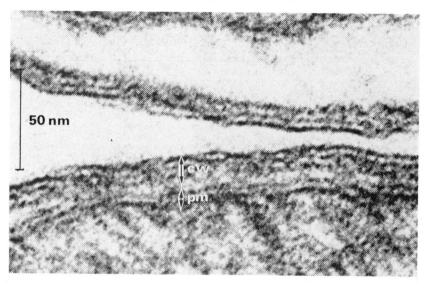

Figure 9. Thin section of cell envelope of *Escherichia coli* showing the five layers of the cell wall (cw) and three layers of the plasma membrane (pm). In the upper cell the wall has become detached from the membrane. Reproduced with permission of S. de Petris. *J. Ultrastructure Res.* **12** (1965) 247.

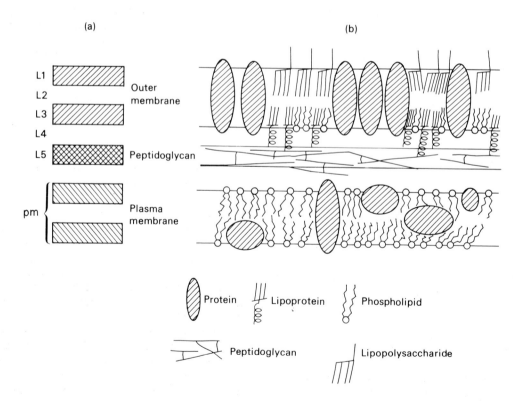

Figure 10. Representation of the cell envelope of *Escherichia coli* based on electron microscopy of thin sections (a), chemical and enzymic analysis of the isolated layers and the ability of the macromolecules to function as antigenic determinants or as receptor sites (b).

Figure 11. Negatively-stained preparation of the cell wall of *Spirillum serpens* showing regular array of subunits. Reproduced with permission of R. G. E. Murray, in *The Bacterial Cell Wall*, by M. R. J. Salton, p. 74, Fig. 26d. Elsevier, London.

The layered arrangement of the polymers of Gram-negative walls can also be deduced from experiments in which components of the wall are selectively removed by chemical or enzymic treatments followed by analyses of the products and by the number of fracture planes obtained in the freeze-etching procedure. The cytoplasmic membrane usually cleaves along the centre of the bimolecular lipid leaflet. This rarely happens with the outer membrane in which most of the phospholipid is believed to be positioned in the inner half together with lipoprotein, while lipopolysaccharide molecules are located in the outer half and proteins span the thickness of the membrane (Figure 10). The rigid layer of peptidoglycan (L5 in Figure 10) has a well-characterized lipoprotein bound covalently to it: the lipid portion of the lipoprotein is in the hydrophobic inner layer of the outer membrane so this results in the outer membrane being linked covalently to the rigid layer.

Halobacteria

These organisms form a special group in that they grow in high concentrations of salts (e.g. 4 M NaCl). Removal of salt results in dissolution of the wall which explains why this structure was not seen in earlier electron micrographs. The wall structure is preserved if salt is included in the fixative before sectioning (Figure 13). The wall appears amorphous, rather like that of Gram-positive bacteria but it differs

in that it does not contain the typical peptidoglycan constituents which make up 50% or more of the weight of the Gram-positive wall: instead, it is built of glycoprotein subunits (Figure 13 inset).

Isolation of cell walls

Before cell walls can be studied chemically they have to be isolated and freed from other components of the cell. The technique used most frequently is that of mechanical disintegration of the cells followed by several cycles of differential centrifugation to separate the walls from the heavier unbroken bacteria and the lighter ribosomes and other cytoplasmic constituents. The particular method of disintegration that is employed affects the final size of the cell wall fragments. Thus, mechanical agitation with glass beads may result in only one tear in the wall and the consequent production of one large fragment per cell, while sonic oscillation produces much smaller fragments. Cell walls often appear flattened when shadowed preparations are examined but retain the overall shape of the organism (Figure 6b) suggesting that it is the wall which contains the rigid components which determine the shape of the bacterium. In general the walls of Gram-positive bacteria are easier to separate and purify than those from Gram-negative organisms which tend to remain associated with the cytoplasmic membrane: consequently

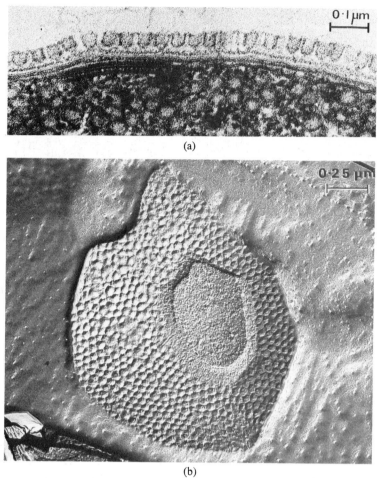

(a)

(b)

Figure 12. Cell wall of *Chromatium buderi*. (a) Thin section showing the multi-layered nature of the cell wall with the outermost layer corresponding to 'wine-glass' shaped subunits. (b) Freeze-etched preparation showing the outer surface composed of tightly-packed cup-shaped units. Reproduced with permission of C. C. Remsen, S. W. Watson and H. G. Truper. *J. Bacteriol.* **103** (1970) 254.

preparations from Gram-negative organisms are generally referred to as the 'cell envelope fraction'. However, the peptidoglycan can be digested with lysozyme resulting in a preparation of osmotically-sensitive spheroplasts. The outer membrane has a higher density than the cytoplasmic membrane and can be separated from it by equilibrium density gradient centrifugation of a lysed spheroplast preparation.

Chemistry

Chemical analysis of isolated cell walls has revealed the complexity and heterogeneity of the individual components. In general the walls of Gram-negative bacteria have a high lipid and low amino sugar content (10–20% and 2–5% of the dry weight respectively) and contain the full range of amino acids found in proteins, while those from Gram-positive organisms have little or no lipid, have a high content of amino sugar (15–20%), but sometimes only a limited range of amino acids. Until it was realized that proteins were present in the walls of many Gram-positive bacteria, purification of the isolated walls often included treatment with a proteolytic enzyme which explains why the full range of amino acids was not found in hydrolysates. Protein where present forms only a low percentage of the total weight of the wall; consequently the amino acids present in peptidoglycan are present at several times the concentration of the other amino acids. The principal polymers found in cell walls are shown in Table 1.

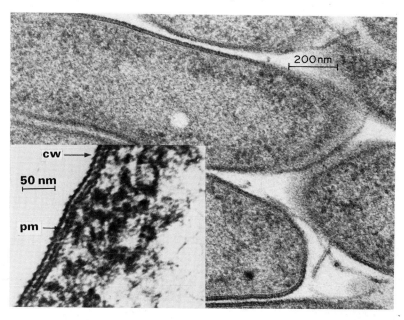

Figure 13. Thin section of *Halobacterium halobium* showing the amorphous appearance of the cell wall. Reproduced with permission of W. Stoeckenius and R. Rowen. *J. Cell. Biol.* **34** (1967) 365.

Inset. Thin section of cell wall (cw) and membrane (pm) showing the glycoprotein subunit nature of the cell wall. Reproduced with permission of K. Y. Cho, C. H. Doy and E. H. Mercer. *J. Bacteriol.* **94** (1967) 196.

Table 1. Principal components of bacterial cell walls

Component	Gram-positive cell wall	Gram-negative cell envelope
Peptidoglycan	+	+
Teichoic acid and/or Teichuronic acid	+	−
Polysaccharide	+	+
Protein	± (not all)	+
Lipid	−	+
Lipopolysaccharide	−	+
Lipoprotein	−	+

Peptidoglycan (mucopeptide, murein)

Structure

Although peptidoglycan is present in all bacteria living in hypotonic environments the proportion of the dry weight of the wall which it forms varies considerably. The cell wall itself may comprise 20% of the organism and the amount of peptidoglycan in it may range from 50–80% in most Gram-positive bacteria (e.g. Bacilli, Staphylococci) to 1–10% in Gram-negative organisms (e.g. Salmonellae, *Escherichia coli*). A characteristic pattern of amino sugars and amino acids is found in hydrolysates of purified peptidoglycan and of these *N*-acetylmuramic acid, diaminopimelic acid (DAP) and the D-isomers of glutamic acid and alanine are found uniquely in this structure. Consequently their presence in bacteria, blue-green algae and Streptomyces spp. has been used as evidence of the occurrence of peptidoglycan in cell walls of these micro-organisms and of their close evolutionary relationship. Structural studies on peptidoglycans from a number of bacteria indicate that the polymer is made up of a polysaccharide backbone (glycan) to which short peptide chains are linked: some or all of these peptide chains are joined together either directly or by other short peptide chains.

Two amino sugars are present in the glycan chains, *N*-acetylglucosamine and its 3-*O*-D-lactyl derivative, *N*-acetylmuramic acid, in alternating sequence with all the linkages β,1–4. It is this portion of the polymer which is sensitive to lysozyme, an enzyme which hydrolyses the β,1–4 link between *N*-acetylmuramic acid and *N*-acetylglucosamine (Figure 14) and the backbone is therefore split into disaccharide units.

The peptide chains linked to the carboxyl group of muramic acid contain four amino acid residues and

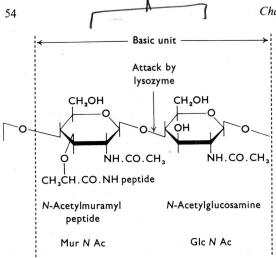

Figure 14. Polysaccharide backbone of cell wall peptido-glycan showing linkage attacked by lysozyme.

have the following sequence —L-Ala—D-Glu—L-R—D-Ala. Exceptions to this include the substitution of L-serine or glycine for L-alanine in some bacteria. The amino acid R can be *meso-* or LL-diaminopimelic acid, L-lysine, L-ornithine, L-diaminobutyric acid or L-homoserine; it is invariably the L-centre of the amino acid which takes part in the peptide link and with the exception of homoserine all the amino acids at this position have two amino groups. The structure of the tetrapeptide subunit is given in Figure 15. The α-COOH group of the D-glutamic acid residue is

sometimes present as the amide or can be combined with glycine in some Micrococci.

The variation in structure of peptidoglycans in different species arises from the manner and degree of cross-linking of tetrapeptide chains. There are thought to be four types of cross-linking but the

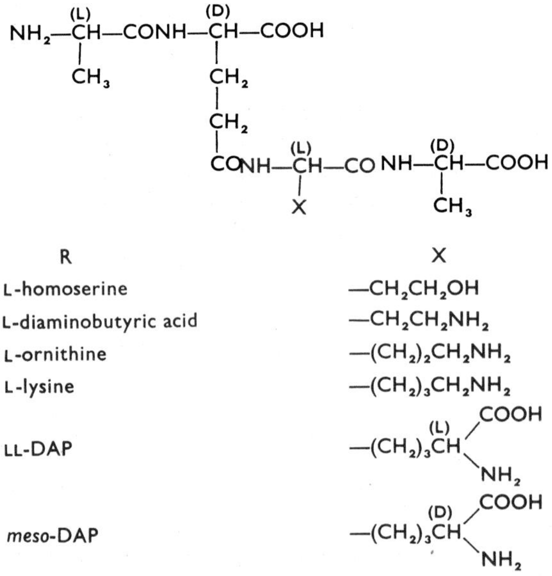

$$NH_2-\underset{|}{\overset{(L)}{CH}}-CONH-\underset{|}{\overset{(D)}{CH}}-COOH$$
$$CH_3 \qquad CH_2$$
$$CH_2$$
$$CONH-\underset{|}{\overset{(L)}{CH}}-CO\,NH-\underset{|}{\overset{(D)}{CH}}-COOH$$
$$X \qquad\qquad CH_3$$

R	X
L-homoserine	—CH_2CH_2OH
L-diaminobutyric acid	—$CH_2CH_2NH_2$
L-ornithine	—$(CH_2)_2CH_2NH_2$
L-lysine	—$(CH_2)_3CH_2NH_2$
LL-DAP	—$(CH_2)_3\overset{(L)}{CH}\diagup\overset{COOH}{}\diagdown NH_2$
meso-DAP	—$(CH_2)_3\overset{(D)}{CH}\diagup\overset{COOH}{}\diagdown NH_2$

Figure 15. General structure of the tetrapeptide subunit L-Ala—D-Glu—L-R—D-Ala. Redrawn from J. M. Ghuysen and M. Leyh-Bouille, *FEBS Symposium* **20** (1970) 59

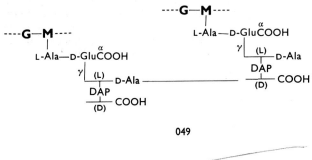

049

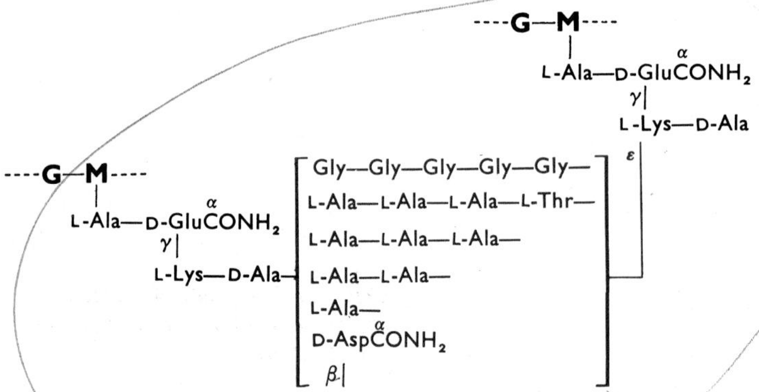

Figure 16. Peptidoglycan of *Escherichia coli* (Type 1). A direct cross-link (in red) joins the tetrapeptides. Redrawn from J. M. Ghuysen and M. Leyh-Bouille. *FEBS Symposium* **20** (1970) 59.

Figure 17. Peptidoglycan of many Gram-positive bacteria (Type 2). A short peptide chain or a single amino acid links the tetrapeptides. Redrawn from J. M. Ghuysen and M. Leyh-Bouille. *FEBS Symposium* **20** (1970) 59.

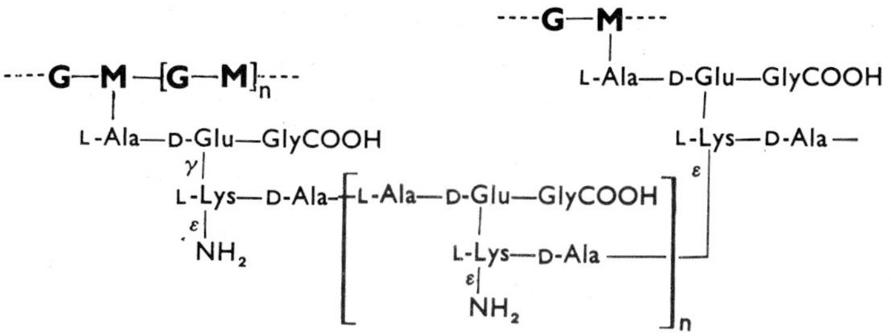

Figure 18. Peptidoglycan of *Micrococcus lysodeikticus* (Type 3). The peptide chain linking the tetrapeptides has the same composition as the tetrapeptide. Redrawn from J. M. Ghuysen and M. Leyh-Bouille. *FEBS Symposium* **20** (1970) 59.

terminal D-alanine of one chain is always involved. The first type may be common to all Gram-negative bacteria and the Gram-positive bacilli and consists of a direct linkage from D-alanine to the amino group on the D-carbon atom of *meso*-DAP in another chain (Figure 16). The second type of cross-linkage is present in the majority of Gram-positive bacteria examined and involves a short peptide or a single amino acid extending from the D-alanine of one chain to the free amino group of a diamino acid in another chain (Figure 17). The pentaglycine cross-link found in *Staphylococcus aureus*, one of the first organisms whose wall structure was carefully investigated, comes into this category. The walls of *Micrococcus lysodeikticus* have a third type of cross-link, again extending from a C-terminal D-alanine to the ε-amino group of a lysine residue as in some type 2 linkages, but comprising varying amounts of a short peptide of the same composition as the peptide attached to the muramic acid (Figure 18). Since the ratio of muramic acid to D-glutamic acid in the peptidoglycan is unity it follows that some of the muramic acid residues are unsubstituted with peptide chains: this suggests that the formation of a

cross-link involves the prior action of an enzyme which catalyses the cleavage of the muramyl-L-alanine bond. When no di-amino acid is present in the tetrapeptide the cross-link is between the D-alanine and the α-COOH group of glutamic acid and consists of a di-amino acid (either D-lysine or D-ornithine (Figure 19)).

Because the degree of cross-linking in the peptido-

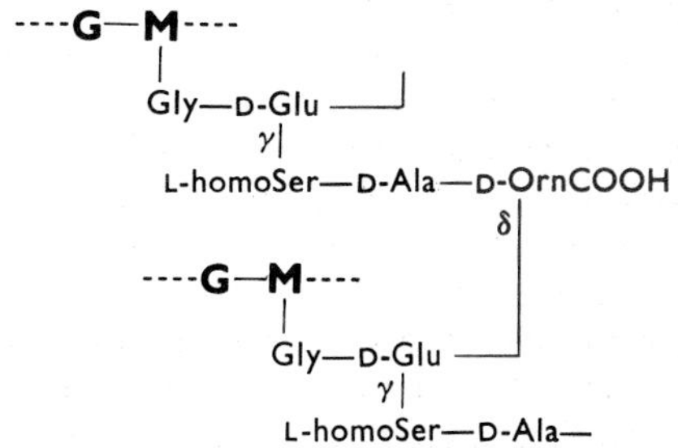

Figure 19. Peptidoglycan of *Corynebacterium poinsettiae* (Type 4). The tetrapeptide does not contain a dibasic amino acid; consequently the cross-link is between two free carboxyl groups. Redrawn from J. M. Ghuysen and M. Leyh-Bouille. *FEBS Symposium* **20** (1970) 59.

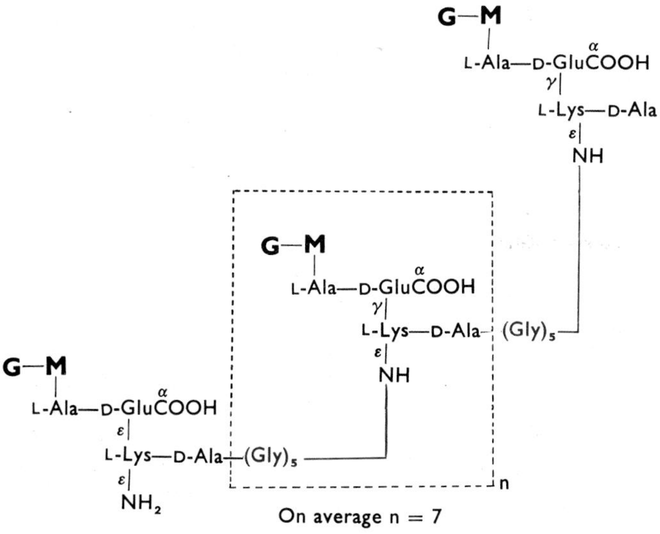

Figure 20. Fragments of Staphylococcal wall obtained after digestion with lysozyme.

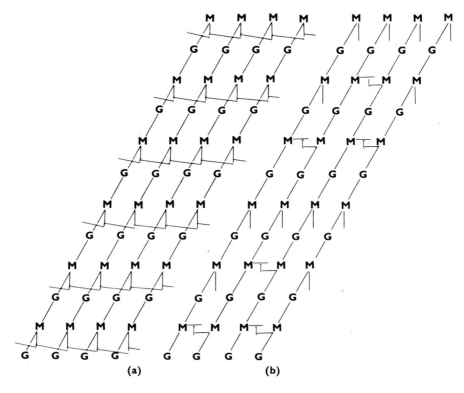

Figure 21. Diagrammatic representation of a single layer of peptidoglycan from (a) *Staphylococcus aureus* and (b) *Escherichia coli*. (a) Redrawn from J. M. Ghuysen, J. L. Strominger and D. J. Tipper, in *Comprehensive Biochemistry*, **26A** (1968) 53. Eds. M. Florkin and E. H. Stotz. Elsevier, Amsterdam, London and New York. (b) Redrawn from J. M. Ghuysen, *Bact. Rev.* **32** (1968) 425.

glycan affects the size of fragments resulting from digestion of the polymer with lysozyme, the types and proportions of such fragments give an insight into the structure of the intact polymer. A high proportion of monomers (disaccharide-tetrapeptide) and dimers (two monomers cross-linked as in Figure 16) is indicative of the low degree of cross-linkage found in bacilli and most Gram-negative bacteria: the larger fragments obtained from Staphylococcal walls (Figure 20) suggest that cross-linkage is almost complete. A single layer of peptidoglycan of *S. aureus* and *E. coli* might have the structure shown in Figure 21, although uncertainties exist concerning the length of the individual glycan chains, the distribution of peptides which are not cross-linked and the regularity of peptide cross-linking between adjacent chains, particularly when more than one layer of peptidoglycan is present. Figure 21 demonstrates the potentiality for size and rigidity of the polymer and it is possible that there is only one molecule per cell, hence the name 'murein sacculus'. Measurements from electron micrographs of the thickness of the peptidoglycan layer are consistent with the idea that Gram-negative envelopes contain a single layer of peptidoglycan while the amount in Gram-positive organisms is indicative of several layers being present.

Biosynthesis

The difficulty of synthesizing such a large polar structure outside a relatively impermeable plasma membrane is overcome by assembling the precursors of the peptidoglycan in the cytoplasm, transporting them through the hydrophobic membrane linked together as a wall subunit attached to a lipid carrier, polymerizing the subunits outside the cell membrane and linking the polymer to the existing wall to make it insoluble.

In detail the process is as follows. *N*-acetyl glucosamine-1-phosphate reacts with uridine triphosphate with the formation of UDP-*N*-acetylglucosamine, one of the two precursors. UDP-*N*-acetylmuramic acid is formed by the reaction of phospho-enolpyruvate with UDP-*N*-acetylglucosamine and reduction of the UDP-*N*-acetyl-3-*O*-enolpyruvylglucosamine so formed. The peptide chain attached to the lactic acid residue of UDP-N-acetylmuramic acid is built up by the sequential addition of amino acids. Each amino acid is added by a specific enzyme, with the exception that the terminal dipeptide, D-alanyl-D-alanine, is added as a single unit (Figure 22); the enzymic reactions are dependent on Mn^{2+} and use ATP as an energy source. Organisms containing DAP have a DAP-adding enzyme which will not add lysine if it replaces DAP in the cell-free assay system: similarly, the enzyme from cells having lysine in the wall will not add DAP to UDP-MurNAc-L-ala-D-glu.

The stepwise build-up of peptidoglycan has been studied in cell-free systems from a number of species and the postulated pathway is shown in Figure 23. The most active cell-free preparations consist of wall-membrane fragments obtained by differential centri-

fugation of bacteria that have been disintegrated mechanically, for example by grinding with alumina or shaking with glass beads. It is believed that these techniques best preserve the *in vivo* arrangement of the enzymes, carrier and wall acceptor. Phosphoryl-MurNAc-pentapeptide is first transferred with the release of UMP to an acceptor present in the membrane fragments. This has been identified as the monophosphate derivative of a C_{55} polyisoprenoid alcohol (undecaprenyl phosphate, Figure 24). A transglycosylation reaction in which N-acetylglu-cosamine is transferred from UDP-GlcNAc results in the production of a disaccharide-pentapeptide derivative of the undecaprenyl pyrophosphate. It is clear that this material (in some cases slightly modified) represents the subunit structure of the peptidoglycan. Modifications, such as the amidation

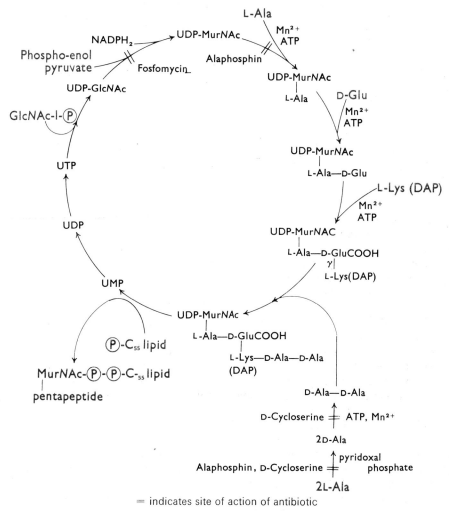

= indicates site of action of antibiotic

Figure 22. Biosynthesis of peptidoglycan. Formation of nucleotide precursors.

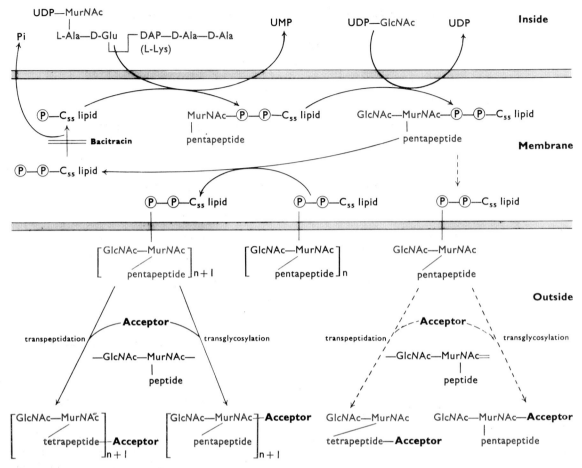

Figure 23. Biosynthesis of peptidoglycan. Incorporation of precursors into linear glycan chains by membrane-bound enzymes. The polymeric glycan may be synthesized while still attached to the C_{55} lipid and be transferred to the peptidoglycan acceptor in the wall either by transglycosylation or transpeptidation (Figure 25). Alternatively the individual cell wall monomer units may be transferred by transglycosylation to the acceptor glycan in the wall or by transpeptidation to an available peptide. These two alternative pathways are denoted by dotted lines.

of the α-COOH group of glutamic acid or the insertion of amino acids which are found in the cross-linking peptide generally occur at this stage. It is interesting that these amino acids are generally inserted from the corresponding amino acyl-tRNA although ribosomes and messenger-RNA are not involved. The same C_{55} phospholipid or one closely related to it has been shown to be involved in the synthesis of the 'O' specific chains of lipo-polysaccharides (p. 71), a mannan polymer in *Micrococcus lysodeikticus*, a capsular polysaccharide, and some wall teichoic acids (p. 63). These observations emphasize the importance of this compound in the synthesis of polymers outside the cytoplasmic membrane. The polymerization of the disaccharide-pentapeptide subunits to form an un-crosslinked

nascent peptidoglycan occurs in some Gram-positive bacteria by insertion of each new unit at the reducing end of a growing peptidoglycan chain still attached to the membrane by a linkage that is labile to mild acid hydrolysis. It is suggested that the glycan chain

$$CH_3C=CH-CH_2(CH_2C=CH-CH_2)_9CH_2C=CH-CH_2$$

Figure 24. C_{55}-polyisoprenoid alcohol phosphate, undecaprenyl phosphate.

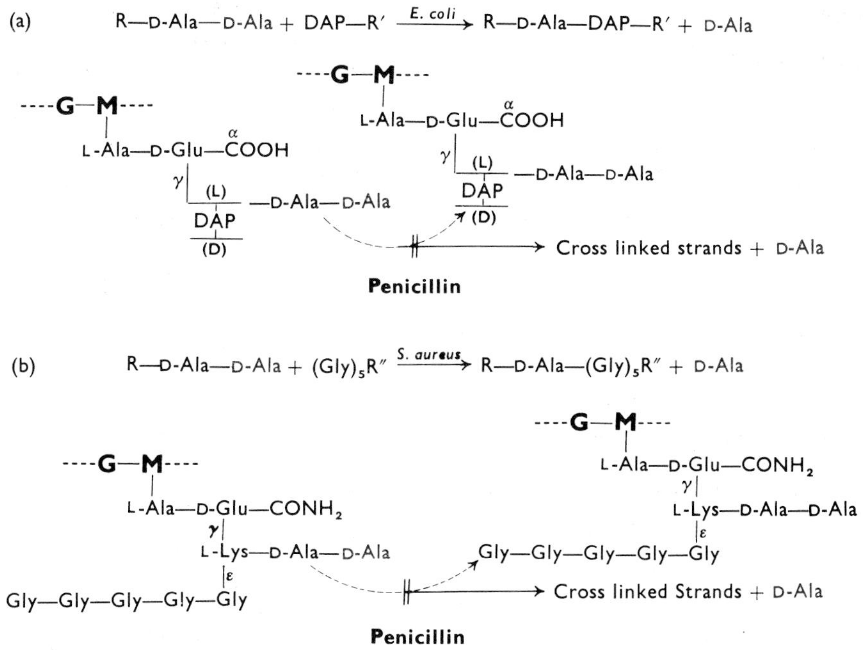

(a) $\quad$ R—D-Ala—D-Ala + DAP—R' $\xrightarrow{\text{E. coli}}$ R—D-Ala—DAP—R' + D-Ala

Figure 25. Biosynthesis of peptidoglycan. The bridge closure reaction in (a) *Escherichia coli* and (b) *Staphylococcus aureus*.

is held in the membrane by the undecaprenyl phosphate. This mechanism of transglycosylation would allow the growing chain to extend away from the external surface of the plasma membrane and permit greater freedom in relation to the positioning of the peptide chains for the final cross-linking reactions. Each transglycosylation reaction releases a molecule of C_{55} undecaprenyl pyrophosphate which is dephosphorylated to complete the cycle of reactions in the membrane.

The last stage in the formation of an insoluble molecule attached to the existing peptidoglycan is a transpeptidation reaction in which a free amino group of a peptide side chain in the wall is joined to the penultimate D-alanine of an adjacent peptide side chain in the nascent peptidoglycan (Figure 25). The cross-linking reaction occurs outside the membrane where ATP is not available to provide energy for the synthesis of a peptide bond: instead, energy for the reaction is provided by hydrolysis of the terminal D-alanyl—D-alanine bond in the pentapeptide chain and one D-alanine is released for every cross-link formed. Few organisms have a degree of peptide cross-linkage greater than 30–50% (*S. aureus* with 90% is an exception) yet isolated peptidoglycans contain mainly tetrapeptides: another enzyme (D,D-carboxypeptidase) cleaves the terminal D-alanine

molecules from uncross-linked pentapeptide chains.

Importance of peptidoglycan

The fact that the cell walls of all bacteria (except halobacteria and mycoplasmas) contain peptidoglycan and that it is often the major component of the wall, amounting to as much as 80% of the dry weight, is strong circumstantial evidence that this component is virtually indispensable. The semipermeable cytoplasmic membrane lying immediately beneath the wall is so fragile that it has to be protected by the cell wall against the osmotic forces exerted within the protoplast. The osmotic pressure may be as much as 15 atmospheres in a Gram-positive coccus and about 5 in a Gram-negative rod. The wall must be rigid and there is abundant evidence that the peptidoglycan component confers this property on the cell wall as a whole.

Enzymic attack by lysozyme: formation of protoplasts

Lysozyme, obtained from egg-white (and found also in a number of natural secretions including tears and sweat) hydrolyses β, 1–4 links between *N*-acetylmuramic acid and *N*-acetylglucosamine. The presence of 6-*O*-acetyl groups on muramyl residues protects some bacteria from the effects of lysozyme but other bacteria (e.g. *Bacillus megaterium*) are lysed by this enzyme unless the treatment is carried out in hyper-

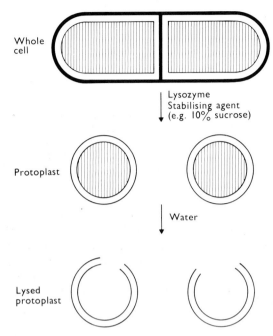

Whole
cell

Lysozyme
Stabilising agent
(e.g. 10% sucrose)

Protoplast

Water

Lysed
protoplast

Figure 26. Formation of protoplasts from *Bacillus mega-terium* by treatment with lysozyme. The cell wall surrounding the photoplast membrane has been removed by the enzymic digestion.

tonic media. Under these conditions protoplasts are formed (Figure 26) which lyse on dilution with water. Degradation of the peptidoglycan therefore results in the disappearance of the cell wall and the loss of rigidity. The cytoplasm, still surrounded by the membrane, becomes spherical and is referred to as a *protoplast*. These structures do not have any of the wall components adhering to them. As far as can be ascertained they are capable of carrying out most of the biosynthetic processes occurring in the intact bacteria including the synthesis of nascent peptidoglycan chains but only under very special circumstances do they revert to the parental shape. Good preparations of protoplasts incorporate radioactive precursors into nucleic acid and protein at rates similar to those of intact cells and can synthesize inducible enzymes. They support the growth of bacteriophages but are not infected by phage since they do not possess the bacteriophage receptors. It has proved difficult to obtain genuine protoplasts of Gram-negative bacteria: treatment of organisms that have been plasmolysed with lysozyme and a metal-chelating agent (which affects the permeability of the outer membrane) results in the production of osmotically-sensitive spheroplasts which still possess the outer membrane component of the cell envelope.

Antibiotics affecting cell wall synthesis

A chemotherapeutic agent which selectively inhibits synthesis of the peptidoglycan is potentially useful in the treatment of bacterial infections since this component of the wall is found uniquely in bacteria and closely related microorganisms. Many antibiotics were recognized as cell wall inhibitors when it was shown that they inhibited the incorporation of typical cell wall amino acids and caused the concomitant accumulation of UDP-MurNAc-peptides. D-Cycloserine bears a striking similarity to D-alanine and it competitively inhibits two enzymes which convert L-alanine to a racemic mixture and then synthesize the D-alanyl-D-alanine dipeptide. Since D-cycloserine binds considerably more strongly to the enzyme than does D-alanine it is a very effective inhibitor at relatively low concentrations. Fosfomycin and alaphosphin are two other antibiotics which inhibit the production of the cytoplasmic precursors (Figure 22).

Penicillin, bacitracin and vancomycin affect peptidoglycan synthesis at some stage after the formation of UDP-MurNAc-pentapeptide. Vancomycin inhibits the polymerization of disaccharide-pentapeptide units resulting in an accumulation of lipid-linked disaccharide-peptides, while bacitracin prevents the regeneration of the C_{55} phospholipid from its pyrophosphate derivative by inhibiting the dephosphorylation reaction. Therefore both antibiotics bring the cycle of reactions shown in Figure 23 to a halt. Whether the bacteria are lysed under these circumstances depends on the activity of autolytic enzymes. Penicillins and cephalosporins (β-lactam antibiotics) inhibit the cross-linking reaction outlined in Figure 25, and their action in growing organisms results in the formation of a defective wall and eventual cell lysis. A number of membrane-bound proteins which bind penicillin covalently have been investigated both biochemically and genetically and tentative roles have been proposed for some of them in cell elongation, cell division and the determination of cell shape.

If an antibiotic affects only peptidoglycan synthesis the organism should not be unduly damaged if the treatment is carried out in protective hypertonic media: this has been found to be the case. Rod-shaped organisms treated with penicillin in the presence of a stabilizing concentration of sucrose lose the shape conferred on them by the rigid cell wall and tend to become spherical (Figure 27). Part of the cell wall remains adhering to the protoplast membrane and because of this the resulting structure is termed a *spheroplast* (*cf.* protoplast). If penicillin is then removed from the medium some of the spheroplasts

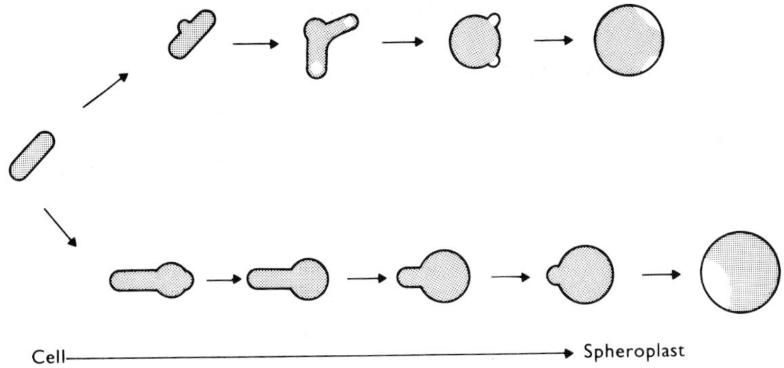

Cell ⟶ Spheroplast

Figure 27. Formation of spheroplasts. These forms can be induced by growth of organisms in the presence of penicillin or by growing DAP-requiring cells in the absence of the amino acid. The cells may bulge either in the centre or terminally.

revert to rod-shaped organisms. A similar sequence of events occurs if a typical peptidoglycan amino acid that cannot be synthesized by the organisms (e.g. DAP in a DAP-requiring organism) is omitted from the medium in which the cells are cultured.

L-forms

L-forms of bacteria are generally regarded as protoplasmic elements not having a defined shape and lacking the rigid component of the cell wall. The loss of shape can be correlated with the loss of peptidoglycan constituents. L-form cultures of many bacterial species have been obtained by growing organisms on agar plates in the presence of pencillin. Although the majority of the parent bacteria are killed, a few survive and after several sub-cultures in the presence of penicillin may not revert to bacteria when they are grown in its absence. They are then termed *stable* L-*forms*. The appearance of the cells of L-forms is similar irrespective of the parental bacterial shape.

Growth of the peptidoglycan polymer

The peptidoglycan is thought to be mainly if not totally responsible for the rigid properties of bacterial cell walls. This cylindrical or spherical 'molecule' is analogous to a bag made of wire netting (chicken wire); the 'strength' of the molecule is dependent on the arrangement of the interpeptide bonds linking the glycan strands which may themselves be several hundred disaccharide-peptide units in length. If the bacterium is to increase in length or diameter it follows that some linkages in the rigid polymer must be broken in order to insert new material. Cutting some of the strands in wire-netting does not lead to collapse of the entire framework. Similarly, controlled hydrolysis of some of the bonds either in the glycan backbone or in the peptide cross-linking chains will not result in much loss of rigidity if bonds are re-formed following the addition of new material. Con-

sequently growth of the polymer involves the controlled activity of a number of autolytic enzymes together with the insertion of new subunits. It is not known how these processes are regulated but there are indications that both synthetic and autolytic processes are localized and do not occur all over the bacterial surface.

Teichoic acids

The name 'teichoic acids' has been given to a group of polymers rich in glycerol phosphate or ribitol phosphate residues and that are associated with the cell wall and/or cytoplasmic membrane of Gram-positive bacteria. Wall teichoic acids can be extracted by treatment of walls or intact cells with dilute acid or alkali: they are linked covalently to peptidoglycan, and are diverse in nature having a large range of glycerol phosphate- and ribitol phosphate-containing structures. Their presence is dependent upon the conditions of growth and they do not occur in all species of Gram-positive bacteria. Membrane-associated teichoic acids can be extracted from whole organisms or membrane fractions with hot 45% aqueous phenol or by hot water extraction of lipid-depleted, freeze-dried bacteria: they invariably contain a chain of glycerol phosphate units linked covalently to a glycolipid component of the cytoplasmic membrane and are present in all Gram-positive bacteria that have been investigated irrespective of the growth conditions. The term 'lipoteichoic acids' has been applied to this class of polymers. The main types of glycerol-containing wall teichoic acids are illustrated in Figure 28. Variations of the type shown in Figure 28a include glycerol phosphate residues linked 1–2 rather than 1–3 and in this instance the R substituent is on position 3. Polymers from the same organism are found which differ in both the substituent and in the degree of

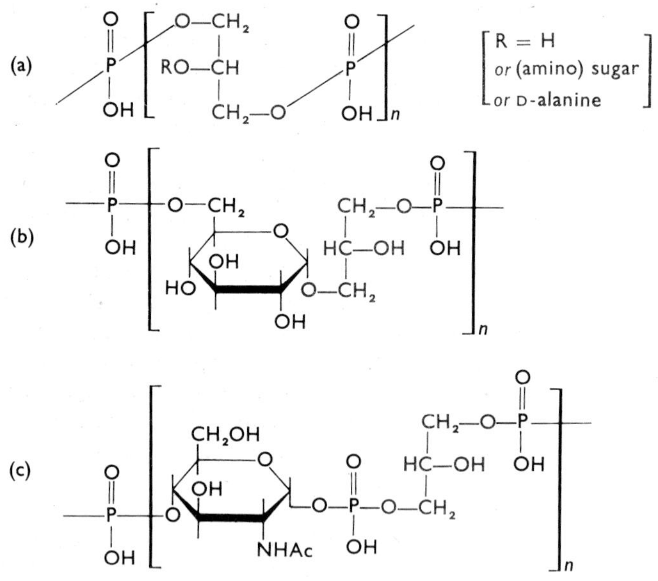

Figure 28. Glycerol teichoic acids. (a) The repeating unit is glycerol phosphate and the sugar or alanine substituent of the glycerol is not in the backbone chain. (b) The repeating unit is glucosylglycerol phosphate. (c) The repeating unit is N-acetyl-glucosamine-l-phosphate-glycerol phosphate. In (b) and (c) the sugar molecules are present in the backbone chain.

substitution. If a sugar or amino sugar is present in the polymer it can be found either as a substituent of the glycerol (Figure 28a) or as a component of the backbone chain (Figure 28b, c) the difference between these two structures being in the ratio of glycerol to phosphate in the polymer chain.

The general structure of the other main type of wall teichoic acid (ribitol teichoic acid) is given in Figure 29; this is typical of *S. aureus* and various bacilli. The position of the ester-linked D-alanine is still uncertain: it is present on the C_2 position of the ribitol in a mutant of *S. aureus* lacking a sugar substituent, but in Bacilli it may be present on the C_2 or C_3 position while the sugar residue is on position C_4. Chain length is generally 30–35 residues. As with the wall glycerol teichoic acids, considerable variation may occur within a single species in respect to degree and type of substitution and the actual linkage of the sugar residues (α or β). More complicated teichoic acids are found in the walls of pneumococci: these

may contain as many as 4 different sugar residues in the chain linking two ribitol phosphate units.

Lipoteichoic acids consist of a chain of 1–3 linked glycerol phosphate residues (about 30 units long) attached to a glycolipid through a phosphodiester

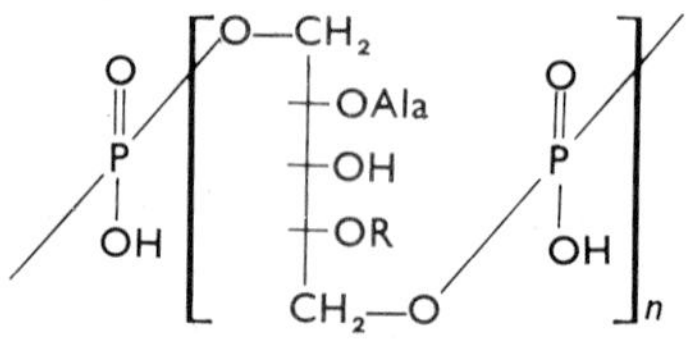

Figure 29. Structure of 1,5-poly(ribitol phosphate) wall teichoic acids. R = glycosyl.

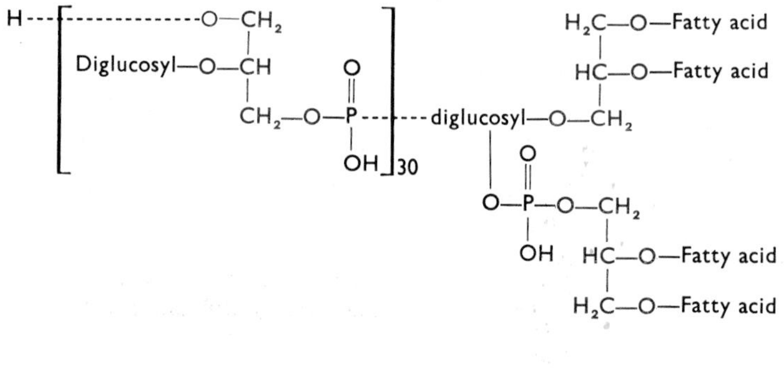

bond. The glycolipid of *Streptococcus faecalis* is believed to be a substituted diglucoside diglyceride: one possible structure is given in Figure 30.

Figure 30. Tentative structure of lipoteichoic acid of *Streptococcus faecalis*. The diglucosyl residues consist of kojibiose (glucose α, 1–2 glucose) units.

Biosynthesis

The discovery of teichoic acids was preceded by the isolation of cytidine diphosphate ribitol (CDP-ribitol) and cytidine diphosphate glycerol (CDP-glycerol). The biosynthetic role of these nucleotide derivatives was predicted before the corresponding polymers were themselves isolated and characterized but it was many years before a direct demonstration of polymer synthesis was achieved. Particulate enzyme preparations were obtained which catalysed the synthesis of a polyglycerophosphate polymer in the following manner:

D-glycerol-1-phosphate + CTP $\longrightarrow$ CDP-glycerol + PPi

CDP-glycerol + (glycerolphosphate)$_n$ $\longrightarrow$
$$CMP + (glycerophosphate)_{n+1}$$

Similar preparations composed of membrane fragments catalysed the synthesis of polyribitol phosphate using CDP-ribitol as a precursor. The monosaccharide and D-alanine substituents are added to the polyol-phosphate backbone chain, the (amino)sugar components being transferred from the corresponding nucleotide precursors. Solubilization and partial purification of the membrane-bound enzymes has revealed a requirement for lipoteichoic acid as an acceptor of the glycerol phosphate or ribitol phosphate units from the nucleotide derivatives. The more highly purified ribitol phosphate polymerase also has a requirement for membrane phospholipid. This requirement for both an acceptor and for a specific spatial arrangement involving phospholipid is also apparent in studies of *in vitro* systems of lipopolysaccharide biosynthesis (page 72). The completed chains of ribitol phosphate or glycerol phosphate units containing glycosyl and/or D-alanine substituents can then be transferred from their membrane anchorage point to the peptidoglycan, though an additional linkage group is present between the two polymers. The more complex polymers which contain glycosyl residues within the polymer chain involve a different membrane attachment site. Early studies with cell-free systems using cytoplasmic membrane fragments as enzyme and acceptor established that the polymer shown in Figure 28b could be synthesized from UDP-glucose and CDP-glycerol. The sugar is transferred first to a phospholipid acceptor in the membrane (probably the same undecaprenyl phosphate as is involved in peptidoglycan biosynthesis), followed by the addition of glycerol phosphate to form the repeating unit which is then transferred from the lipid intermediate to the growing polymer chain (Figure 31). Competition experiments suggest that there is a pool of undecaprenyl phosphate which can function, at least *in vitro*, in either teichoic acid or peptidoglycan biosynthesis.

Linkage to wall polymers

Digestion of cell walls with enzymes that catalyse hydrolysis of the peptidoglycan results in the liberation of teichoic acid chains with small fragments of peptidoglycan attached. Since some muramic acid-6-phosphate was found in the hydrolysates of peptidoglycan from which teichoic acid had been removed by extraction with mild acid it was supposed that wall teichoic acids were attached to the peptidoglycan. It has now been established by detailed chemical analysis and by biosynthetic evidence that a linkage unit consisting of three glycerol phosphates and one N-acetylglucosamine residue occurs between the terminal phosphate of a ribitol phosphate chain and the 6-position of muramic acid (Figure 32). Biosynthesis of the linkage group is achieved by sequential transfer of N-acetylglucosamine-1-phosphate and three glycerol phosphate units to an undecaprenyl-phosphate carrier in the membrane. Presumably the wall teichoic acid is transferred from the lipoteichoic acid carrier to the linkage group and the entire structure is added to the nascent peptidoglycan which

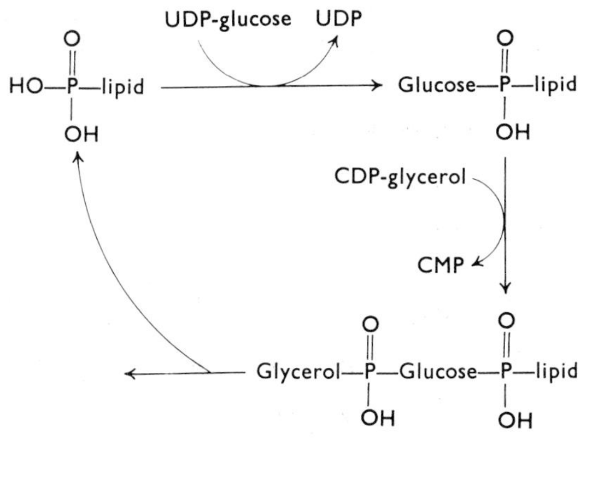

Figure 31. Biosynthesis of poly-(glucosyl-glycerol) phosphate.

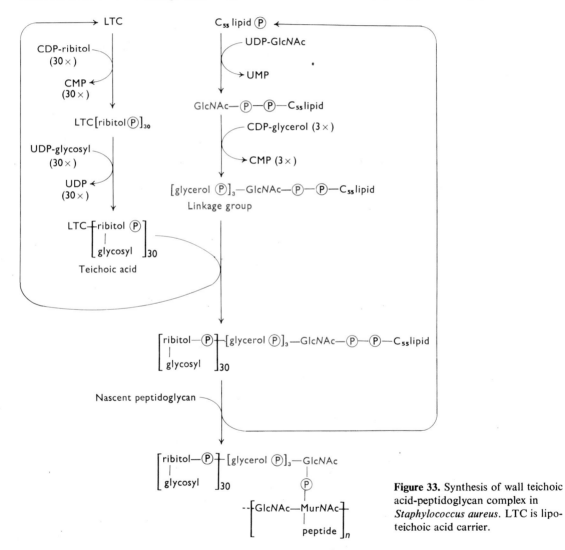

Figure 32. Structure of the linkage unit joining the wall teichoic acid to the peptidoglycan of *Staphylococcus aureus*.

Figure 33. Synthesis of wall teichoic acid-peptidoglycan complex in *Staphylococcus aureus*. LTC is lipo-teichoic acid carrier.

is then inserted into the cell wall (Figure 33). It follows that covalent attachment of teichoic acid to the wall should be dependent on peptidoglycan biosynthesis: this is so in most, but not all, of the systems that have been investigated.

Function

Wall teichoic acids may account for 50% of the weight of the wall (10% of the dry mass of the organism) and Gram-positive bacteria must divert a substantial proportion of their metabolic activity

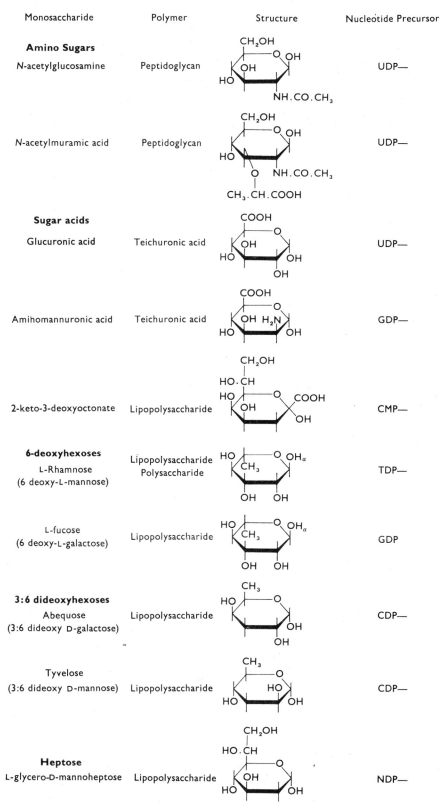

Monosaccharide	Polymer	Structure	Nucleotide Precursor
Amino Sugars			
N-acetylglucosamine	Peptidoglycan		UDP—
N-acetylmuramic acid	Peptidoglycan		UDP—
Sugar acids			
Glucuronic acid	Teichuronic acid		UDP—
Amihomannuronic acid	Teichuronic acid		GDP—
2-keto-3-deoxyoctonate	Lipopolysaccharide		CMP—
6-deoxyhexoses L-Rhamnose (6 deoxy-L-mannose)	Lipopolysaccharide Polysaccharide		TDP—
L-fucose (6 deoxy-L-galactose)	Lipopolysaccharide		GDP
3:6 dideoxyhexoses Abequose (3:6 dideoxy D-galactose)	Lipopolysaccharide		CDP—
Tyvelose (3:6 dideoxy D-mannose)	Lipopolysaccharide		CDP—
Heptose L-glycero-D-mannoheptose	Lipopolysaccharide		NDP—

Figure 34. Structure of carbohydrate components occurring in cell wall polymers.

into their synthesis. The amount of polymer is dependent on the medium: under conditions of Mg^{2+}-limitation they are produced in abundance; when phosphate is limited the polymer is to a large extent replaced with another anionic polymer, teichuronic acid, consisting of alternating residues of *N*-acetylgalactosamine and glucuronic acid (Figure 34; see also p. 122). It seems likely that a negatively-charged polymer in the wall is essential as there are indications that it functions to bind cations in the wall and to provide the correct cationic environment at the cell membrane (many of the wall polymer biosynthetic enzymes require a high concentration of Mg^{2+} ions for optimal activity *in vitro*). Wall teichoic acids are also receptor sites for some bacteriophages: the sugar residues are important though a rather low degree of stereospecificity is found in relation to the adsorption process. Isolated teichoic acid is incapable of inactivating phages so the true receptor site in the wall is probably a peptidoglycan-teichoic acid complex.

The lipoteichoic acids are synthesized even under conditions of phosphate deprivation and apart from their role in the biosynthesis of wall teichoic acids it is believed that they may regulate the activity of wall-bound autolytic enzymes. The flexible glycerol chain of these teichoic acids ensures that they can extend from their attachment site in the membrane to the external surface of the wall (they are antigenic even in whole cells) and it has been found that acylated lipoteichoic acid inhibits the activity of enzymes that hydrolyse linkages in the peptidoglycan. The de-acylated lipoteichoic acid has no inhibitory effect. This may provide part of a control mechanism for ensuring that balanced growth occurs.

Polysaccharides and amino-sugar polymers

The presence of sugars and amino-sugars in teichoic acids complicates studies of polysaccharide polymers: nevertheless it has been shown that many Gram-positive bacteria contain polymers of sugar or sugar phosphate residues. A polymer consisting of poly *N*-acetylglucosamine-1-phosphate comprises 40% of the weight of the wall of *Staph. lactis* 2102: more complex polymers with repeating units of glucose-*N*-acetylgalactosamine-1-phosphate have been found in the walls of a Micrococcus and a Bacillus. The teichuronic acids referred to earlier in connection with their ability to replace wall teichoic acids contain alternating residues of a uronic acid and an amino sugar while walls of *Micrococcus lysodeikticus*

are rich in aminomannuronic acid which may be present in an analogous structure. The possible function of these anionic polymers has been discussed in the section dealing with the functions of teichoic acids (see p. 64). The polysaccharides present in the walls of Gram-negative organisms are present as lipopolysaccharides and are discussed on page 68.

Proteins

In many species of bacteria the external layer of the wall is composed of regularly-arranged protein sub-units (Figures 7, 11, 12). In some bacteria these can be detached and dissociated into the individual subunits. Under specified conditions these can re-associate into regular arrays with the same dimensions as the pattern in intact bacteria. This external layer is not present in all bacteria or even in all strains of the same species so that the function of this protein layer remains obscure.

Isolated envelope preparations of Gram-negative bacteria contain the cytoplasmic membrane which is rich in protein. The outer and inner membranes can be separated by equilibrium density-gradient centrifugation and the outer membrane has been shown to contain proteins in addition to phospholipid and lipopolysaccharide. Five major and approximately thirty minor proteins are present and there is extensive protein-protein interaction (as revealed by cross-linking reagents) in addition to interaction with phospholipids and peptidoglycan. One of the major outer membrane proteins is a lipoprotein, one third of which is bound covalently to the underlying peptidoglycan thus linking the rigid layer of the wall to the outer membrane (Figure 35). The linkage is from the ε-NH_2 group of the C-terminal lysine residue of the lipoprotein to a diaminopimelic acid residue in the peptidoglycan. There is approximately one molecule of lipoprotein for every ten disaccharide peptide subunits in the peptidoglycan. The remaining two thirds of the lipoprotein occurs in the free form in the outer membrane. The lipoprotein is synthesized as a polypeptide precursor bound to the cytoplasmic membrane by an *N*-terminal sequence of 20 amino acids, most of which are hydrophobic. Before translocation to the outer membrane and attachment to peptidoglycan, the hydrophobic polypeptide must be cleaved and a fatty acid and diglyceride residue added to the new *N*-terminal cysteine residue (Figure 36).

The other four major outer membrane proteins of *E. coli* can act as phage adsorption sites indicating that

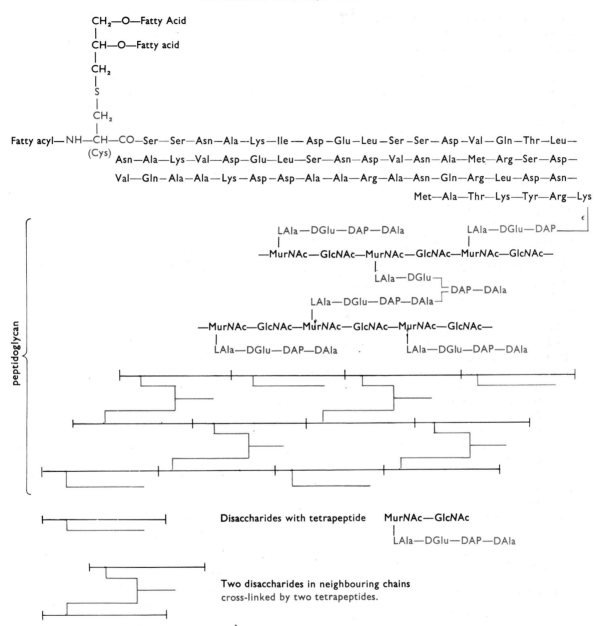

Figure 35. Structure of the lipoprotein-peptidoglycan complex of *Escherichia coli.*

they are available on the outer surface of the membrane. Other experiments indicate that some of them span the whole thickness of the membrane and interact with the peptidoglycan. Many of the minor proteins are apparently multi-functional. In addition to any structural role they may have in determining the three-dimensional architecture of the membrane which is a permeability barrier to hydrophilic molecules with molecular masses greater than 500–600, some of the minor proteins, which were originally detected as phage and colicin receptor sites, have been shown to be important in the transport of Vitamin B_{12} and of high molecular weight ferric iron complexes. Bacteriophage- and colicin-resistant strains lacking the receptor proteins are unable to take up the iron complexes and Vitamin B_{12}.

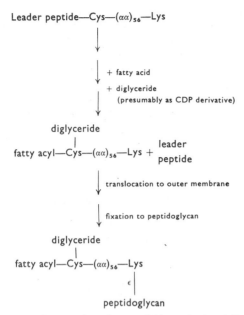

Figure 36. Proposed pathway of biosynthesis of lipo-protein-peptidoglycan complex.

Lipids

A conspicuous difference in lipid content is found between cell walls isolated from Gram-positive and -negative bacteria. Gram-positive organisms have very little extractable lipid in the wall with the exception of the mycobacteria which have as much as 60% of the dry weight of the wall composed of wax, phosphatides and bound lipids. The high lipid content of the outer membrane of Gram-negative bacteria is due to the lipid A of lipopolysaccharides (see p. 70), the diglyceride and fatty acid residues attached to the *N*-terminal cysteine of the lipoprotein (page 67) and phospholipids (mostly phosphatidyl-ethanolamine). The lipoprotein and phospholipids are believed to occur predominantly in the inner half of the outer membrane while lipid A is present in the outer half.

Lipopolysaccharides

The monosaccharides in the cell walls of Gram-negative bacteria such as *Salmonella typhimurium* and

Figure 37. Sequence of the three structural regions of lipopolysaccharides.

Escherichia coli bear a striking similarity to those found in the immunologically specific lipopolysac-charides isolated by extraction of whole cells with phenol/water. The resulting aqueous phase contains nucleic acids and polysaccharides in addition to lipopolysaccharide which can be sedimented by centrifugation at 100 000 × g. Thin-section micro-scopy shows that bacteria that have undergone this extraction procedure have lost the outer 3-layered membrane of the cell envelope but still retain the cytoplasmic membrane and the rigid layer of the wall.

Chemical investigation of a large number of lipo-polysaccharides indicates the presence of the general structure shown in Figure 37. Lipid A is found in all lipopolysaccharides and the core oligosaccharide region is attached to it. There are a number of mutant strains which lack part of this region and it is through analysis of the 'deficient' lipopolysaccharides in these strains that the detailed chemical structure has been elucidated. The structure of the core is thought to be similar if not identical in closely related strains. The outermost region of the lipopolysac-charide comprises the 'O' specific chains, the structures of which are highly species specific.

Lipopolysaccharides from all wild-type strains of salmonellae contain glucose, galactose, *N*-acetyl-glucosamine, 2-keto-3-deoxyoctonate (KDO) and L-glycero-D-mannoheptose (core region) together with a number of other sugars which may include hexoses (galactose, mannose), pentoses (ribose, xylose), 6-deoxyhexoses (rhamnose, fucose) and 2:6-dideoxy-hexoses (abequose, tyvelose etc.). These sugars are found in the 'O' specific chains. One such chain from *Salmonella typhimurium* contains galactose, mannose, rhamnose and abequose and its structure is shown in Figure 38. It contains several repeating units of a specific sequence of sugars.

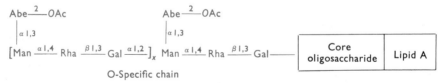

Figure 38. Structure of the 'O' specific chains of the lipopolysaccharide of *Salmonella typhimurium*. The repeating unit imparts antigenic activity to the lipopolysaccharide; the sugar composition is strain-specific.

The order of sugar residues within the oligosaccharide chain of the core region has been determined by the use of mutants blocked in the synthesis of nucleotide-sugar precursors or in the transfer of sugar or phosphate residues. The lipopolysaccharides formed under such conditions lack certain sugar residues but this, apparently, has no effect on the ability of the mutants to survive under laboratory conditions. The compositions of the lipopolysaccharides from a number of mutant classes are shown in Table 2 and the composite structure derived from these and other more detailed chemical studies is given in Figure 39. It is thought that all closely related strains possess this structure. The 'O' specific chains are attached to the glucose II residue.

The core oligosaccharide is in turn linked through one of the KDO residues to lipid A the structure of which has been determined (Figure 40). It contains two glucosamine residues, fully substituted with long chain fatty acids, β-hydroxymyristic acid and phosphate.

It is apparent that the molecular mass of the

Table 2. Structure of the lipopolysaccharides of *Salmonella* mutants unable to synthesize the complete core region

Mutant type	Deficiency	Components present
1	Addition of heptose	$[KDO]_3 \rightarrow$ lipid A $\uparrow$ (P)—Ethanolamine
2	Addition of second heptose	Hep $\rightarrow [KDO]_3 \rightarrow$ lipid A $\uparrow$ (P)—Ethanolamine
3	Synthesis of UDP—Glc	Hep $\rightarrow$ Hep $\rightarrow [KDO]_3 \rightarrow$ lipid A $\uparrow$ (P)—Ethanolamine
4	Transfer of phosphate	Glc $\rightarrow$ Hep $\rightarrow$ Hep $\rightarrow [KDO]_3 \rightarrow$ lipid A $\uparrow$ (P)—Ethanolamine
5	Synthesis of UDP—Gal	Hep $\downarrow$ Glc $\rightarrow$ Hep $\rightarrow$ Hep $\rightarrow [KDO]_3 \rightarrow$ lipid A $\uparrow$ $\uparrow$ $\uparrow$ (P) (P) (P)—Ethanolamine $\uparrow$ (P)—Ethanolamine
6	Synthesis of UDP—GlcNAc	Gal Hep $\downarrow$ $\downarrow$ Glc $\rightarrow$ Gal $\rightarrow$ Glc $\rightarrow$ Hep $\rightarrow$ Hep $\rightarrow [KDO]_3 \rightarrow$ lipid A $\uparrow$ $\uparrow$ $\uparrow$ (P) (P) (P)—Ethanolamine $\uparrow$ (P)—Ethanolamine
7	Synthesis of 'O' specific chain	GlcNAc Gal Hep $\downarrow$ $\downarrow$ $\downarrow$ Glc $\rightarrow$ Gal $\rightarrow$ Glc $\rightarrow$ Hep $\rightarrow$ Hep $\rightarrow [KDO]_3 \rightarrow$ lipid A $\uparrow$ $\uparrow$ $\uparrow$ (P) (P) (P)—Ethanolamine $\uparrow$ (P)—Ethanolamine

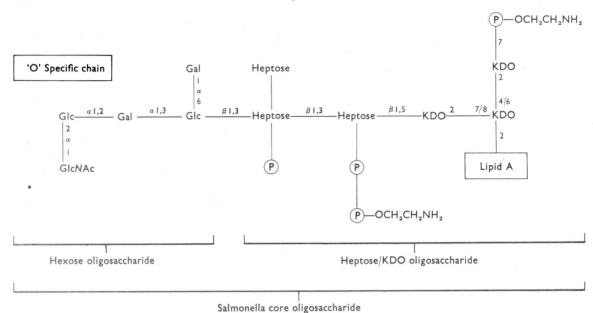

Figure 39. Structure of the 'core' oligosaccharide of *Salmonella spp*. The structure has been deduced from the analysis of partial acid hydrolysates of lipopolysaccharides isolated from normal strains and from mutants unable to synthesize, the complete structure.

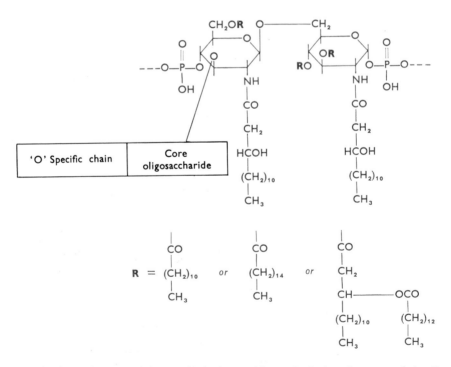

Figure 40. Structure of lipid A in *Salmonella* spp. The fatty acids on the hydroxyl groups of the disaccharide are present in the ratio of 1:1:1.

complex, lipid A—core oligosaccharide—'O' specific chain, is several thousand and the availability of phosphate residues in the lipid A may result in pyrophosphoryl linkages between individual subunits with increase in the size of the polymer. These large molecules are important constituents of the outer membrane of the cell envelope because they contain both hydrophobic and hydrophilic residues.

Biosynthesis

The 'O specific chains are synthesized in a manner analogous to peptidoglycans—again an extracellular polymeric component being made from small precursors. The repeating unit is first assembled as an oligosaccharide linked to a C_{55} polyisoprenoid alcohol phosphate (Figure 24) each sugar being added from its nucleotide precursor. In *Salmonella typhimurium* this involves four different bases in the nucleotides, *U*DP-galactose, *T*DP-rhamnose, *G*DP-mannose and *C*DP-abequose (Figure 41). The completed oligosaccharide is then polymerized to form the 'O' specific chain still attached to the phospholipid in

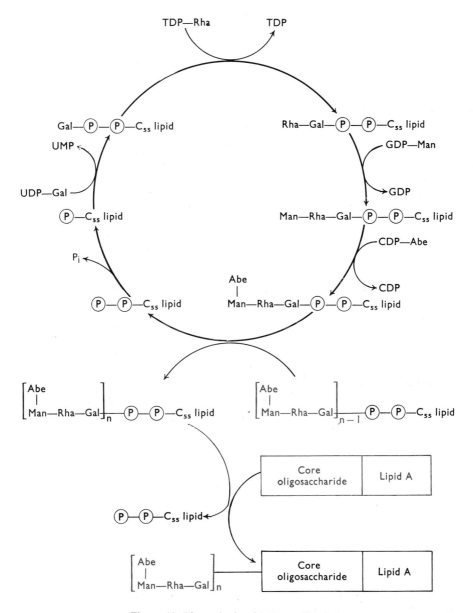

Figure 41. Biosynthesis of 'O' specific chain.

①   UDP-Glc + Glucose-deficient lipopolysaccharide (LPS) ⟶ UDP + Glc-LPS

②   UDP-Gal + Glc-LPS ⟶ UDP + Gal-Glc-LPS

③   UDP-Glc + Gal-Glc-LPS ⟶ UDP + Glc-Gal-Glc-LPS

④   UDP-GlcNAc + Glc-Gal-Glc-LPS ⟶ UDP + GlcNAc-Glc-Gal-Glc-LPS

Figure 42. Biosynthesis of core of lipopolysaccharide. This has been elucidated using mutants which are unable to synthesize either UDP-glucose or UDP-galactose. The former mutant contains an incomplete lipopolysaccharide to which glucose can be added, the second, one to which galactose can be added. Further additions, catalysed by particulate enzyme preparations, are then possible.

the cytoplasmic membrane. The chain extends by the addition of the repeating unit to the reducing end of the growing chain. The completed chain is then transferred to the core of the lipopolysaccharide.

The core polysaccharide is built up sequentially by the addition of sugars to the incomplete core. Little is known of how the heptose residues are transferred since the nucleotide precursor has not been identified. The addition of glucose, galactose and *N*-acetylglucosamine has been studied using cell envelope preparations from mutant organisms as a source of enzymes and as the receptor. In addition to the particulate enzyme preparations, soluble enzymes have been isolated which catalyse the addition of glucose and galactose to boiled cell envelope preparations from suitable mutant organisms. The product of the first reaction is used as the substrate for the second as is illustrated in the scheme for the synthesis of part of the core oligosaccharide (Figure 42). The cell envelope has been fractionated in an attempt to find the material which accepts the monosaccharides. The active material is a complex of lipopolysaccharide and phospholipid since the purified lipopolysaccharide alone does not function as an acceptor, while a mixture of lipopolysaccharide and phospholipid is active if it is prepared under specified conditions of heating and slow cooling. A complex containing phospholipid, lipopolysaccharide and the soluble enzyme which catalyses the addition of galactose to the incomplete lipopolysaccharide has been obtained both by density gradient centrifugation and by monolayer experiments. In the latter a monolayer of phospholipid is first prepared and galactose-deficient lipopolysaccharide acceptor injected below it into the aqueous phase. An interaction between phospholipid and lipopolysaccharide is indicated by a rise in pressure and analysis of the monolayer indicates that it contains both phospholipid and lipopolysaccharide. If purified galactosyl transferase is then injected below the surface of the film there is a further increase in pressure and

the enzyme becomes incorporated into the film forming a ternary complex. This complex catalyses the addition of galactose from UDP-galactose to the galactose-deficient lipopolysaccharide. It seems that the phospholipid is important in maintaining the correct spatial arrangement of the acceptor-enzyme complex.

Synthesis of the core and complete 'O' antigen regions followed by attachment to each other is catalysed by enzymes located in the cytoplasmic membrane. The resulting lipopolysaccharide is exported to the outer membrane by an irreversible step whose mechanism is unknown. Incomplete lipopolysaccharide molecules made in the absence of an essential component and that are already present in the outer membrane cannot be completed when the essential nutrient is added back to the growth medium, although complete new chains are synthesized and exported. It is believed that a limited number of export sites are available (200 per cell) as revealed by labelling newly-synthesized lipopolysaccharide with ferritin-conjugated antibody, but within a short time the 'label' is distributed all over the cell surface presumably due to movement of the lipopolysaccharide in the fluid environment of the membrane.

Function

Since mutants containing only the lipid A and the KDO region of the lipopolysaccharide grow in the laboratory just as well as wild-type strains it is clear that the 'O' specific chains and monosaccharides found in the outer core structures are dispensible. However, only wild-type strains containing the complete 'O' specific chain are pathogenic so these components appear to be essential for successful invasion of the host. Futher the presence of charged phosphate groups may permit the polymer to function in the binding of ions and possibly to maintain a defined ionic environment in this area of the wall (*cf.* the function of teichoic acids in Gram-positive organisms, page 64).

Site of cell wall formation and cell division

The biosynthesis of the cell wall and cell division involve the controlled and integrated deposition of several different polymers which form a recognized structure when assembled. As the cell wall is very thick in molecular terms it follows that new polymers being synthesized by membrane-bound enzymes and laid down on the inside of the cell wall will not necessarily be in the same location by the time they reach the outside of the wall and appear on the cell surface. Therefore it may be misleading to investigate sites of deposition of the wall polymers using methods that rely on the interaction of a marker with the outside of the cell wall. Nevertheless the labelling of new wall with fluorescent or ferritin-labelled antibodies,

or with bacteriophages, has yielded valuable information on the presumed sites of synthesis of the cell wall polymers as have other methods including the location of newly incorporated radio-isotopes by autoradiography and the segregation of radioactive or other markers of the cell surface (transport systems, flagella). When morphological markers are present (e.g. the raised wall bands in *Streptococcus faecalis* and *B. subtilis*) the biochemical observations can be checked by an alternative method. In the majority of Gram-positive organisms that have been investigated the peptidoglycan and polymers linked to it covalently are unstable and an appreciable fraction of the wall (as much as 50%) may be lost in the course of a single generation: this complicates the interpretation of 'labelling' experiments. A

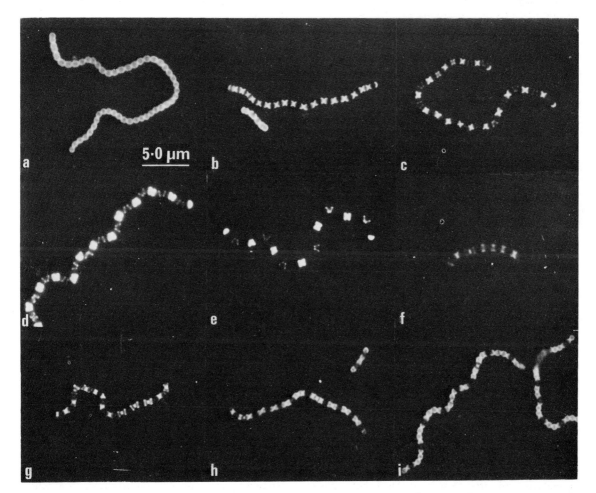

Figure 43. Ultra-violet photomicrographs of *Streptococcus pyogenes* showing cell wall immunofluorescence patterns. In photographs a–e, taken at 15 min intervals, new wall appears as dark patches between the light semi-circles. In photographs f–i, also taken at 15 min intervals, newly-synthesized wall is labelled with the fluorescent antibody. Reproduced with permission of R. M. Cole. *Science N.Y.* **135** (1962) 722.

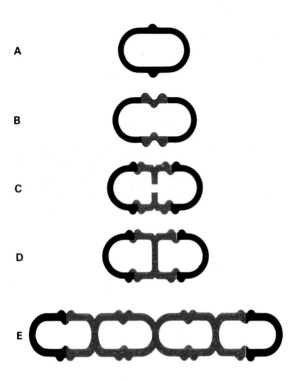

Figure 44. Representation of growth of cell wall and of septa in *Streptococcus faecalis*. The sequence represents successive stages in growth and is based on electron micrographs.

In (A) the wall and the equatorial wall band have been synthesized in the preceding generation and are 'old' (black). Deposition of new material (red) occurs in the region of the band and a notch develops. There is concomitant separation of the wall band areas (B). In (C) formation of the septum continues and the wall bands move further apart. In (D) the septation is complete, the wall bands have reached the equators of the daughter cells, and new septa are being laid down. In (E) the whole process has been repeated: there are four cells and only the poles of the outermost pair consist of old wall.

further difficulty encountered in Gram-negative organisms is that the outer membrane is a dynamic and relatively fluid structure and hydrophobic polymers deposited in a relatively few areas may become randomly distributed over the entire cell surface in a short time.

The peptidoglycan is, however, stable in the Gram-positive streptococci: hence the results of biochemical and morphological investigations are likely to be reliable. The site of cell wall formation was examined first using fluorescent antibodies. Intact streptococci or isolated cell walls were injected into rabbits and after several weeks antibodies were prepared from the serum. The antibody molecules were treated with fluorescein isothiocyanate to give a fluorescent derivative which still reacted with cell wall material. Bacteria were grown in the presence of this fluorescent antibody for a certain length of time: the antibody was then removed and the bacteria were re-incubated in fresh medium. Samples were examined microscopically by ultra-violet illumination: old wall appeared brilliantly fluorescent while new wall was dark. The results were checked by the reverse technique: organisms were exposed initially to unlabelled antibody to prevent subsequent reaction of old wall with fluorescent antibody followed by growth in fresh medium. Bacterial smears taken from the re-incubated culture were stained on slides with fluorescent antibody to show the formation of new wall, or alternatively, labelled antibody was included in the medium used for re-incubation. Results obtained by the use of this method had to be interpreted with caution since not all wall components are necessarily replicated simultaneously and at the same sites. When *Streptococcus pyogenes* was stained with antibodies specific for two different wall components (protein and polysaccharide) essentially the same pattern of cell wall growth was revealed (Figure 43). Cell wall growth was initiated equatorially and synthesis of the cross-wall and peripheral wall occurred simultaneously. In this manner the new halves of the daughter cocci are formed back to back. The 'old' ends of the cocci are preserved intact but are gradually pushed further apart as shown in Figures 43 and 44. When this type of experiment is repeated with Gram-negative organisms new wall is apparently incorporated in a diffuse manner but this probably results from the mobility of the relevant antigens in the outer membrane.

Longitudinal sections of *Streptococcus faecalis* reveal the presence of a raised wall band whose position can be followed throughout the growth of a

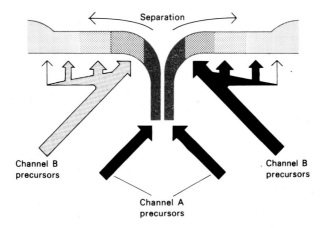

Figure 45. Model illustrating the channels of incorporation of cell wall precursors in order to achieve enlargement of the cell surface and cell division in *Streptococcus faecalis*. Channel A precursors are involved in cross wall formation and are fed in at constant rate throughout the growth cycle. Channel B precursors thicken the peripheral wall formed from the separating halves of the septum at its base. The precise sites of incorporation of the precursors are unknown.

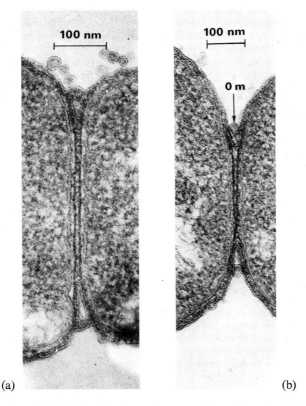

(a)   (b)

Figure 46. Sections of *Escherichia coli* illustrating two stages in septum formation and cell division. In (a) the septum is composed of the plasma membrane and two lamellae of peptidoglycan. In (b) cell division is effected by ingrowth of the outer membrane (Om) between the two layers of peptidoglycan. Reproduced with permission of I. D. J. Burdett and R. G. E. Murray. *J. Bacteriol.* **119** (1974) 1039.

cell: at the beginning of a new round of growth, wall material is deposited immediately beneath the band to initiate a new septum. A notch then appears in the wall band and the two halves of the band are separated as new peripheral wall is synthesized between them (Figure 44). It is believed that there is a single growth zone and that peripheral and septal wall are synthesized by the controlled deposition of wall precursors arising from two channels (Figure 45). One set of precursors (channel A) is involved solely in the synthesis of the new cross wall and three-dimensional reconstructions of whole cells from thin sections suggest that the flow of precursors through this channel into the cross wall is constant until the cross wall closes. Precursors flowing through channel B thicken the cross wall at its base as it separates into the two layers of peripheral wall and expand the surface area of the peripheral wall. A gradual reduction in the flow of precursors through this second channel as the growth cycle proceeds results in an increasing curvature of the new cell pole and eventual cell division (Figure 44). It follows that material deposited initially in the cell septum becomes part of the peripheral wall as the cross wall separates at its base as a result of controlled autolytic action. The actual location of the site of incorporation of the precursors is unknown as modifications and re-modelling of the peptidoglycan may occur after the initial incorporation event. This technique of using measurements taken from thin sections of cells to reconstruct a cycle of cell wall assembly is currently being applied to the bacilli in which small raised wall bands have recently been observed: it should then be possible to compare the growth of the wall in rods and cocci.

In Gram-positive organisms the septum is composed of a double thickness of wall, and cell division apparently involves the action of autolytic enzymes. In Gram-negative organisms the process is more complicated. In thin sections septum formation can be seen to be initiated as an ingrowth of the cytoplasmic membrane together with the peptidoglycan (rigid layer). Eventually a complete septum is formed consisting of two lamellae of peptidoglycan separated by an electron-transparent gap but the characteristic three-layered outer membrane is absent from the septum (Figure 46). After the septum is complete the outer membrane grows inwards, a process which eventually results in the separation of the two daughter cells.

CAPSULES

Capsules are manufactured by certain Gram-positive and Gram-negative bacteria and form the outermost layer of the cell. They are not indispensible structural elements and their occurrence is subject to cultural conditions. The capsule is not shown up easily in electron micrographs (Figure 5, p. 47) but its presence can be demonstrated immunologically or by dispersing the cells in Indian ink (Figure 47).

Before it was known that some capsules were immunologically active, it was believed that they contained homogeneous accumulations of amorphous material surrounding the cell wall. However, many of the capsules are heterogeneous and the presence of polypeptide and polysaccharide material has been demonstrated in the capsule of *Bacillus megaterium* strain M after exposure of organisms to antibodies reacting with these two components. Striated structures within a capsular matrix have been detected in

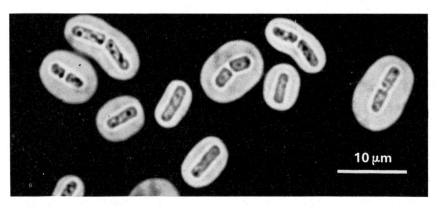

Figure 47. Capsule of *Bacillus megaterium* demonstrated by dispersing the cells in Indian ink. Reproduced with permission of C. F. Robinow. Unpublished.

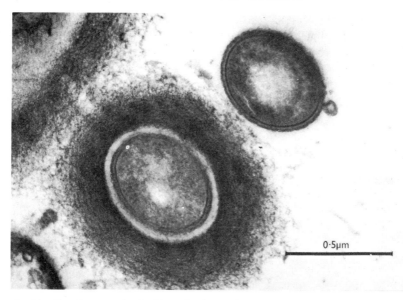

Figure 48. Polysaccharide slime layer, stained with ruthenium red, showing a concentric arrangement of the strands. Reproduced with permission of H. C. Jones, I. L. Roth and W. M. Sanders. *J. Bacteriol.* **99** (1969) 316.

Escherichia coli strain Lisbonne, while the capsule of other organisms may not be of even thickness over the whole surface of the organism.

Many organisms do not produce a well-defined capsule but secrete a loose slime or extracellular gum composed of polysaccharide. This material may be loosely adsorbed on the cell surface and can frequently be removed by washing. Occasionally it may be laid down in an ordered manner (Figure 48).

Sugars and amino sugars are found most frequently in capsules: uronic acids also occur and distinguish the capsular polysaccharide from cell wall polysaccharides in which they are seldom detected. The capsules of a number of bacilli contain polypeptide material in addition to polysaccharide: the peptide is composed of D-glutamic acid units linked together through the α-amino and γ-carboxyl groups. The α-carboxyl groups are apparently in the amide form—reminiscent of the D-glutamyl residues of the cell wall peptidoglycan.

It is not clear what holds the capsule in position. The fact that material related to wall polymers has

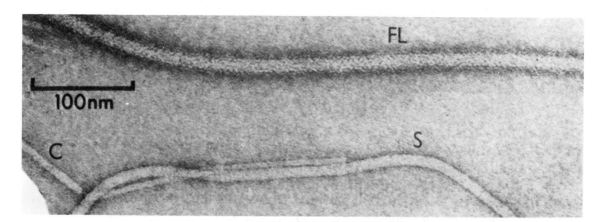

Figure 49. Negatively-stained common I pilus (C), F-like sex pilus (S) and flagellum (FL) of *Escherichia coli*. Reproduced with permission of A. M. Lawn. *Bacterial ultrastructure* (1976) p. 73. Edited by R. Fuller and D. W. Lovelock. Academic Press.

been found in some capsules (e.g. teichoic acid in pneumococci; peptidoglycan components in *Bacillus anthracis*) suggests that some wall substances may protude from the wall, and it is conceivable that covalent bonds link these protrusions with the true capsular materials.

SURFACE APPENDAGES

Pili (Fimbriae)

Two types of filamentous appendages can protrude from the bacterial surface: pili and flagella. Pili (hair-like structures) have been found in Gram-negative bacteria only: they are invariably shorter and thinner than flagella being approximately 7–9 nm in diameter and having an axial hole 2–2.5 nm in diameter. Several types of pili exist (at least six have been identified) and two have been examined in detail, the common I pilus of *Escherichia coli* and the F-like sex pilus distinguishable by size difference (Figure 49). I pili are composed of a protein (M_r 17 000) arranged in a right-handed helix of pitch 2.4 nm. The sex pilus consists of two parallel protein rods, 8.5 nm in diameter and has a phosphoglycoprotein as its monomeric subunit. The structures of pili are better characterized than their function. I pili are numerous and evenly distributed over the cell surface (Figure 50). They are believed to function in adhesion, are antigenic, and might be involved in pathogenicity since only the piliated forms of some organisms are virulent. Sex pili are of at least two types, the I-type and the F-type which can be distinguished from each other and from the more numerous I pili by differential labelling either with antisera or a male-specific RNA bacteriophage (Figure 51). Type F pili are assembled rapidly from preformed F pilin and become detached from the cell and appear in the medium when they have become longer than 1–2 μm: this may have the effect of limiting the number to between one and three per cell. They function in the transfer of nucleic acid between mating cells of *Escherichia coli*. It should be noted that the protein of which the pili are composed must have two other properties: that of assembling into pili and that of triggering the release of RNA from male-specific phages.

Flagella

The majority of motile bacteria have flagella which enable them to travel at speeds as great as 50 μm (i.e.

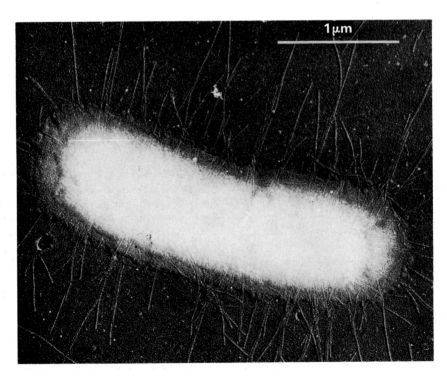

Figure 50. *Escherichia coli* with attached pili. Reproduced with permission of C. Brinton. Unpublished.

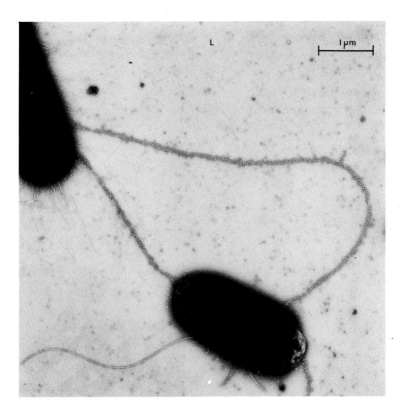

Figure 51. F pili, 'stained' with MS–2 phage, forming a link between an Hfr cell and a F⁻ cell of *Escherichia coli*. Reproduced with permission of R. Curtiss, L. G. Caro, D. P. Allison and D. R. Stallions. *J. Bacteriol* **100** (1969) 1091.

fifty times their length) per second and to respond to attractants and repellants in the medium (chemotaxis). In structure and mechanism they are completely different from eukaryotic flagella or cilia. The bacterial flagellum is composed of a filament, a hook and a basal structure (Figure 52a). The filament is a long helical structure containing 11 parallel strands composed of the protein subunit flagellin. Each filament arises from a hook structure which also consists predominantly of a single subunit while the remainder of the basal structure contains at least nine polypeptides. The basal structure is held in the cell envelope by a series of rings, two in Gram-positive and four in Gram-negative bacteria where the two innermost rings are embedded in the cytoplasmic membrane and the two outer are attached to the rigid layer and outer membrane (Figure 52b). Bacteria may have a single flagellum, a tuft arising from one or both poles of the cell, or many flagella, evenly distributed over the surface (peritrichate,

Figure 53). Until recently it was believed that bacterial flagella moved by means of a wave propagated along the filament. It is now clear that flagella are semi-rigid and that the bacterium is moved by the flagella rotating around the point of insertion in a counter-clockwise direction when viewed from behind. This was demonstrated by tethering bacteria having only a single flagellum to a glass slide by means of flagella antibodies fixed to the slide, and observing them microscopically. It is believed that the basal structure has three functions: it anchors the flagellum to the cell envelope; it is in contact with the plasma membrane from which it can accept energy; and it is the motor which drives rotation of the flagellum.

At least thirty genes influence motility, many more than are needed to code for the actual polypeptides comprising the three portions of the flagellum: therefore isolated flagella do not function *in vitro*. In bacteria that have many flagella, counter-clockwise

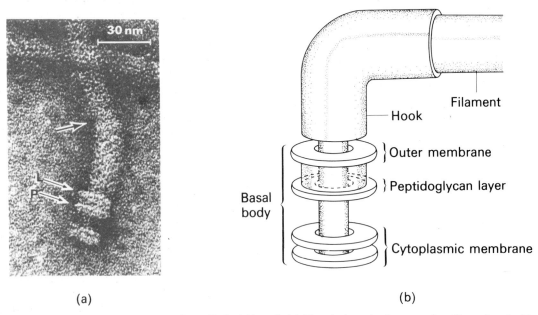

(a)   (b)

Figure 52. The basal end of the flagellum from *Escherichia coli.* (a) Negatively-stained preparation. Reproduced with permission of M. L. De Pamphilis and J. Adler. *J. Bacteriol* **105** (1971) 384. (b) Model constructed from measurements taken from large numbers of electron micrographs. The two lower rings of the basal body are embedded in the plasma membrane and the two upper rings in the outer membrane and peptidoglycan.

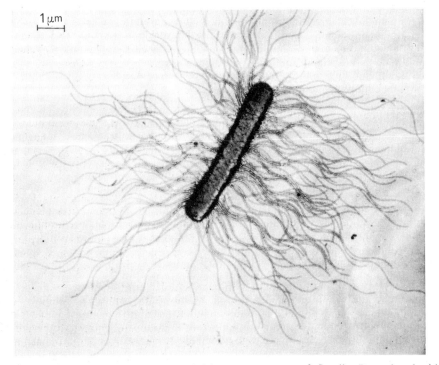

Figure 53. *Salmonella typhimurium* showing the peritrichate arrangement of flagella. Reproduced with permission of J. Hoeniger. *J. Gen. Microbiol.* **40** (1965) 29.

rotation of the left-handed helical filaments results in the formation of a bundle. Movement of bacteria is made up of 'runs' (smooth motion in a straight line) and 'tumbles'. Tumbling motion is achieved by reversal of the direction of rotation of the flagella. Monoflagellate organisms simply back away but in peritrichate organisms the bundle of flagella flies apart, the morphology of the flagellar helix alters (it becomes right-handed and has half the wavelength of the normal state) and the individual filaments become tangled. A tumble may last for a fraction of a second and the bacteria then run in a new, randomly chosen direction. The frequency of tumbling is increased by chemical repellents and decreased by attractants with the result that bacteria move towards an attractant and away from a repellent. Bacteria are equipped with specific chemo-receptors (at least 20 different proteins in *Escherichia coli*) that detect the molecule involved and activate a general chemotaxis pathway. Signals are transmitted to the flagellar motor which regulates the frequency of tumbling. Transport of the chemical is not necessary for chemotaxis although the same chemoreceptors may function as components of the machinery for transporting compounds into bacteria (p. 87).

Energy to rotate flagella is provided not by direct hydrolysis of ATP but by the discharging of proton gradients formed across the cell membrane (p. 87). Thus, mutants deficient in the ATPase protein can swim only when allowed to respire (i.e. they cannot use ATP generated by anaerobic glycolysis). It is not known how the flagellum rotates or how the proton circulation is coupled to rotation. Presumably one of the rings which links the flagellum to the cytoplasmic membrane is fixed to the rod at the base of the flagellar filament but can rotate freely in the membrane while the other rings act as bushes in which the flagellum can turn. In addition to the proton gradient for rotation of the flagellum, chemotaxis also has a direct requirement for ATP, possibly as an intermediate in the formation of S-adenosyl-methionine which seems to play an important role in the process.

BACTERIAL CYTOPLASM

In general bacterial cytoplasm appears to lack the extensive membrane systems that are such a feature of plant and animal cells but has a high overall density resulting from the presence of large numbers of ribosomes. Improved fixation and thin sectioning techniques have revealed details of the fine structure which will be presented in this section.

MEMBRANES
Cytoplasmic membrane
(protoplast membrane, plasma membrane)

Structure
The first direct indication that protoplast membranes and cell walls were separate structures resulted from the isolation of protoplasts of *Bacillus megaterium*: a limiting semi-permeable membrane surrounds these spherical bodies and is preserved intact when the protoplasts are maintained in a hypertonic medium. If such protoplast suspensions are diluted with water the protoplasts lyse and the membrane fraction can be isolated by differential centrifugation for chemical characterization. Treatment of Gram-negative bacteria with lysozyme leaves the outer membrane intact; lysis of the resulting spheroplasts and centrifugation of the particulate material gives rise to a mixture of cytoplasmic and outer membranes which may be separated by centrifugation in a sucrose density gradient. It is impossible to state whether any membrane preparation is 'pure' since other cell components may be strongly adsorbed to it after breakage: conversely, some 'true' membrane components may be removed by the stringent washing procedures. The preparations that are obtained consist almost entirely of lipid and protein and account for approximately 10% of the dry weight of the bacteria: other substances found associated with the membrane include carbohydrate, RNA and DNA. The ratio of protein to lipid varies from 2:1 to 4:1, values much higher than in eukaryotic membranes with the exception of the inner mitochondrial membrane.

Characterization
The main class of lipid present in bacterial cytoplasmic membranes is phospholipid: Gram-positive bacteria contain several types within this class while Gram-negative organisms possess mainly phosphatidylethanolamine (Table 3). The types of phospholipid found and the amount of each present depend on the bacterial strain and on the growth conditions, particularly the pH value of the medium: for this reason the percentage composition cannot be included in Table 3. The structure of the various phospholipids is shown in Figure 54. The other main classes of lipid found predominantly in Gram-positive organisms are glycolipids and glycophospholipids. Minor components include ubiquinones (which function in the respiratory chain) and carotenoids.

Gram-positive and -negative bacteria differ markedly in the fatty acid composition of the membrane lipids. Branched-chain fatty acids are characteristic

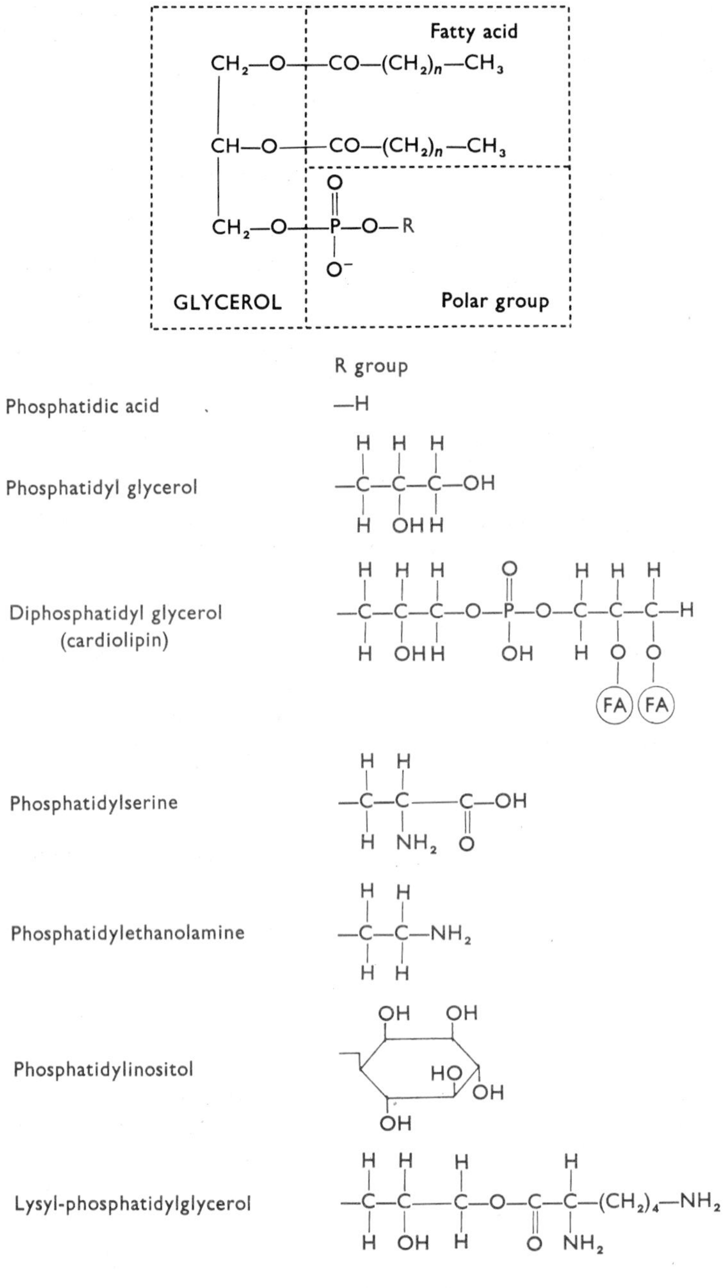

Figure 54. Phospholipids found in bacteria.

of Gram-positive organisms (e.g. *Micrococcus lyso-deikticus* contains 90% of its fatty acid residues as branched-chain C_{15}) and virtually no unsaturated fatty acids are found. Gram-negative bacteria contain a mixture of saturated and unsaturated fatty acids in which chain lengths of C_{16} and C_{18} predominate, and also cyclopropane acids. The ratio of saturated to unsaturated fatty acids is dependent on the growth temperature. Thus in cultures growing at low temperatures (e.g. 20°) unsaturated fatty acids predominate but at higher temperatures (e.g. 37°) some of the unsaturated fatty acids are replaced by saturated derivatives and the ratio of the two types approaches unity. This effect is even more marked at

Table 3. Principal classes of phospholipids found in the membrane of Gram-positive bacteria and the cell envelopes of Gram-negative organisms

Gram-positive	
Bacillus megaterium	PE, PG, di-PG, lys-PG
Bacillus subtilis	PE, PG, di-PG, αα-PG
Bacillus licheniformis	PE, PG
Streptococcus faecalis	PA, PG, di-PG, αα-PG
Staphylococcus aureus	PA, PG, di-PG, αα-PG
Micrococcus lysodeikticus	PG, di-PG, PI
Gram-negative	
Azotobacter spp.	PE
Escherichia coli	PE (PG, PS, αα-PG)
Abbreviations	
PE	Phosphatidylethanolamine
PG	Phosphatidylglycerol
di-PG	Di-phosphatidylglycerol (cardiolipin)
lys-PG	Lysyl-phosphatidylglycerol
αα-PG	Aminoacyl-phosphatidyl-glycerol
PA	Phosphatidic acid
PI	Phosphatidylinositol
PS	Phosphatidylserine

the two extremes of the range of growth temperature. Thus in psychrophilic organisms growing at 2°, practically all of the fatty acids are unsaturated whereas in thermophilic bacteria growing at 60° they are all saturated. This response to temperature is important in maintaining the fluidity of the lipids in the membrane.

The major component of the cytoplasmic membrane is protein (60–80%) but it is unusual for any one protein to be present in large amounts. Polyacrylamide gel electrophoresis of the proteins obtained by dissolving the membrane in the detergent, sodium dodecylsulphate, indicates that at least 50–70 polypeptide species are normally present, some of them as glycoproteins. Multi-subunit proteins, however, will yield several polypeptides (5 for the ATPase) and consequently the number of proteins will be less than the number of bands detected by polyacrylamide gel electrophoresis. Some proteins are easily removed by aqueous washing and are apparently loosely bound to the surface (*extrinsic* or *epi-proteins*) while others are only extracted by organic solvents or by detergents which dissociate the membrane structure and are presumed to penetrate the hydrophobic core of the membrane (*intrinsic, integral* or *endo-proteins*). Proteins which have been studied most extensively include ATPase, a complex protein with a headpiece forming the inside

surface of the cytoplasmic membrane and a 'stalk' which may be involved in channelling protons through the membrane, and bacteriorhodopsin, the light-driven proton pump in the purple membrane of halobacteria.

Models of membrane structure

In thin sections the cytoplasmic membrane appears as a three-layered structure (Figure 9, p. 50) with two electron-dense layers separated by an electron-transparent zone. This observation, together with X-ray diffraction data led Robertson to propose the 'unit-membrane' hypothesis which is an extension of the ideas elaborated by Danielli and Davson from physical studies involving measurements of permeability, surface tension and conductivity. The 'unit membrane' is rigidly defined as a bimolecular lipid leaflet in which the hydrocarbon chains of the phospholipids are close-packed and oriented perpendicular to the surface of the leaflet (Figure 55a). The 'core' of the membrane, therefore, has hydrophobic properties and it is this region which is considered to be electron-transparent in micrographs of thin sections. The polar groups of the phospholipids are on the outside of the leaflet and are linked electrostatically to charged groups of the proteins or glyco-proteins. These hydrophilic regions were considered to be equivalent to the electron-dense areas on both sides of the membrane and it was further proposed that the proteins were in the extended β-configuration. This model explains many of the properties of cytoplasmic membranes, particularly their relative impermeability to many simple substances such as amino acids, sugars and even ions, and various electrical properties such as resistance and capacitance. However, the rigid definition of the model as presented here is not in accord with more recent experimental data.

A great deal of the evidence for the 'unit membrane' hypothesis was obtained with myelin, a highly specialized membrane with an extremely low protein:lipid ratio and no enzymic activity. Since bacterial membranes have a protein:lipid ratio of between two and four it could be argued that the structure of myelin has little bearing on bacterial membrane structure. However, the important consideration is not the ratio of protein to lipid but whether the total amount of lipid is sufficient to cover the surface of the membrane in the form of a bimolecular lipid leaflet. For *Acholeplasma laidlawii* only 62% of the surface area could be covered by a bilayer of the phospholipid in the membrane but it is not known whether this is a typical figure. It has now been demonstrated that proteins are held in the membrane by predominantly hydrophobic interactions with each other and with phospholipids. Much of the protein is in the α-helix configuration, and this is not in accord with Robertson's unit membrane hypothesis. Studies of the interactions between membrane proteins and lipids led to the suggestion that the proteins could contribute to the overall structure of the membrane and to the concept of specific lipo-protein subunits with both structural and functional roles. However, no exclusively structural membrane protein common to a variety of membranes has yet been found and there is no reliable evidence for such a model.

Most physical determinations using modern techniques (circular dichroism, optical rotatory dispersion, electron spin resonance, differential scanning colorimetry) suggest a more liberal interpretation of the unit membrane hypothesis. These studies show that membranes consist largely of a lipid bilayer in which the fatty acid side-chains are not arranged in a semi-rigid manner but are in constant motion. The *Mycoplasma* are particularly useful in these investigations since these organisms incorporate into membrane lipids whatever fatty acids are present in the growth medium. The effect of varying the fatty acid composition on such properties as permeability and lipid mobility has been measured. Since membranes can be prepared very easily from this group of organisms (they have no wall components) it is possible to study intact cells, membranes derived from them, and model lipid bilayer systems in which the same phospholipids are present as in the cytoplasmic membrane. The three different systems yield almost identical results which support the existence of a lipid bilayer over 60–80% of the area covered by the cytoplasmic membrane.

Proteins are believed to penetrate this fairly fluid bilayer to a lesser or greater extent, as shown in Figures 55b and 56, causing limited disturbance of the fatty acid side-chains of the phospholipids and interacting with other molecules by predominantly hydrophobic interactions. The protein molecule may pass completely through the membrane or may exist in a hydrophobic environment in the middle of the bilayer. The location of each protein molecule is determined by its amino acid sequence and by the location of the specific phospholipids with which it associates. This model has become known as the *fluid mosaic model of membrane structure* and was proposed by Singer and Nicolson in 1972.

The vectorial movement of many different types of molecule across membranes suggests that neither proteins nor phospholipids (see below) are arranged symmetrically within its framework. Chemical labelling studies with intact protoplasts and with purified cytoplasmic membranes support this hypothesis: some proteins could be labelled from the outside and others from the inside of the membrane while others were labelled from either side. More recently the enzymic and antigenic architecture of the membranes of Gram-positive and -negative bacteria have been examined with the expected result that both the antigenic and enzymic activities are distributed asymmetrically.

Freeze-etch electron micrographs of membranes support the idea that proteins are located within the hydrophobic core of the membrane. There is evidence that the fracture plane of membrane structures passes through the hydrophobic centre thus enabling replicas of the two internal surfaces to be examined. The concave fracture surface is virtually free of particles but has occasional depressions while the convex fracture surface is extensively covered with particles 5–10 nm in diameter. In general the bacterial cytoplasmic membrane would be expected to have much

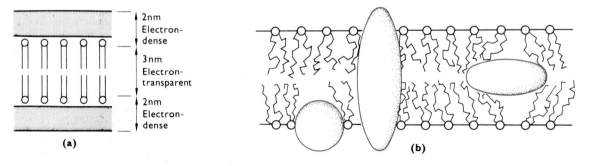

Figure 55. Diagrammatic representation of cross-section through the cytoplasmic membrane. (a) Unit membrane with layers of protein attached to the outside of the bimolecular lipid leaflet as proposed by Danielli and Davson and modified by Robertson. (b) Fluid mosaic model with protein molecules penetrating the phospholipid bilayer to varying extents.

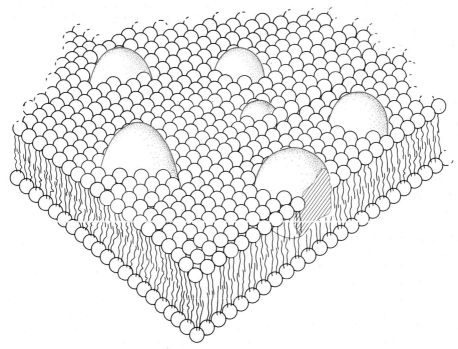

Figure 56. The fluid mosaic model of membrane structure proposed by Singer and Nicolson. The phospholipid molecules adjacent to globular protein molecules are relatively fixed in position; those next to other phospholipid molecules are very mobile.

more protein on both its surfaces than is indicated in Figures 55b and 56 in view of the protein: lipid ratio and this is borne out by electron micrographs of negatively-stained preparations (Figure 57a) and also by micrographs of freeze-etched bacteria (Figures 6a and 7, pp. 48–9).

Function

The bacterial cytoplasmic membrane has a central functional role in the prokaryotic cell which is devoid of the specialized membrane-bounded organelles (e.g. mitochondria, chloroplasts, endoplasmic reticulum, Golgi apparatus) found in eukaryotic cells. The membrane is a multifunctional structure that is involved in energy generation, biosynthesis, transport and secretion in addition to acting as a relatively impermeable barrier.

The chemiosmotic hypothesis of Mitchell (see Chapter 3, p. 147) predicts that the proton-motive force which performs osmotic, chemical and mechanical work is generated by the vectorial movement of protons across a sealed membrane. Other than mesosomes which are largely devoid of enzymic activity (page 90) and specialized membranes in photosynthetic bacteria, the plasma membrane is the only membranous organelle in bacteria. It contains the components of the respiratory chain (cytochromes, flavins, ubiquinones) and several of the enzymes feeding reducing equivalents into it (succinate, malate, NADH dehydrogenases) as well as the associated ATPase which is located on the inside surface. The working of the respiratory chain or the hydrolysis of ATP by the ATPase results in the production across the membrane of a gradient of protons which can drive various processes (Figure 58 and page 147). The bacterial cytoplasmic membrane is therefore analogous to the mitochondrial inner membrane.

The cytoplasmic membrane is also the obvious location for at least some of the enzymes and carriers involved in the biosynthesis of extracellular polymers present in bacterial cell walls and capsules. It therefore participates in the biosynthesis of peptidoglycan, teichoic acids, teichuronic acids, lipopolysaccharides and polysaccharides, in addition to anchoring some of the growing polymers to the membrane via the polyisoprenoid alcohol or the lipoteichoic acid. It is also the site of synthesis of its own phospholipids and glycolipids. The role of the membrane in the synthesis of DNA, RNA and protein is not clear.

It is difficult to prove whether substances adhering to the membrane after several washings are con-

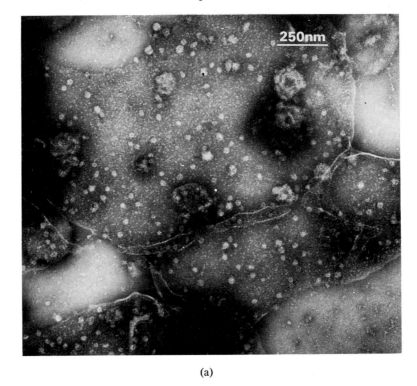

(a)

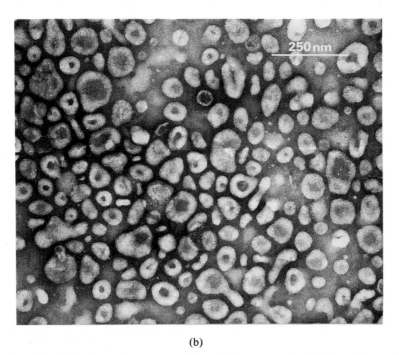

(b)

Figure 57. Negatively-stained isolated membrane preparations of *Micrococcus lysodeikticus*. (a) cytoplasmic membrane showing ATPase particles (b) mesosomal vesicles. Reproduced with permission of M. R. J. Salton. *Symp. Soc. Gen. Microbiol.* **28** (1978) 201. Edited by R. Y. Stanier, H. J. Rogers and J. B. Ward. Cambridge University Press.

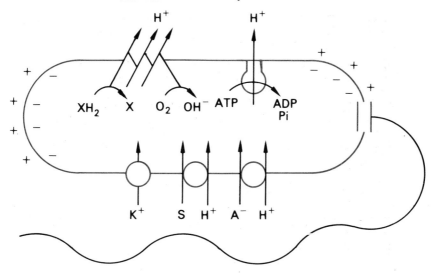

Figure 58. Secondary coupling of transport carriers and of the flagellar motor to the proton circulation according to the chemiosmotic hypothesis. The proton gradient is generated either by the action of the respiratory chain or by the ATPase working in the hydrolytic direction. K^+ represents a cationic (lysine, K^+), S an uncharged (e.g. lactose, proline) and A^- an anionic substance (e.g. glutamate, phosphate), which are transported into the cell in response to the proton and electrical gradients. An ingoing neutral molecule or anion is accompanied by H^+.

taminants or are membrane components. The converse is also true—some truly membrane components are easily washed off. Nevertheless it seems likely that membrane-bound polysomes are active in protein synthesis though it has not been established whether they are concerned only in synthesis of certain classes of proteins (e.g. extracellular proteins). It appears that DNA is attached to the cytoplasmic membrane either directly or through the mesosomal membranes (p. 90) since *membrane* fractions have been obtained containing a small percentage of the total membrane together with most of the DNA (free DNA would not be obtained in this fraction). It is possible that membranes are connected in some way with DNA metabolism either in respect to DNA synthesis or to the separation of the replicating strands.

As the cytoplasmic membrane is virtually impermeable it follows that essential parts of the transport machinery will be located within the hydrophobic framework. The work of transport is extremely important to a bacterium living in an aqueous environment and it is estimated that the percentage of the total energy produced by the cell that is directed towards transport may be as much as 50%. Transport systems for supplying nutrients to the cell (i.e. excluding those involved in the export of polar molecules or in the uptake of DNA) can be separated into three categories: those in which transport is coupled to proton movement, those which are directly depen-

dent on ATP and those which involve group translocation.

The classical lactose permease of *Escherichia coli* is one of the many systems that now appear to be driven by the circulation of protons. The general principles and predictions linking transport of anionic, cationic and unchanged substances to the proton circulation are outlined in Figure 58. The primary proton pump (respiratory chain or ATPase) is vectorial and generates two gradients, one of pH value and one of electrical potential. A cationic metabolite (e.g. the amino acid lysine or potassium ions) could be translocated by responding to the membrane potential (interior negative). Anionic substances (glutamate, phosphate) are transported with protons as the uncharged acids and accumulated in response to the pH gradient. Uncharged metabolites such as proline and some sugars may be carried inwards in the same direction as protons (a symport mechanism) and accumulated in response to the total proton-motive force. Dissipation of the gradients with the class of chemicals known as ionophores (which artificially conduct protons across the membrane) results in loss of the ability to transport metabolites in this way. Although it is clear that membrane proteins act as carriers of the substances to be transported the molecular mechanism of the transport process is still uncertain.

The second category of transport mechanism

involves ATP directly. Gram-negative bacteria contain several proteins localized in the 'space' between the cytoplasmic and outer membranes. These proteins recognize and bind certain sugars, amino acids and ions and are released into the medium when cells receive an osmotic shock. They are referred to as *periplasmic binding proteins* and are believed to determine the affinity and specificity of the overall transport system. Other protein components are involved in the transport process and are located in the cytoplasmic membrane. A characteristic of all transport systems involving binding proteins is that they are resistant to proton conductors (ionophores): in addition, mutants which lack the ATPase can support transport only by glycolysis and not by respiration. Transport occurs only in the inward direction whereas those systems driven by the proton circulation can transport substances in either direction depending on which side of the membrane is made artificially alkaline or electrically negative.

The third type of transport system involves group translocation: in this process the substance that is transported is also chemically modified. The best characterized system in both Gram-positive and -negative bacteria is the *phosphotransferase system*. It relies on the interaction of one or more proteins arranged in the membrane (which bind a specific sugar molecule on the outside of the membrane) with a series of cytoplasmic proteins that transfer phosphate from phosphoenolpyruvate to the sugar. All the systems studied involve two non-specific cytoplasmic proteins: *enzyme 1* which is phosphorylated by phosphoenolpyruvate and a low molecular weight, histidine-containing protein (HPr) which is phosphorylated by enzyme 1. The transport process also involves sugar-specific components (as many as 10 different complexes exist in *Escherichia coli*) consisting either of a membrane-bound component IIB′ and a cytoplasmic component III (Figure 59a) or two membrane-bound components IIB and IIA (Figure 59b). Component III or IIA is phosphorylated by HPr and component IIB or IIB′ catalyses the

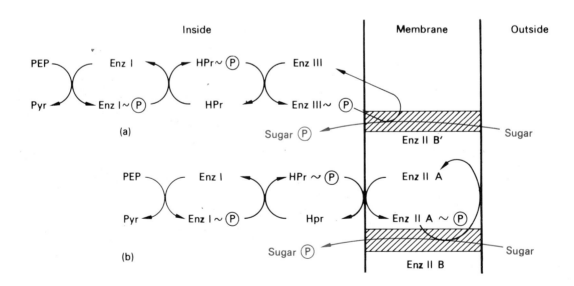

Figure 59. Group translocation by the phosphotransferase system. Diagrammatic representation of the reactions involved in the transfer of phosphate to the sugar as it is transported into the cell. Two alternative schemes are shown. In (a) the two sugar-specific proteins consist of a cytoplasmic component III and a membrane-bound IIB′. In (b) both components IIA and IIB are membrane-bound.

transfer of phosphate to the sugar as it is transported into the cell. The complete system is shown diagrammatically in Figure 59 and the result of the process is that the sugar is deposited inside the cell in a phosphorylated form, immediately available for further metabolism.

The membrane must also function in the secretion of proteins. This process has been studied in detail in eukaryotic organisms and in many instances it has been demonstrated that this class of proteins is synthesized on membrane-bound ribosomes associated with the endoplasmic reticulum. The primary translation product of the messenger RNA contains a hydrophobic peptide at the amino terminus of the polypeptide that enables the growing chain to be secreted through the membrane as it is being synthesized. The hydrophobic 'leader' sequence is subsequently cleaved and the protein assumes its stable, three-dimensional, configuration in the lumen of a vesicle that eventually fuses with the plasma membrane to release the protein from the cell. Prokaryotic organisms do not possess an endoplasmic reticulum or Golgi membranes so it is imagined that the secretory process is analogous to the actual synthetic stage in eukaryotes, i.e. synthesis on membrane-bound ribosomes and passage through the membrane into the medium. Evidence for such a process in bacteria is slight: the only system studied in detail has been the production of penicillinase in *Bacillus licheniformis*. This organism produces large amounts of the protein which exists in both membrane-bound and exocellular forms, the only difference in structure being that the membrane-bound species contains an additional N-terminal sequence which anchors the protein in the membrane. The additional sequence is cleaved by a proteolytic enzyme to release the penicillinase into the medium. A protein similar, if not identical, to the *in vivo* membrane-bound enzyme has been synthesized by an *in vitro* system from the organism but it remains a possibility that the *initial* translation product contains some additional amino acids as compared with the membrane-bound enzyme.

The phospholipids in membranes do not function simply as a permeability barrier. They are arranged assymetrically as shown by differential labelling techniques and by selective digestion with specific phospholipases. In these ways 66% of the total phosphatidylethanolamine in *Bacillus megaterium* has been shown to be located in the inner leaflet facing the cytoplasm and 33% in the outer leaflet. Some of the phospholipid molecules function as essential components of enzyme systems. Phosphatidylethanolamine is essential for the function of the

purified galactosyl transferase in lipopolysaccharide synthesis (page 72) and for wall teichoic acid synthesis (page 63). Our modern concept of membranes is therefore of dynamic and fluid structures in which there is considerable lateral mobility of both phospholipid and protein molecules. The proteins and phospholipids are located in specific regions so that enzyme systems involved in transport and biosynthesis of macromolecules can work vectorially. It seems likely that the same general features of membrane structure and function elaborated here will apply to all types of non-specialized biological membranes.

Mesosomes

Structure
The majority of Gram-positive and Gram-negative bacteria contain membranous organelles formed by involution of the protoplast membrane and located in the region of the bacterial DNA or at the site of cell division or spore formation. The term *mesosomes*

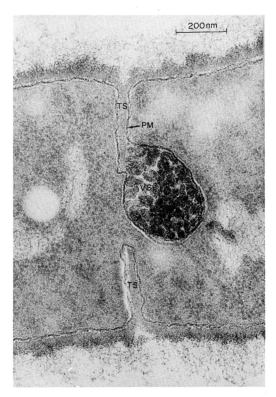

Figure 60. Thin section of *Bacillus megaterium* showing the close association between the transverse septum (TS), the plasma membrane (PM) and the mesosomal vesicles (VS). Reproduced with permission of D. J. Ellar, D. G. Lundgren and R. A. Slepecky. *J. Bacteriol.* **94** (1967) 1189.

has been applied to these structures, which can be visualized in intact cells by negative staining. The mesosome shown in Figure 60 is associated with an early stage of septum formation and thin sections showing later stages of the process demonstrate that mesosomes are present at this site throughout its entire construction. Different arrangements of the membranes in mesosomes have been observed; vesicles, probably interconnected, are seen in most sections (Figure 60) while whorls of membranes are seen in others. It has been suggested that the appearance is dependent on the method of fixation prior to sectioning as mesosomes are seen much less frequently in unfixed cells. Micrographs obtained by the freeze-fracturing procedure emphasize the close association between the mesosomes and the cytoplasmic membrane and give the impression that the internal membranes are present as vesicles (Figure 61). Fixation with glutaraldehyde prior to freeze-fracturing results in mesosomes being seen with much greater frequency than in unfixed cells. It is possible that mesosomes are seen only as a result of physical perturbation of the fluid mosaic nature of the membrane system and are therefore artifactual.

Mesosomes form a pocket in which membranous components are contained and it is these membranous vesicles that are released when intact cells are converted to protoplasts. They can be recovered from the medium by centrifugation and their structure and enzymic activities determined. Negative staining reveals smooth vesicles that are different in appearance from the particle-covered cytoplasmic membrane preparations (Figures 57a and b). Mesosomes of Gram-negative bacteria are much less prominent than those of Gram-positive organisms: they are present mainly as small infoldings of the cytoplasmic membrane and are frequently associated closely with the nuclear material.

Function

It has been suggested that mesosomes may act as a site for cell respiration and energy production, a specific site for cross-wall formation in Gram-positive bacteria, a control centre for orderly cell division, an organ of attachment for the bacterial nucleus during replication, and a site for DNA uptake during transformation. The location of mesosomes at the site of septum formation provides strong support for some role in its synthesis—whether as a site of assembly of the building blocks or by directing and controlling autolytic enzymes that are necessarily involved in cell division has yet to be proved. Another suggested function concerns anchorage of the chromatinic body during replication of the

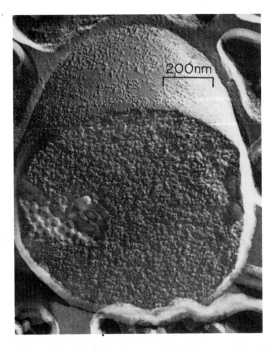

Figure 61. Freeze-fractured *Bacillus subtilis* showing a mesosome in close contact with the septum. Reproduced with permission of N. Nanninga. *J. Cell. Biol.* **39** (1968) 251.

bacterial nucleus: this is based on analysis of thin sections which show that contact is maintained between the mesosome and the replicating DNA. The cytoplasmic membrane can be separated from the internal membranes of mesosomes for study of the distribution of enzymes between the two types of membrane. The mesosomal membranes are virtually devoid of enzymes and carriers involved in respiration and are not specific sites for the synthesis of phospholipid or peptidoglycan. They contain some of the antigens expressed on the outer surface of the cytoplasmic membrane and are enriched in lipomannan (which replaces lipoteichoic acid in *Micrococcus lysodeikticus*) but are otherwise apparently lacking in activity. It should be emphasized, however, that the mesosomal membrane fraction may not contain the bounding membrane of the mesosomal sac as this probably becomes part of the protoplast membrane: therefore absence of an enzyme does not prove that the mesosome is not involved in a certain process.

Photosynthetic membranes

Purple and green bacteria have the ability to carry out photophosphorylation, i.e. they are photosynthetic organisms. The purple bacteria contain membrane-bounded structures arranged as lamellae or as vesicles. The vesicles or 'chromatophores' of *Rhodospirillum rubrum* (Figure 62) can be obtained by differential centrifugation of extracts of photosynthetically-grown organisms that have been mechanically disrupted. The particles are approximately 60 nm in diameter, they are chemically similar to membrane preparations, and they catalyse a light-dependent process resulting in the production of energy. The cellular content of vesicles is paralleled by the chlorophyll concentration: as the light intensity under which the bacteria are grown is increased so

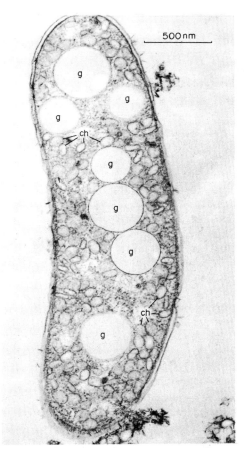

Figure 62. Longitudinal section of an intact cell of *Rhodospirillum rubrum* showing chromatophores (ch) and granules of poly-β-hydroxybutyrate (g). Reproduced with permission of E. S. Boatman. *J. Cell Biol.* **20** (1964) 297.

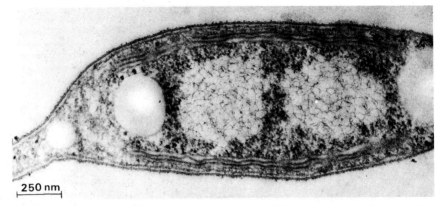

Figure 63. Transverse section of *Rhodomicrobium vannielii* showing the symmetrical stacking of lamellar membranes on either side of the cell. Reproduced with permission of W. C. Trentini and M. P. Starr. *J. Bacteriol.* **93** (1967) 1699.

both the cellular content of 'chromatophores' and the amount of chlorophyll decrease and *vice versa*. Because the membrane fraction of *Rhodospirillum rubrum* prepared from osmotically-lysed organisms contains the entire content of chromatophores it may be that the vesicles are attached to or arise from the protoplast membrane. Thin sections of lysed organisms demonstrate that the protoplast membrane is continuous with the membrane surrounding some of the vesicles or lamellar structures. It is believed that the vesicular structures obtained by physical fractionation techniques are artifacts resulting from detachment of the photosynthetic membranes from the cytoplasmic membrane and the subsequent formation of vesicles. Many other purple bacteria contain lamellar structures arranged in stacks near the periphery of the cell (Figure 63) and attached to the cytoplasmic membrane. Although it has not been definitely established that the lamellae are the sites of photosynthesis the circumstantial evidence is strong; the amounts of membrane and of chlorophyll alter in parallel on changing light intensity, temperature or oxygen tension.

Until recently it was believed that the chlorophyll pigments in green bacteria were not located in any special structure. However, improved fixation and embedding techniques have revealed large vesicles lying immediately beneath the cytoplasmic membrane. These vesicles have been isolated and shown to be enriched in the photosynthetic pigments.

Gas vacuoles

Gas vacuoles are found in many non-motile bacteria living in aquatic environments. They are present in green and purple bacteria and in some non-photosynthetic organisms (e.g. halobacteria). By electron microscopy gas vacuoles have been shown to consist of clusters of small, regularly-shaped gas vesicles, those of *Halobacterium halobium* being $0.3-0.4$ μm in length. The vesicles have a thin, proteinaceous wall only 2 nm thick which is freely permeable to gases and which appears striated in freeze-etched preparations of cells (Figure 64a) or negatively-stained preparations of isolated vesicles (Figure 64b). Gas vesicles probably function as buoyancy tanks thus keeping the micro-organisms containing them close to the interface of environments that are poor in oxygen. The vesicles are deflated by sudden increases in pressure with a consequent loss in buoyancy of the organisms. Another possibility is that they might perform a light-shielding function.

BACTERIAL CHROMOSOMES (CHROMATINIC BODIES, DNA-PLASM)

The bacterial chromosome is not surrounded by a membrane, has no defined shape and is seen in electron micrographs as an irregularly-shaped, electron-transparent area within the cytoplasm. Sometimes this is called a 'nucleoid'. It has been calculated that the genetic information of an *Escherichia coli* cell is contained in approximately 1.6×10^7 nucleotides (*c.* 3% of the dry weight of the cell). DNA preparations from lysed spheroplasts spread on thin protein films support the view that one or very few molecular strands are present (see Figure 2, p. 187) in agreement with the autoradiographic studies of Cairns (page 212). The amount of DNA per cell varies

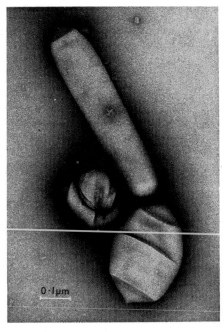

(a)   (b)

Figure 64. Gas vesicles of *Halobacterium halobium* demonstrated by (a) freeze-etching of intact cells and (b) negative staining of isolated vesicles. Reproduced with permission of (a) W. Stoeckenius and W. H. Kunau. *J. Cell Biol.* **38** (1968) 337 and (b) G. Cohen-Bazire, R. Kunisawa and N. Pfennig. *J. Bacteriol.* **100** (1969) 1049.

considerably depending upon the rate at which the cell is growing. Under optimum conditions it takes approximately 40 min to replicate the complete bacterial chromosome, which means that in rapidly growing cells the initiation of one round of DNA replication occurs before the previous round is complete. Consequently, rapidly dividing cells contain 2-4. chromosomes while slow growing cells have 1–2. Cytological observations suggest that nuclear separation occurs almost immediately after DNA replication is completed.

It is obvious that the double strand of DNA, 2000 μm long, has to be folded several hundred times in order to be compressed into a bacterium perhaps 2 μm long. The state of the DNA in the intact cell can be inferred from what is seen in thin sections of bacteria that have been fixed by the Ryter-Kellenberger (R-K) method. In this procedure fixation with osmium tetroxide is carried out in the presence of calcium ions and amino acids followed by treatment with an aqueous solution of uranyl ions. This technique prevents shrinkage during the subsequent dehydration step and results in a homogeneous appearance of the nucleus (Figure 65). Normal fixation methods not involving the use of Ca^{2+} and amino acids result in side-to-side aggregation of the DNA fibrils. The main disadvantage of the R-K technique

is the difficulty in recognizing the orientation of the DNA fibres but the micrographs obtained using this method show that the DNA-plasm is present in the hydrated state. It is possible that the high degree of hydration is particularly favourable for the penetration of enzymes and DNA-precursors between the strands during the replication process and for protein synthesis to occur as RNA is being transcribed on the DNA template but this has not been determined experimentally.

The physical processes involved in unwinding the original DNA helix and separating the two daughter helices so that one passes into each of the progeny are incompletely understood. It is possible that mesosomes (page 89), or growth of the cytoplasmic membrane *between* attachment points of the two daughter DNA strands to it, may be involved in the separation of the nuclei and distribution to the two new cells.

RIBOSOMES (RIBONUCLEOPROTEIN PARTICLES)

The cytoplasm of rapidly growing bacteria is filled with darkly staining granules approximately 10–20 nm in diameter (Figure 5, p. 47) and grouped in clusters. These granules are equivalent to the ribosomes which can be obtained by high-speed centrifugation of disrupted bacteria. It is difficult to investigate the state in which ribosomes exist *in vivo*, but the available data suggest a high degree of organization within the cytoplasm, a view that is consistent with the great metabolic activity of bacteria. They are linked together transiently by messenger RNA to form polyribosomes which function in protein synthesis (Figure 66). The detailed structural and functional organization of the RNA and proteins in ribosomes of *Escherichia coli* has been investigated by chemical and immunochemical techniques and by reconstitution experiments with the purified constituent molecules and is considered in Chapter 6.

CYTOPLASMIC INCLUSIONS

Reserve materials may accumulate in the cytoplasm when certain bacteria are incubated under specific conditions: among these products can be included glycogen, lipid droplets, polymerized inorganic metaphosphate and sulphur. Of these, the first three are generally absent during active growth of the organism while sulphur granules seem to be essential as an energy reserve for sulphur bacteria.

Glycogen granules (granulose)

These appear as electron-transparent spheres without a limiting membrane; they may account for as much as 50% of the dry weight of the organism and tend to accumulate under conditions of nitrogen starvation in *Escherichia coli* when the organisms are provided with a carbon source. If nitrogen is supplied the accumulated glycogen is broken down and used as a carbon source in the same way as glucose.

Lipid granules

Poly-β-hydroxybutyric acid is accumulated in large granules—easily visible by phase contrast microscopy—by a wide variety of bacteria including bacilli and *Rhodospirillum rubrum*. It is a storage product and may occupy a considerable proportion of the cytoplasm in cells of old cultures (Figure 62). Large numbers of lipid droplets may be found in bacteria: these can be stained with lipid-soluble dyes such as Sudan Black.

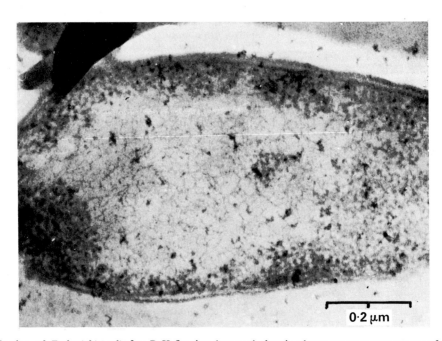

Figure 65. Section of *Escherichia coli* after R-K fixation (see text) showing homogeneous appearance of the chromosome. The black deposits are artifacts of staining. Reproduced with permission of G. Wolfgang Fuhs. *Bact. Rev.* **29** (1965) 277 and Pergamon Press, Oxford.

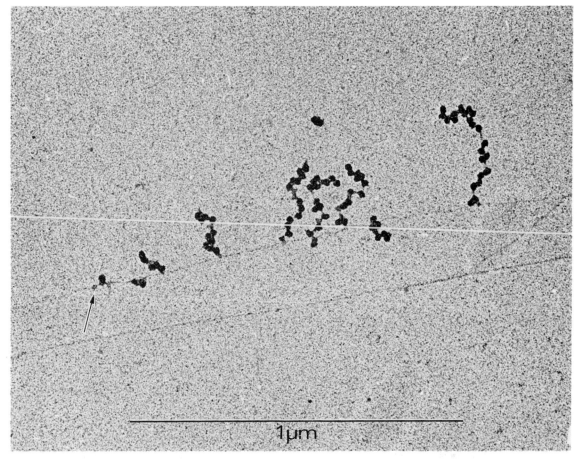

Figure 66. Complex of ribosomes and messenger RNA (polyribosomes) linked to *Escherichia coli* DNA. The arrow indicates the presumptive initiation site for transcription of messenger RNA from the DNA. Reproduced with permission of B. A. Hamkalo and O. L. Miller. *Ann. Rev. Biochem.* **42** (1973) 379.

Metachromatic granules (volutin granules, polymetaphosphate granules)

In the majority of organisms which contain these granules their appearance and the concomitant accumulation of polyphosphate tends to occur towards the end of the growth cycle. They are composed in part of inorganic metaphosphate in a polymer of high molecular mass and are readily stained with toluidine blue (Figure 67), hence the name metachromatic. The granules from *Micrococcus lysodeikticus* are atypical in that they accumulate during the log phase of growth and disappear in the stationary phase. The main constituents of the granules are lipid and protein in addition to the phosphate which may represent more than 50% of the total phosphate in the cell. The electron-dense phosphate is present in a rosette arrangement aound the periphery of the granule.

Sulphur droplets

Prominent deposits of sulphur (Figure 68) are found in the large purple sulphur bacteria growing in the presence of sulphide. The sulphur arises from the oxidation of sulphide during photosynthesis and is in turn oxidized to sulphate when the cells have been depleted of sulphide, and so the droplets disappear. The sulphur is probably in an unstable form since the stable orthorhombic allotopic form does not exist as wet spherical droplets. The small purple sulphur bacteria and green sulphur bacteria do not store sulphur within the cytoplasm but deposit it outside the cell.

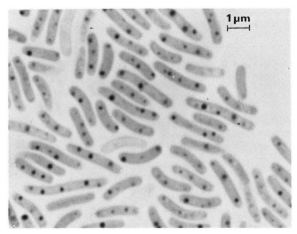

Figure 67. Metachromatic granules of *Spirillum* revealed by staining with toluidine blue. Photographed by Dr. J. P. Truant and reproduced with permission of R. G. E. Murray. See *The Bacteria*, Volume I (1960), Plate I, Fig. 6, p. 41. Eds. R. Y. Stanier and I. C. Gunsalus, Academic Press. New York and London.

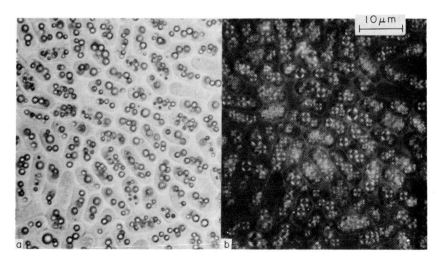

Figure 68. Photomicrograph of *Chromatium okenii* containing sulphur granules. (a) Bright-field microscopy, (b) polarizing microscopy. Reproduced with permission of G. J. Hageage, E. D. Eanes and R. L. Gherna. *J. Bacteriol.* **101** (1970) 464.

FURTHER READING

A Cell walls

General Reviews

1 Braun V. and Hantke K. (1974) Biochemistry of bacterial cell envelopes. *Ann. Rev. Biochem.* **43**, 89.

2 Hussey H. and Baddiley J. (1976) Biosynthesis of bacterial cell walls. *The enzymes of biological membranes*, **2**, 227. Edited by A. Martonosi. Wiley, London, New York, Sydney and Toronto.

Peptidoglycan

3 Schleifer K. H. and Kandler O. (1972) Peptidoglycan types of bacterial cell walls and their taxonomic significance. *Bacteriol. Revs.* **36**, 407.

Teichoic acids

4 Archibald A. R. (1974) The structure, biosynth asindes function of teichoic acid. *Advances in Microbial Physiology* **11**, 53.

5 Coley J., Tarelli E., Archibald A. R. and Baddiley J. (1978) The linkage between teichoic acid and peptidoglycan in bacterial cell walls. *FEBS Letters* **88**, 1.

Cell wall growth

6 Shockman G. D., Daneo-Moore L. and Higgins M. L. (1974) Problems of cell wall and membrane growth enlargement and division. *Annals of the New York Acad. Sci.* **235**, 161.

7 Higgins M. L. and Shockman G. D. (1976) Study of a cycle of cell wall assembly in *Streptococcus faecalis* by three dimensional reconstructions of thin sections of cells. *J. Bacteriol.* **127**, 1346.

Flagella

8 Adler J. (1975) Chemotaxis in bacteria. *Ann. Rev. Biochem.* **44**, 341.

B Membranes

9 Hamilton W. A. (1975) Energy coupling in microbial transport. *Advances in Microbial Physiology* **12**, 1.

10 Harold F. M. (1977) Membranes and energy transduction in bacteria. *Current topics in Bioenergetics* **6**, 83.

11 Salton M. R. J. (1974) Membrane associated enzymes in bacteria. *Advances in Microbial Physiology* **11**, 213.

12 Salton M. R. J. and Owen P. (1976) Bacterial membrane structure. *Ann. Rev. Microbiol.* **30**, 451.

Mesosomes

13 Higgins M. L., Tsien H. C. and Daneo-Moore L. (1976) Organisation of mesosomes in fixed and unfixed cells. *J. Bacteriol.* **127**, 1519.

C Books of general interest

14 Fuller R. and Lovelock D. W. eds. (1976) *Microbial Ultrastructure*. Academic Press. London, New York and San Francisco.

15 Leive L. ed. (1973) *Membranes and walls of bacteria*. Dekker, New York.

16 Quinn P. J. (1976) *The molecular biology of cell membranes*. MacMillan, London.

17 Stanier R. Y., Rogers H. J. and Ward J. B. eds. (1978) *Relations between structure and function in the Prokaryotic cell*. Symp. Soc. Gen. Microbiol. **28**. Cambridge University Press. This book contains review articles on several topics covered in this chapter.

Chapter 2
Growth: Cells and Population

The growth of bacteria can be considered either in terms of what is happening in the individual cells or in terms of the growth characteristics of large cell populations. At the level of individual cell behaviour we are concerned with the initiation and duration of events such as chromosome replication and separation, the synthesis and insertion of new cell-wall material (see Chapter 1, p. 73) and the signals that co-ordinate chromosome replication with cell division. The study of population behaviour is concerned with the kinetics of growth, the factors that affect the mean generation time, the environmental limits to growth, etc. We shall begin by considering individual cells.

THE CELL CYCLE

Cell cycles are usually described in terms of a series of identifiable landmarks which are supposed to follow one another in an unalterable sequence, the fulfillment of each event being dependent on the completion of the preceding one. Thus in a typical eukaryotic cell it is possible to map out the sequence of events shown in Figure 1a. The landmarks are the onset and end of DNA synthesis S, mitosis and cell division M. The gaps between these landmarks are called G_1 and G_2. One of the few generalizations that can be made about eukaryotic cell cycles is that when the cycle time (growth rate) is altered by enrichment or depletion of the growth medium the time intervals between the landmarks do not change proportionately. S and G_2 tend to be little altered. The major variable is G_1.

Until about a decade ago bacterial cell cycles appeared to be remarkably featureless. There is no identifiable mitosis, and DNA synthesis goes on continuously in rich growth media—the S period thus seems to occupy the whole cell cycle. Subsequent studies, particularly with *E. coli*, have revealed a more complex and interesting situation. At slow growth rates (i.e. doubling times greater than one hour) the cell cycle is superficially similar to that of a eukaryotic cell. There is now a defined S period (except that it is called C (Figure 1b and 1c). It is followed by a gap prior to cell division (called D and corres-

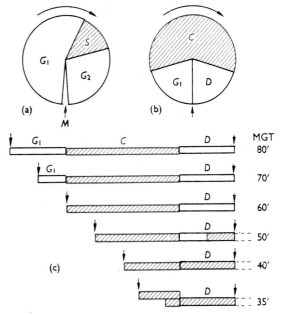

Figure 1. The cell cycle. (a) Representation of the cell cycle in a eukaryote (see text). (b) and (c) Alternative representations of the cell cycle in a prokaryote. C is the period of chromosome replication; it is followed by a period, D, at the end of which the cell divides (↑); there is then a gap, G_1, preceding further DNA synthesis. When cells are grown more rapidly, G_1 is reduced and may be absent so that C and D may overlap (c).

ponding to $G_2 + M$) and may be preceded by another gap which has not been named but which would correspond to G_1. Extensive studies have shown that, as with eukaryotic cells, C and D do not vary in proportion to the growth rate and that the major variable is G_1. For the purpose of illustration let us assume that C and D are totally unaffected by the growth rate (although current evidence suggests that they do lengthen somewhat as the cell cycle lengthens) and examine how the cell cycle would have to change to accommodate a progressive increase in growth rate.

In Figure 2, C and D are assumed to be 40 minutes and 20 minutes (the approximate experimental values found for *E. coli* at fast growth rates). When the doubling time is greater than $C + D$ minutes,

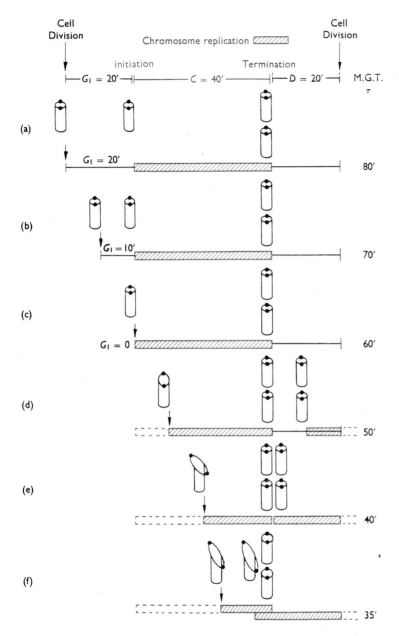

Figure 2. Effect of growth rate on timing of chromosome replication. The chromosome form during a single cycle of *Escherichia coli* beginning immediately after cell division is illustrated for each of six growth rates (τ, mean generation time, 80–35 min). It is assumed that the time C required for replication is 40 min and that it is followed by a period D of 20 min at the end of which the cell divides. C and D are assumed to be constant whatever the growth rate.

When the generation time exceeds $C + D$ (> 60 min) an interval G_1 precedes replication as in (a) and (b); when it is equal to or less than $C + D$, then G_1 is zero as in (c)–(f).

When the generation time is less than $C + D$, e.g. 50 min (d), replication begins during the preceding D period; when it is equal to the replication time of the chromosome (40 min), synthesis of DNA is continuous throughout the growth cycle (e). Further reduction in generation time leads to dichotomous replication, a second initiation occuring before completion of the first round of replication (f).

the G_1, C, and D periods will be clearly demarcated (Figure 2a). As the cycle time shortens it will do so at the expense of G_1 (Figure 2b). When the cycle time is equal to $C + D$, rounds of replication will commence immediately after cell division and there will be no G_1 period (Figure 2c). If the doubling time is less than $C + D$ minutes, rounds of replication which end in one cell cyle must start in the D period of the previous cycle (Figure 2d) and when the cycle time shortens to C minutes the D period will

no longer be detectable. Replication will now be continuous and the cell cycle will be featureless at a superficial level of analysis. At even shorter cycle times, new rounds of replication must start before previous ones are terminated (Figure 2e). During this overlapping period, chromosomes are described as replicating dichotomously.

An alternative way of thinking about the pattern of events shown in Figure 2 is to consider the onset of the C period, the end of the C period, and cell division as an unalterable sequence of events which can be called the $C + D$ cycle. When doubling times are less than $C + D$ minutes, successive $C + D$ cycles overlap. The noteworthy feature of the cell cycle when $C + D$ cycles overlap is that the chromosomes which are distributed to daughter cells have already embarked on a new round of replication.

Initiation of chromosome replication

We can deduce the average time of initiation of rounds of chromosome replication in the cell cycle provided we know the average length of C and D and the doubling time. It is simply necessary to count backwards $C + D$ minutes from cell division. The cell faces a different problem. It cannot count backwards in this way in order to determine when it should have started a new round of replication. Nor is it apparent how it can count forwards from the previous cell division because the time of initiation bears no simple relationship to the preceding cell division. This suggests that the timing of initiation is determined by a different mechanism. An important clue to this mechanism is provided by observations in *E. coli*, which suggest that the ratio of the number of origins to the cell mass at the time of initiation is the same at all growth rates. This ratio is, in effect, the concentration of chromosome origins and it thus seems possible that cells can detect the concentration of origins and trigger a round of replication when it reaches a critical value. The immediate consequence of initiation is to double the concentration of origins, and initiation will not recur until this has fallen to the critical value again as a result of further growth in cell mass. This control mechanism would constitute a biological clock with mass, or volume, or some other related parameter as its time base. Its importance in the cell cycle stems from the fact that initiation of new rounds of chromosome replication is the rate-controlling step in DNA synthesis. Thus, a culture in which the number of cells doubles every hour must double its number of chromosomes every hour. If it initiates a round of replication once per hour, then it will terminate a round once per hour under equilibrium conditions and consequently make two chromosomes from one in an hour irrespective of the rate at which replication forks traverse the chromosome.

The underlying biochemical mechanisms which control initiation of chromosome replication have been the subject of much speculation based on little hard fact. Initiation of replication has been shown

to take place at a unique site on the chromosome in a number of species of bacteria. This site is called the *origin* of replication, or chromosome origin, and in *E. coli* it is located very near to the *ilv* locus (Chapter 7). In *E. coli* and *B. subtilis* replication forks traverse the chromosome in both directions and meet at approximately 180° from the origin. The stretch of DNA containing the origin in *E. coli* has been sequenced but little is known about the biochemistry of initiation. Nevertheless, the process of initiation is clearly distinct from replication itself. Thus, genes have been identified whose products are needed for initiation but not for ongoing replication. In addition, inhibition of RNA synthesis or protein synthesis by addition of rifampicin or chloramphenicol or by removal of an amino acid required by an amino acid auxotroph, prevents initiation but allows rounds of replication that are in progress to continue to completion. Likewise, cells of cultures which stop growing naturally as they enter stationary phase end up with completed chromosomes.

This common response to cessation of growth is what would be expected if the frequency of initiation is determined by the rate of mass increase as suggested earlier.

Methods for studying the cell cycle

Several distinct methods have been used to study the cell cycle in bacteria. One is to observe the growth of single cells microscopically (e.g. by time-lapse photography) under conditions designed to maintain a constant environment. These studies are of limited value with organisms as small as bacteria. There, are additional problems such as disturbances introduced when transferring cells from liquid cultures to solid surfaces on which they can be photographed. Nevertheless, for some questions, this kind of analysis is the only one available. For example, one important finding from such studies is that the variation in cell size at division in *E. coli* is substantially less than the variation in cell age. Thus, there appears to be an important size factor that controls the cell cycle and this might be the initiation mass.

A second method is to fractionate a steady-state exponential culture on the basis of size or age. One example of size fractionation is illustrated in Figure 3. The advantage of this kind of analysis is that one is, in effect, examining an undisturbed culture. A more recent method of size fractionation which has considerable promise is by sedimentation in an

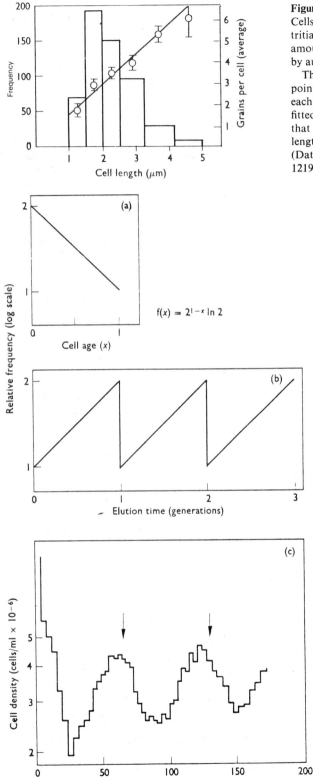

Figure 3. Rate of protein synthesis during the cell cycle. Cells of *Salmonella typhimurium* were labelled with tritiated leucine for 1/20 of a generation time and the amount of label incorporated into protein was determined by autoradiography.

The histogram shows the distribution of cell sizes. The points represent the average number of grains per cell for each of the blocks of the histogram. The straight line is fitted to the data by the method of least squares and shows that the rate of protein synthesis is proportional to cell length (i.e. to volume) and hence increases with cell age. (Data from Ecker and Kokaisl (1969) *J. Bacteriol.* **98**, 1219.)

$$f(x) = 2^{1-x} \ln 2$$

(a) The age distribution of the cells. For every 'old' cell about to divide there are two that have just been born. In a semilogarithmic plot, the two extreme points are joined by a straight line.

(b) Theoretical elution pattern. When the filter is inverted and irrigated with growth medium, the original cells remain attached, grow, and divide; the newborn cells are eluted. The cells at each peak are those descended from those that were newborn when the culture was deposited on the membrane. (see text).

(c) An experimental elution curve obtained with *Escherichia coli* B/r (data of P. Meacock).

Figure 4. Production of synchronized culture by elution of cells from nitrocellulose membrane. Cells from a culture in exponential growth are placed on the membrane.

H_2O/D_2O density gradient. It has the advantage of being applicable to any species. If particular size fractions are to be cultured subsequently (see below) it may be important that the gradient, unlike the more commonly used sucrose gradient, imposes no osmotic stress. Fractionation on the basis of age is more complex and has so far been successful with only very few strains. A brief description of the method used is necessary because it has been so useful in elucidating important features of the cell cycle in *E. coli*.

Cells of *E. coli* adhere to nitrocellulose membrane filters. If cells from a steady-state exponential culture are so deposited, the filter inverted in a suitable apparatus, and warm growth medium passed through from above, then with at least one group of strains (*B/r*), cells are eluted into this medium in a characteristic oscillatory pattern (Figure 4). It seems that after an initial wash-off period mainly newly-divided cells are eluted from the membrane. The oscillation in numbers is predicted because in the exponential culture that was deposited

on the filter there are twice as many newborn cells as cells about to divide and there is an exponential fall in the relative number of cells between these two extreme ages. The theoretical age distribution is shown in Figure 4 and an experimental elution curve is compared with the theoretically predicted one.

On this hypothesis the cells which wash off the membrane at the peak of the elution curve are the progeny of cells that were newborn at the time they were deposited on the filter. Those at the trough are the progeny of cells that were about to divide when they were deposited. If the parental culture is pulse-labelled with radioactive thymidine before the cells are deposited on the filter, the amount of label incorporated in the eluted cells will be proportional to the amount incorporated in cells of that age in the parental culture.

Figure 5a shows the predicted result of such an experiment if the culture had a doubling time of 65 minutes and average values of *C* and *D* of 40 and 20 minutes. Figure 5b shows the results obtained

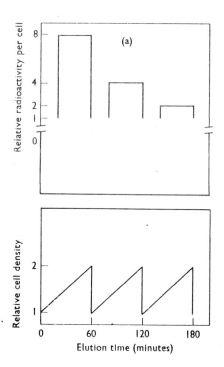
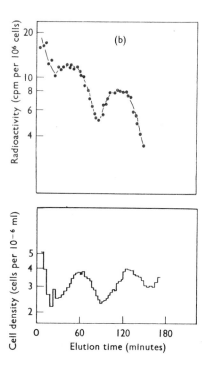

Figure 5. Elution curve for a culture pulse-labelled with [³H]thymidine (see text). (a) The theoretical behaviour of a culture having *C* and *D* periods of 40 and 20 minutes, respectively. Cells which were in the *D* period during the pulse would have incorporated no thymidine and would be eluted first. Those cells which were in the *C* period would all have incorporated equal amounts of radioactivity and would be eluted next. Note that the specific radioactivity of successive generations falls by half. (b) Experimental results for a culture with doubling time of 65 minutes.

in an experiment in which the doubling time was 65 minutes. The results are broadly consistent with those predicted. Indeed it was from experiments of this sort that the values of *C* and *D* were first estimated in bacteria. A limitation of the technique is that it is restricted to a few strains of one species of bacteria. Another is that the elution profiles are 'noisy' and also rapidly depart from the theoretical elution curve with successive elution cycles. At slow growth rates the noisiness of the elution pattern makes a quantitative interpretation of the data difficult and subjective. This may be why recent evidence does not support the constancy of *C* and *D* that was initially proposed on the basis of experiments using this technique.

The third way to study the cell cycle is to obtain synchronous cultures. New-born cells can be collected by elution in the case of *E. coli B/r* and then cultivated. A more widely applicable technique is to sediment cells of an exponential culture through a sucrose gradient. Cells in the uppermost fractions are removed and grow synchronously. Figure 6 illustrates the degree of synchrony that can sometimes be obtained and also the variability of the results. In the experiment shown the first cell cycle is longer than the average doubling time of the culture from which it was taken, and the second cycle is shorter. It is often assumed that the long first cell cycle in this kind of experiment is due to physiological disturbances produced by the sedimentation. A more likely explanation is that since there is inevitably a distribution of cell sizes about the mean cell size at birth then the more effectively one manages to skim off the smallest size fraction (and

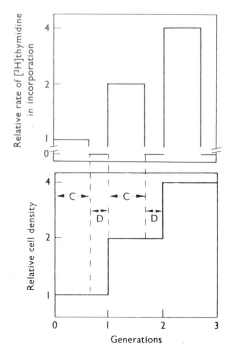
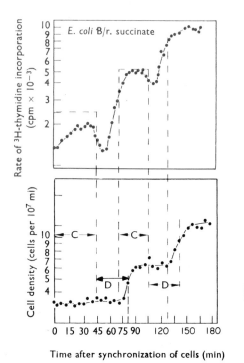

Figure 6. Determination of *C* and *D* periods in a synchronous culture. (a) The theoretical behaviour of a homogeneous population of newborn cells growing with a doubling time of *C* + *D*. The black curve represents cell numbers and the red, the rates of DNA synthesis. (b) A culture obtained by selecting from an exponential culture those cells which sedimented most slowly in a sucrose gradient (Gundas and Pardee (1974) *J. Bacteriol.* **117**, 1216). The fraction was incubated in growth medium and successive samples were pulse-labelled with [³H]-thymidine to measure the rate of DNA synthesis.

The *C* period was estimated from the mid-points of the rises and falls in incorporation (assumed to indicate mean times of initiation and termination of rounds of DNA replication).

The *D* period was estimated as the time from termination to the mid-point of the rise in numbers due to division.

Note the differences in the durations of successive cell cycles and in the timing of initiation, that two initiation events appear to take place within the first cell cycle, and that the estimates of *C* vary.

consequently the better the synchrony) the more one is studying a population of cells whose size at birth was less than the average birth size. Such cells will be expected to have a longer cycle time and to initiate rounds of DNA synthesis later in the cell cycle than an average cell.

Finally, the fact that deprivation of a required amino-acid blocks initiation, as do inhibitors of protein or RNA synthesis, can be used to obtain cultures with synchronous chromosome replication. Cultures treated in this way will complete rounds of replication but will not start new ones. If the block is removed, initiation does not begin in all cells simultaneously, presumably because they did not all have the appropriate ratio of cell mass to number of chromosome origins. However, if DNA synthesis is prevented then, after growth recommences, the concentration of chromosome origins will fall to the critical value, permitting initiation in an increasing proportion of the cell population. If DNA synthesis is allowed to resume after about one mass doubling all the cells immediately embark on a new round of replication. Such cultures show *synchronised chromosome replication* but it is important to realise that in no sense are these *synchronous cultures* in which cells are of uniform age and size.

Segregation of chromosomes and plasmids

Regular distribution of bacterial chromosomes to sister cells implies the existence of an efficient segregation mechanism. In the absence of an apparatus like the spindle of eukaryotic cells it is generally assumed that chromosomes are attached to the cell membrane and that growth of the membrane between them draws sister chromosomes apart. Likewise, plasmids are seldom lost from growing laboratory cultures. Even in the case of low copy-number plasmids like F, the incidence of plasmidless cells is less than $1:10^3$. This suggests that they too must have a segregation mechanism.

The notion of specific attachment sites has also been invoked as a control mechanism which determines the copy number and incompatibility properties of plasmids. Thus it has been widely supposed that the frequency of replication is determined by the availability of attachment sites or, in other words, that the copy number is determined by the number of attachment sites. Likewise an incoming conjugative plasmid is supposed to be unable to replicate in cells containing the homologous plasmid because all attachment sites are occupied. There is some evidence that plasmids and chromosomes are attached to cell membranes and also that the attachment is necessary for replication. The notion that the cell envelope fulfils the role of a mitotic spindle is therefore an appealing and plausible one. On the other hand, it is no longer thought that attachment sites if they exist, have a key role in determining copy number or plasmid incompatibility.

Control of cell division

The simplest interpretation of the $C + D$ cycle is that termination of a round of replication acts as the signal that triggers the septation process which results in cell separation D minutes later. It is equally plausible to suppose that septation is triggered at the time of initiation and that the septation process takes $C + D$ minutes. Most likely is the probability that septation and its relationship to cell growth and chromosome replication are more complex than simple models of this type imply. Thus when thermosensitive mutants of *S. typhimurium* and *E. coli* (which cannot start new rounds of replication at elevated temperatures) are incubated at these high temperatures, they continue to grow and divide long after chromosome replication has stopped, producing normal sized anucleate cells. Clearly each act of septation does not depend on a preceding initiation or termination of chromosome replication.

The process of septation and its relationship to growth of the envelope is better understood in Gram-positive organisms like *Streptococcus faecalis* than it is in Gram-negative bacteria like *E. coli*. It would appear that new wall material is inserted centrally at a single annular growth zone and that the newly synthesized envelope is drawn outwards to accommodate the increase in volume of the growing cell. This interpretation is consistent with studies on the distribution of old and newly synthesized components (see Chapter 1, Figure 44). It is still not clear what determines whether the additional material inserted at the growth zone should close the annulus to form a septum or be drawn out to form additional peripheral wall. An attractive idea is that whereas mass increases exponentially, envelope increases linearly. If this were so, there would have to be a discrete doubling in the rate of envelope synthesis once in every cell cycle to accommodate an exponential increase in the amount of envelope. This sudden doubling in the rate of wall synthesis at the central growth zone might cause the newly synthesized envelope to close off the septum instead of forming peripheral wall. On this model the timing of septum formation is a consequence of the varying relative rates of mass and wall synthesis.

Certain mutants of *E. coli* produce small chromosomeless cells, 'mini-cells', as a result of occasional aberrant septum formation very near to the cell pole. This shows that the site of septum formation is genetically determined and it has been suggested that in these mutants the normal inactivation of growth sites is aberrant. However, at present the mode of growth of cell envelope and the control of septum formation in the Gram-negative bacteria —including *E. coli*—is unclear.

Growth rate and cell size

The richer the growth medium the larger is the size of the bacterial cells growing in it. This is a fact that has been known for many years. The important variable is the growth medium. By contrast, if growth rates are altered by cultivating bacteria at different temperatures in the same growth medium there is little change in average cell size (see below).

The increase in average cell size can be understood by considering the implications of the occurrence of initiation at a constant average origin concentration followed by cell division $C + D$ minutes later. Consider a cell with a doubling time of $C + D$ (60 min). Initiation will occur at the time of cell division and the cells have a mass M (the initiation mass = 1/initiation concentration) at this time in the cell cycle (Figure 7). The mass (dry wt) of individual cells probably increases exponentially and the cell divides $C + D$ minutes later at a mass of $2M$. Now imagine that it were possible to pick out such a cell from the culture at birth and drop it into a richer medium permitting a faster growth rate (mass doubling time, 35 min). Suppose also that the rate of mass increase adjusted immediately to the new rate of growth (Figure 7). Note firstly, that following this shift-up (see below) the time of the next cell division would not be altered if the $C + D$ rule holds, but the cell would be larger because the amount of new mass accumulated in $C + D$ minutes will be greater. Secondly, note that since the cell mass doubles in 35 minutes instead of 60 minutes the initiation of a new round of replication will begin before cell division occurs. In other words, C and D cycles will now start to overlap. Thirdly, note that after one cell cycle at the faster growth rate a new equilibrium is reached with a larger cell size and an

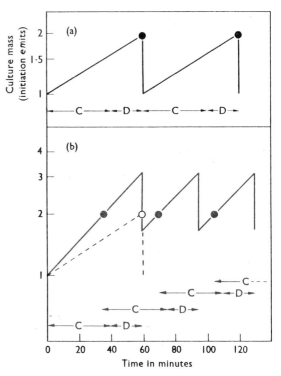

Figure 7. Effect of growth rate on cell size and timing of initiation of chromosome replication. The C and D periods are assumed to be constant and to be 40 min and 20 min respectively. (a) Mass doubling time 60 min. Growth and division of a single cell with a size of one initiation mass. The time of initiation of a new round of chromosome replication is indicated (●). (b) Mass doubling time 35 min. Rounds of chromosome replication overlap because the doubling time of the culture is less than C. The dashed curve reproduces the growth curve in (a) for comparison.

altered time of initiation. The greater the increase in growth rate the greater will be the difference in cell size and it can be shown that for exponential cultures in balanced growth the average cell size ($\overline{M}$) will be given by the following expression:

$$\overline{M} = k2^{(C + D/\tau)}$$

where τ is the doubling time of the culture. The shape of the curve relating $\overline{M}$ and growth rate will depend on whether C, D and k are invariant or not. Only in the simple (and probably too simple) case assumed here for illustrative purposes will a plot of log $\overline{M}$ versus growth rate be a straight line.

This analysis of the cell cycle also helps to explain the kinetics with which various cell parameters respond to a shift-up (see below) and the way they vary during the evolution of a culture through lag, exponential and stationary phases (Figure 8; see also Figure 15 p. 118).

After a shift-up the rate of mass increase adjusts almost immediately to the new rate. The rate of DNA synthesis reaches the new rate after C minutes. The rate of cell division equilibrates to the new rate after $C + D$ minutes have elapsed. This illustrates in another way why cell size increases when the growth rate increases. It also shows that the concentration of DNA (DNA/mass) decreases when the growth rate increases (see below).

During the growth of a culture from an inoculum of stationary phase cells (Figure 8) there is invariably an appreciable lag before cell number starts to increase. The delay before mass begins to increase is shorter. The effect of this time difference is that average cell size increases as cells emerge from stationary phase. If a steady-state of exponential growth is reached all measurable parameters will increase exponentially and in parallel. As the culture enters stationary phase a reversal of the sequence of events takes place because the rate of initiation decelerates $C + D$ minutes ahead of cell division.

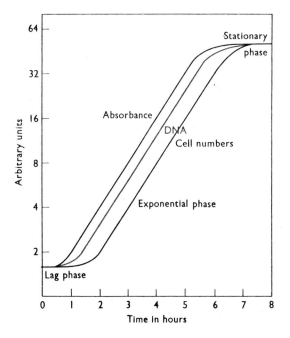

Figure 8. Representation of the growth cycle of a bacterial culture. If the frequency of initiation of new rounds of DNA replication is determined by the rate of accumulation of new mass, then the curves for absorbance and cell numbers will not become parallel for at least $C + D$ minutes (see text).

Because cell division continues exponentially while the mass increase is declining, the cells become progressively smaller. This sequence of events, shown in Figure 8, is predictable from what we know of the bacterial cell cycle. The figure illustrates the point that a culture which is in the exponential phase of growth for one measured parameter, say cell number, may not be in exponential growth for another, e.g. absorbance. Indeed batch cultures of bacteria probably never reach steady-state conditions unless started from very low inocula or repeatedly diluted at low absorbance values.

Growth rate and concentration of DNA

The fall in DNA concentration with increasing growth rate can be expressed mathematically as follows:

$$\bar{G}/\bar{M} = \frac{\tau}{kC \ln 2} \cdot (1 - 2^{-C/\tau})$$

Where $\bar{G}$ is the average amount of DNA per cell in genome equivalents and k is a constant related to the initiation concentration. The relationship is illustrated diagrammatically in Figure 9, where the curves represent the growth of one initiation mass unit of bacterial cytoplasm with time. The fact that the cytoplasm is packaged into cells which divide is ignored. One initiation mass unit will contain one chromosome which has just begun a round of replication. This is represented by a horizontal bar of length C minutes and whose height on the vertical axis indicates the mass at which new rounds of replication commence. In Figure 9a a mass doubling time of 70 minutes is assumed. There will consequently be a gap between rounds of replication. By the time the culture mass has increased 3-fold it will contain 4 fully replicated chromosomes.

If, at time zero, the same cell had been dropped into a richer medium permitting a doubling time of 40 min (Figure 9b) a new round of replication would have started immediately the preceding one was completed. There would have been no gap between rounds of replication. More important, by the time the culture mass had increased 3 units the second round of replication would not have been completed and there would have been two half replicated chromosomes present instead of four fully replicated chromosomes—the DNA concentration would have fallen.

If the cell were dropped into an even richer

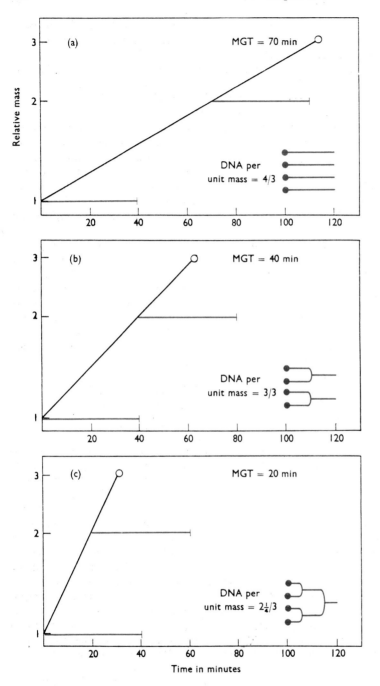

Figure 9. The effect of growth rate on the DNA concentration and on the configuration of an 'average' chromosome. Three different mass doubling times are illustrated: (a) 70 min, (b) 40 min and (c) 20 min. The transit time of a replication form is indicated by a red horizontal line and is assumed to be 40 min. The diagram in the bottom corner of each panel shows the chromosome configuration in a cell mass three times the initiation mass.

medium permitting a doubling time of 20 min, a second round of replication would commence before the first was completed. Replication would now be dichotomous. Moreover, when the culture mass reached 3 there would be a single chromosome with three replication forks present (Figure 9c)—the DNA concentration would have fallen even further.

Figure 9 illustrates the effect of growth rate on the relative number of copies of genes located at different positions on a chromosome. In a random exponential culture there will always be more copies of genes near the origin than of genes near the terminus. The steepness of this gradient of *relative gene dose* increases with increasing growth rate. It also depends on the replication time (*C*). Measurements of relative gene dose were used to determine the relative positions of genes on the chromosome of *Bacillus subtilis* (Chapter 7 p. 308). Similarly, such measurements have been used to obtain an estimate of *C* in undisturbed exponential cultures of *Escherichia coli* in which the position of genes on the chromosome has been determined by other methods.

It is often assumed that dichotomous replication leads to an increase in the number of genes located near the chromosome origin. This is not so. We have already noted that the concentration of DNA falls as the growth rate of a culture increases. Figure 9 shows that this fall affects different genes to different extents. Those near the chromosome terminus suffer the greatest fall in concentration. There will be no change in the concentration of genes near to the chromosome origin if, as we are assuming, the initiation concentration is the same at all growth rates. The average concentration of genes at intermediate positions will fall by intermediate amounts. Thus an increase in the *relative number* of genes near the origin is perhaps more informatively described as a decrease in the *relative concentration* of genes near the terminus. The special feature of bacterial cell cycles, like that discussed here, in which the replication time of the chromosome can be long relative to the cell cycle time, is that gene concentrations cannot be maintained at constant values for all genes over a range of growth rates. It also means that the concentration of one gene relative to another can vary up to four-fold during a single cell cycle (see Figure 9). The differential effect of growth rate on gene concentration may provide a reason for non-random gene order. Thus it might be expected that genes whose products are normally rate limiting for growth would be located close to the chromosome origin. In this location the gene concentration and hence the concentration of its product would be the maximum possible at all growth rates. This may be why the six copies of the gene coding for the ribosomal RNA in *E. coli*, which account for the bulk of the RNA being synthesized at any instant, are clustered around the chromosome origin.

Effects of growth rate on composition of bacteria

Some of these effects have been mentioned and others will occur later. Since the growth of a population reflects what is happening in the individual cells it will be useful to summarize here those effects of growth rate that are more immediately relevant.

In general, when bacteria grow faster they are larger and they contain more RNA, most of which is ribosomal. The experimental observations, made originally in coliform bacteria, show that the growth rate varies linearly with the ribosome content. The interpretation placed on this is that ribosomes are rate-limiting in protein synthesis. In fact, the RNA per cell varies with growth rate over a 10-fold range, and in very rapidly growing cells the RNA may account for 30% of the mass. This means that almost half of the dry weight of the cells consists of ribosomes.

The DNA per cell also increases with the growth rate but to a lesser extent than the cell size. Hence the DNA decreases in relation to the cell mass.

The cell envelopes are usually of constant thickness and it is found that the proportion of a bacterium which is wall plus membrane decreases with growth rate and thus with cell size. This is because the ratio of surface to volume decreases as the size increases.

GROWTH OF CELL POPULATIONS

Exponential growth

Bacteria multiply by binary fission but the two daughter cells may not grow at identical rates or divide synchronously. Hence, in a growing population, there is spread of individual generation times and consequently the term 'doubling time' of the population is often preferred as a measure rather than the mean generation time. In spite of the spread in generation times, populations of bacteria generally increase with a relatively constant doubling time provided that the environmental conditions do not vary markedly. Thus, a population of organisms (number *N*) will give rise to 2*N* progeny after one culture doubling time (t_d) and in turn to 2^2N after a similar period of time has elapsed. This progression can be represented as an exponential series:

$$2^0N_0 \rightarrow 2^1N_0 \rightarrow 2^2N_0 \rightarrow 2^3N_0 \rightarrow 2^4N_0 \rightarrow 2^nN_0$$

where N_0 is the initial population density, and *n* the number of culture doublings that have occurred after some period of time (*t*). Thus:

$$n = t/t_d$$

It follows that the number of organisms present after *n* culture doublings (N_t) will be related to the initial number of organisms (N_0) by the general expression:

$$N_t = N_0 \cdot 2^{t/t_d}$$

which in the logarithmic form is:

$$\ln N_t - \ln N_0 = \left(\frac{t}{t_d}\right) \cdot \ln 2 = t(0 \cdot 693/t_d) \qquad (1)$$

Therefore, plotting the logarithm of the number of organisms against the time of incubation gives a straight line the slope of which is $0 \cdot 693/t_d$.

The specific growth rate constant, μ

An alternative expression for growth can be derived on the assumption that the rate of increase in cell numbers (dN/dt) at any time (t) is proportional to the total number of cells (N) already present. Therefore:

$$\frac{dN}{dt} = \mu N \qquad (2)$$

where μ is a constant, usually referred to as the '*specific*' or '*instantaneous*' growth rate constant. It can be defined as $1/N \cdot dN/dt$ and, upon integration, equation (2) gives:

$$\ln N_t - \ln N_0 = \mu t$$

or

$$\ln \frac{N_t}{N_0} = \mu t$$

or

$$N_t = N_0 \cdot e^{\mu t}$$

The relationship between μ and the doubling time (t_d) can be easily established. Thus, for a doubling of the culture, $t = t_d$ and $N_t = 2N_0$, and so $\ln (2N_0/N_0) = \mu \cdot t_d$. Therefore:

$$\mu = \frac{\ln 2}{t_d} = \frac{0 \cdot 693}{t_d} \qquad (3)$$

Since biomass can generally be determined more accurately than the number of organisms, the basic growth equations are usually expressed in terms of mass rather than number. Conventionally, the biomass concentration is given the symbol 'x', and the specific growth rate (μ) is therefore defined by

$$\frac{1}{x} \cdot \frac{dx}{dt}$$

The specific growth rate constant (μ) is greatly influenced by environmental conditions. Thus, in natural ecosystems the specific growth rate may vary widely. In the laboratory, cultures can be grown at a constant rate for long periods.

Batch culture: the bacterial growth cycle

When a small population of bacteria is inoculated into a suitable medium growth generally does not start immediately; there is a variable *lag period*. Moreover, once growth commences, an exponential increase in population density is observed only over a relatively short period (often less than ten doubling times, depending on the inoculum size). Growth then ceases and the culture enters the so-called *stationary phase*, after which the cells slowly die and lyse. Thus, with these closed *batch cultures*, at least three well-defined phases of growth can be defined (*lag*; *exponential*; *stationary*) which together constitute the *bacterial growth cycle*. It is appropriate to consider this cycle in more detail and to distinguish it from the cell division cycle.

The lag phase

When a batch of fresh medium is inoculated with cells taken from a culture that has been grown into the stationary phase, a significant period of time may elapse before growth resumes. This is due to the fact that, during the stationary phase, cells undergo marked changes in chemical composition (see later). If, however, the inoculum is taken from a culture that is still in the exponential phase, the lag period generally will be much shorter or even absent altogether. Moreover, if cells from a culture are inoculated into a medium of substantially different composition then the lag may be prolonged, even when an inoculum of actively growing cells is used. Indeed, such a period of metabolic adjustment is often observed with a culture in which two different sugars are present as carbon and energy sources (see Figure 3, Section 2, p. 17). The organisms use the sugars sequentially, and there is a short lag period between exhaustion of the first sugar and growth on the second. This pattern of response, first described by Monod, is termed *diauxie* (see Chapter 8).

Many other factors influence the length of the lag period, either by impeding the process of adaptation or of growth, or both. These factors tend to be more effective with small inocula than with large ones. For example, certain aerobic nitrogen fixers, like *Azotobacter* spp., are highly sensitive to oxygen when fixing nitrogen. A small inoculum of such cells in fresh, oxygen-saturated medium free of fixed nitrogen, might fail to grow whereas cells in a larger inoculum might consume sufficient of the dissolved oxygen to enable nitrogen fixation, and therefore growth, to proceed. Similarly, even heterotrophic organisms have a substantial requirement for carbon dioxide but usually this compound is generated sufficiently fast by catabolism. However, if a low concentration of organisms is vigorously aerated, and if the pH value is below neutrality, so much carbon dioxide may be swept from the aqueous environment that CO_2-requiring reactions, and thereby growth, are retarded. In such cases low concentrations of TCA-cycle intermediates (such as fumarate, succinate or malate) may 'spark' growth.

As expected, the length of the lag period is usually diminished if organisms are transferred from a medium that is nutritionally demanding to one that is less so. Conversely, the lag period is usually prolonged when organisms are transferred from a rich medium to a poor one. This is because biosynthetic pathways that were repressed by components of the rich medium have to be regenerated before growth can resume. Such considerations do not invariably apply, particularly with autotrophic organisms in which the presence of complex organic nutrients may inhibit growth.

The exponential phase

The lag phase may be looked upon as a period of metabolic adjustment which eventually leads to a balanced synthesis of the cells' polymeric constituents (i.e. growth) and cell division. As mentioned previously, the specific growth rate is not an inherent and invariant property of the organisms; but is influenced markedly by the prevailing environmental conditions, particularly by the complexity of the medium and the nature of the major carbon and energy source. In general, heterotrophic bacteria grow faster in media containing a mixture of complex organic nutrients (amino acids, purines, pyrimidines, and vitamins) than they do in simple salts media. Again, with simple salts media, bacteria grow faster when supplied with some carbon substrates than with others. Thus, cultures of *Escherichia coli* grow more rapidly on glucose than on any other single carbon substrate.

With facultatively anaerobic bacteria, growth in the presence of oxygen is usually faster than in its absence. This presumably reflects the greater efficiency with which organisms can generate ATP aerobically. Further, substances like nitrate, whose assimilation imposes extra energetic demands upon the organisms, usually support growth at a lesser rate than those, like ammonia, whose assimilation requires less energy input.

The exponential growth rate may be influenced by the pH value of the culture and by its osmolarity.

One factor, however, which most consistently affects the growth rate of bacteria is the temperature. Invariably, the growth rate increases with temperature up to some well defined optimum value, beyond which it decreases sharply to zero. Thus the difference between the optimum temperature for growth and the maximum that can be tolerated is small as compared with the difference between the optimum and minimum temperatures. The lower limit of temperature at which growth will occur is not easily defined, but generally speaking the temperature range within which observable growth occurs seldom exceeds 35°C (Figure 10a). The optimum temperature varies widely between different species, and the terms *thermophile, mesophile* and *psychrophile* are used to refer, respectively, to species which grow optimally at high, intermediate and low temperatures.

In order to interpret the influence of temperature on microbial growth rate it is necessary to consider its effect on the rates at which chemical reactions proceed. The relationship between reaction rate and temperature can be described by the Arrhenius equation:

$$\log_{10} v = \frac{-\Delta H}{2.303 \, RT} + C$$

where v represents the reaction rate, $-\Delta H$ the activation energy of the reaction, R the gas constant and T the absolute temperature (in K). This predicts a linear relationship between the logarithm of the reaction velocity and the inverse of the absolute temperature. However, a plot of this kind for bacterial growth shows marked deviations from linearity (Figure 10b). At higher growth temperatures this can be attributed to protein denaturation, with concomitant loss of enzyme activity. Not all the cellular proteins are equally thermolabile, and inactivation of only one key enzyme will cause growth to cease. This is clearly shown by temperature-sensitive mutant strains. In some of these an enzyme essential for growth is altered so that it is more susceptible to heat denaturation than is the wild-type enzyme. It is therefore possible by a single mutation to *decrease* the maximum temperature at which the organism can grow; on the other hand it is by no means as easy to *increase* the maximum growth temperature by mutation.

The factors which affect growth of an organism at low temperatures are less well defined. It has been suggested that the temperature minimum is determined by the control mechanisms of the cell which may be more sensitive to low temperatures than are the metabolic sequences which they control. This view is based on the considerable sensitivity of

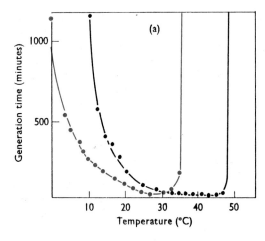
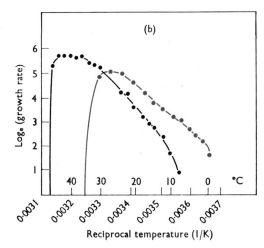

Figure 10. Temperature and growth rate. (a) Effect of temperature on the generation time of a mesophile and a psychrophile. (b) Arrhenius plot of the relationship between growth rate and temperature for the same two species. Growth rate is expressed in terms of generations per hour.

allosteric enzymes (see Chapter 8) to environmental conditions. Other factors are also important, since it is known that mesophilic organisms (those with a maximum growth rate in the range 30°–40°C) respond to a lowered incubation temperature by increasing the proportion of unsaturated fatty acyl residues in their membrane lipids. Lipids containing unsaturated fatty acids melt at lower temperatures than those containing the corresponding saturated acids. Thus, the minimum temperature may be prescribed by the limit to which the chemical composition of the plasma membrane can be adjusted. Similar considerations may also play a role in setting the maximum temperature at which bacteria may grow. Brief exposures to high temperatures may cause an efflux of low molecular weight 'pool' constituents from cells, presumably by impairing membrane function.

The stationary phase

As growth proceeds, the concentrations of essential nutrients diminish and those of end-products increase exponentially. In particular, if growth ceases because of the depletion of an essential nutrient, its concentration will fall from one-half of the initial concentration to zero over the period of the final doubling of biomass (Figure 11). For organisms growing in a simple salts medium containing limiting amount of a single carbon and energy source, the transition from exponential phase to stationary phase is abrupt. For cultures growing in a complex medium such as nutrient broth, the transition is often greatly extended, with ancillary carbon

substrates (principally amino acids) becoming depleted in sequence.

The word 'stationary' does not imply that all biosynthetic activity has ceased. Indeed, the onset of the stationary phase frequently induces extensive changes. In spore-forming bacteria such as *Bacillus subtilis*, sporulation is initiated by depletion of certain nutrients, and a succession of morphological changes ensues leading to the formation of a highly resistant dormant spore within the mother cell.

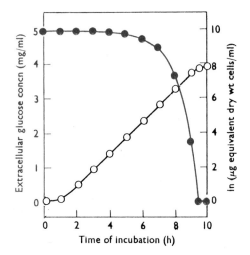

Figure 11. Growth limitation by a specific nutrient. Growth of a culture in a simple glucose/salts medium at 37°. Note that the concentration of bacteria (○) is plotted as its logarithm, whereas that of glucose (●) is on an arithmetic scale.

Cells undergoing sporulation are therefore far from static, metabolically, but are synthesizing many enzymes and new compounds in a sequential way (see Chapter 9).

Changes also occur in non-sporulating bacteria as they enter the stationary phase. DNA synthesis continues for some time after net increase in mass has ceased, and the organisms divide to give cells small in size and low in RNA content. Many proteins, including extracellular enzymes, are synthesized during the stationary phase. When growth is restricted by lack of exogenous carbon substrate or utilizable nitrogen source, the intermediates needed for RNA and protein synthesis cannot be formed *de novo*. However, in starved cells they can be, and generally are, provided by hydrolysis of inessential RNA and proteins, mediated by intracellular nucleases and proteinases the synthesis or activation of which follows the onset of starvation. This recycling is known as *turnover*, and is an important feature enabling cells to survive and adapt to altered environments (see Chapter 8).

The stationary phase is not always caused by depletion of an essential nutrient. Growth generally alters the pH value and occasionally this may inhibit further cell synthesis. Even with organisms growing on neutral sugars and producing carbon dioxide as the sole end-product of catabolism, there may be a substantial shift in the pH value owing to the disproportionate uptake of cations (NH_4^+, K^+ and Mg^{2+}) over anions (PO_4^{3-}, SO_4^{2-}). For example, *Bacillus subtilis* may contain about 12% of its dry matter as nitrogen, and 5% as potassium, whereas the cellular phosphorus content is about 3·5%, and sulphur 0·5% of the cells' dry weight. Thus, the molar ratio of assimilated ($NH_4^+ + K^+$) to ($PO_4^{3-} + SO_4^{2-}$) would be approximately 7, and the charge ratio about 2·5. A similar imbalance occurs with Gram-negative bacteria, although the cellular potassium and phosphorus requirements are substantially lower. Not surprisingly, therefore, the growth of organisms with, say, glucose as the sole carbon source, produces a progressive fall in the pH value unless the initial medium is heavily buffered. Conversely, growth on organic acids such as acetate, lactate or succinate causes the pH value to rise because of the greater uptake of anions over cations, and growth often ceases due to the increase in pH value.

In general, then, growth causes the environment to change until ultimately it can no longer support, or permit, further cell synthesis; the culture then enters the stationary phase. It follows that the sequence of events routinely observed in a batch culture, and referred to as the 'growth cycle', is in no way an inherent property of the organisms. However, if the changes in the environment occurring as a consequence of growth were to be continuously corrected, then growth would continue indefinitely. In fact, this is easily achieved and organisms can then be cultured continuously.

CONTINUOUS CULTURE

From a consideration of the events occurring during the exponential phase of a batch culture, it is clear that the growth of organisms could be prolonged indefinitely by providing fresh medium at a rate sufficient to keep the population density constant at some submaximal value. With a constant population the specific rates of substrate utilization would also be constant. Moreover, the concentrations of the end-products would no longer increase since these would be diluted out as fast as they were accumulating. The culture is kept at the same volume by an overflow device. This, then, is one simple form of continuous culture device, called a

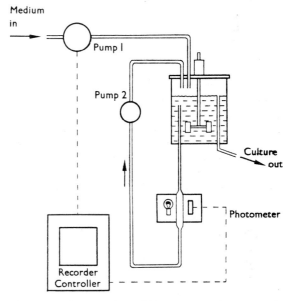

Figure 12. Diagram of a turbidostat. A small amount of culture is continuously circulated (by pump 2) through an external loop around which is housed a device for monitoring the optical density. The output from the photocell is connected to a recorder–controller which activates the medium pump (1) whenever the optical density rises above some set value. The culture volume is maintained constant by means of an overflow tube.

turbidostat, and its essential components are outlined in Figure 12.

Theoretically, a turbidostat represents one of the simplest forms of continuous culture but in practice it is the most difficult to operate. It has the advantage that it allows organisms to grow at their maximum rate in any particular medium and is, therefore, useful in studying factors that either stimulate or depress the growth. The main problem associated with its use stems from the tendency of bacteria to adhere to surfaces of the growth vessel. This gives false turbidity readings and thus false estimates of the growth rate.

The chemostat

The second, and operationally more reliable, method of growing bacterial cultures continuously is the *chemostat* technique. This can be most easily understood by considering what would happen to a turbidostat culture if the optical control were disconnected and the pump set to deliver medium at a rate which was less than that needed to maintain growth at the maximum rate. Since the rate at which organisms grew would be greater than the rate at which the culture was being diluted with fresh medium, their concentration would increase. However, this increase could not continue indefinitely because of the growth-associated changes in the environment. On the other hand, growth would not cease completely, as in a batch culture, since fresh medium is continuously being added. In fact, after some time, the culture would come into a steady state in which the rate is proportional to the rate of addition of fresh medium. More precisely, the rate of growth is equal to the *dilution rate* (*D*; the flow rate of the medium divided by the total volume of culture). Although, with this apparatus, fresh medium is added at a constant rate, the organisms grow exponentially (but at a sub-maximal rate), since the dilution rate of the culture is an *exponential* function of time. That this is so can be seen by considering what happens when a culture in a chemostat is sterilized by adding a small volume of formalin. The instantaneous rate at which the biomass concentration (x) in the culture vessel of volume V will change is proportional to the flow rate (f) of medium entering the vessel, and the concentration of organisms present:

$$\frac{dx}{dt} = -\left(\frac{f}{V}\right)x = -D(x) \qquad (4)$$

where the dilution rate $D = f/V$. Integration gives the exponential expression:

$$x = x_0 \cdot e^{-Dt} \qquad (5)$$

(This equation is formally identical to that which describes the decay of a radio-active element.) Of course, with a chemostat culture, it applies only to the special case where further growth is inhibited. When cells are growing at a rate such that their concentration is not changing with time (that is, when the culture is in a steady state) it is clear that the growth rate must be numerically equal to the washout rate (or dilution rate, D).

In a chemostat culture (in contrast to a turbidostat culture) organisms are growing at a submaximal rate that is determined by the dilution rate. Nevertheless, just as in a turbidostat their growth follows exponential kinetics since, at steady state, the specific growth rate must equal the dilution rate which is an exponential function. Thus, irrespective of the composition of the medium, organisms can be grown at a wide variety of different exponential rates in a chemostat-type culture; or grown continuously at their maximum rate in a turbidostat-type culture. When organisms are grown in a complex medium in a chemostat, the environmental changes that cause the culture to come into a steady state are not known precisely. But with a defined simple salts medium it is easy to arrange for one essential nutrient to be present in a sufficiently low concentration to limit the population density. In a chemostat culture the concentration of the limiting nutrient in the reservoir determines the *population density*: its rate of supply (i.e. rate of addition of medium) specifies the *growth rate*. Any compound or element for which the organism has an absolute growth requirement can be made the growth-limiting nutrient: thus, cultures can be carbon substrate-limited, ammonia-limited, phosphate-limited, etc. Further, if an organism has a particular auxotrophic requirement, that substance can be made the growth-limiting nutrient.

The chemostat therefore allows a choice of bacterial density and also of growth rate from close to the maximum to very low values. Again, by choosing different growth-limiting components it is possible to gain considerable insight into the physiology of microbial growth.

Theoretical aspects

Dependence of growth rate on the concentration of a limiting substrate

From the above it is clear that, with chemostat cultures, it is the concentration of growth-limiting nutrient in the reservoir medium that determines the population density. However, it is not the dilution rate *per se* that regulates the growth rate but the concentration of growth-limiting nutrient in the *culture fluid* (which varies with dilution rate). Monod showed empirically that growth rate was an approximate hyperbolic function of the concentration of the growth-limiting nutrient (s). This relationship could be expressed mathematically in a form similar to that used to define the rate of an enzyme-catalysed reaction in relation to its substrate concentration, *viz*:

$$\mu = \mu_{max} \left(\frac{s}{K_s + s} \right) \qquad (6)$$

where μ_{max} is the maximum (unconstrained) growth rate, and K_s a saturation constant that is numerically equal to the growth-limiting substrate concentration allowing growth to proceed at one-half its maximum rate.

This formulation adequately describes what is found with many real chemostat cultures at equilibrium (i.e. when steady state conditions prevail). However, it does not accurately predict transient changes in growth rate following changes in limiting substrate concentration. Here it is often found that the response is not immediate, as predicted, but that a slow adjustment occurs during which time many components of the cell change quantitatively (see p. 118).

Growth rate and rate of uptake of nutrient
Net growth requires the uptake of essential nutrients, and in steady state chemostat cultures, growth rate (μ) is determined by, and proportional to, the rate of uptake of the growth-limiting nutrient (q). That is:

$$q \propto \mu = Y \cdot q = Y \left(\frac{1}{x} \cdot \frac{ds}{dt} \right) \qquad (7)$$

where Y is the proportionality factor or *yield value* (normally expressed as grams of organisms formed/mole of substrate consumed).

Again, this relationship applies strictly only to steady state conditions; q can vary widely during a transient state without concomitant changes in μ. This is particularly true for nutrients such as the carbon and energy source.

Characteristics of the steady state
For a complete and quantitative statement of the behaviour of bacterial cultures in a chemostat, it is important to specify more precisely the effect of dilution rate on the concentration of growth-limiting nutrients (s) in the *culture*, and on the biomass concentration (x).

In a chemostat culture, changes in the concentration of organisms depend upon the relationship between the growth and washout. Thus:

$$\begin{array}{l} \text{change in cell} \\ \text{biomass in the culture} \end{array} = \text{growth} - \text{washout}$$

$$\frac{dx}{dt} = \mu.x - D.x$$

$$= x(\mu - D)$$

Whenever $\mu > D$, the concentration of organisms will increase with time, and whenever $\mu < D$ it will decrease. Hence, at any fixed value of D, changes in the concentration of organisms will depend solely on μ which, itself, is critically dependent on s (the *culture* growth-limiting substrate concentration) as specified by equation (6). That is:

$$\frac{dx}{dt} = x \left[\mu_{max} \left(\frac{s}{K_s + s} \right) - D \right] \qquad (8)$$

Similarly, changes in the concentration of the growth-limiting nutrient will depend on the balance between the input (supply) rate and both the rate of loss in the effluent culture and the rate of consumption by the bacteria. Thus:

$$\begin{array}{l} \text{change in} \\ \text{substrate} \\ \text{concentration} \end{array} = \begin{array}{l} \text{input with} \\ \text{medium} \end{array} - \begin{array}{l} \text{loss in} \\ \text{effluent} \end{array} - \begin{array}{l} \text{consumption} \\ \text{by organisms} \end{array}$$

$$\frac{ds}{dt} = D.S_r - D.s - q \cdot x$$

$$= D(S_r - s) - \frac{\mu x}{Y}$$

where x is the concentration of organisms (mg equivalent dry wt/ml) and S_r is the concentration of growth-limiting substrate in the *influent medium*. Rearranging, and substituting for μ (equation 6):

$$\frac{ds}{dt} = D(S_r - s) - \frac{\mu_{max} \cdot x}{Y} \left(\frac{s}{K_s + s} \right) \qquad (9)$$

When a culture is in a steady state, both $dx/dt = 0$ and $ds/dt = 0$, and unique values exist for the concentration of both organisms (x) and growth-limiting substrate (s). These steady state values are identified by the symbols $\tilde{x}$ and $\tilde{s}$, respectively. Thus:

$$\tilde{x} = Y(S_r - \tilde{s}) \qquad (10)$$

and

$$\tilde{s} = K_s \left(\frac{D}{\mu_{max} - D} \right) \qquad (11)$$

Substituting for $\tilde{s}$ in equation (10) gives:

$$\tilde{x} = Y \left[S_r - K_s \left(\frac{D}{\mu_{max} - D} \right) \right] \qquad (12)$$

It follows from equation (11) that the main effect of changing the dilution rate is to change the concentration of growth-limiting nutrient in the culture, thereby effecting a change in the specific growth rate (equation 6). Furthermore, consideration of equations (6), (8) and (9) reveals that, provided that the dilution rate is set at a value below that permitting cells to grow at their maximum (unconstrained) rate, steady state conditions invariably will become established. The culture is self-regulating, and any transient changes in the steady state will set up reactionary forces that 'steer' the culture (often rapidly) back to equilibrium.

At equilibrium, and provided that μ_{max}, K_s and Y are true constants, the steady state concentrations of cells and limiting substrate should vary with dilution rate as shown in Figure 13. In practice, such a pattern of change is rarely found, irrespective of the growth limitation. This is mainly because the yield value is not a true constant, but often is markedly dependent on the growth rate. In Figure 14a the rate of glucose consumption in a glucose-limited chemostat culture of *Klebsiella aerogenes* is shown as a function of the specific growth rate (= dilution rate). Clearly, for the yield value to be a constant, the plot of q versus μ (Figure 14a) must pass through the origin; and if the organism has a relatively high affinity for carbon-substrate, $\tilde{x}$ should marginally increase as the dilution rate is lowered

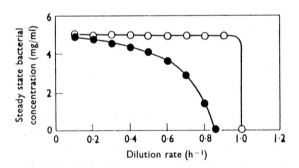

Figure 13. Theoretical plot of steady state bacterial concentration as a function of growth rate. The data apply to a glucose-limited, chemostat culture in which the yield value (Y) is 90 (g cells formed/mole glucose assimilated); the input concentration of glucose (S_r) is 10 g glucose/l (or 55·6 mM); and the maximum specific growth rate (μ_{max}) is 1·0. In the upper plot ($\bigcirc$), the saturation constant (K_s) is taken to be 0·1 mM, and in the lower plot ($\bullet$), it is taken to be 10 mM. The points are calculated according to equation 11:

$$\tilde{s} = K_s \frac{D}{\mu_{max} - D}$$

towards zero (see Figure 13). Generally, neither condition is met: as the growth rate is decreased, the specific rate of carbon-substrate consumption invariably tends to some finite value at $D = 0$. Hence, since $Y = \mu/q \ (= D/q$, at steady state), the yield value must tend to zero as $D \rightarrow 0$; such is clearly evident in Figure 14b.

Energetic aspects of bacterial growth

In order to explain the observed variations in yield value with growth rate in carbon-substrate limited cultures, it has been suggested that a portion of the substrate must be oxidized to provide energy for growth-unassociated 'maintenance' functions (which have been assumed to include maintenance of solute gradients, motility, and turnover of macromolecules). It is not clear whether, and to what extent, the energy diverted into these processes varies with growth rate, and this makes it difficult to formulate a mathematical relationship between growth rate and yield. Nevertheless, if it is assumed that the maintenance energy requirement does not vary with growth rate, then the rate of substrate uptake can be estimated by extrapolation to zero growth rate and so provide a direct measure of it. Hence, if this rate of substrate uptake is deducted from the actual rate expressed at each growth rate, then the substrate requirement associated solely with growth can be evaluated. Thus:

$$q_{actual} = q_{growth} + q_{maintenance}$$

Dividing by the growth rate ($\mu = D$, at steady state)

$$q/\mu = q_g/\mu + q_m/\mu$$

But since q/μ is the reciprocal of the yield value, then:

$$1/Y = 1/Y_g + q_m \cdot 1/\mu \qquad (13)$$

where Y_g is the 'true' growth yield constant (that is, the yield value corrected for maintenance losses). It follows, therefore, that plotting the reciprocal of the observed yield value $(1/Y)$ against the reciprocal of the growth rate $(1/\mu)$ should produce a straight line with a slope (equal to q_m, the maintenance rate) that intersects the ordinate at a value of $1/Y_g$. This is shown in Figure 14c.

While this approach does have success in predicting deviations in the steady state cell concentration at low growth rates, in carbon-substrate limited chemostat cultures, it should not be accepted uncritically. There is no *a priori* reason to suppose either that the maintenance rate or the growth-

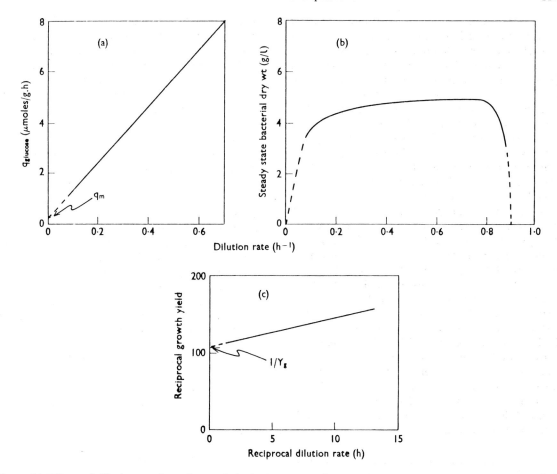

Figure 14. Effects of dilution rate in a glucose-limited, chemostat culture. *Klebsiella aerogenes* was grown in a simple glucose/salts medium at 35°C and pH 6·8. (a) Effect of dilution rate on the specific rate of glucose consumption. (b) Effect of dilution rate on the steady state bacterial concentration (note that $\mu_{max} = 0.9$ h^{-1}). (c) Double reciprocal plot of the growth yield versus the dilution rate. The intercept on the ordinate where $1/D = 0$ (i.e. where the cells are growing infinitely fast) is a measure of $1/Y_g$, according to equation 10: $\bar{x} = Y(S_r - s)$.

associated substrate requirements do not vary with growth rate; in fact evidence exists to suggest that they do.

For a clearer assessment of the energetics of bacterial growth, a more rational approach would be to determine the relationship between the specific growth rate and the specific rate of ATP synthesis (i.e. q_{ATP}). This, however, creates problems since, in general, no direct measure can be made of the rate of ATP synthesis in growing bacteria. The best that can be done is to measure the rate of change in some other parameter which is linked to ATP synthesis. For example, with organisms growing anaerobically on glucose and producing CO_2 and ethanol as end-products, the q_{ATP} should be equal to the $q_{ethanol}$ since each mole of glucose fermented

gives a net yield of two moles of ATP and two moles of ethanol. With a bacterial culture growing anaerobically in a nutritionally complex medium with a single fermentable carbonsubstrate energy source, yield values of about 10 (g organisms formed/mole ATP generated) have been obtained which, when corrected for maintenance energy losses, give a Y_{ATP}^{max} value of about 14.

For many organisms growing aerobically, the bulk of the ATP is generated by oxidative phosphorylation; and since the complete oxidation of one mole of glucose consumes six moles of oxygen and can generate, maximally, 38 moles of ATP, it is possible to assess the Y_{ATP} value from the oxygen yield value (Y_0). In practice, this gives low values (e.g. about 8 for *Klebsiella aerogenes*) which, though possibly realistic, may be an underestimate since the respiratory chain in many bacteria (including *K. aerogenes*) may contain fewer proton translocating loops than do mitochondria (see Chapter 3, p. 149). Again, it is still not certain that the transfer of electrons to oxygen along the respiratory chain in bacteria is coupled stoichiometrically to ATP synthesis.

Transient states: unbalanced growth in batch cultures

Exponential growth cannot continue for long periods either in batch cultures or in natural environments and, as growth ceases, a large number of interrelated events ensue.

We can begin by considering some of the events in a controlled experiment when cultures are shifted from one steady state to another. This can be done by adding richer nutrients or, in the case of a chemostat culture, by increasing the rate of flow of medium. This is called *shift-up*. Conversely, the slowing down of growth is called *shift-down*. This can be carried out by washing cells grown in a rich medium and transferring them to a nutritionally poorer medium.

Since cells growing at the faster rate are larger (see Figure 7) and are richer in RNA, it is pertinent to ask how, during a *shift-up*, small cells which are relatively poor in RNA acquire the characteristics of fast-growing cells. As shown schematically in Figure 15, the rate of RNA synthesis accelerates almost immediately, is followed by an increase in the rate of protein synthesis and, a few minutes later, by an increase in the rate of DNA synthesis. Cell division continues at the original rate until the cells have acquired the larger size characteristic of the faster growth. Thus the characteristic composition of cells growing in the new medium is established very rapidly. This sequence during *shift-up* is expected because the rate of protein synthesis is proportional to the cells' ribosome content (see above).

Cells in a rich medium have available a variety of nutrients. These cause wide-spread repression of many enzyme systems concerned with their synthesis. Bacteria growing in nutrient broth do not synthesize amino acids, purines, pyrimidines, etc. Indeed, this sparing action is probably the reason why they grow faster. If they are 'shifted-down' to a poorer medium, containing a single source of carbon and energy, they lack many of the enzymes now necessary. The result is a long lag before cell mass

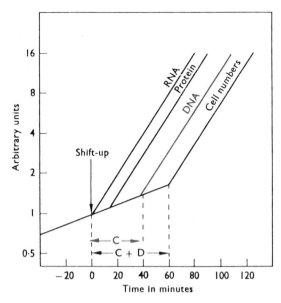

Figure 15. Representation of a shift-up. A culture growing exponentially in synthetic medium is shifted at time zero to nutrient broth where the growth rate is three doublings per hour. Since growth is balanced, the semilogarithmic plot of the data before the shift includes in a single line the *relative* increments in the various components. After the shift, the synthesis of RNA, protein, and DNA change to the new faster rate in that order. Subsequently, the rate of cell division increases to the same new value, but only after $C + D$ minutes. The change in the rate of DNA synthesis occurs C minutes after the shift-up.

increases. However DNA synthesis and cell division continue for some time, resulting in small uninucleated cells, with a low RNA content (Figure 16). During this period the cells synthesize the enzymes they lack. Since there is no net synthesis of RNA and protein, new enzymes must be made from existing cell material (see protein and RNA turnover, Chapter 8). Gradually, net protein synthesis is resumed and after many hours the cells acquire the characteristics appropriate for growth in the poorer medium.

During transition following *shift-up* and *shift-down*, the culture is in a condition of *unbalanced growth* during which synthesis of one or more species of macromolecules may be dissociated from the synthesis of others. Examples of typical manipulations which lead to the uncoupling of synthetic activities are shown in Table 1.

The synthesis of DNA is particularly easy to dissociate from other synthetic activities. Thus, synthesis can be prevented without affecting other

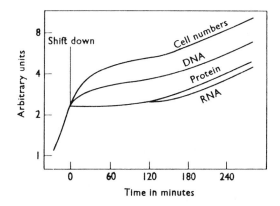

Figure 16. Representation of a shift-down. A culture growing exponentially in nutrient broth at a rate of three doublings per hour is washed by centrifugation and, at zero time, is put in synthetic medium where the rate is only 1·2 doublings per hour. As in Figure 15, a single line denotes the relative increment of various cell components before the shift. After shift-down, cell division and DNA synthesis continue for some time, rapidly resulting in smaller cells. Net protein and RNA synthesis, however, are resumed only gradually and the size and composition characteristic of the new medium are only obtained after three to four hours. During the initial lag, extensive turnover of RNA takes place.

synthetic processes for a considerable time, as, for instance, when thymine is removed from a mutant that requires it. Conversely, DNA synthesis may continue in the absence of RNA or protein synthesis. This is observed in a *shift-down* or under conditions of starvation, such as when amino acids are removed from stringent mutants requiring them (Table 1). The amount of DNA made corresponds to that required for the completion of existing rounds of replication (see above).

The synthesis of RNA and of protein, although readily dissociated from the synthesis of DNA, are closely connected. Thus, if RNA synthesis is inhibited, for instance by the antibiotics actinomycin D or rifamycin, protein synthesis will proceed at a rapidly diminishing rate which parallels the rate of decay of messenger RNA.

When protein synthesis is inhibited by the addition of a variety of antibiotics (chloramphenicol, the tetracyclines, etc.), the synthesis of RNA continues at a fast rate and proceeds until a two-or three-fold increment has occurred. Depriving the cells of a single amino acid, however, not only causes immediate cessation of protein synthesis, but, in the case of some strains, of ribosomal RNA synthesis as well. These are called *stringent*; they differ from

Table 1. Examples and consequences of unbalanced growth

Cause of inhibition	Synthesis inhibited	Residual synthesis				Some consequences of unbalanced growth
		DNA	Messenger RNA	Ribosomal RNA	Protein	
Removal of thymine from requiring mutants	DNA	0	+	+	+	Cells die, sometimes due to induction of prophages
Addition of low concentrations of actinomycin D	RNA	+	0	0	+	Enzymes continue to be made on pre-existing messenger RNA
Removal of amino acids from 'stringent' mutants	Protein and ribosomal RNA	+	+	?	0	Each chromosome finishes its round of replication; messenger RNA synthesis is balanced by its breakdown, thus, no net RNA accumulates
Removal of amino acids from 'relaxed' mutants	Protein	+	+	+	0	Cells accumulate protein-poor ribosomal particles
Addition of penicillin	Peptidoglycen	+	+	+	+	Cells lyse due to continued cytoplasmic synthesis and lack of cell wall synthesis

relaxed mutants which continue to synthesize ribosomal RNA as well as messenger RNA, in the absence of the required amino acids. This suggests that the synthesis of messenger RNA and ribosomal RNA are controlled by different regulatory mechanisms (see Chapter 5).

Unbalanced growth can also arise as the result of inhibition of the synthesis of components of cell envelopes. For instance, formation of cell wall peptidoglycan can be inhibited by penicillin. Under these conditions other cell components such as nucleic acids and proteins may continue to be made for many hours. However, this leads to the weakening of the cell envelope and to extrusion of the protoplasm (Chapter 1).

Steady states: balanced growth in chemostat cultures

The relationship between bacterial growth rate and macromolecular composition is best studied with cultures in a chemostat. For example, with a glycerol-limited culture of *Klebsiella aerogenes* the RNA content increased from about 8% (w/w) at a dilution rate (D) of 0.1 h^{-1} to about 18% (w/w) at $D = 0.9$ h^{-1}, and there was a concomitant decrease in the DNA and protein (Figure 17). However, cell size is also markedly dependent on growth rate, and increased at least three-fold. Hence it is clear that the amounts of DNA and protein *per cell* do not decrease; indeed, they increase significantly, but not as extensively as that of RNA. The variation of DNA content per cell with growth rate is an important feature of the cell division cycle and reflects the co-ordination of DNA synthesis with cell division (see above).

Effects of temperature

Changes in temperature may affect batch cultures and chemostat cultures differently. In a batch culture, changes in temperature do not provoke changes in the content of RNA and ribosomes but do alter the rate of protein synthesis and consequently the growth rate. This can be understood if one assumes that the ribosome content is maximal during exponential growth and that a decrease in the temperature would decrease all metabolic processes more or less equally. However, in a chemostat culture if the growth rate is maintained at a submaximal value and the temperature is decreased, the ribosome content will increase to allow protein synthesis to continue at the rate fixed by the dilution rate (see Table 2). These changes are similar to those occurring when the growth rate is varied and the temperature held constant.

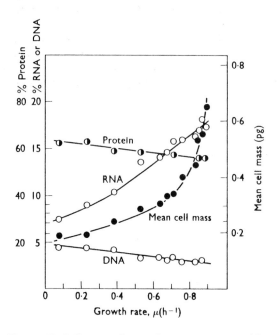

Figure 17. Influence of growth rate on composition. Effect of growth rate on macromolecular composition and mean cell size of populations of *Klebsiella aerogenes* growing in a glycerol-ammonium-salts medium with glycerol as the growth-limiting factor. Data from: Herbert, D. (1961). In *Microbiol Reactions to Environment*. Symp. Soc. Gen. Microbiol. **11**, 391–416.

Specific nutrient limitations

The regulatory mechanisms of the cell generally ensure efficient use of nutrients. Hence, one might expect organisms to restrict to a minimum those components derived from a growth-limiting nutrient. For example, proteins and nucleic acids both contain substantial amounts of nitrogen, whereas polysaccharides contain only a little (if amino-sugars are present). Not surprisingly, therefore, when some bacteria are cultured in a nitrogen-limited medium the content of protein, RNA and DNA decreases but substantial amounts of glycogen accumulate (Table 3). Further, under these conditions, some organisms synthesize large amounts of extracellular polysaccharides (slime).

Table 2. Influence of growth temperature on the macromolecular composition of *Klebsiella aerogenes* growing at a fixed dilution rate ($0.2 \, h^{-1}$) in Mg^{2+}-limited and glycerol-limited chemostat cultures. Values expressed as g component/100 g dried bacteria.

Temperature of growth (°C)	Protein	DNA	RNA	Carbohydrate	Sum
			Mg^{2+}-limited		
25	68·9	2·8	14·9	4·3	90·9
30	67·5	2·9	12·4	2·6	85·4
35	70·0	2·9	10·7	2·2	85·8
40	69·0	2·9	9·4	2·1	83·4
			Glycerol-limited		
25	59·4	2·7	11·8	8·9	82·8
30	67·5	2·6	11·0	3·6	84·7
35	68·4	3·3	10·1	3·3	85·1
40	74·1	2·4	8·6	2·9	88·0

Note the progressive increase in cellular RNA content as the growth temperature is decreased. This is indicative of a progressive increase in cellular ribosome content.

When the carbon-substrate is not limited, other intracellular storage polymers may accumulate. Thus when *Azotobacter beijerinckii* fixes atmospheric nitrogen in an oxygen limited environment, 65% of the cells, dry mass may be poly β-hydroxybutyrate. Similar results have been obtained with an oxygen-limited culture of *Hydrogenomonas eutropha* growing in the presence of excess hydrogen. Other types of growth limitation cause the accumulation of other storage polymers. For example, sulphate-limitation in some strains of *Klebsiella aerogenes* causes polyphosphates to accumulate.

In Gram-negative bacteria, phosphate-limitation may lead to some accumulation of glycogen, and the cellular phosphorus content decreases slightly. However, in some Gram-positive bacteria (e.g.

Table 3. Influence of growth rate and growth limitation on the macromolecular composition of *Klebsiella aerogenes* growing in chemostat culture (35°C; pH 6·8). Values are expressed as g component/100 g dried bacteria.

Dilution rate (h^{-1})	Protein	DNA	RNA	Carbohydrate	Sum
			Mg^{2+}-limited		
0·1	74·5	3·5	8·4	2·3	88·7
0·2	75·5	3·6	11·0	2·4	92·5
0·4	72·5	3·2	15·1	2·8	93·6
0·8	69·0	2·7	16·4	4·3	92·4
			NH_4^+-limited		
0·1	61·5	3·8	7·5	21·7	94·5
0·2	64·2	3·8	10·0	16·0	94·0
0·4	69·4	4·0	13·6	9·8	96·8
0·8	69·3	4·0	18·3	4·6	96·2

Note that the magnesium-limited cells have a generally similar macromolecular composition to glycerol-limited cells (Table 2) but that, by comparison, nitrogen-limited cells contain increased amounts of carbohydrate (glycogen) particularly when growing at the lower rates.

Bacillus subtilis), growth in a phosphate-limited environment results in a substantial decrease in the phosphorus content. In these, unlike Gram-negative species, the walls are rich in *teichoic acid* (a polyol-phosphate polymer) which may account for almost 50% of the cell-bound phorphorus. When growth is limited by the availability of phosphate, the wall-bound teichoic acid is no longer synthesized and a non-phosphorus-containing polymer (*teichuronic acid*) is deposited instead (Table 4).

Table 4. Influence of the growth-limiting component of the medium on the composition of the cell walls of *Bacillus subtilis* var *niger*. Cultures were grown in a chemostat at a dilution rate of 0.2 h^{-1} (35°C, pH 7.0). Values are expressed as g component/100 g dried cell walls.

Component	Phosphate-limited organisms	Magnesium-limited organisms
Phosphorus	0·2	6·0
Glucose	< 1	28
Glucuronic acid	22	< 2
Galactosamine	14	< 2

The chemostat provides a highly selective environment, and it might be argued that the changes in wall composition result not from phenotypic change, but from the selection of mutants. Thus, under phosphate-limitation those mutants might be selected that either had a greater affinity for phosphate, or were devoid of wall-bound teichoic acid. That the changes involved are phenotypic is shown by the kinetics of the transition between one type of growth-limitation and another.

If, during the transition, some component of the cells ceased to be synthesized but was not destroyed, its concentration in the culture would decrease exponentially according to the exponential washout equation:

$$x_t = x_0 e^{-Dt} \qquad (5)$$

Conversely, if some other component were absent from the growing cells but, following a stimulus, started to be synthesized in proportion to cell growth, its concentration in the culture would increase at a rate such that:

$$p_t/p_s = 1 - e^{-Dt} \qquad (14)$$

where p_s represents the final steady state concentration of the component, and p_t its concentration at any time (t) after its synthesis started. The disappearance of teichoic acid and the synthesis of teichuronic acid follow these equations fairly well and the transition is far too rapid to be attributable to the selection of mutants.

Growth of mixed microbial populations

Pure cultures are rarely found outside the laboratory except under extreme environmental conditions in which other species cannot grow. In concluding this chapter, it is useful to consider briefly the physiological and environmental factors that promote or impede the growth of mixed bacterial populations.

In most natural environments the growth of microbes is limited by the availability of essential nutrients. This is analogous to conditions in a chemostat culture and one might start by considering the likely fate of a contaminant, following the theoretical principles outlined in the earlier sections.

Assuming that the Monod equation (6) applies to the relationship between growth-limiting substrate concentration and growth rate, and that no·interaction occurs between organisms other than competition for growth-limiting substrate, the outcome of contamination of the culture can be predicted by considering the respective relationships between μ and s for both the original and the contaminating organisms. Figure 18a shows the relationship defined by equation 6 for two different bacterial species (X and Y) that have different maximum growth rates and affinities for a common growth-limiting nutrient. Suppose that initially the culture contains only organism Y, is growing at a dilution rate D, and that at some moment it becomes contaminated with organism X. Since, at that time, the growth-limiting substrate concentration is s_y, contaminant organisms will grow at a rate equal to μ_x. Because this growth rate is greater than the dilution rate, the concentration of X will increase, thereby lowering the growth-limiting substrate concentration to the value s_x. At this substrate concentration, Y will be able to grow only at a rate of μ_y. Since this is less than the dilution rate (D), Y will be progressively washed from the culture. Figure 18a represents the outcome when the kinetic properties of the two competing species are markedly different. Such differences are often found and the theory is then borne out in practice (see Figure 18b).

This is an example of how a simple ecological experiment can be carried out under controlled

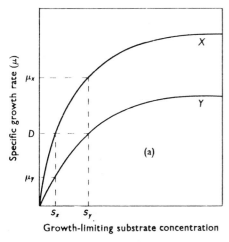 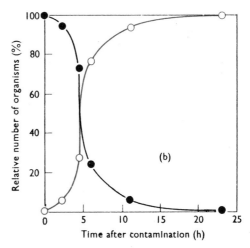

Figure 18. Growth of mixed populations. (a) Hypothetical saturation curves for two species (X and Y) growing in magnesium-limited, chemostat cultures. At dilution rate D_1, the concentration of growth-limiting substrate in a culture of X would be s_x and that in a culture of Y would be s_y. It is assumed that the maximum growth rate ($\mu_{\max}$) for X is greater than that for Y and that the saturation constants (K_s values) have a similar relationship. Data of Meers J. L. and Tempest D. W. (1968) *J. Gen. Microbiol.* **52**, 309–317. (b) Changes in the proportion of *Bacillus subtilis* (●) and *Klebsiella aerogenes* (○) following contamination of a magnesium-limited culture of the former with a low concentration of the latter. Initially, the contaminant population amounted to only 0·5% of the total population but after 24 h, almost no *B. subtilis* could be detected in the culture. Data of Tempest D. W., Dicks J. W. and Meers J. L. (1967) *J. Gen. Microbiol.* **49**, 139–147.

conditions. It is also possible to obtain stable mixed populations in chemostats when the species cross-feed one another. Such systems can be used as models for the study of symbiosis.

FURTHER READING

Brock T. D. (1971) Microbial growth rates in nature *Bacteriol. Rev.* **35**, 39-58.

Churchward G. & Bremer H. (1977) Determination of deoxyribonucleic acid replication time in exponentially growing *Escherichia coli* B/r. *J. Bact.* **130**, 1206–1213.

Helmstetter C. E. & Cooper S. (1968) DNA synthesis, during the division cycle of rapidly growing *Escherichia coli* B/r. *J. Mol. Biol.* **31**, 507–518.

Helmstetter C. E., Pierucci O., Weinberger M., Holmes M. & Tang M-S. (1978) Control of cell division in *Escherichia coli*. In *The Bacteria*, Vol. VII, pp 517–579. Eds. L. N. Ornston & R. Sokatch. Academic Press, New York.

Maaløe O. & Kjeldgaard N. O. (1966) *Control of Macromolecular Synthesis*. W. A. Benjamin, New York.

Monod J. (1949) The growth of bacterial cultures. *Annual Review of Microbiology* **3**, 371–394.

Pirt S. J. (1975) *Principles of Microbe and Cell Cultivation*. Blackwell Scientific Publications, Oxford.

Pritchard R. H. (1974) On the growth and form of a bacterial cell. Review Lecture. *Phil. Trans. Roy. Soc. London* **267**, 303–336.

Pritchard R. H. (1978) Control of DNA replication in bacteria. In *DNA Synthesis Present and Future*. Eds. I. Molineux & M. Kohiyama. Plenum Press, New York & London.

Tempest D. W. (1970) The continuous cultivation of micro-organisms. 1. Theory of the chemostat. In *Methods in Microbiology* **2**, 259–276. London and New York: Academic Press.

Veldkamp H. (1976) *Continuous culture in Microbial Physiology and Ecology*. Meadowfield Press Ltd, Durham, England.

Chapter 3
Class I Reactions: Supply of Carbon Skeletons

Most bacteria can use any of a number of compounds as carbon sources. The pseudomonads for instance can grow on various hydrocarbons, phenols, aliphatic amides, etc. but most other types of bacteria are more restricted in their biochemical potentialities. However, almost all can grow on glucose and it will be useful to consider the ways in which this substance is metabolized to secure carbon skeletons and energy for biosynthesis.

By a series of enzymically catalysed reactions the hexose molecule is degraded, usually after the introduction of inorganic phosphate, to smaller molecules which can then be utilized as building blocks for the synthesis of new cell material. In the course of these reactions energy in the form of ATP is produced for utilization in the subsequent energy-requiring synthetic reactions. At present, four major metabolic pathways for glucose catabolism have been discovered and, as we shall see, these are usually related to the particular mode of life of the organism concerned. The differences in these metabolic sequences reside in the presence or absence of specific enzymes catalysing particular reaction sequences and, in fact, all four pathways utilize an identical series of enzymes for converting triose phosphate to pyruvate. It is possible, therefore, to visualize a central area of metabolism common to all pathways of glucose metabolism, with characteristic differences lying outside this zone.

According to species, bacteria are able to live either in the presence or absence of oxygen; some are intolerant of one or other of these conditions and are classified as strict aerobes or anaerobes, but many bacteria can tolerate both conditions and are referred to as facultative anaerobes. The pathway for glucose metabolism which operates in bacteria must therefore be related to the conditions of life under which the organism can exist. Under aerobic conditions oxygen will be the ultimate hydrogen acceptor but under anaerobiosis, products of glucose metabolism must serve as hydrogen acceptors in a series of balanced oxidation-reduction reactions if glucose degradation and growth are to occur. These aspects of glucose metabolism will become apparent in the following discussions of metabolic pathways. First, however, it is necessary to consider some fundamental aspects of energetics before describing the ways in which bacteria secure their energy.

FUNDAMENTAL ASPECTS OF ENERGETICS

When a sytem of free energy G_1 undergoes change to a system of free energy G_2, the change, ΔG, which is equal to $G_2 - G_1$, will be a measure of the work which can be done in association with that change. If the system moves from a high to a low free energy, i.e. ΔG is negative, then the reaction will be spontaneous in the thermodynamic sense. Spontaneous, that is feasible, reaction is possible only if there is a *decrease* in free energy; such reactions are termed *exergonic*. If an *increase* in free energy is associated with the reaction, then work must be put into the system to bring about the change and the reaction is *endergonic*. By convention, a decrease in free energy is denoted by a negative sign and an increase by a positive sign. It should be noted, however, that a reaction with a high negative free energy change does not necessarily occur at a measurable rate and that a catalyst may be essential for significant reaction to occur.

For a given reaction

$$a\text{A} + b\text{B} \rightleftharpoons c\text{C} + d\text{D}$$

where a, b, c and d are the respective number of molecules of A, B, C and D participating in the reaction, the free energy change ΔG, at constant temperature and pressure, is given by the expression

$$\Delta G = -RT \ln K_{eq} + RT \ln \frac{(C)^c(D)^d}{(A)^a(B)^b} \qquad (1)$$

where R is the gas constant, T the absolute temperature in kelvins and K_{eq} the equilibrium constant for the reaction. (A) and (B) are the activities of the reactants and (C) and (D) the activities of the products present in solution. Consider now the situation when all the reactants and products are in their *standard states*. The standard state is a convenient reference condition in which, by convention, the activities of all components in aqueous solution are taken as unity and gases as 1 atmosphere, at a fixed temperature of 25°C (298 K). Consequently, in this situation, the second of the right hand terms in equation (1) becomes zero and the free energy change is now equal to $-RT \ln K_{eq}$. This particular value of ΔG is defined as the *standard free energy change* of the reaction and is denoted by $\Delta G°$. Thus equation (1) may be rewritten as

$$\Delta G = \Delta G° + RT \ln \frac{(C)^c(D)^d}{(A)^a(B)^b} \qquad (2)$$

where $\Delta G° = -RT \ln K_{eq}$. Values of $\Delta G°$ are additive so that if values of $\Delta G°$ for two reactions are known that of a third coupled reaction with a common intermediate may be calculated. Note that since $\Delta G°$ refers to the standard state of unit activity of hydrogen ions, it refers to pH 0, which is of limited application to living organisms. Consequently the value of $\Delta G°$ at a specified pH other than zero (usually pH 7·0) is generally taken; this is designated $\Delta G°'$ and the pH and temperature must be specified. Note that it is ΔG and not $\Delta G°$ which determines whether or not spontaneous reaction occurs.

As the reaction proceeds towards equilibrium, the free energy of the system will decrease and it is in this process that the reaction will be able to do work at constant temperature and pressure. At equilibrium the free energy of the system is at its minimum for the given conditions, no further net chemical change occurs, ΔG is zero and the system is incapable of doing work. At equilibrium, therefore, since $\Delta G = 0$, equation (1) becomes

$$\Delta G° = -RT \ln \frac{(C)^c(D)^d}{(A)^a(B)^b}$$

At this point, it is important that the distinction between ΔG and $\Delta G°$ be clearly understood, for it is frequently a matter of some confusion, especially when the factors which determine the direction in which a reaction proceeds are considered. The key to this situation is the value of ΔG as defined by equation (2), namely as the sum of $\Delta G°$ and the term $RT \ln (C)^c(D)^d/(A)^a(B)^b$. The value of the latter term relative to the value of $\Delta G°$ ($= -RT \ln K_{eq}$) will determine the direction in which a reversible reaction proceeds. Clearly, if the ratio of activities of products to those of reactants is less than unity, the term is negative. Consequently if this ratio of

activities is less than the value of the equilibrium constant K_{eq}, ΔG for the overall reaction must be negative and the reaction will proceed from left to right. Only when this ratio is sufficiently large in comparison with K_{eq} to change the negative value of ΔG to a positive one will the reaction proceed in the reverse direction.

Sometimes a reaction which will not of itself proceed because of the increase in free energy involved, can be made to do so by coupling it with a reaction which has a suitable decrease in free energy. For this to occur there must be an intermediate common to both reactions. Consider now the following three reactions.

(1) $A + B \rightleftharpoons C + D$ $\Delta G_1°$ small positive

(2) $D + N \rightleftharpoons P + Q$ $\Delta G_2°$ large negative

(3) $A + B + N \rightleftharpoons C + P + Q$ $\Delta G_3°$ moderately negative

Since reaction (1) has a small positive value of $\Delta G°$, the reaction can proceed from left to right but will lead at equilibrium to little conversion of A and B to C and D. When coupled through the common intermediate D to reaction (2) for which $\Delta G°$ is large and negative, the overall reaction (3) has $\Delta G°$ moderately negative and so will result in extensive conversion of A and B to products. It is important to note, however, that kinetic factors may also be involved and for A and B to be converted to products the intermediate D must participate in reaction (2) at an appropriately rapid rate.

In the case of hydrolytic reactions of the type

$$A + H_2O \rightleftharpoons C + D$$

since the concentration of water is 55·5 M, the term $(C)(D)/(A)(H_2O)$ is very small and ΔG is consequently large and negative. Thus for many hydrolytic reactions $K_{eq} \gg 1$ and attempts to reverse the reaction will lead to very low values of (A) because the term (H_2O) is so large (55·5 M). However, under appropriate conditions, net synthesis of A from C and D can occur. The requisite conditions would be that A participates in a coupled reaction and, since (A) is very low, the enzyme concerned with the utilization of A either has a very low Michaelis constant (K_m) for A, i.e. a high affinity for A, or is present in the cell in large amounts. Additionally, as already mentioned, kinetic factors can assume major importance, namely the *rate* at which A is utilized in the coupled reaction.

Many examples of coupled reactions occur in living systems and of especial significance are those involving the so-called high energy compounds, of which high energy phosphates are of great importance. The hydrolysis of most phosphate esters involves a standard free energy change of about -8.37 kJ (-2 kilocalories) per mole, whereas the hydrolysis of compounds such as adenosine triphosphate (ATP), adenosine diphosphate (ADP), arginine phosphate and creatine phosphate involves a much greater standard free energy change of around -29.3 to 50.2 kJ (-7 to -12 kilocalories) per mole. Compounds which exhibit this greater $\Delta G^{\circ\prime}$ (pH 7) on hydrolysis are termed 'high energy' compounds. Certain other compounds of metabolic importance also display high standard free energies of hydrolysis and can be similarly designated as high energy compounds, e.g. the thiol esters of coenzymes such as acetyl coenzyme A and succinyl coenzyme A. Their importance resides in the ability of living systems to transform the available free energy of metabolic processes into the chemical energy characteristic of compounds containing these bonds and which manifests itself as a high standard free energy of hydrolysis. High energy compounds thus serve as 'storehouses' of energy for the organism and play a vital role in the economy of the living cell.

It is now considered that all the chemical energy derived from the oxidation of foodstuffs must first be converted to high energy phosphate compounds before it can be utilized to drive biosynthetic reactions or for the performance of mechanical work. Likewise, the radiant energy utilized in photosynthesis must be converted to chemical energy in the form of ATP by reactions taking place in the chloroplast.

The key reactions are thus

$$ATP + H_2O \rightleftharpoons ADP + P_i$$
$$\Delta G^{\circ\prime} = -30.5 \text{ kJ or } -7.3 \text{ kcal mole}^{-1}$$

$$ADP + H_2O \rightleftharpoons AMP + P_i$$
$$\Delta G^{\circ\prime} = -30.5 \text{ kJ or } -7.3 \text{ kcal mole}^{-1}$$

where P_i represents inorganic phosphate and the $\Delta G^{\circ\prime}$ values refer to pH 7.0 and 25°C. The first of these reactions is catalysed by an important enzyme called *adenosine triphosphatase* (ATPase). However, in many reactions ATP undergoes pyrophosphate cleavage to give AMP and inorganic pyrophosphate PP_i, thus

$$ATP + H_2O \rightleftharpoons AMP + PP_i$$
$$\Delta G^{\circ\prime} = -41.8 \text{ kJ or } -10.0 \text{ kcal mole}^{-1}$$

and the $\Delta G^{\circ\prime}$ value is substantially greater than in the corresponding orthophosphate hydrolysis of ATP and ADP, a factor which is valuable in certain reactions such as the activation of fatty acids and amino acids, e.g.

$$R.COOH + CoASH + ATP \rightleftharpoons$$
$$RCOSCoA + AMP + PP_i$$

$$RCHNH_2COOH + ATP \rightleftharpoons RCHNH_2CO.AMP + PP_i$$

AMP differs from ATP and ADP in not being a high energy compound.

The regeneration of ATP from AMP and pyrophosphate necessitates two additional enzymes, *inorganic pyrophosphatase* and *adenylate kinase*. The former enzyme catalyses the reaction

$$PP_i + H_2O \rightleftharpoons$$
$$2P_i \qquad \Delta G^{\circ\prime} = -19.2 \text{ kJ or } -4.6 \text{ kcal mole}^{-1}$$

and this free energy change can sometimes ensure that biosynthetic reactions proceed essentially to completion, i.e. equilibrium is far to the right, as for example in the activation of fatty acids referred to above. The inorganic orthophosphate produced can then participate in the regeneration of ATP from ADP.

The rephosphorylation of AMP is achieved by adenylate kinase

$$AMP + ATP \rightleftharpoons 2ADP$$

which, as we shall see, is a very important enzyme in the energetic economy of the cell.

The energy state of a living cell depends upon the balance between the concentrations of the adenine nucleotides ATP, ADP and AMP. When most of the adenylate is present as ATP the cell will have a maximum energy level and conversely when most of it is present as AMP the energy level will be minimal.

To express quantitatively the energy state of a cell Atkinson has introduced the concept of the *energy charge* of the adenylate system, somewhat analogous to the charge of a storage battery. The reaction

$$ATP + 2H_2O \rightleftharpoons AMP + 2P_i$$

formally represents energy acceptance and donation by the adenylate system and the degree of charge is proportional to the amount of phosphate added to AMP, the addition of 2 moles of phosphate per mole of adenylate fully charges the system which then exists solely as ATP. For convenience, to enable the charge to range from 0 to 1 instead of 0 to 2, Atkinson defines energy charge as half the number of anhydride-bound phosphate groups per adenine moiety, i.e. in terms of concentrations

$$\text{Energy charge} = \frac{[ATP] + 0.5[ADP]}{[ATP] + [ADP] + [AMP]}$$

This single parameter enables the energy level of the cell to be defined and is of great assistance when the energy-linked control of metabolic systems is considered, on account of the heterogeneity of response to the individual adenine nucleotides encountered with certain regulatory enzymes, e.g. some respond primarily to the individual concentration of ATP, ADP or AMP, whereas others respond either to [ATP]/[ADP] or [ATP]/[AMP]. The concept of the energy charge thus unifies the treatment of these cases. The principal limitations of this parameter are, however, that it does not afford information concerning the turnover of ATP and of the total adenylate pool of an organism, which can vary significantly at a fixed charge.

The generation of high-energy phosphate compounds occurs by three major metabolic processes, substrate-level phosphorylation, oxidative phosphorylation, and photophosphorylation, characteristic of anaerobic, aerobic, and photosynthetic conditions, respectively. Anaerobic glycolysis employs only substrate-level phosphorylation while the tricarboxylic acid cycle involves principally oxidative phosphorylation but also one substrate-level phosphorylation step, as will be emphasized in the subsequent discussion of these processes.

PATHWAYS OF GLUCOSE CATABOLISM

The four routes of glucose catabolism currently recognized are: (1) Embden–Meyerhof glycolysis, (2) pentose phosphate cycle, (3) Entner–Doudoroff pathway, and (4) phosphoketolase pathway. The distinguishing features of these will now be discussed.

The Embden–Meyerhof glycolytic pathway

The Embden–Meyerhof glycolytic pathway (Figure 1) consists of some ten enzymes which effect the conversion of glucose (C_6) to pyruvate (C_3). The system can operate under both aerobic and anaerobic conditions. Aerobically it usually functions in conjunction with the tricarboxylic acid cycle which can oxidize pyruvate to CO_2 and H_2O. Anaerobically, pyruvate, or products of its further (anaerobic) metabolism, must be reduced, e.g. with the formation

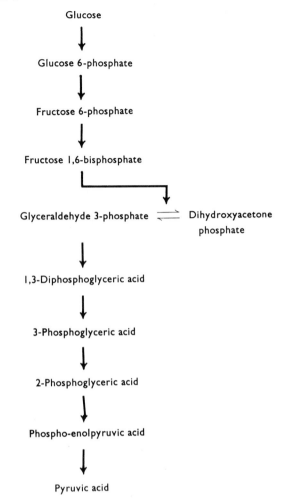

Figure 1. Embden–Meyerhof glycolytic pathway of glucose degradation.

of lactate or ethanol. The details of this sequence of reactions will now be considered.

Glucose must first be phosphorylated in the 6-position and the mechanism by which this occurs depends on the type of organism. Aerobic bacteria including many pseudomonads employ the enzyme *hexokinase* which requires Mg^{2+} for activity and utilizes 1 molecule of ATP. This essentially irreversible reaction yields glucose 6-phosphate and ADP.

Facultative organisms such as *Escherichia coli* effect phosphorylation of glucose and other sugars during transport into the cell (see p. 88). This process is more widespread than was once thought.

Next follows an isomerization of glucose 6-phosphate to fructose 6-phosphate catalysed by

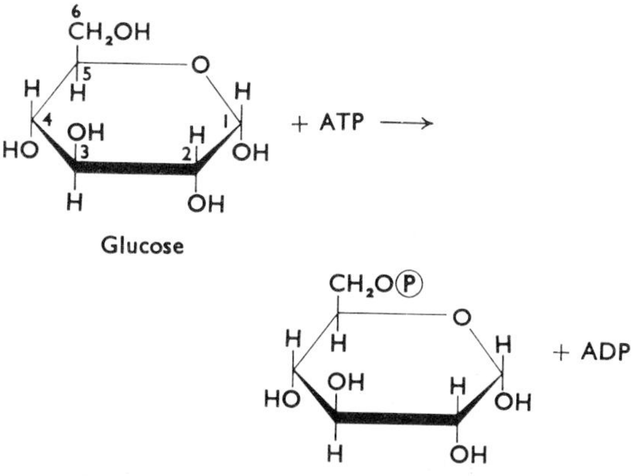

Glucose

+ ATP ⟶

Glucose 6-phosphate

+ ADP

phosphohexose isomerase, which has no cofactor requirement. At equilibrium some 70% of the glucose form is present.

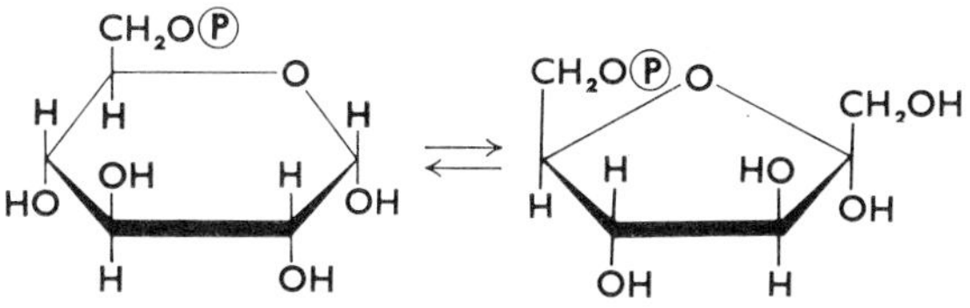

Glucose 6-phosphate ⇌ Fructose 6-phosphate

A second phosphate group is now introduced into the 1-position of the fructose by the action of *phosphofructokinase*, an enzyme characteristic of this pathway of glycose metabolism, and the presence of which in a given type of bacterium is taken as good evidence for the operation of glycolysis. Like hexokinase, it requires ATP and Mg^{2+} and is irreversible. Up to this stage of glycolysis two moles of ATP have been utilized.

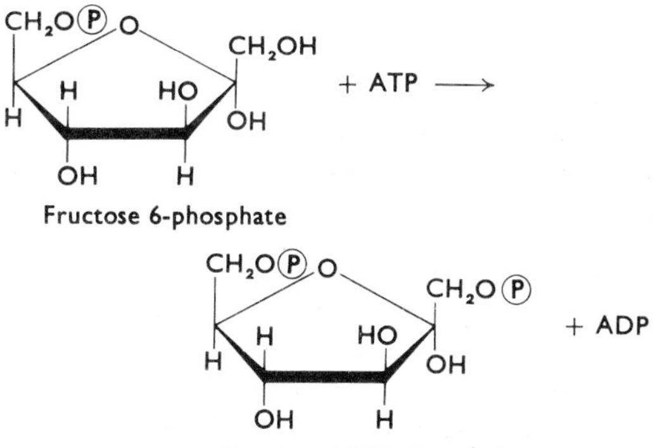

Fructose 6-phosphate

+ ATP ⟶

Fructose 1,6-bisphosphate

+ ADP

The hexose molecule is now split by *fructose bisphosphate aldolase* into two C_3 units—the triose phosphates, glyceraldehyde 3-phosphate and dihydroxyacetone phosphate.

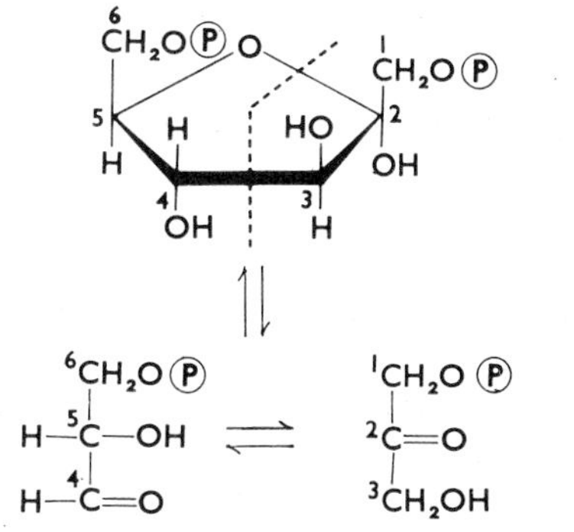

Glyceraldehyde 3-phosphate

Dihydroxyacetone phosphate

The triose phosphates are produced in equivalent amounts but due to the action of *triose phosphate isomerase*, which catalyses their interconversion, they are brought to an equilibrium which is about 95% in favour of dihydroxyacetone phosphate. On account of this, dihydroxyacetone phosphate was originally believed to be the sole product of aldolase action. Dihydroxyacetone phosphate has been found to inhibit fructose bisphosphate aldolase competitively. The fission of the hexose occurs in such a way that dihydroxyacetone phosphate is derived from carbon atoms 1, 2 and 3 and glyceraldehyde 3-phosphate from atoms 4, 5 and 6 of the fructose molecule.

The triose phosphate stage of glycolysis marks the point of linkage with fat metabolism because of the possibility of interconversion of dihydroxyacetone phosphate and glycerol. In some bacteria glycerol is first phosphorylated by a kinase and ATP (in *Escherichia coli* this reaction is the pacemaker for glycerol metabolism), and then the α-glycerophosphate formed undergoes oxidation to dihydroxyacetone phosphate catalysed by α-glycerophosphate dehydrogenase.

α-Glycerophosphate + NAD^+ ⇌ Dihydroxyacetone phosphate + NADH + H^+

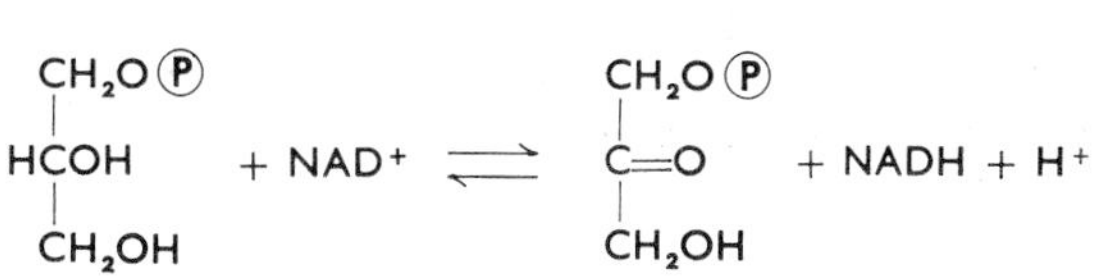

Despite the fact that the equilibrium of triose phosphate isomerase greatly favours dihydroxyacetone phosphate, glyceraldehyde 3-phosphate is the substrate for the next step in glycolysis—an oxidation to 1,3-diphosphoglyceric acid. The enzyme responsible, *triose phosphate dehydrogenase*, requires NAD^+ and inorganic phosphate, which is incorporated into the product:

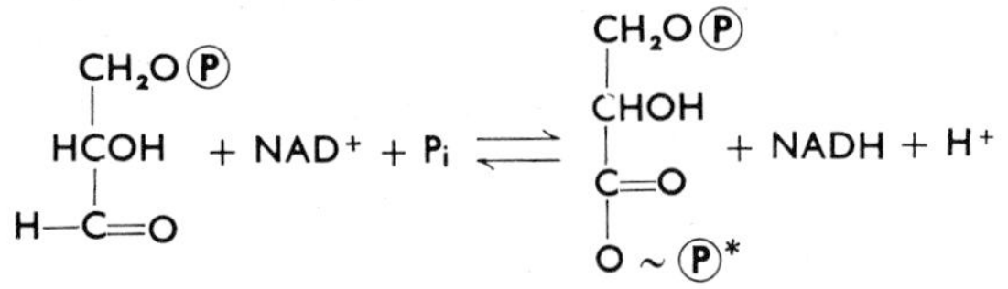

$$
\begin{array}{c}
\text{CH}_2\text{O}\ \textcircled{P} \\
|\\
\text{HCOH} \\
|\\
\text{H—C}{=}\text{O}
\end{array}
+ \text{NAD}^+ + \text{P}_i \rightleftharpoons
\begin{array}{c}
\text{CH}_2\text{O}\ \textcircled{P} \\
|\\
\text{CHOH} \\
|\\
\text{C}{=}\text{O} \\
|\\
\text{O} \sim \textcircled{P}^*
\end{array}
+ \text{NADH} + \text{H}^+
$$

Glyceraldehyde 3-phosphate   1,3-Diphosphoglyceric acid

$*\sim$ Indicates a high-energy phosphate bond (cf. p. 127).

This reaction represents the first and only oxidation in the sequence of glycolysis leading to pyruvic acid. The NADH must be reoxidized for the reactions to continue. Aerobically this occurs via the cytochrome system (see Appendix B). Anaerobically it is coupled with reduction of an organic compound—often one produced later in the glycolysis pathway, namely pyruvate. Thus pyruvate oxidizes NADH and is itself reduced to lactate.

It should be noted that the product 1,3-diphosphoglycerate contains a high-energy bond (the acyl phosphate in the 1-position). The next reaction is the transfer of this phosphate group to ADP by the action of *phosphoglycerate kinase* leaving 3-phosphoglycerate as the other product: Mg^{2+} is a cofactor for this enzyme, as with other kinases. The energy made available by the dehydrogenation process is thus coupled to ATP synthesis. This is the first energy-yielding step of glycolysis and the process is one of substrate-level phosphorylation as opposed to electron transport phosphorylation associated with the respiratory chain (see p. 147).

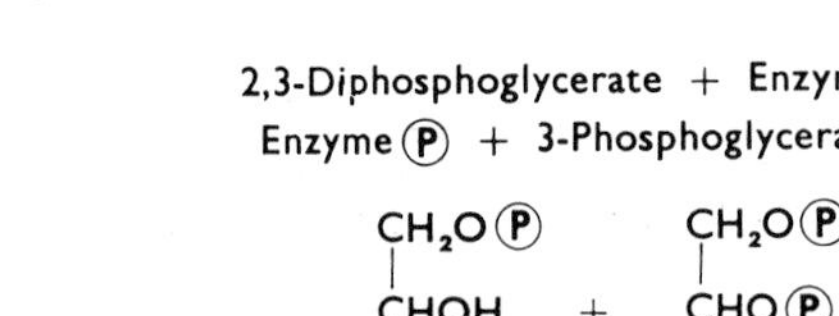

$$
\begin{array}{c}
\text{CH}_2\text{O}\ \textcircled{P} \\
|\\
\text{CHOH} \\
|\\
\text{C}{=}\text{O} \\
|\\
\text{O} \sim \textcircled{P}
\end{array}
+ \text{ADP} \rightleftharpoons
\begin{array}{c}
\text{CH}_2\text{O}\ \textcircled{P} \\
|\\
\text{CHOH} \\
|\\
\text{COOH}
\end{array}
+ \text{ATP}
$$

1,3-Diphosphoglyceric acid   3-Phosphoglyceric acid

3-Phosphoglycerate now undergoes conversion to

2-phosphoglycerate by the action of the enzyme *phosphoglyceromutase*. This reaction requires a trace of 2,3-diphosphoglycerate and the mechanism is believed to involve a phosphorylated enzyme intermediate (as shown at the bottom of this page).

It will be noticed that the donor molecule, by transfer of its phosphate group in the 3-position, becomes the product of the reaction and the original substrate appears as the diphosphate which can then in turn act as a donor molecule.

Water is now removed to yield phospho-enol-pyruvic acid by the action of *enolase* which requires a divalent metal ion such as Mg^{2+}, Mn^{2+} or Zn^{2+}:

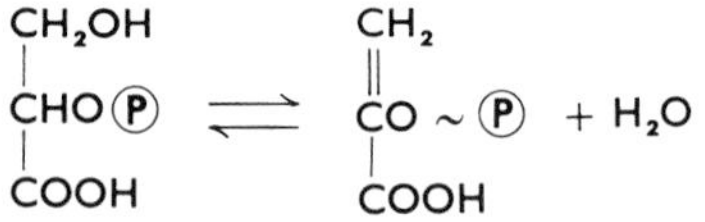

$$
\begin{array}{c}
\text{CH}_2\text{OH} \\
|\\
\text{CHO}\ \textcircled{P} \\
|\\
\text{COOH}
\end{array}
\rightleftharpoons
\begin{array}{c}
\text{CH}_2 \\
\|\\
\text{CO} \sim \textcircled{P} \\
|\\
\text{COOH}
\end{array}
+ \text{H}_2\text{O}
$$

2-Phosphoglyceric acid   Phospho-enolpyruvic acid

Enolases from organisms which require Mg^{2+} are susceptible to inhibition by fluoride, due to the formation of magnesium fluorophosphate, whereas those requiring Mn^{2+} are not because the manganese salt, unlike the magnesium salt, is soluble. Inhibition by fluoride results in the accumulation of phosphoglycerate and for many years the detection of phosphoglycerate under these conditions was taken as evidence for the operation of glycolysis. However it is now known that enolase is part of the central area of glucose metabolism common to the different pathways.

The effect of dehydration is to produce another high-energy phosphate bond, this time an enol-phosphate, which is now transferred to ADP by a further enzyme, *pyruvate kinase*. Thus the energy yielded by the dehydration step is coupled to the synthesis of another molecule of ATP; this is the second energy-yielding step of glycolysis and again it is an example of substrate-level phosphorylation. The equilibrium of the kinase reaction greatly favours ATP formation and the reverse reaction is

2,3-Diphosphoglycerate + Enzyme $\rightleftharpoons$ 2-Phosphoglycerate + Enzyme $\textcircled{P}$

Enzyme $\textcircled{P}$ + 3-Phosphoglycerate $\rightleftharpoons$ Enzyme + 2,3-Diphosphoglycerate

$$
\begin{array}{c}
\text{CH}_2\text{O}\ \textcircled{P} \\
|\\
\text{CHOH} \\
|\\
\text{COOH}
\end{array}
+
\begin{array}{c}
\text{CH}_2\text{O}\ \textcircled{P} \\
|\\
\text{CHO}\ \textcircled{P} \\
|\\
\text{COOH}
\end{array}
\rightleftharpoons
\begin{array}{c}
\text{CH}_2\text{O}\ \textcircled{P} \\
|\\
\text{CHO}\ \textcircled{P} \\
|\\
\text{COOH}
\end{array}
+
\begin{array}{c}
\text{CH}_2\text{OH} \\
|\\
\text{CHO}\ \textcircled{P} \\
|\\
\text{COOH}
\end{array}
$$

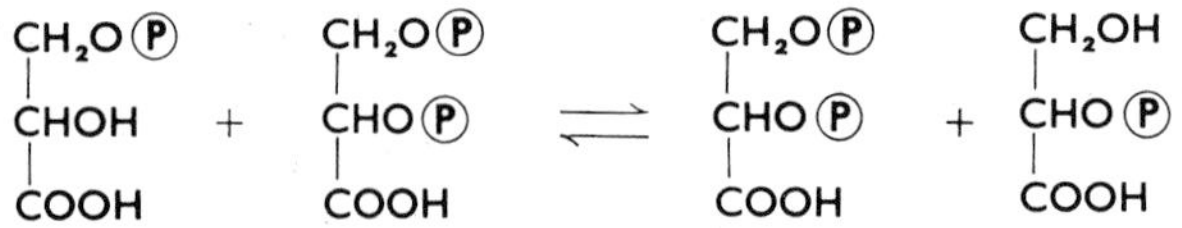

3-Phosphoglycerate   2,3-Diphosphoglycerate   2,3-Diphosphoglycerate   2-Phosphoglycerate

extremely difficult to demonstrate:

$$
\begin{array}{c}
CH_2 \\
\| \\
C\!-\!O \sim \circled{P} + ADP \\
| \\
COOH
\end{array}
\longrightarrow
\begin{array}{c}
CH_3 \\
| \\
C\!=\!O + ATP \\
| \\
COOH
\end{array}
$$

Phospho-enolpyruvic   Pyruvic acid
acid

Fermentation products

Bacteria differ from yeast in that the majority of them give rise to a diversity of products by the further metabolism of pyruvate. These reactions permit an overall oxidation-reduction balance to be preserved under anaerobic conditions and the products formed are characteristic of the particular species. Some of these reactions can give rise to additional energy by

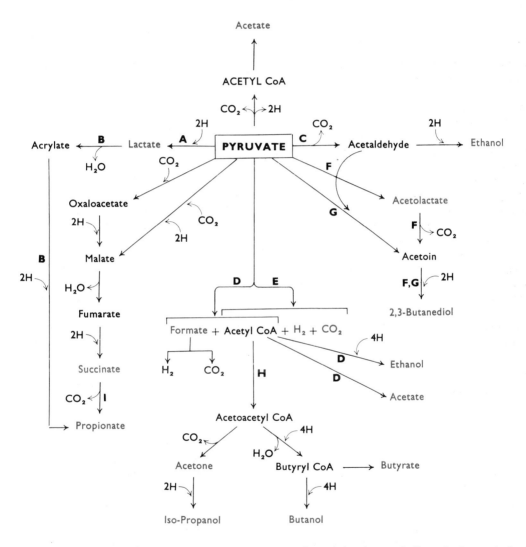

Figure 2. Bacterial fermentation products of pyruvate. Pyruvate formed by the catabolism of glucose is further metabolized by pathways which are characteristic of particular organisms and which serve as a biochemical aid to identification. End products of fermentations are indicated in red.

A   Lactic acid bacteria (*Streptococcus, Lactobacillus*)
B   *Clostridium propionicum*
C   Yeast, Acetobacter, Zymomonas, *Sarcina ventriculi, Erwinia amylovora.*
D   Enterobacteriaceae (coli-aerogenes)

E   Clostridia
F   Klebsiella
G   Yeast
H   Clostridia (butyric, butylic organisms)
I   Propionic acid bacteria

...vel phosphorylation reactions, for ex-
...e conversion of an acyl coenzyme A
...e to the free acid

CoA + ADP + P$_i$ $\rightleftharpoons$ Acetate + ATP + CoA

Butyryl CoA + ADP + P$_i$ $\rightleftharpoons$ Butyrate + ATP + CoA.

but, in general, the energy yield is poorer than that derived from the conversion of glucose to pyruvate. The fermentation reactions of pyruvate carried out by various micro-organisms are summarized in Figure 2.

The full sequence of reactions from glucose to pyruvate has now been described and it will be noted that, since each fructose 1,6-bisphosphate molecule gives rise to two triose phosphates, four moles of ATP are produced per mole of fructose bisphosphate fermented. As one mole of ATP is used for the phosphorylation of glucose and a second for the phosphorylation of fructose 6-phosphate, the net yield is two moles of ATP per mole of glucose fermented.

The fate of pyruvate produced by bacteria varies tremendously, depending on the species of organism and environmental conditions, such as pH, so that products of pyruvate metabolism can be used as an aid to classification. Homolactic acid bacteria (i.e. those which produce lactate as the major product of glucose fermentation, as opposed to heterolactic bacteria which produce substantial amounts of other products in addition to lactate) resemble muscle in yielding two moles of lactate per mole of glucose fermented. The enzyme involved in the reduction of pyruvate, *lactate dehydrogenase*, utilizes the NADH produced by the triose phosphate dehydrogenase reaction, as previously discussed, thus permitting anaerobic oxidation of glyceraldehyde 3-phosphate. The linked reactions are:

in such a way that the carboxyl groups of the resulting pyruvate molecules are derived from carbon atoms 3 and 4 of the original glucose molecule. Consequently, in the lactic fermentation the carboxyl groups of lactate are similarly derived from 3-C and 4-C. In the alcohol fermentation, however, these carbon atoms are eliminated as carbon dioxide and the acetaldehyde and ethanol are derived from 1-C, 2-C and 5-C, 6-C as shown at the top of the next page.

Using specifically labelled [^{14}C]glucose the distribution of the carbon atoms of glucose in intermediates and products of different metabolic pathways can be established and this enables deductions to be made concerning the occurrence and quantitative significance of these pathways (see top of p. 133).

Control of glycolysis

The principal functions of glycolysis are two-fold, to provide energy (directly and via intermediates which are oxidized in the tricarboxylic acid cycle) and to furnish carbon skeletons for the biosynthesis of cellular components. The term *amphibolic* has been applied to such dual function pathways, to distinguish them from the strictly catabolic and anabolic sequences of metabolism (see also Chapter 8, p. 340). The ATP produced in the exergonic reactions is utilized in the endergonic synthetic pathways with the generation of ADP or AMP and, as previously discussed (p. 127), the energy state of a bacterial cell depends on the balance between the concentration of the adenine nucleotides ATP, ADP and AMP.

A key enzyme in the interconversion of adenylate nucleotides is *adenylate kinase* which requires Mg^{2+}

Glyceraldehyde 3-phosphate + NAD$^+$ $\rightleftharpoons$ 3-Phosphoglycerate + NADH + H$^+$

CH$_3$CO.COOH + NADH + H$^+$ $\rightleftharpoons$ CH$_3$CHOH.COOH + NAD$^+$

Pyruvic acid   Lactic acid

Coupled reactions of this type, involving nicotinamide nucleotides as the common reactant, are encountered frequently in the anaerobic metabolism of bacteria since, under these conditions, reoxidation of the reduced coenzymes cannot employ oxygen as the terminal acceptor. Instead, a reaction involving the generation of a reduced coenzyme is linked to a second reaction which utilizes it as a reductant.

A feature of the Embden–Meyerhof mechanism is that fission of fructose 1,6-bisphosphate occurs

and catalyses the reaction

ATP + AMP $\rightleftharpoons$ 2ADP

The importance of Mg^{2+} ions in the reactions of adenylate nucleotides must here be emphasized, especially since the affinity of ATP for Mg^{2+} is much greater than that of ADP or AMP.

When the energy charge is low there must be some means of increasing the rate of glycolysis and, conversely, when the energy charge is high of de-

$$\overset{1}{C}-\overset{2}{C}-\overset{3}{C}-\overset{4}{C}-\overset{5}{C}-\overset{6}{C}$$

$$\overset{1}{C}H_3\overset{2}{C}HOH.\overset{3}{C}OOH \longleftarrow \overset{1}{C}H_3\overset{2}{C}O.\overset{3}{C}OOH \qquad HOO\overset{4}{C}.\overset{5}{C}O.\overset{6}{C}H_3 \longrightarrow HOO\overset{4}{C}.\overset{5}{C}HOH.\overset{6}{C}H_3$$

$$\overset{3}{C}O_2 \qquad\qquad \overset{4}{C}O_2$$

$$\overset{1}{C}H_3\overset{2}{C}HO \qquad\qquad OHC.\overset{5\ 6}{C}H_3$$

$$\overset{1}{C}H_3\overset{2}{C}H_2OH \qquad\qquad HO.H_2\overset{5}{C}.\overset{6}{C}H_3$$

Distribution of carbon atoms of glucose and products of metabolic pathways.

creasing the rate of ATP production. This is achieved by modulating the rates of two key reactions in glycolysis, the phosphorylation of fructose 6-phosphate to fructose 1,6-bisphosphate catalysed by phosphofructokinase, and the conversion of phospho-enolpyruvate to pyruvate effected by pyruvate kinase. Phosphofructokinase is activated by ADP or AMP and inhibited by phospho-enolpyruvate, while pyruvate kinase is activated by AMP and fructose 1,6-bisphosphate. Energy-linked control of this type is clearly of great value for the regulation of catabolic sequences which, in general, are not subject to end-product or negative feedback control of the type encountered in biosynthetic pathways (see Chapter 4, pp. 159–160). However, in the case of glycolysis the dual function of energy generation and production of intermediates for biosynthesis imposes upon it a form of regulation rather more complicated than that encountered with a strictly catabolic pathway involving only energy-linked control. An additional control feature, apparently unique to amphibolic pathways, is *precursor activation* which, in a sense, is the opposite of feedback control where the last metabolite of a pathway inhibits the first enzyme of the sequence. In precursor control the first metabolite of the sequence activates the last enzyme of that sequence. It is not possible here to enter into a detailed discussion of precursor activation and one example must suffice. Fructose 1,6-bisphosphate may be regarded as the first metabolite of a metabolic sequence in glycolysis which leads to pyruvic acid, the final enzymic step being the conversion of phospho-enolpyruvate to pyruvate, catalysed by a pyruvate kinase. As already mentioned, fructose 1,6-bisphosphate activates this pyruvate kinase thus functioning in a positive 'feedforward' manner.

Figure 3 illustrates the regulatory mechanisms currently known to operate in glycolysis and the tricarboxylic acid cycle.

Glucogenesis

When bacteria grow on pyruvate or other C_3 or C_4 compounds as the sole source of carbon it is essential that they synthesize glucose from these substrates, a process termed glucogenesis (or gluconeogenesis). As three reactions of glycolysis are virtually irreversible on account of their highly exergonic nature, namely those catalysed by hexokinase, phosphofructokinase and pyruvate kinase, it is not possible for the cell simply to reverse the glycolytic sequence in order to synthesize glucose, and alternative means have evolved to circumvent these irreversible steps.

In *Escherichia coli* it has been found that two enzymes which enable the organism to grow on C_4 compounds such as succinate or malate are phospho-enolpyruvate carboxykinase and malate enzyme:

$$HOOC.CH_2CO.COOH + ATP \xrightarrow{\overset{\text{PEP carboxy}}{\text{kinase}}}$$

$$CH_2{:}CO \sim \textcircled{P}.COOH + CO_2 + ADP$$

$$HOOC.CH_2CHOH.COOH + NADP^+ \xrightarrow{\overset{\text{malate}}{\text{enzyme}}}$$

$$CH_3CO.COOH + CO_2 + NADPH + H^+$$

Loss of the carboxykinase by mutation completely prevents growth on C_4 compounds and acetate, yet such mutants are still able to grow on C_3 compounds such as pyruvate or lactate. This observation was explained by the discovery of an enzyme *phospho-enolpyruvate synthase*, induced during growth on C_3

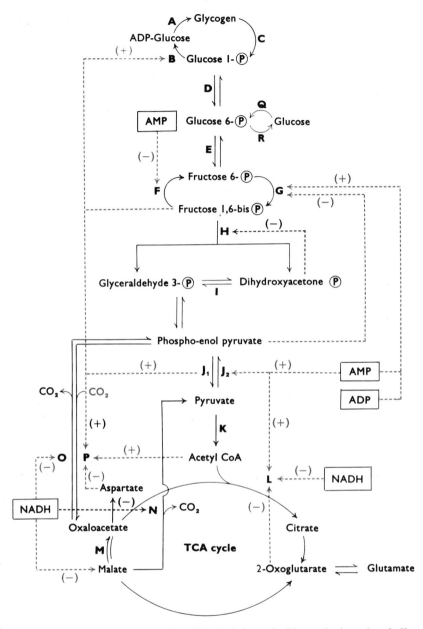

Figure 3. Regulation of glycolysis and tricarboxylic acid cycle in bacteria. Plus and minus signs indicate activation and inhibition respectively of the appropriate enzymes by various metabolites. Note that the controls indicated do not all necessarily apply to a single species of organism.

A   Glycogen synthase
B   ADP-glucose pyrophosphorylase
C   Phosphorylase
D   Phosphoglucomutase
E   Phosphohexose isomerase
F   Fructose 1,6-bisphosphatase
G   Phosphofructokinase
H   Fructose bisphosphate aldolase
I   Triosephosphate isomerase
J   Pyruvate kinase (there are two pyruvate kinases in

enteric bacteria, one J_1 activated by fructose 1,6-bisphosphate and the other J_2 by AMP)

K   Pyruvate dehydrogenase
L   Citrate synthase
M   Malate dehydrogenase
N   Malate enzyme
O   Phospho-enolpyruvate carboxykinase
P   Phospho-enolpyruvate carboxylase
Q   Hexokinase
R   Glucose 6-phosphatase

compounds, which requires Mg^{2+} and effects the reaction:

$$CH_3COCOOH + ATP \longrightarrow$$
$$CH_2{=}CO \sim \circled{P}.COOH + AMP + P_i$$

In this way phospho-enolpyruvate is synthesized from pyruvate at the expense of two of the high energy bonds of ATP, one of which is preserved in the phospho-enolpyruvate formed. In passing, it should be noted that growth on C_3 compounds demands that the organism be able to synthesize C_4 compounds from C_3 and CO_2 in order that the tricarboxylic acid cycle operate to produce both energy and intermediates for biosynthesis (see p. 146). Studies with mutants of the Enterobacteriaceae, unable to grow on glucose, glycerol, pyruvate or their precursors unless the medium is supplemented with utilizable intermediates of the tricarboxylic acid cycle, have revealed that the key enzyme is *phospho-enolpyrurate carboxylase*, which catalyses the reaction

$$CH_2{=}CO \sim \circled{P}.COOH + CO_2 \longrightarrow$$
$$HOOC.CH_2CO.COOH + H_3P_i$$

This enzyme therefore fulfils a replenishing role by permitting net synthesis of C_4 compounds and belongs to the category of anaplerotic enzymes (see p. 151).

The irreversibility of the phosphofructokinase reaction is overcome by the enzyme fructose bis-phosphatase which hydrolyses phosphate from the 1-position, thus

Fructose 1,6-bisphosphate + $H_2O \longrightarrow$
$$Fructose\ 6\text{-phosphate} + P_i$$

The necessity for this reaction in glucogenesis is illustrated by the fact that mutants of *Escherichia coli* devoid of the enzyme are unable to grow on substrates such as acetate, succinate or glycerol and have an absolute requirement for hexoses.

By utilizing these reactions in conjunction with the reversible steps of glycolysis the bacterial cell is able to synthesize glucose 6-phosphate from C_3 and C_4 compounds. Studies with mutants of *Escherichia coli* lacking triosephosphate isomerase have demonstrated that this enzyme is essential for glucogenesis, thus lending support to the pathway proposed.

Glycogenesis

It has been mentioned that bacteria such as *Escherichia coli* accumulate substantial amounts of glycogen under conditions where the nitrogen of the medium becomes exhausted in the presence of excess carbon sources. Glycogen synthesis requires glucose 1-phosphate as its substrate and this is formed from glucose 6-phosphate by the enzyme *phosphoglucomutase*, which undergoes phosphorylation and de-phosphorylation during the reaction and involves glucose 1,6-bisphosphate as an intermediate.

Glucose 6-phosphate   Glucose 1,6-bisphosphate
+ $\rightleftharpoons$ + $\rightleftharpoons$
Phosphoenzyme   Dephosphoenzyme

Glucose 1-phosphate
+
Phosphoenzyme

Catalytic amounts of glucose 1,6-bisphosphate are essential for the reaction to proceed and these are formed by the action of phosphoglucokinase on glucose 1-phosphate and ATP.

Glucose 1-phosphate reacts with ATP under the influence of ADP-glucose pyrophosphorylase to yield ADP-glucose and inorganic pyrophosphate

Glucose 1-phosphate + ATP $\longrightarrow$ ADP-glucose + PP_i

ADP-glucose then transfers its glucosyl moiety to a glycogen primer (having $\alpha[1 \rightarrow 4]$ bonds) in a reaction catalysed by *glycogen synthase*

ADP-glucose + (Glucose)$_n$ $\longrightarrow$ ADP + (Glucose)$_{n+1}$

Glycogen breakdown to glucose 1-phosphate occurs by a different route, involving the enzyme *phosphorylase*

(Glucose)$_n$ + P_i $\longrightarrow$ (Glucose)$_{n-1}$ + Glucose 1-$\circled{P}$

These differences of synthesis and degradation permit the regulation of glycogen synthesis which, in bacteria, occurs by modulation of the activity of ADP-glucose pyrophosphorylase. This enzyme is strongly activated by fructose 1,6-bisphosphate and NADPH and, to a lesser extent, by glyceraldehyde 3-phosphate and phospho-enolpyruvate. It would seem that the signals for glycogen synthesis are the concentrations in the intracellular pools of these glycolytic intermediates and the accumulation of NADPH in the cell under circumstances where they are not being utilized for biosynthetic reactions.

Mammalian glycogen synthesis differs from that in bacteria and plants in employing UDP-glucose as the glucosyl carrier and by being regulated at the levels of glycogen synthase and phosphorylase. To date, no allosteric control of bacterial phosphorylase has been observed.

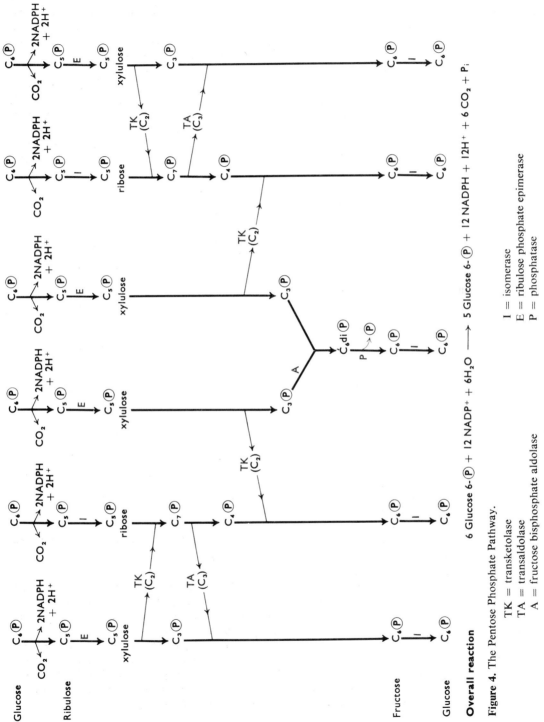

Figure 4. The Pentose Phosphate Pathway.

TK = transketolase
TA = transaldolase
A = fructose bisphosphate aldolase

I = isomerase
E = ribulose phosphate epimerase
P = phosphatase

Overall reaction

6 Glucose 6-$\textcircled{P}$ + 12 NADP$^+$ + 6H$_2$O $\longrightarrow$ 5 Glucose 6-$\textcircled{P}$ + 12 NADPH + 12H$^+$ + 6 CO$_2$ + P$_i$

The pentose phosphate cycle

For many years Embden–Meyerhof glycolysis was considered to be the sole pathway for glucose catabolism, but the process referred to as the *pentose phosphate cycle* has now been shown to be of major importance to living organisms, including many bacterial species. In recent years, however, the pentose cycle, as originally described by B. L. Horecker, has been shown to occur only in adipose tissue. J. F. Williams and his colleagues have found that a more complex cycle, involving the participation of octulose, an eight-carbon sugar, operates in other mammalian tissues. To date, there is no report of this so-called *new pentose cycle* having been sought or found in bacteria and so the following discussion is confined to the original formulation of the process. The overall effect of the cycle may be summarized as follows:

6 Glucose 6-(P) + 12 NADP$^+$ + 6H$_2$O $\longrightarrow$

5 Glucose 6-(P) + 6CO$_2$ + 12 NADPH + 12H$^+$ + P$_i$

This is illustrated diagrammatically in Figure 4 and it will be appreciated that the cyclic involvement of six molecules of glucose 6-phosphate is equivalent to the complete oxidation of one of them to CO$_2$ and water, with the regeneration of five molecules of glucose 6-phosphate. The sequence involves the oxidative decarboxylation of a hexose to a pentose followed by anaerobic rearrangements of the carbon skeletons of pentose so produced. The mechanism of these reactions will now be considered.

The pentose phosphate pathway diverges immediately from Embden–Meyerhof glycolysis in that oxidation of glucose 6-phosphate occurs to yield 6-phosphogluconolactone, catalysed by *glucose-6-phosphate dehydrogenase*. This lactone is then hydrolysed by a *lactonase* to yield 6-phosphogluconate which is subsequently oxidatively decarboxylated to give the pentose phosphate, ribulose 5-phosphate. Experiments with isotopes have clearly shown that it is the 1-C atom of glucose which is eliminated as CO$_2$ (Figure 5).

Ribulose 5-phosphate, under the influence of the enzymes *pentose phosphate isomerase* and *xylulo-epimerase*, forms an equilibrium mixture of ribose 5-phosphate, ribulose 5-phosphate and xylulose 5-phosphate. Now rearrangements of the carbon skeletons of these pentose monophosphates occur

under the influence of the enzymes *transketolase* and *transaldolase*. The intermediate formation of C$_7$ and C$_4$ sugars occurs in a series of reactions which does not require the presence of oxygen. First, one molecule each of the ribose 5-phosphate and xylulose 5-phosphate react to form the seven-carbon sugar, sedoheptulose 7-phosphate and the triose, glyceraldehyde 3-phosphate; the reaction is catalysed by transketolase which requires thiamine pyrophosphate (TPP) as cofactor. In keeping with the known mechanism of action of TPP (see Appendix B) it is believed that an 'active' glycolaldehyde complex is formed with the enzyme-coenzyme by fission of xylulose 5-phosphate into glyceraldehyde 3-phosphate and a two-carbon fragment. The active glycolaldehyde then becomes attached to ribose 5-phosphate forming sedoheptulose 7-phosphate, thus:

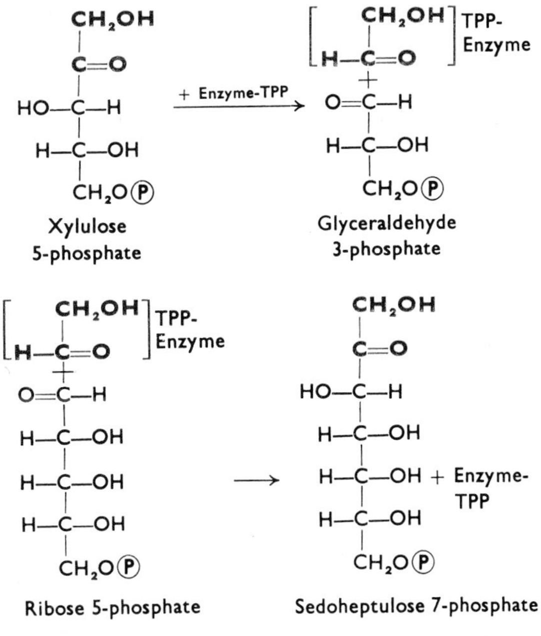

A further transfer reaction between sedoheptulose 7-phosphate and glyceraldehyde 3-phosphate is catalysed by the enzyme transaldolase, and by its action fructose 6-phosphate and the four-carbon sugar erythrose 4-phosphate are formed. The reaction resembles that catalysed by transketolase but in this case the active moiety transferred is dihydroxyacetone and no cofactor requirement has been demonstrated (see top of next page).

```
        CH₂OH
         |
         C=O
         |
   HO—C—H
         |
    H—C—OH          H—C=O
         |              |
    H—C—OH    +    H—C—OH      ⟶
         |              |
    H—C—OH          CH₂O℗
         |
        CH₂O℗

  Sedoheptulose      Glyceraldehyde
  7-phosphate        3-phosphate
```

```
                                   CH₂OH
                                    |
                                    C=O
                                    |
        O=C—H             HO—C—H
            |                      |
       H—C—OH            H—C—OH
            |                      |
       H—C—OH     +      H—C—OH
            |                      |
          CH₂O℗             CH₂O℗

      Erythrose            Fructose
      4-phosphate          6-phosphate
```

Transketolase now catalyses a reaction between erythrose 4-phosphate and xylulose 5-phosphate such that another mole of fructose 6-phosphate and one of glyceraldehyde 3-phosphate are formed:

```
                            CH₂OH
                             |
     O=C—H                C=O
         |                   |
    H—C—OH    +    HO—C—H          ⟶
         |                   |
    H—C—OH            H—C—OH
         |                   |
       CH₂O℗              CH₂℗

   Erythrose            Xylulose
   4-phosphate          5-phosphate
```

```
            CH₂OH
             |
             C=O
             |
       HO—C—H            O=C—H
             |               |
        H—C—OH    +    H—C—OH
             |               |
        H—C—OH            CH₂O℗
             |
           CH₂O℗

      Fructose            Glycer-
      6-phosphate         aldehyde
                          3-phosphate
```

The fructose 6-phosphate molecules, by the action of phosphohexose isomerase, are converted to glucose 6-phosphate and may then re-enter the cycle. The overall effect of the cycle at this stage is therefore:

6 Glucose 6-phosphate   ⟶   4 Glucose 6-phosphate +

2 Glyceraldehyde 3-phosphate + 6CO₂

Triose phosphate isomerase converts one molecule of glyceraldehyde 3-phosphate to dihydroxyacetone phosphate and this is condensed with another molecule to yield fructose 1,6-bisphosphate by a reversal of the aldolase reaction:

Glyceraldehyde 3-℗

⥯

Dihydroxyacetone ℗

+   ⇌   Fructose 1,6-bis ℗

Glyceraldehyde 3-℗

The hexose bisphosphate is dephosphorylated and isomerized to give glucose 6-phosphate. This yields a fifth molecule of glucose 6-phosphate, i.e. six molecules of hexose have been converted to five and six molecules of carbon dioxide have been formed.

It will be seen in Figure 5 that two reactions result in formation of NADPH so that 12 molecules have to be reoxidized for each turn of the cycle, since six molecules of hexose are involved. In bacteria, according to species, either two or three ATP molecules can be formed from the oxidation of each NADPH. The ATP production is thus 24 or 36 molecules but one is expended in the phosphorylation of glucose to glucose 6-℗ hence the overall reactions may be written as:

Glucose + 6 O₂   ⟶

6 CO₂ + 6 H₂O + Energy [= 23 or 35 ATP]

These reactions illustrate how the pentose cycle yields energy in effect by oxidizing glucose completely to carbon dioxide and water. However, this oxidative function is neither the sole nor the major role of the cycle. Thus it is also evident that NADPH formed in the pentose cycle serves as an important, and probably the major, source of reducing power for fatty acid biosynthesis in a series of reactions which effects the condensation and reduction of acetate units to yield long chain fatty acids, as described in Chapter 4, pp. 179–183. In this sense, then, NADPH may be regarded as an alternative form of energy and one which is utilized for fatty acid synthesis. But it must be appreciated that, besides producing energy for biosynthetic reactions and cellular maintenance, the pentose cycle can also serve to furnish the intermediates required for biosynthesis, and compounds such as pentose, erythrose and triose phosphates will be drained from the cycle for this purpose.

It should be noted that since the anaerobic rearrangements of the carbon skeletons in the pentose cycle are reversible, it is possible for five molecules

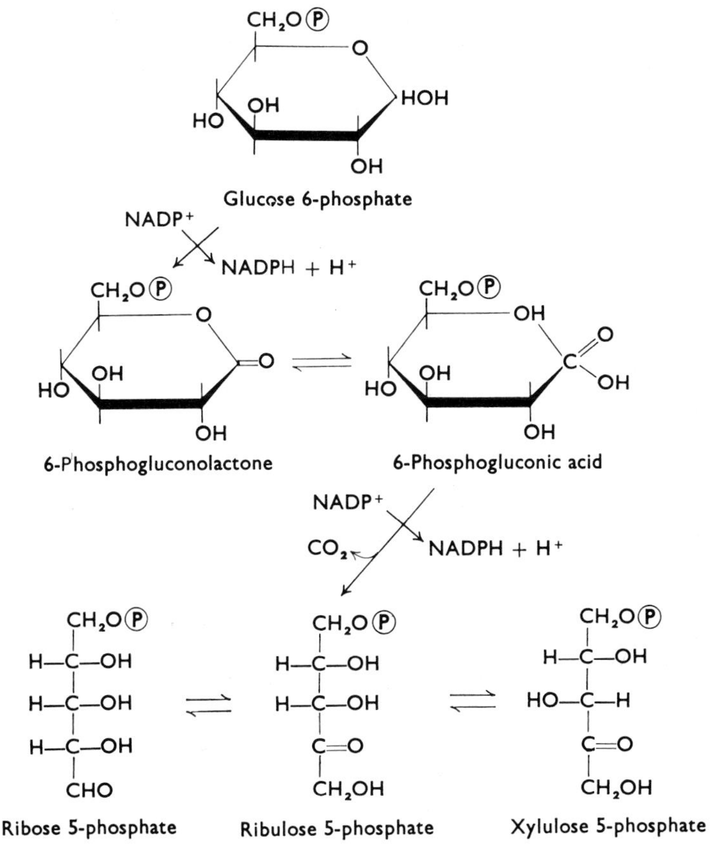

Figure 5. Oxidative decarboxylation of glucose 6-phosphate.

of hexose phosphate to give rise to six molecules of pentose phosphate by nonoxidative reactions, as opposed to their direct oxidation to pentose phosphate and carbon dioxide.

These reactions assume particular importance in organisms such as some pseudomonads which lack the enzyme 6-phosphogluconate dehydrogenase and so are unable to generate pentose phosphate by the oxidative decarboxylation of glucose 6-phosphate. In these circumstances fructose 6-phosphate and glyceraldehyde 3-phosphate can initiate the reverse, anaerobic sequence leading to pentose phosphate.

The Entner–Doudoroff pathway

A new pathway of glucose metabolism was discovered by Entner and Doudoroff who observed that *Pseudomonas saccharophila* preferentially released $^{14}CO_2$ from [1-^{14}C]glucose and that the pyruvate produced was labelled in a manner different from that characteristic of the Embden–Meyerhof route;

the carboxyl of one pyruvate was derived from 1-C and the other from 4-C:

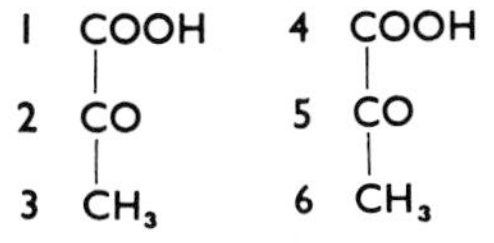

1	COOH	4	COOH
2	CO	5	CO
3	CH$_3$	6	CH$_3$

The isolation of a new intermediate, 2-oxo-3-deoxy-6-phosphogluconate and the discovery of two new enzymes, one a dehydratase for 6-phosphogluconate and the other an aldolase specific for the new intermediate, enabled them to formulate the metabolic sequence shown in Figure 6.

The dehydratase removes the elements of water from 6-phosphogluconate and requires Fe^{2+} ions and glutathione for maximal activity. No cofactors have been demonstrated for the aldolase which splits the resulting 2-oxo-3-deoxy-6-phosphogluconate to glyceraldehyde 3-phosphate and pyruvate; the former can be converted to a second molecule of pyruvate by the reactions common to glycolysis.

This sequence of reactions, commonly referred to as the Entner–Doudoroff pathway, has been found to occur in a number of Gram-negative species, particularly among *Pseudomonas* and *Azotobacter* spp., but it can also be induced by growing the Gram-positive, lactate-producing organism *Streptococcus faecalis* on gluconate. Studies with labelled substrates show that glucose is metabolized exclusively by the Entner–Doudoroff pathway in *Pseudomonas*

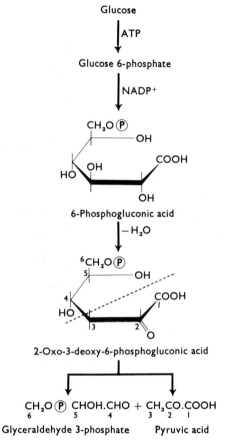

Figure 6. Entner-Doudoroff pathway of glucose degradation.

saccharophila, Zymomonas mobilis (Pseudomonas lindneri) and *Zymomonas anaerobia* and is the major pathway in other pseudomonads.

Energetically, the anaerobic operation of the Entner–Doudoroff pathway is only half as efficient as anaerobic glycolysis, for the net yield of ATP per mole of glucose is 1 mole instead of 2, because only 1 mole of triose phosphate is produced and oxidized. This gives 2 moles of ATP but the net yield is only one because of the need to phosphorylate glucose.

Phosphoketolase pathway

Some kinds of bacteria carry out a fermentation of glucose which yields lactate together with other major products such as CO_2, acetate or ethanol. They also ferment pentoses with the formation of lactate and acetate. An example is *Leuconostoc mesenteroides* which ferments glucose according to the equation:

$$Glucose \longrightarrow Lactate + Ethanol + CO_2$$

The enzymes phosphofructokinase, aldolase and triose phosphate isomerase are absent, indicating a departure from Embden–Meyerhof glycolysis. This was confirmed by the use of [1-^{14}C]- and [3,4-^{14}C$_2$]-glucose which indicated the following derivations:

1	†C		1	†CO$_2$
2	C		2	CH$_3$
3	*C	$\longrightarrow$	3	*CH$_2$OH
4	*C		4	*COOH
5	C		5	CHOH
6	C		6	CH$_3$

These results suggest that the hexose is decarboxylated to a pentose which then undergoes fission, a conclusion supported by the presence of glucose 6-phosphate dehydrogenase and 6-phosphogluconate dehydrogenase in the organism. These enzymes bring about the oxidative decarboxylation of glucose 6-phosphate to ribulose 5-phosphate and are NAD$^+$-rather than NADP$^+$-dependent.

The substrate for the fission is xylulose 5-phosphate and *phosphoketolase* cleaves this pentose phosphate to acetyl phosphate and glyceraldehyde 3-phosphate. The enzyme requires TPP, inorganic phosphate, Mg^{2+} and a thiol compound for activity:

$$
\begin{array}{ccc}
\begin{array}{c}
CH_2OH \\
| \\
C=O \\
| \\
HO-C-H \\
| \\
H-C-OH \\
| \\
CH_2O\,\text{\textcircled{P}}
\end{array} + P_i \longrightarrow &
\begin{array}{c}
CH_3 \\
| \\
\text{\textcircled{P}} \sim O-C=O \\
\text{Acetyl phosphate}
\end{array} \\
\\
Xylulose \\ 5-phosphate &
\begin{array}{c}
O=C-H \\
| \\
H-C-OH \\
| \\
CH_2O\,\text{\textcircled{P}} \\
Glyceraldehyde \\ 3-phosphate
\end{array}
\end{array}
$$

The purified enzyme is specific for xylulose 5-phosphate and does not attack fructose 6-phosphate or sedoheptulose 7-phosphate although a similar enzyme from *Acetobacter xylinum* does split fructose

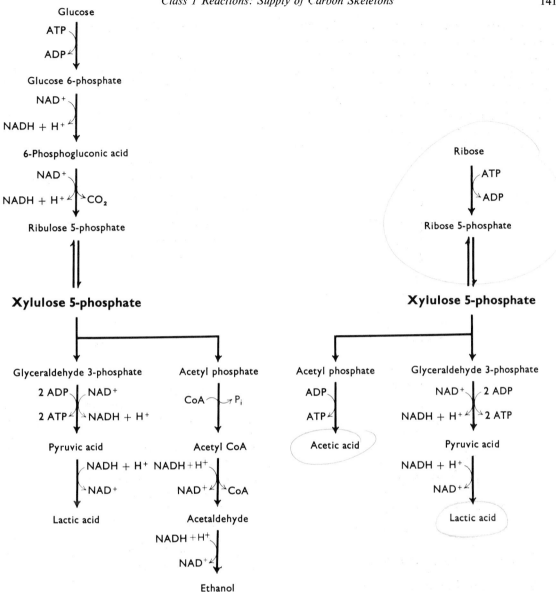

Figure 7. Phosphoketolase pathway. Fermentation of glucose to lactic acid, ethanol and CO_2 and of ribose to lactic acid and acetic acid.

$$\text{Glucose} + \text{ADP} + P_i \longrightarrow \text{Lactic acid} + \text{ethanol} + CO_2 + \text{ATP}$$

$$\text{Ribose} + 2\text{ADP} + 2P_i \longrightarrow \text{Lactic acid} + \text{acetic acid} + 2\,\text{ATP}$$

6-phosphate to yield erythrose 4-phosphate and acetyl phosphate. In the presence of ADP, cell extracts catalyse the formation of glyceraldehyde 3-phosphate, acetate and ATP so that the cleavage yields energy to the organism via substrate-level phosphorylation. Again in this fermentation the triose phosphate is converted to pyruvate by the glycolytic enzymes and the pyruvate is reduced to

lactate. The other ultimate product of the fission, i.e. acetate or ethanol, depends on the oxidation-reduction balance of the system which, in turn, depends on whether the substrate is a hexose or a pentose. Figure 7 reveals that conversion of glucose to xylulose 5-phosphate involves two oxidation steps, with the formation of 2NADH, or the equivalent of 4H. Metabolism of glyceraldehyde 3-phosphate to pyruvate yields a further 2H making 6H in all. Two of these are utilized for the reduction of pyruvate to lactate and, to balance the overall reaction, the remaining 4H are used to reduce acetyl phosphate to ethanol, via acetyl coenzyme A and acetaldehyde. However, when pentose is the substrate the two initial oxidation steps are eliminated and the reducing power formed, equivalent to only 2H, is utilized for lactate formation, leaving acetyl phosphate as the other product.

This difference in products is also reflected in the energetics of the fermentations for whereas there is a net yield of 2 moles of ATP per mole of pentose fermented, the yield from glucose is only one as a consequence of the loss of the high energy bond of acetyl phosphate by its reduction to ethanol. This observation underlines the general principle that alcohol formation by any route other than pyruvate decarboxylation and acetaldehyde reduction is energetically wasteful to the organism.

CATABOLISM OF LIPIDS

Two of the most important kinds of lipids are triglycerides and phospholipids. The former are long-chain fatty acyl esters:

$$CH_2O.OC(CH_2)_pCH_3$$
$$CHO.OC(CH_2)_qCH_3$$
$$CH_2O.OC(CH_2)_rCH_3$$

where p, q and r may be the same or different, are usually even numbers, and are frequently about 16. Phospholipids have the third acyl group replaced by phosphate linked to ethanolamine, serine or inositol. Lipids occur as components of membranes and can also serve as reserves of carbon and energy. When they are catabolized they are first hydrolysed to yield the free fatty acids. As shown in Figure 8 these are then degraded stepwise to acetyl coenzyme A which is further metabolized in the tricarboxylic acid cycle.

The process of β-oxidation yields energy via respiratory chain phosphorylation as one molecule each of reduced flavoprotein and NADH, generated in each sequence leading to the release of acetyl coenzyme A, are re-oxidized via the electron transport system (see p. 147). Additional energy accrues as the acetyl coenzyme A is oxidized via the tricarboxylic acid cycle. In mitochondria the re-oxidation of NADH yields 3ATP molecules and that of reduced flavoprotein furnishes 2ATP, whereas in

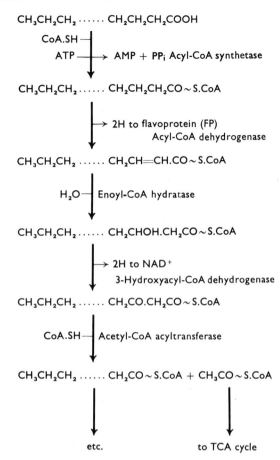

Figure 8. β-Oxidation of fatty acids.

many bacteria the corresponding energy yields are 2ATP and 1ATP molecules respectively (see p. 149).

THE TRICARBOXYLIC ACID (KREBS) CYCLE

We have seen how the various pathways of glucose metabolism lead to the formation of pyruvate. Under aerobic conditions pyruvate is then oxidized by a cyclic process termed the *tricarboxylic acid cycle*. Entry to this is gained after pyruvate has been converted to a C_2-unit by loss of carbon dioxide (Figure 9). The tricarboxylic acid cycle effects the oxidation of two-carbon units to carbon dioxide and water and constitutes the most important single mechanism for the generation of ATP in aerobic organisms. Additionally, it is of importance for the production of carbon skeletons for synthetic reactions, particularly those leading to the synthesis of amino acids, e.g.

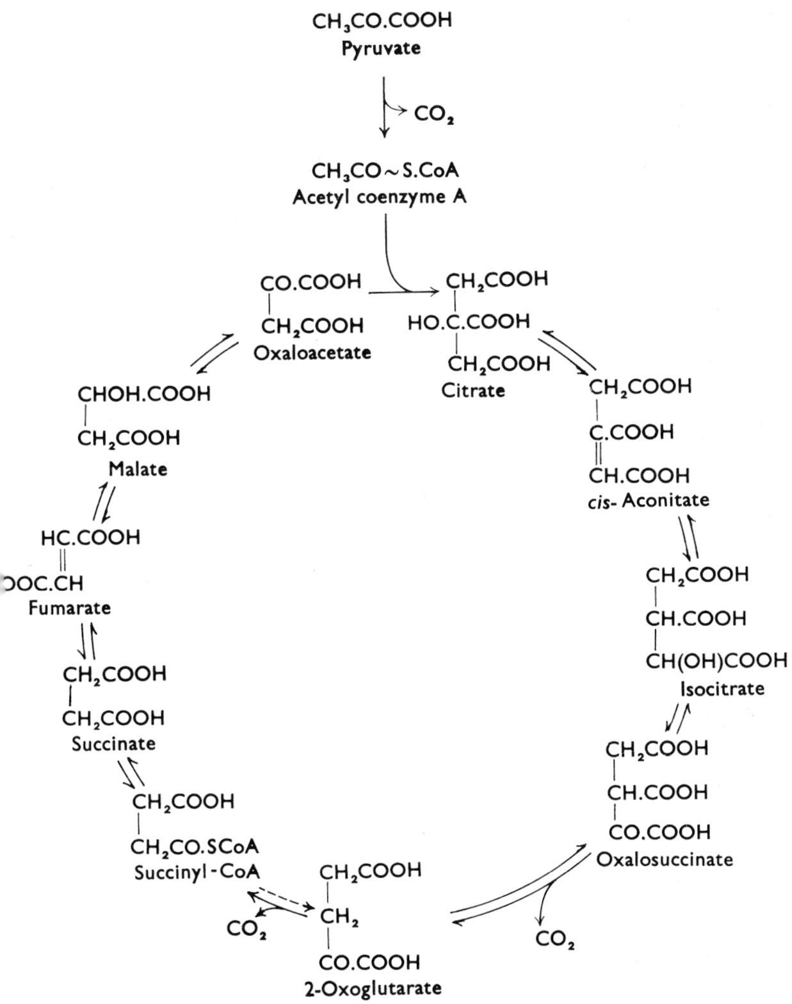

CH$_3$CO.COOH
Pyruvate

→ CO$_2$

CH$_3$CO∼S.CoA
Acetyl coenzyme A

CO.COOH
|
CH$_2$COOH
Oxaloacetate

CH$_2$COOH
|
HO.C.COOH
|
CH$_2$COOH
Citrate

CHOH.COOH
|
CH$_2$COOH
Malate

CH$_2$COOH
|
C.COOH
‖
CH.COOH
cis- Aconitate

HC.COOH
‖
ϽOC.CH
Fumarate

CH$_2$COOH
|
CH.COOH
|
CH(OH)COOH
Isocitrate

CH$_2$COOH
|
CH$_2$COOH
Succinate

CH$_2$COOH
|
CH.COOH
|
CO.COOH
Oxalosuccinate

CH$_2$COOH
|
CH$_2$CO.SCoA
Succinyl-CoA

CO$_2$

CH$_2$COOH
|
CH$_2$
|
CO.COOH
2-Oxoglutarate

CO$_2$

Figure 9. The tricarboxylic acid cycle.

aspartic and glutamic acids. The relative importance of these two functional roles of the cycle depends to a large extent on whether or not the cells are growing. This aspect of metabolism is considered subsequently on page 146. Before entering the tricarboxylic acid cycle, pyruvate is converted to acetyl coenzyme A. This oxidative decarboxylation is brought about by a multi-enzyme complex which comprises three enzymes, five different cofactors and requires Mg^{2+} ions. The cofactors include thiamine pyrophosphate (TPP), coenzyme A (CoA), lipoic acid, flavin adenine dinucleotide (FAD) and NAD (see Appendix B). Coenzyme A combines with substrates via its thiol group; for this reason it is usually denoted as CoA.SH. Combination with acyl groups, e.g. acyl-S.CoA, gives rise to C—S bonds which have a high free energy of hydrolysis and such compounds therefore belong to the class of high-energy compounds.

Lipoic acid (thioctic acid) is a dithiol compound which, in its coenzyme form, is linked covalently *via* its carboxyl group to a protein sidechain amino group, i.e. lipoamide (for convenience, the acid is shown here). The open chain form can be reversibly oxidized to a disulphide, five-membered ring form; these are denoted by lip(SH)$_2$ and lipS$_2$ respectively:

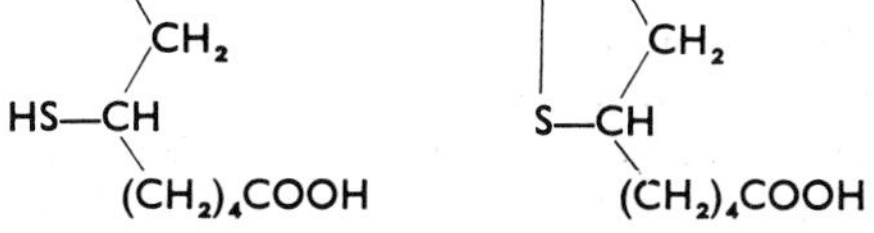

HS—CH$_2$
　　CH$_2$
HS—CH
　　(CH$_2$)$_4$COOH

Reduced

S—CH$_2$
　　CH$_2$
S—CH
　　(CH$_2$)$_4$COOH

Oxidized

Lipoic acid

FAD is a hydrogen carrier which undergoes reversible reduction and oxidation by accepting and relinquishing two hydrogen atoms (see Appendix B):

$$FAD + 2H \rightleftharpoons FADH_2$$

The first step in the oxidative decarboxylation of pyruvate is catalysed by pyruvate dehydrogenase which requires TPP and Mg^{2+} as cofactors. An enzyme-bound 'active acetaldehyde'-TPP complex, which has been shown to be 2-hydroxyethyl-2-thiamine pyrophosphate, is produced and this is then transferred to one of the sulphur atoms of the di-sulphide form of lipoic acid, being oxidized to acetyl lipoate.

$$CH_3CO.COOH + TPP \longrightarrow [CH_3CHO]TPP + CO_2$$

      Pyruvate   Active acetaldehyde

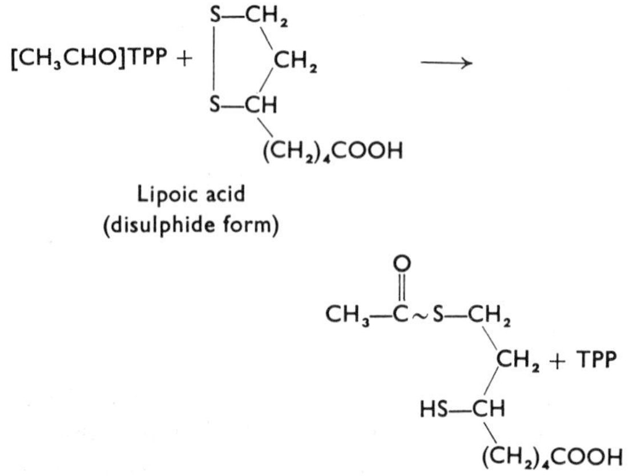

$$[CH_3CHO]TPP +$$
S—CH₂
CH₂
S—CH
(CH₂)₄COOH
$\longrightarrow$

Lipoic acid
(disulphide form)

O
‖
CH₃—C~S—CH₂
CH₂ + TPP
HS—CH
(CH₂)₄COOH

Acetyl lipoic acid

The lipoic acid is covalently bound to the second enzyme of the complex, *dihydrolipoyl transacetylase*. The C~S bond in the acyl lipoates resembles that in acyl CoA compounds and has a high free energy of hydrolysis, i.e. acyl lipoates are high-energy

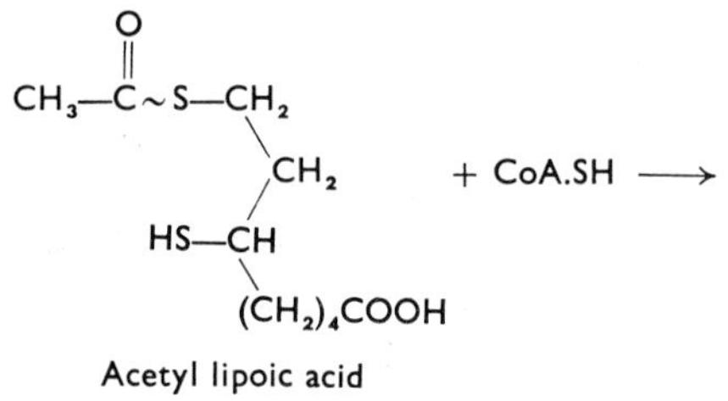

O
‖
CH₃—C~S—CH₂
CH₂   + CoA.SH $\longrightarrow$
HS—CH
(CH₂)₄COOH

Acetyl lipoic acid

O
‖
CH₃—C~S.CoA + HS—CH₂

Acetyl CoA
CH₂
HS—CH
(CH₂)₄COOH

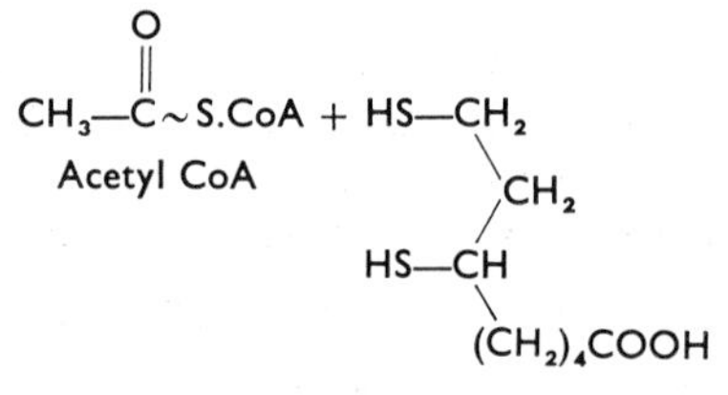

Reduced lipoic acid

compounds. The TPP is liberated and is thus able to act catalytically. The acetyl lipoate now serves as an acyl donor for coenzyme A, resulting in the formation of free acetyl coenzyme A and reduced lipoic acid bound to the transacetylase enzyme (see foot of left column).

Reduced lipoyl transacetylase is then re-oxidized by the third enzyme of the complex, *dihydrolipoyl dehydrogenase*, which contains tightly bound FAD; this becomes reduced and is in turn re-oxidized in a mechanism involving NAD.

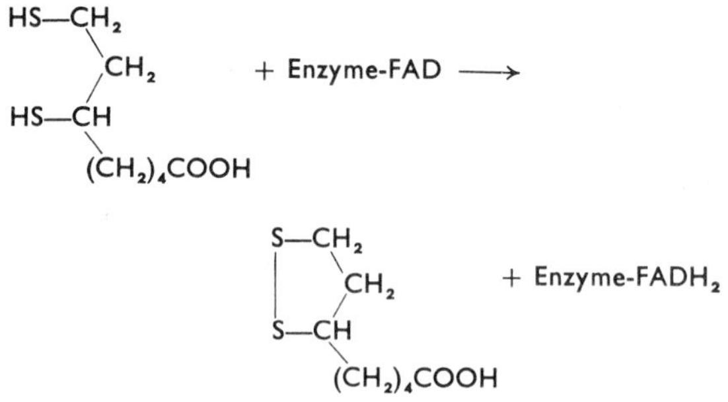

HS—CH₂
CH₂   + Enzyme-FAD $\longrightarrow$
HS—CH
(CH₂)₄COOH

S—CH₂
CH₂   + Enzyme-FADH₂
S—CH
(CH₂)₄COOH

$$Enzyme\text{-}FADH_2 + NAD^+ \longrightarrow$$
$$Enzyme\text{-}FAD + NADH + H^+$$

The overall reaction for the oxidation of pyruvate may be written as:

$$Pyruvate + CoA.SH + NAD^+ \longrightarrow$$
$$Acetyl \sim S.CoA + NADH + H^+ + CO_2$$

The pyruvate dehydrogenase complex has been isolated from *Escherichia coli* and found to have a relative molecular mass of $4\cdot6 \times 10^6$. It consists of 24 molecules of pyruvate dehydrogenase each binding one molecule of TPP, one molecule of dihydrolipoyl transacetylase containing 24 polypeptide chains each possessing one molecule of lipoic acid, and 12 molecules of dihydrolipoyl dehydrogenase each containing one molecule of FAD. The pyruvate de-hydrogenase units and dihydrolipoyl dehydrogenase units are located on the outside of the transacetylase core.

Under certain circumstances energy may be made available by the conversion of acetyl CoA to acetate and CoA with concomitant formation of a mole of ATP.

Quantitatively, however, the most important reaction which acetyl CoA undergoes in metabolism is condensation with oxaloacetate to form citrate and thus to gain entry to the tricarboxylic acid cycle. The reaction is catalysed by *citrate synthase*.

Acetyl CoA
$CH_3CO\sim S.CoA$
+
$$O=\underset{\underset{CH_2COOH}{|}}{\overset{\overset{CH_2COOH}{|}}{C}}-COOH \quad \rightleftharpoons \quad HO\underset{\underset{CH_2COOH}{|}}{\overset{\overset{CH_2COOH}{|}}{C}}-COOH + CoA.SH$$

Oxaloacetic acid 　　Citric acid

The formation of citrate is followed by isomerization to D-isocitrate via *cis*-aconitate. This is catalysed by the enzyme *aconitate hydratase (aconitase)* which is specific for *cis*-aconitate and D-isocitrate and which, therefore, catalyses two distinct dehydration reactions, one involving a hydroxyl group attached to a tertiary carbon atom and the other a secondary hydroxyl attached to a secondary carbon atom. This seemed so remarkable that it was originally assumed that two different aconitases existed, one for each reaction. All attempts to resolve the enzyme into two such components have failed and the relative activities of the two reactions remain constant throughout purification:

$$\underset{\underset{CH_2COOH}{|}}{\overset{\overset{CH_2COOH}{|}}{HO.C.COOH}} \quad \underset{\pm H_2O}{\overset{Aconitase}{\rightleftharpoons}} \quad \underset{\underset{CH_2COOH}{|}}{\overset{\overset{CH.COOH}{||}}{C.COOH}} \quad \underset{\pm H_2O}{\overset{Aconitase}{\rightleftharpoons}}$$

Citric acid 　　　*cis*-Aconitic acid*

$$\underset{\underset{CH_2COOH}{|}}{\overset{\overset{CH(OH)COOH}{|}}{HC.COOH}}$$

Isocitric acid

*There is some uncertainty concerning the role of *cis*-aconitate as an obligatory intermediate. It has been suggested that an enzyme-bound carbonium ion is the true intermediate and *cis*-aconitate is produced from this ion as a side product.

In the next step of the cycle isocitrate undergoes oxidative decarboxylation to 2-oxoglutarate. This is catalysed by *isocitrate dehydrogenase* for which both NAD and NADP requirements have been established, depending upon the source of the enzyme:

$$\underset{\underset{CH_2COOH}{|}}{\overset{\overset{HOCH.COOH}{|}}{CH.COOH}} + NADP \rightleftharpoons$$

$$\underset{\underset{CH_2COOH}{|}}{\overset{\overset{O=C-COOH}{|}}{CH_2}} + NADPH_2 + CO_2$$

2-Oxoglutaric acid

Oxalosuccinic acid has been postulated as an intermediate in this reaction and it has been shown that highly purified isocitrate dehydrogenase will catalyse the decarboxylation of this compound.

The overall oxidative decarboxylation reaction is reversible and, therefore, enables carbon dioxide fixation to occur. It also is one of the energy-yielding steps of the cycle since reduction of nicotinamide nucleotide occurs.

2-Oxoglutarate is now oxidized to succinate and carbon dioxide by the action of *2-oxoglutarate oxidase*, a reaction similar to pyruvate oxidation which has already been discussed:

$$\underset{\underset{\underset{COOH}{|}}{CH_2}}{\overset{\overset{\overset{O=C-COOH}{|}}{CH_2}}{|}} + CoA.SH + NAD^+ \longrightarrow$$

2-Oxoglutaric acid

$$\underset{\underset{\underset{COOH}{|}}{CH_2}}{\overset{\overset{\overset{O=C\sim S.CoA}{|}}{CH_2}}{|}} + CO_2 + NADH + H^+$$

Succinyl CoA

Succinyl CoA is then cleaved to succinic acid and CoA with the concomitant formation of a mole of ATP. The reaction is catalysed by *succinyl CoA synthetase (succinate thiokinase)* and affords another example of substrate-level phosphorylation.

$$\underset{\underset{\underset{COOH}{|}}{CH_2}}{\overset{\overset{\overset{O=C-S.CoA}{|}}{CH_2}}{|}} + ADP + P_i \rightleftharpoons$$

Succinyl CoA

$$\underset{\underset{\underset{COOH}{|}}{CH_2}}{\overset{\overset{\overset{O=C-OH}{|}}{CH_2}}{|}} + ATP + CoASH$$

Succinic acid

In mitochondria the corresponding reaction involves GDP in place of ADP and the GTP formed then transfers its terminal phosphate to ADP under the influence of the enzyme *nucleoside diphosphokinase*.

$$GTP + ADP \rightleftharpoons GDP + ATP$$

Some bacteria, however, are capable of utilizing either ADP or GDP.

Succinate next undergoes oxidation to fumarate by *succinate dehydrogenase*:

$$
\begin{array}{c}
\text{CH}_2\text{COOH} \\
| \\
\text{CH}_2\text{COOH}
\end{array}
+ \text{FP} \xrightarrow[\text{dehydrogenase}]{\text{Succinate}}
$$

Succinic acid

$$
\begin{array}{c}
\text{HC---COOH} \\
\| \\
\text{HOOC---CH}
\end{array}
+ \text{FPH}_2
$$

Fumaric acid

Fumarate is reversibly hydrated with insoluble particles which contain also the necessary enzymes for transferring electrons to molecular oxygen, it is often referred to as the succinoxidase system. Succinate dehydrogenase is a flavoprotein (FP) which is oxidized *via* the cytochrome system.

Fumarate is reversibly hydrated to malate by the enzyme *fumarate hydratase* (*fumarase*) which therefore carries out a similar type of reaction to aconitase. Only the L-isomer of malate is formed and only fumarate, not maleate, can serve as the substrate:

$$
\begin{array}{c}
\text{HC---COOH} \\
\| \\
\text{HOOC---CH}
\end{array}
+ \text{H}_2\text{O} \xrightarrow{\text{Fumarase}}
\begin{array}{c}
\text{HOCH---COOH} \\
| \\
\text{CH}_2\text{COOH}
\end{array}
$$

Fumaric acid   L-Malic acid

The last step of the cycle is the oxidation of L-malate to regenerate oxaloacetate by a nicotinamide nucleotide-requiring *malate dehydrogenase*. At neutral pH values the equilibrium is greatly in favour of malate formation. However, under the normal conditions of operation of the tricarboxylic acid cycle, oxaloacetate is removed at a rapid rate and therefore malate oxidation proceeds readily:

$$
\begin{array}{c}
\text{HOCH---COOH} \\
| \\
\text{CH}_2\text{COOH}
\end{array}
+ \text{NAD}^+ \xrightarrow[\text{dehydrogenase}]{\text{Malate}}
$$

L-Malic acid

$$
\begin{array}{c}
\text{O}{=}\text{C---COOH} \\
| \\
\text{CH}_2\text{COOH}
\end{array}
+ \text{NADH} + \text{H}^+
$$

Oxaloacetic acid

In passing, it may be noted that another enzyme for the oxidation of malate is known, namely *malate dehydrogenase* (*decarboxylating*). This so-called 'malate enzyme' catalyses an oxidative decarboxylation to pyruvate and CO_2 using NADP as cofactor:

$$
\begin{array}{c}
\text{HOCH---COOH} \\
| \\
\text{CH}_2\text{COOH}
\end{array}
+ \text{NADP}^+ \rightleftharpoons
$$

L-Malic acid

$$
\begin{array}{c}
\text{O}{=}\text{C---COOH} \\
| \\
\text{CH}_3
\end{array}
+ \text{CO}_2 + \text{NADPH} + \text{H}^+
$$

Pyruvic acid

There is no evidence that oxaloacetate is an intermediate and since this compound must be formed to permit the cycle to function, the malate enzyme is not believed to play a role in the tricarboxylic acid cycle. Present views assign to this enzyme the functions of producing pyruvate for glucose synthesis (glucogenesis) when growth occurs on dicarboxylic acids such as succinate and malate, and the formation of NADPH for reductive biosynthesis. Enzymes which catalyse the decarboxylation of oxaloacetate to pyruvate and CO_2 and require only a divalent metal ion as cofactor are also known. Again, these enzymes are not strictly relevant to the present discussion of the cycle.

We have now seen how a two-carbon fragment produced in metabolism is condensed with oxaloacetate and, by undergoing a series of reactions, regenerates oxaloacetate and is itself oxidized to two molecules of carbon dioxide and two molecules of water. Oxaloacetate is thus enabled to combine with another molecule of acetyl CoA and participate in another cycle. It must be emphasized that the carbon atoms of the regenerated oxaloacetate are not identical with those of the molecule of oxaloacetate which initiated the cycle.

Significance of the tricarboxylic acid cycle

The tricarboxylic acid cycle fulfils two major roles, the production of energy and the provision of intermediates for biosynthesis, and their relative importance depends on whether or not the cells are growing. By isotopic experiments it has been shown that during exponential growth of *Escherichia coli* in glucose ammonium salts medium the principal function of the tricarboxylic acid cycle is to provide intermediates while glycolysis furnishes energy. When growth ceases, a switch of roles occurs, the cycle now generates energy and glucose is used for glycogenesis, leading to a deposition of glycogen in the cells.

From our previous discussion it is apparent that the tricarboxylic acid cycle is important in the total oxidation of carbohydrates. However, it is also important in the oxidation of many other substances. For instance, fatty acids are degraded stepwise to acetyl-CoA, many amino acids are converted to the keto acids, pyruvic, 2-oxoglutaric and oxaloacetic acids, and indeed intermediates of the cycle such as citric and succinic acids may be directly oxidized by this pathway.

Furthermore, as outlined in Chapter 4, these same substances may be precursors in the biosynthesis of amino acids and nucleotides.

Regulation of the tricarboxylic acid cycle

Control of the tricarboxylic acid cycle is exerted principally at the citrate synthase stage, i.e. on the first enzyme of the cycle. Weitzman has found that the citrate synthases of 18 genera of Gram-negative bacteria (including Azotobacter and Pseudomonas) were inhibited by NADH whereas the enzyme from aerobic Gram-positive bacteria was unaffected. Some of these enzymes are, however, inhibited by ATP. Within the NADH—susceptible group, subgroups could be distinguished on the basis of the ability of AMP to reverse the NADH inhibition. It has been suggested that only those organisms with a relatively simple mesosomal structure are susceptible to inhibition by NADH. Thus under conditions which result in an accumulation of reducing power and ATP, e.g. when biosynthesis is curtailed, the overall operation of the tricarboxylic acid cycle will be inhibited in susceptible bacteria, thus conserving energy and carbon. The citrate synthase of *coli* has also been shown to be inhibited by 2-oxoglutarate.

Although the oxidation of pyruvate to acetyl CoA is not part of the cycle it does furnish one of the substrates for citrate synthase and is itself also subject to regulation. Acetyl CoA, the product of the oxidation, inhibits pyruvate dehydrogenase, the first enzyme of the complex, by negative feedback inhibition (see Chapter 4, pp. 159). This enzyme is activated by AMP and inhibited when the energy charge is high.

ENERGY GENERATION BY BACTERIA

We have noted that bacteria secure the energy they need for growth by one or more of three general mechanisms, namely substrate-level phosphorylation, oxidative phosphorylation and (in the case of photosynthetic organisms) photophosphorylation. *Substrate-level phosphorylation*, which can occur under both anaerobic and aerobic conditions, may take one of two forms, described by the following equations

(1) Substrate $\sim$ (P) + ADP $\rightleftharpoons$ Substrate + ATP

(2) Substrate $\sim$ X + ADP + P_i $\rightleftharpoons$

Substrate + X + ATP

Examples of reaction (1) are found in glycolysis where 1,3-diphosphoglyceric acid and phosphoenolpyruvic acid serve as phosphate donors, while succinyl coenzyme A, in the tricarboxylic acid cycle, affords an example of reaction (2) [however, in eukaryotes the initial phosphate acceptor is GDP (see p. 145)]. Noteworthy is the fact that relatively few substrate-level phosphorylation reactions have been discovered.

Oxidative phosphorylation occurs when ATP synthesis is coupled to electron transport reactions associated with the oxidation of either organic compounds in the chemo-organotrophs, or inorganic compounds in the chemo-lithotrophs (see p. 153). Electrons are transferred from the substrate to a series of electron carriers, which span the gap between the redox potential of the substrate oxidation reaction and that of oxygen, until they react with two protons, and an atom of oxygen to form water

$$2H^+ + 2e + \tfrac{1}{2}O_2 \longrightarrow H_2O$$

catalysed by a terminal *cytochrome oxidase* enzyme.

Photophosphorylation occurs in photosynthetic bacteria which utilize radiant energy to drive electron trans port. In some cases it may be cyclic and in some noncyclic; the free energy available from such electron transport is coupled to ATP synthesis. These organisms derive the reducing power they require either from the oxidation of inorganic substances in the photolithotrophs, or from organic compounds in the photo-organotrophs (see p. 153).

Although differences in detail exist, the fundamental features of ATP generation coupled to electron transport are remarkably similar in mitochondria, bacteria and photosynthetic systems. In all cases, the enzymes and carriers are located in membranes and they are arranged in an asymmetric manner so that the reactions catalysed are *vectorial*, i.e. directional across the membrane as opposed to the more usual *scalar* or non-directional reactions encountered in soluble phases. The precise mechanism whereby ATP is linked to electron transport remains controversial and in the past three principal hypotheses have held the field. However, in recent years one of these models, the *chemiosmotic theory* of Mitchell, has gained fairly general acceptance because it explains most of the experimental observations and, moreover, furnishes a unifying concept which links the various energy-dependent functions in mitochondria, bacteria, chloroplasts and muscle. For these reasons, attention here will be confined to this theory, although it must be stressed that the fine details of the scheme are still the subject of extensive current research.

The chemiosmotic theory

Membranes are structurally and functionally asymmetric, i.e. the two surfaces of a membrane possess different components and different enzymes. In the chemiosmotic hypothesis, it is postulated that enzymes in the membrane which are associated with oxidative phosphorylation are asymmetrically oriented so that they catalyse vectorial reactions which effect the transport of molecules, chemical groups and ions across the membrane. Further, some of these reactions result in the separation of electric charges within and across the membrane; the recombination of such charges is associated with the performance of chemical, osmotic and mechanical work.

In its essential features, the chemiosmotic theory demands the presence in a membrane of a proton-translocating electron transport chain and a proton-translocating adenosine triphosphatase (ATPase). The latter enzyme catalyses the readily reversible vectorial reaction

$$ATP_{in} + H_2O_{out} + 2H^+_{in} \rightleftharpoons ADP_{in} + P_{i\,in} + 2H^+_{out}$$

where the subscripts 'in' and 'out' represent respectively the phases on the inner and outer sides of the membrane which is essentially impermeable to most ions, including H^+ and OH^-. Then the net result of either electron transport or the hydrolysis of ATP is the generation of gradients of both electrical potential ($\Delta\psi$) and pH (ΔpH) across the membrane, with the soluble phase on one side of the membrane being electrically negative and alkaline with respect to the other. These parameters give rise to a proton motive force (ΔP) defined by the equation

$$\Delta P = \Delta\psi - Z\Delta pH$$

where ΔP is in millivolts, $\Delta\psi$ is the potential difference across the membrane, $Z = 2\cdot3\,RT/F$, where R is the gas constant, T the absolute temperature and F the Faraday ($Z = 59$ mV at 25°C), and ΔpH the pH difference between the inner and outer phases. In order to maintain an electrochemical proton gradient there must be two compartments and consequently such gradients can only be measured with membranes that form topologically closed units, an experimental observation which supports the chemiosmotic hypothesis. ΔP, which, according to the magnitude of the individual terms, may be a function primarily of $\Delta\psi$ or of ΔpH, or a combination of the two, drives various transmembrane energy-dependent processes, such as ATP synthesis by reversal of the proton-translocating ATPase and the accumulation of metabolites *via* their respective permease systems.

The principle of ATP synthesis according to the chemiosmotic theory is illustrated in Figure 10 with reference to a system for translocating two protons. Substrate AH_2 is oxidized by a dehydrogenase which reacts on the inner side of the membrane. Associated with the enzyme are a hydrogen carrier and an electron carrier which are oriented across the membrane and operate in sequence to enable two protons to be released to the exterior and two electrons to be carried back to the inner side of the membrane. There they are taken up simultaneously with two protons by a hydrogen acceptor, B, to yield the reduced form BH_2. This sequence comprises a *two-proton-translocating loop* or *segment*, the net result of which is the release of $2H^+$ on the outer side and the uptake of $2H^+$ on the inner side of the membrane, thus creating a proton-motive force, and with the inside compartment becoming alkaline relative to the outside. The donor and acceptor molecules react with the enzyme complex on the same side of the membrane (although it must be emphasized that alternative formulations are possible in which they react on opposite sides of the membrane).

The associated two-proton translocating ATPase that permits energy coupling is also shown diagrammatically in Figure 10. It comprises two components, a stalk-like hydrophilic unit F_1, which is associated with adenine nucleotide binding and possesses catalytic activity as an ATPase even when

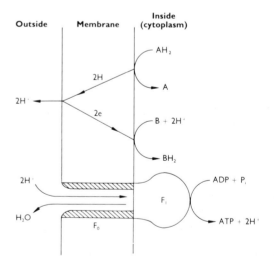

Figure 10. The chemiosmotic theory. Diagrammatic representation of a proton-tranlocating oxidation-reduction loop and a proton-translocating ATPase comprising two moieties, F_0 in the membrane and F_1 projecting into the cytoplasm. The extrusion of two protons is coupled to the synthesis of one ATP molecule. (Note that 2H is equivalent to $2H^+ + 2e$.)

removed from the membrane, and a hydrophobic moiety F_0 which furnishes a pore through the membrane and permits the passage of H^+ and H_2O to and from F_1. The ATPase enzyme complex is reversible so that it can not only synthesize ATP in oxidative phosphorylation but also hydrolyse ATP to generate a protonmotive force. The latter reaction assumes importance in bacteria that can grow under anaerobic conditions since, under these circumstances, the energy generated as ATP by substrate-level phosphorylation must be used to energize the membrane to the essential functions of active transport and reversed electron transport.

In oxidative phosphorylation, the protonmotive force generated across the membrane by electron transport reverses the proton-translocating ATPase so that, in the presence of ADP and P_i in the cytoplasm, ATP is synthesized. Consequently, the extrusion of two protons is associated with the subsequent synthesis of one molecule of ATP as the protons move inwards, as shown in Figure 10. The complete electron transport chain extending from a substrate to oxygen usually embodies more than one proton-translocating segment of the type discussed. Thus in mitochondria oxidizing NADH there are three such loops involved, while only two operate in succinate oxidation. Therefore, the P/O or P/2e ratio for the oxidation of NADH is 3, while that for succinate is 2.

Bacterial respiration and oxidative phosphorylation

During recent years attention has been increasingly focused on bacterial respiration and oxidative phosphorylation. While it is now established that bacteria possess electron-transporting and proton-translocating loops generally similar to those of mitochondria, considerable variations are encountered. These occur in the nature of the oxidants and reductants used, in the much greater diversity of the redox components found in bacterial cytoplasmic membranes, and in the number of energy conservation sites they possess. Consequently it is not possible to draw up a universally applicable chart of energy yields for aerobic bacteria utilizing an energy source such as glucose, since some organisms appear to possess three energy transduction sites and others only two. These sites can be identified with three proton-translocating segments of the chain, referred to as 1, 2 and 3, and the observed differences may be explained by the inability of certain bacteria to synthesize a high redox potential cytochrome c (segment 3) whereas most organisms are able to synthesize

segments 1 and 2 during normal aerobic growth*.

The principal difference between mitochondrial and bacterial respiratory chains resides in the great diversity of cytochrome types found in bacteria. Thus all mitochondria contain identical cytochromes (b, c, c_1 and aa_3) whereas *E. coli* has at least nine cytochromes, not all of which are membrane-associated and involved in oxidative phosphorylation; the cytochrome oxidase (aa_3) of mitochondria is replaced by carbon monoxide-binding cytochromes o, a_1, d and/or $c_{(CO)}$, which may function as oxidases. Further, the redox carriers synthesized by *E. coli* are greatly influenced by the conditions of growth, such as growth phase, carbon source, terminal electron acceptor and the strain of the organism.

Mitochondrial and bacterial electron transport systems also differ in relation to their quinonoid components; some bacteria resemble mitochondria in possessing ubiquinones (coenzyme Q) while Gram-positive bacteria contain mainly naphtha-quinones (menaquinones, MK, which are vitamin K derivatives) and facultative anaerobic bacteria have both.

There is one bacterium, however, *Paracoccus denitrificans*, which has an electron transport chain very similar to that of mitochondria (Figure 11) and this observation has led to the suggestion that the inner membrane of the present-day mitochondrion evolved from the plasma membrane of an ancestral form of *P. denitrificans* by endosymbiosis with a primitive host cell. Figure 11 compares some respiratory chains of bacteria with that of the mitochondrion.

It is possible to replace oxygen as the terminal electron acceptor in the electron transport chain by fumarate or nitrate, and in these circumstances anaerobic growth can occur in the absence of fermentable substrates. Considerable research with *E. coli* has shown that the presence of fumarate allows the organisms to grow anaerobically with glycerol as the carbon source; two enzymes are specifically

---

* A fourth proton-translocating segment, designated 0, has been associated with a membrane-bound transhydrogenase which catalyses the reaction

$$NADPH + NAD^+ \rightleftharpoons NADP^+ + NADH$$

However, in those bacteria which synthesize the transhydrogenase there is little evidence to suggest that the enzyme plays a significant role in energy generation; it seems more likely that it serves as a means of generating NADPH from NADH, although experimental evidence is lacking.

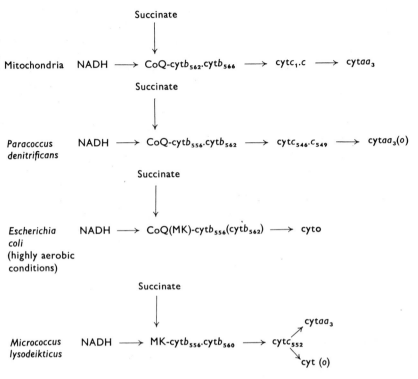

Figure 11. Comparison of mitochondrial and some bacterial electron transport chains. Components which are present in low concentrations or are kinetically inactive are shown in parentheses. MK, menaquinone.

induced under these conditions, namely an anaerobic L-α- glycerophosphate dehydrogenase and a fumarate reductase which is distinct from succinate dehydrogenase. Similarly, anaerobic growth of *E. coli* with nitrate depends on the induction of two membrane-bound components, cytochrome $b_{556}^{NO_3^-}$ and nitrate reductase, which enable various non-fermentable substrates, e.g. D-lactate, to serve as the sole source of carbon. Some aerobic bacteria, such as *Pseudomonas aeruginosa*, can also grow anaerobically if nitrate is added to the medium.

Energy yields as a result of glucose catabolism

The generation of energy in the tricarboxylic acid cycle occurs in the oxidative steps, namely the oxidation of pyruvate, isocitrate, 2-oxoglutarate, succinate and malate. We can consider the total yield of energy in terms of molecules of ATP produced from pyruvate in the course of one turn of the tricarboxylic acid cycle, i.e. in the oxidation of pyruvate to carbon dioxide and water. The energy yielding steps and the ATP generated by mito-

chondria are as follows:

Pyruvate $+ NAD^+ \longrightarrow$ Acetyl CoA $+ NADH + H^+$ 3 ATP

Isocitrate $+ NADP^+ \longrightarrow$
                              2-Oxoglutarate $+ NADPH + H^+$ 3 ATP

α2-Oxoglutarate $+ NAD^+ \longrightarrow$
                              Succinyl CoA $+ NADH + H^+$ 3 ATP

Succinyl CoA $\longrightarrow$ Succinate $+ CoA$ 1 ATP

Succinate $+ FP \longrightarrow$ Fumarate $+ FPH_2$ 2 ATP

Malate $+ NAD^+ \longrightarrow$ Oxaloacetate $+ NADH + H^+$ 3 ATP

                                              Total   15 ATP

Hence the oxidation of pyruvate by the cycle in mitochondria yields 15 molecules of ATP. In bacteria, however, the yield is not always so high because many organisms possess two, not three energy conservation sites, as discussed above. Thus the comparable yield from one turn of the cycle would be some 10 molecules of ATP. Frequently when glucose is metabolized by the Embden–Meyerhof pathway and the tricarboxylic acid cycle in bacteria, there is a net yield of either 38 or 26 molecules of ATP (Figure 12). This may be compared with anaerobic glycolysis or yeast fermentation:

Glucose + 2ADP + 2P$_i$ ⟶ 2 Lactate + 2 ATP

Glucose + 2ADP + 2P$_i$ ⟶

$$2 \text{ Ethanol} + 2 CO_2 + 2 \text{ ATP}$$

when the much greater efficiency of the aerobic process in terms of energy yield is apparent.

The energy which is biologically useful to an organism is that which is obtained in the form of ATP, and is not necessarily identical with the free energy change of the process. When the energy yield is measured in terms of the molar growth yield (grams dry weight of organism produced per mole of glucose utilized as the energy source) the values reflect the yield of ATP. The measurement of molar growth yields thus represents a valuable experimental technique for the assessment of the energy (ATP) derived from a specified carbon source by a given type of bacterium, and can be used in deducing the metabolic pathways involved. For example, the anaerobic fermentation of glucose by the Embden–Meyerhof, Entner–Doudoroff and phosphoketolase pathways yields respectively 2, 1 and 1 moles of ATP per mole of glucose fermented; consequently the amount of growth supported per mole of glucose by the Embden–Meyerhof pathway will be double that obtained with either of the other two types of metabolism.

THE GLYOXYLATE CYCLE

As mentioned above, intermediates of the tricarboxylic acid cycle may be continuously drawn off and used as precursors of amino acids, etc. They must be replenished if the cycle is to go on functioning. This usually happens by carboxylation of pyruvate or phospho-enolpyruvate (PEP) to yield oxaloacetate. The following reactions bring this about:

CH$_3$CO.COOH + CO$_2$ + ATP $\xrightarrow{\text{pyruvate carboxylase}}$

HOOC.CH$_2$CO.COOH + ADP + P$_i$

CH$_2$:CO∼℗.COOH + CO$_2$ + H$_2$O $\xrightarrow{\text{PEP-carboxylase}}$

HOOC.CH$_2$CO.COOH + P$_i$

The carboxylation of PEP by PEP-carboxylase appears to be the essential reaction for *E. coli in vivo* since only mutants which lack this enzyme fail to grow on pyruvate or its precursors unless tricarboxylic acid cycle intermediates are added to the growth medium. PEP-carboxylase is absent, however, from pseudomonads and Arthrobacter which

employ the ATP-dependent pyruvate carboxylase to produce oxaloacetate.

However, many micro-organisms are able to utilize acetate as the sole carbon source and if under these conditions the tricarboxylic acid cycle is to continue to provide energy and intermediates for biosynthesis, there must be an alternative method of replenishing its intermediates from C$_2$-units. A pathway which does this is the *glyoxylate cycle* (Figure 13) and pathways having this function are sometimes referred to as 'anaplerotic' (from the Greek for 'filling up'), since they replenish intermediates drained off for biosynthesis. Effectively the net result of this cycle is the condensation of two molecules of acetate to give one of succinate. This is made possible by two additional enzymes that we have not so far discussed. One of these is *isocitrate lyase* and the other

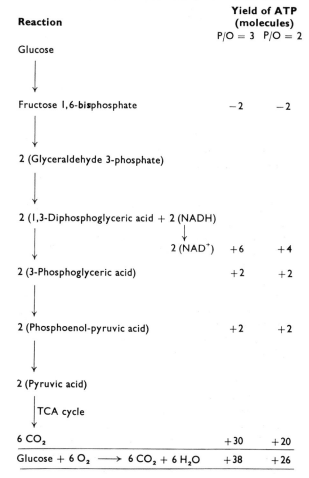

Reaction	Yield of ATP (molecules)	
	P/O = 3	P/O = 2
Glucose		
↓		
Fructose 1,6-bisphosphate	−2	−2
↓		
2 (Glyceraldehyde 3-phosphate)		
↓		
2 (1,3-Diphosphoglyceric acid + 2 (NADH)		
↓ 2 (NAD$^+$)	+6	+4
2 (3-Phosphoglyceric acid)	+2	+2
↓		
2 (Phosphoenol-pyruvic acid)	+2	+2
↓		
2 (Pyruvic acid)		
↓ TCA cycle		
6 CO$_2$	+30	+20
Glucose + 6 O$_2$ ⟶ 6 CO$_2$ + 6 H$_2$O	+38	+26

Figure 12. Yield of ATP as a result of aerobic catabolism of glucose via Embden–Meyerhof pathway and tricarboxylic acid cycle in bacteria when the P/O ratios are 3 and 2 respectively.

is *malate synthase* and they are found in greatly increased amounts in cells grown on acetate. The reactions then are as follows:

(1)   Acetyl CoA + Oxaloacetate + H_2O ⟶ Citrate + CoA

(2)   Citrate ⇌ Isocitrate

(3)   Isocitrate ⇌ Succinate + Glyoxylate

(4)   Acetyl CoA + Glyoxylate + H_2O ⟶ Malate + CoA

(5) $\underline{\text{Malate} + \tfrac{1}{2}O_2 \longrightarrow \text{Oxaloacetate} + H_2O}$

2 Acetyl CoA + $\tfrac{1}{2}O_2$ + H_2O ⟶ Succinate + 2 CoA

It will thus be seen that the glyoxylate cycle interlocks with the tricarboxylic acid cycle and that some enzymes and intermediates are common to both. The tricarboxylic acid cycle can then operate normally for the oxidation of acetate and the production of biosynthetic intermediates, being replenished by the succinate produced via the glyoxylate cycle.

The essential requirement of isocitrate lyase for growth on acetate was demonstrated by the inability

of mutants devoid of the enzyme to grow on this substrate, although they were still able to oxidize it. Such mutants oxidized acetate to completion, as opposed to the wild type organism which oxidized acetate with the simultaneous assimilation of 20–30% of the carbon, thus demonstrating both the role of the anaplerotic enzymes in biosyntheses from acetate and also the independence of these enzymes from those concerned with the provision of energy.

The glyoxylate cycle appears to be regulated by inhibition of isocitrate lyase. In *Escherichia coli* this enzyme is powerfully inhibited by phospho-enol-

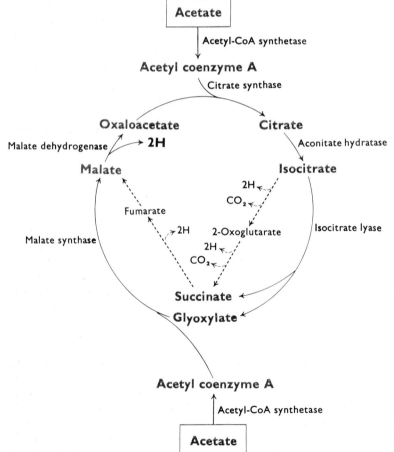

Figure 13. The glyoxylate cycle. The dotted lines indicate reactions of the tricarboxylic acid cycle.

pyruvate in a non-competitive manner, which may be regarded as a feedback mechanism. Accumulation of phospho-enolpyruvate would thus inhibit the key enzyme of the anaplerotic sequence leading to phospho-enolpyruvate formation. The operation of this mechanism has been verified by the use of mutants of *Escherichia coli* which lack phospho-enol-pyruvate carboxylase (p. 135) and consequently, although able to form phospho-enolpyruvate from pyruvate *via* the synthase reaction (p. 133), are unable to remove it. The addition of pyruvate to cultures of the mutant arrests growth on acetate because it causes accumulation of phospho-enolpyruvate.

In other micro-organisms compounds such as succinate, glycollate and pyruvate have also been found to inhibit isocitrate lyase.

AUTOTROPHS AND HETEROTROPHS

Most species of bacteria use organic carbon compounds and from them make their own cellular components. They derive energy from exergonic catabolic reactions of these organic substrates and are called *heterotrophs*. The *autotrophs* can use carbon dioxide as their sole source of carbon. The energy they require to do this may be supplied by light, as in *photosynthesis*, or from exergonic inorganic oxidation reactions as occurs in the *chemosynthetic* bacteria. Because photosynthetic organisms can utilize organic compounds other than carbon dioxide, a modified terminology has now been adopted somewhat replacing the older autotroph/heterotroph and photosynthetic/chemosynthetic categories. The *lithotrophs* use carbon dioxide and reduce it to organic compounds. This requires energy and reducing power. The latter is supplied by inorganic substances, the former by radiation in the *photolithotrophs* and by inorganic oxidation in the *chemolithotrophs*. The organotrophs use organic substrates and may be *photo-organotrophs* using light energy for assimilation of carbon dioxide and organic substances or may be *chemo-organotrophs* using oxidation or fermentation of organic compounds. The reducing power needed in the photo-organotrophs is again supplied by organic material.

Bacterial photosynthesis closely resembles the process found in plants and blue-green algae (cyanobacteria) but differs in utilizing somewhat different pigment systems and in not producing oxygen. This is due to the possession by bacteria of only a single light-absorbing reaction, whereas oxygen evolution requires two such steps.

Phototrophic bacteria

Photosynthesis comprises an extremely complex series of reactions and it is only within recent years that the nature of these has been elucidated. There are two essential, closely integrated yet quite distinct processes involved, namely the absorption of light energy by the photosynthetic pigments and its conversion to the chemical bond energy of ATP, and the biosynthetic reactions leading from CO_2 or organic compounds to cell materials and utilizing the chemical bond energy derived from the light.

Whereas the first process is dependent on light and unique to photosynthetic organisms, the reactions leading to synthesis of cell material can occur in the dark and are also found in many non-photosynthetic bacteria. Even the key reactions by which CO_2 fixation is achieved in photosynthetic bacteria are found in chemolithotrophic organisms.

The main difference between plant and bacterial photosynthesis lies in the nature of the ultimate electron donor. Plants utilize water as the electron donor and release oxygen: isotopic studies with water enriched with ^{18}O have revealed that the oxygen is derived from the water, so that the overall series of reactions may be designated as:

$$CO_2 + 2H_2^{18}O \xrightarrow{\text{Light}} (CH_2O) + {}^{18}O_2 + H_2O$$

Photosynthetic bacteria do not use water but rather reduced sulphur compounds or organic compounds or even molecular hydrogen as electron donors, the particular donor depending on the species. The reaction is anaerobic and involves only one light-dependent reaction.

Photosynthetic bacteria fall into two principal groups, referred to as the green and purple bacteria and distinguished by their photosynthetic pigments. The green *Chlorobium* genus contains one of two different chlorophylls together with alicyclic carotenoids, while the purple bacteria always contain one particular type of chlorophyll, bacteriochlorophyll, together with various aliphatic carotenoids. In consequence, organisms of the two groups absorb light of different wavelengths for the process of photosynthesis.

The green bacteria are usually photolithotrophic and strictly anaerobic organisms which utilize CO_2 as the carbon source and H_2S as an electron donor. The purple bacteria can be classified on a physiological basis as purple sulphur bacteria (Thiorhodaceae) or non-sulphur purple bacteria (Athiorhodaceae). The purple sulphur bacteria resemble the green bacteria in being anaerobic and photolithotrophic whereas the non-sulphur purple bacteria prefer organic compounds as the source of carbon and reducing power for photosynthesis, although they can also reduce CO_2. Unlike most photosynthetic bacteria, some of the non-sulphur purple organisms are able to tolerate oxygen and can grow in the dark, deriving their energy by oxidation of organic compounds and using oxygen as the final electron acceptor, i.e. as chemo-organotrophs.

The green sulphur bacteria and the purple sulphur bacteria use hydrogen sulphide as an exogenous electron donor for cellular synthesis from CO_2. Hydrogen sulphide is oxidized to sulphate in two stages, the first of which is analogous to the overall equation for plant photosynthesis:

(Bacteria)
$$CO_2 + 2\,H_2S \xrightarrow{\text{Light}} (CH_2O) + H_2O + 2\,S$$

cf. (Plants)
$$CO_2 + 2\,H_2O \xrightarrow{\text{Light}} (CH_2O) + H_2O + O_2$$

Elemental sulphur frequently accumulates in the purple sulphur bacteria and then, when the exogenous source of hydrogen sulphide is exhausted, disappears as it is further oxidized to sulphate:

$$3\,CO_2 + 2\,S + 5\,H_2O \xrightarrow{\text{Light}} 3\,(CH_2O) + 2\,H_2SO_4$$

Some purple bacteria, both sulphur and non-sulphur, can use molecular hydrogen and their photosynthetic reaction can be formulated:

$$CO_2 + 2\,H_2 \xrightarrow{\text{Light}} (CH_2O) + H_2O$$

The non-sulphur purple bacteria can use organic substances as electron donors:

$$CO_2 + 2\,CH_3CHOH.CH_3 \xrightarrow{\text{Light}}$$
$$(CH_2O) + H_2O + 2\,CH_3COCH_3$$

but usually can also assimilate the organic substrate. Thus many strains use lower fatty acids. Acetate, for instance, can be assimilated to form the reserve material poly-β-hydroxybutyrate $[(C_4H_6O_2)_n]$. This anaerobic process requires ATP (generated from the light) and reducing power, derived by anaerobic breakdown of some acetate *via* the tricarboxylic acid cycle:

$$CH_3COOH + 2\,H_2O \longrightarrow 2\,CO_2 + 8\,H$$

Pairs of acetate molecules are combined, reduced and polymerized:

$$2\,CH_3COOH + 2\,H \longrightarrow [CH_3CH.CH_2C{=}O] + 2\,H_2O$$
$$\underset{[C_4H_6O_2]}{\overset{|}{O}}$$

The overall reaction is thus:

$$CH_3COOH + 2\,H_2O \longrightarrow 2\,CO_2 + 8\,H$$
$$8\,CH_3COOH + 8\,H \longrightarrow 4\,(C_4H_6O_2) + 8H_2O$$
$$\overline{\quad 9\,CH_3COOH \xrightarrow{\text{Light}} \quad}$$
$$4\,(C_4H_6O_2) + 2\,CO_2 + 6\,H_2O$$

In general, however, photosynthesis results in endergonic reduction of carbon dioxide to carbohydrate.

Fixation of carbon dioxide: photolithotrophs and chemolithotrophs

Experiments with $^{14}CO_2$ show that in extremely short time periods the first labelled compound formed is glycerate 3-phosphate with the isotope in its carboxyl group. The detection also of labelled sedoheptulose 7-phosphate and ribulose 1,5-bisphosphate in the early stages of photosynthesis suggested that some of the reactions of the pentose phosphate cycle might be involved in the process. It is now known that of the fifteen or more reactions involved in photosynthesis only two are specific to photosynthetic and chemolithotrophic organisms. The rest are reactions common to glycolysis, the pentose phosphate cycle and the formation of carbohydrate from non-carbohydrate precursors, i.e. reactions found in non-photosynthetic organisms.

The two specific reactions are the phosphorylation of ribulose 5-phosphate to ribulose 1,5-bisphosphate under the influence of the enzyme *phosphoribulokinase*, and the fission of ribulose 1,5-bisphosphate by carbon dioxide and water, to yield two molecules of glycerate 3-phosphate, catalysed by the enzyme *ribulose bisphosphate carboxylase*. Phosphoribulokinase is analogous in its action to the phosphofructokinase of the Embden–Meyerhof sequence, although differing in its specificity:

The ribulose bisphosphate carboxylase reaction is a complex one in which ribulose 1,5-bisphosphate

2 Glyceraldehyde 3-phosphate $\longrightarrow$ Fructose 6-phosphate + P_i

Fructose 6-phosphate + Glyceraldehyde 3-phosphate $\longrightarrow$ Xylulose 5-phosphate + Erythrose 4-phosphate

Erythrose 4-phosphate + Glyceraldehyde 3-phosphate $\longrightarrow$ Sedoheptulose 7-phosphate + P_i

Sedoheptulose 7-phosphate + Glyceraldehyde 3-phosphate $\longrightarrow$ Ribose 5-phosphate + Xylulose 5-phosphate

Ribose 5-phosphate + Xylulose 5-phosphate $\longrightarrow$ 2 Ribulose 5-phosphate

Xylulose 5-phosphate $\rightleftharpoons$ Ribulose 5-phosphate

---

5 Glyceraldehyde 3-phosphate $\longrightarrow$ 3 Ribulose 5-phosphate + 2 P_i

---

3-Ribulose 5-phosphate + 3ATP $\longrightarrow$ 3 Ribulose 1,5-bisphosphate

---

reacts with carbon dioxide to give two molecules of glycerate 3-phosphate:

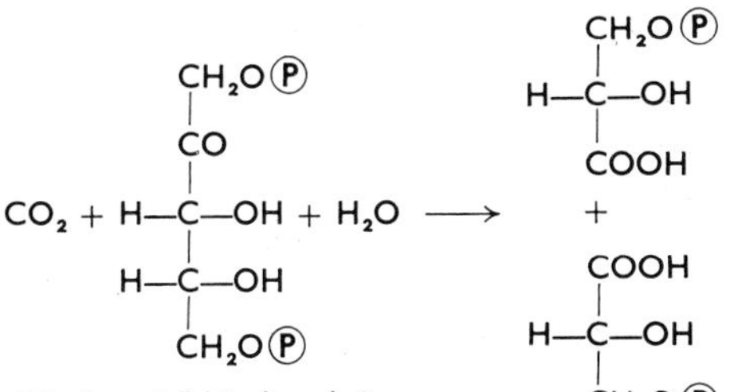

The subsequent reactions of glycerate 3-phosphate are common to glycolysis although operating in the reverse direction and employing $NADP^+$ rather than NAD^+. Glycerate 3-phosphate is thus phosphorylated by ATP in the presence of *phosphoglycerate kinase* to 1,3-diphosphoglycerate and then reduced to glyceraldehyde 3-phosphate by triose phosphate dehydrogenase and NADPH. An equilibrium mixture of glyceraldehyde 3-phosphate and dihydroxyacetone phosphate is produced by the action of triose phosphate isomerase, and one molecule of each then combines with the other under the influence of aldolase to form fructose 1,6-bisphosphate. The phosphate group in the 1-position is then removed by a specific phosphatase and the resulting fructose 6-phosphate undergoes reactions of the pentose phosphate cycle catalysed by transketolase and transaldolase.

Fructose 6-phosphate and glyceraldehyde 3-phosphate are converted to xylulose 5-phosphate and erythrose 4-phosphate by transketolase. Erythrose 4-phosphate and glyceraldehyde 3-phosphate yield sedoheptulose 1,7-bisphosphate in a reaction catalysed by transaldolase, and the latter compound loses a phosphate group to become sedoheptulose 7-phosphate. This reacts with glyceraldehyde 3-phosphate to yield one molecule each of xylulose 5-phosphate and ribose 5-phosphate, again catalysed by transketolase. Pentose phosphate isomerase and xylulo-epimerase convert ribose 5-phosphate and xylulose 5-phosphate to ribulose 5-phosphate and phosphorylation of this yields ribulose 1,5-bisphosphate. This series of reactions is denoted at the top of this page. The fixation of each molecule of CO_2 in photosynthesis thus requires the expenditure of 3 molecules of ATP, two to phosphorylate the two molecules of glycerate 3-phosphate formed and one to phosphorylate ribulose 5-phosphate produced in the regeneration cycle. The overall process leading to hexose synthesis may be represented as shown at the foot of this page The reactions of the photosynthetic or Calvin cycle are depicted schematically in Figure 14. For purposes of clarity only half the number of molecules involved are represented leading to the formation of a 'half molecule' of hexose. There is still doubt as to whether this pathway operates in the green sulphur bacteria.

It must be emphasized that the photosynthetic cycle provides also the intermediary carbon compounds for biosynthetic reactions associated with

Fixation	6 Ribulose 1,5-bisphosphate + 6 CO_2 + 6 H_2O $\longrightarrow$ 12 Glycerate 3-phosphate
Reduction	12 Glycerate 3-phosphate + 12 ATP + 12 NADPH + 12 H^+ $\longrightarrow$
	12 Glyceraldehyde 3-phosphate + 12 ADP + 12 P_i + 12 $NADP^+$ + 12 H_2O
Regeneration	12 Glyceraldehyde 3-phosphate + 6 ATP $\longrightarrow$
	6 Ribulose 1,5-bisphosphate + 6 ADP + 5 P_i + Fructose 6-phosphate
Sum	6 CO_2 + 6 H_2O + 18 ATP + 12 NADPH + 12 H^+ $\longrightarrow$
	Fructose 6-phosphate + 18 ADP + 12 $NADP^+$ + 17 P_i

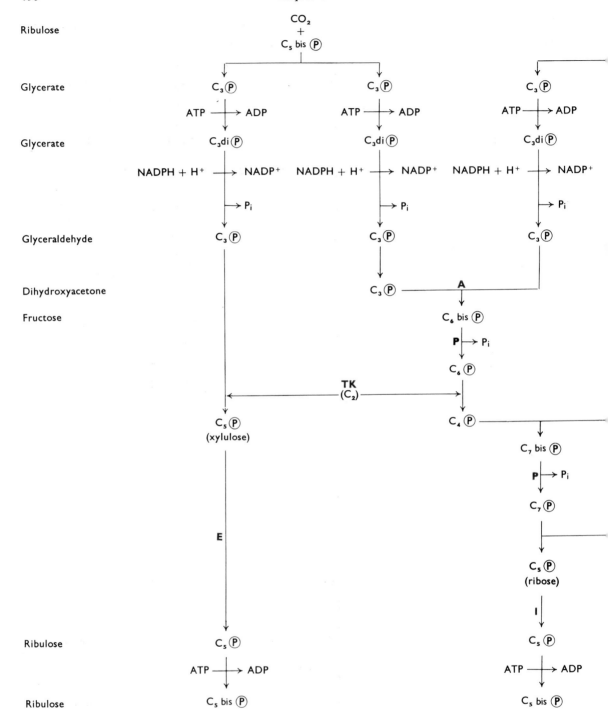

Overall reaction

$$3 CO_2 + 3 H_2O + 9 ATP + 6 NADPH + 6 H^+ \longrightarrow \tfrac{1}{2} \text{Fructose } 6\text{-}\text{P} + 9 ADP + 6 NADP^+ + 8\tfrac{1}{2} P_i$$

Figure 14. The photosynthetic or Calvin cycle.

TK = transketolase, A = aldolase, E = epimerase, I = isomerase, P = phosphatases.

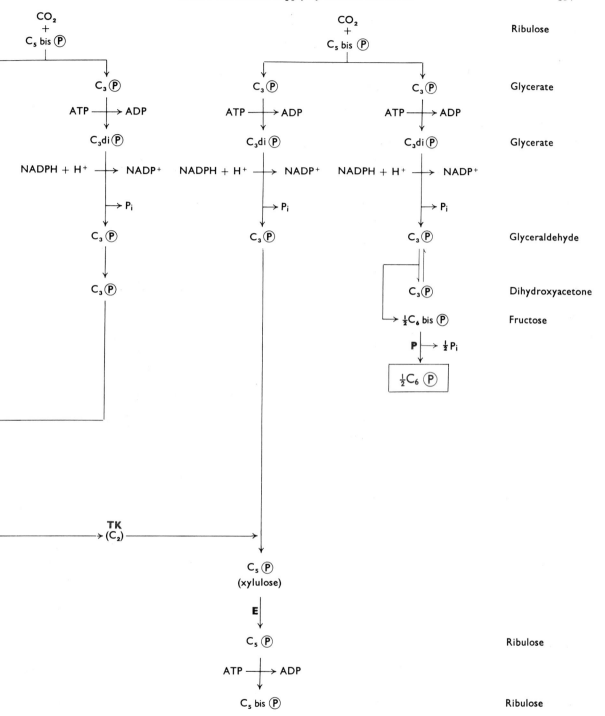

growth, in addition to producing hexose. Thus compounds such as triose phosphates and pentoses will be drained from the cycle to fulfil these requirements.

The same pathway of CO_2 fixation as occurs in photosynthetic bacteria has now been demonstrated in many of the chemolithotrophic bacteria, the difference residing solely in the nature of the energy-yielding mechanisms. Thus the key enzymes ribulose bisphosphate carboxylase and phosphoribulokinase are present and following exposure to $^{14}CO_2$ the label appears rapidly in glycerate 3-phosphate, hexose phosphates, sedoheptulose phosphate and ribulose phosphate.

The exergonic reactions in different organisms may include oxidation of ammonia to nitrite and then to nitrate, of inorganic sulphur compounds (H_2S, S, SO_3^{2-}, etc.), of ferrous to ferric compounds and of molecular hydrogen to water. Usually the oxidizing agent is oxygen but some species are anaerobes using inorganic nitrate as oxidant. Thus *Thiobacillus denitrificans* oxidizes H_2S to sulphate while nitrate is reduced to molecular nitrogen:

$$5\,H_2S + 8\,KNO_3 \longrightarrow$$
$$3\,K_2SO_4 + 2\,KHSO_4 + 4\,N_2 + 4\,H_2O$$

This exergonic reaction is coupled to the formation of ATP from ADP and inorganic phosphate; the electron transport chain involves flavoproteins and cytochromes as in aerobic oxidations (see Appendix B).

Control of autotrophic carbon dioxide fixation

Present evidence indicates that autotrophic CO_2 fixation is regulated by modulation of the activity of phosphoribulokinase. From the preceding discussion it will be apparent that the overall reactions leading to hexose synthesis make heavy demands on the ATP and reducing power (NADPH) of the organism. Phosphoribulokinase has been shown to be sensitive to the energy charge of the cell and the enzyme from several types of chemolithotrophic bacteria is inhibited by AMP. It has also been found recently that NADH activates the phosphoribulokinase of hydrogen bacteria and *Rhodopseudomonas*. Thus regeneration of the CO_2-acceptor appears to be controlled by the energy charge of the autotrophic cell and in such a way that fixation and reduction occur only when the cell is well-endowed with energy.

FURTHER READING

1 Pfennig N. (1967). Photosynthetic bacteria. *Annual Review of Microbiology* **21**, 285.

2 Fraenkel D. G. & Vinopal R. T. (1973) Carbohydrate Metabolism in Bacteria. *Annual Review of Microbiology.* **27**, 69.

3 Kornberg H. L. (1966) Anaplerotic sequences and their role in metabolism. In *Essays in Biochemistry*, Volume 2 (eds. Campbell P. N. and Greville G. D.), p. 1. Academic Press, London.

4 White A. and others (1978) *Principles of Biochemistry.* Sixth ed. McGraw-Hill, New York.

5 Metzler, D. E. (1977). *Biochemistry. The Chemical Reactions of Living Cells* Academic Press, London.

6 Lehninger A. L. (1975) *Biochemistry.* Second ed. Worth, New York.

7 Sanwal B. D. (1970). Allosteric controls of amphibolic pathways in bacteria. *Bacteriological Reviews* **34**, 20.

8 Haddock B. A. & Jones C. W. (1977). Bacterial respiration. *Bacteriological Reviews* **41**, 47.

9 Gregory R. P. F. (1971) *Biochemistry of Photosynthesis.* Wiley-Interscience, London.

10 Hamilton W. A. & Haddock B. A. (eds.) (1977) *Microbial Energetics.* C.U.P. London.

11 Quayle, J. R. (ed.) (1979) *Microbial Biochemistry.* University Park Press, Baltimore.

Chapter 4
Class II Reactions: Synthesis of Small Molecules

BIOSYNTHESIS OF AMINO ACIDS

As outlined in Section 4 and Figure 1, p. 21, the 20 amino acids commonly found in proteins are related to one another in a series of *families*, according to their biosynthetic origin. The families comprise

(a) the aromatic family with three amino acids
(b) the aspartate family with six amino acids
(c) the glutamate family with four amino acids
(d) the serine family with three amino acids
(e) the pyruvate family with three amino acids
(f) histidine, which is not related to any of the other amino acids, but the synthesis of which is closely related to that of the purines.

We shall discuss these in turn. Each family is illustrated by a detailed metabolic scheme with structural formulae. The compounds are numbered and the numbers will be referred to in the text. The enzymes involved are indicated on the schemes by letters which are used when the control of these pathways is being discussed.

Control of biosynthetic reactions is necessary in order to ensure that the cell does not waste energy and carbon synthesizing metabolites which are already in the medium. This topic is discussed further in Chapter 8. Generally the same biosynthetic pathway may be subjected to two kinds of control, represented respectively by alteration of the rate of enzyme synthesis and by modulation of enzyme activity. It will be convenient to consider the two mechanisms together and to recapitulate briefly some of the points made in Section 8.

Reduction in the differential rate of enzyme *synthesis* (see Chapter 8, p. 329) is termed *repression*; reduction in enzyme *activity* is often caused by negative *feedback inhibition*. Usually it is one or more terminal products of a biosynthetic pathway which are active in repression and feedback inhibition (hence the name of the latter, which is also called *retro-inhibition* or *end-product inhibition*). The main impression from recent work on control mechanisms, especially of feedback inhibition, is of the variety of different mechanisms observed in different bacterial species. Consequently, only an outline of some of the better authenticated examples can be given here.

Consider a branched biosynthetic pathway for two metabolites Q and Y in which there is a series of common intermediates:

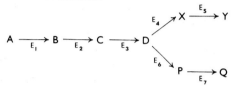

It is obvious that if Y has the property of inhibiting any of the enzymes before the branch point, its presence in the medium will arrest further synthesis of Q as well as its own synthesis. At least six different patterns of feedback inhibition have evolved, which surmount this problem in different ways. These may be briefly summarized as follows:

(i) *Concerted or multivalent feedback inhibition.* In this mechanism Q and Y separately have no effect on enzyme E_1, but together they are potent inhibitors.

(ii) *Co-operative feedback inhibition.* Here Q and Y separately are weakly inhibitory to E_1, but together they exert an effect that is more than the sum of their individual inhibitory effects.

(iii) *Cumulative feedback inhibition.* In this mechanism a given end-product inhibits E_1 by a given percentage irrespective of the presence of other inhibitors. The presence of a second product increases the total inhibition. If all the products are present, the activity of E_1 is completely inhibited.

(iv) *Compensatory antagonism of feedback inhibition.* In this situation in which Q might totally inhibit E_1 in a system in which Y is required to react with an intermediate, Z, from some other pathway, then Z is able to decrease the inhibition produced by Q, so preventing a situation in which Z accumulates because Q has shut off the synthesis of Y.

(v) *Sequential feedback inhibition.* Here Y inhibits enzyme E_4 only and Q inhibits enzyme E_6 only. However this inhibition causes the accumulation of D in either case and D is an inhibitor of E_1.

(vi) *Multiple enzymes with specific regulatory effectors.* In this case there are different but *isofunctional* enzymes catalysing the conversion of A to B and each is inhibited by one of the end-products. Thus if Q inhibits one form of E_1, Y can still be formed by the action of another.

There are also at least four patterns of repression of enzyme synthesis.

(a) In *simple end-product repression* all the enzymes of a pathway are repressed by the presence of the end-product.

(b) In other cases, the end-product represses only the first enzyme, while the others are absent unless induced by the product of the first reaction.

(c) In branched pathways *multivalent repression* of all enzymes by the concerted action of all the end-products may be observed. If any one of the end-products is not present, no repression is observed.

(d) Alternatively there may be *cumulative repression* in which there are several different E_1 enzymes each repressed by the presence of one particular end-product.

Feedback inhibition is a control mechanism that works rapidly and prevents existing enzymes from wasting energy and carbon in making a metabolite already available. It operates within seconds of the addition of a controlling metabolite and on the disappearance of the exogenous metabolite the inhibition is just as quickly relieved.

Repression on the other hand is a much more slowly acting control both in its imposition and in its removal. Thus although addition of a controlling metabolite rapidly prevents further synthesis, the existing enzyme is not usually destroyed, but merely diluted out by growth, the specific activity of the enzyme falling by a factor of 2 per generation time. Similarly, when the repressing metabolite is removed, some time will have to elapse (about 2 generation times) before the specific activity of the enzyme rises to near its fully derepressed level.

The family of aromatic amino acids
(Figures 1a and 1b)

The three aromatic amino acids phenylalanine, tyrosine and tryptophan are derived from a seven-carbon straight chain compound 3-deoxy-7-phospho-D-arabinoheptulosonic acid (1). This arises by condensation of a C_4 compound D-erythrose 4-phosphate (which comes from glucose via the pentose phosphate cycle)

with the C_3 compound phospho-enolpyruvate (a derivative of glycolysis) with the elimination of a single molecule of orthophosphate. The variant pathway involving pretyrosine (8a) occurs in the cyanobacteria (blue-green algae) and may be an alternative route in other organisms such as the Pseudomonads.

Thus, of the three aromatic amino acids, phenylalanine and tyrosine derive their aromatic rings from 3-deoxy-7-phospho-arabinoheptulosonate with the loss of one C atom as CO_2, and their C_3 side chains from phospho-enolpyruvate. Tryptophan derives its aromatic ring from the same source, its heterocyclic N atom from glutamine, its remaining ring C atoms from C-1 and C-2 of ribose, and its C_3 side chain from serine. Other aromatic compounds (e.g. menaquinone) also arise from chorismic acid.

Control

At least six different control patterns have been recognized in the aromatic family of amino acids. Only the mechanisms which have been studied in detail will be mentioned here.

In *Bacillus subtilis*, the end-products respectively inhibit the first enzyme in their terminal branch of the pathway. Thus tyrosine (12) inhibits prephenate dehydrogenase (L), phenylalanine (10) inhibits prephenate dehydratase (J) and tryptophan (17) inhibits anthranilate synthase (N). These inhibitions cause accumulation of the substrates of the enzymes, i.e. prephenate (8) and chorismate (7), and these compounds in turn inhibit the first enzyme of the whole pathway, phospho-2-keto-3-deoxy-heptonate aldolase (A). This is an example of sequential feedback.

In *Escherichia coli* and *Salmonella typhimurium*, in contrast, the control by feedback inhibition is quite different (Figure 2). The first reaction in the pathway is catalysed by three separate *isofunctional* phospho-2-keto-3-deoxyheptonate aldolases (A). One is repressed and inhibited by tyrosine, one by phenylalanine and the third is repressed and inhibited by tryptophan. In addition, there are two separate chorismate mutases (enzyme H). One of these is physically associated with prephenate dehydrogenase (L) and both enzymes of the complex are repressed by tyrosine which also inhibits enzyme L. The other chorismate mutase is associated with prephenate dehydratase (J) and these enzymes are both repressed and inhibited by phenylalanine.

Although less work has been done with the aerobic pseudomonads, it appears that enzyme A is inhibited only by tyrosine. This may however only indicate that there are other isoenzymes which are either very labile or repressed.

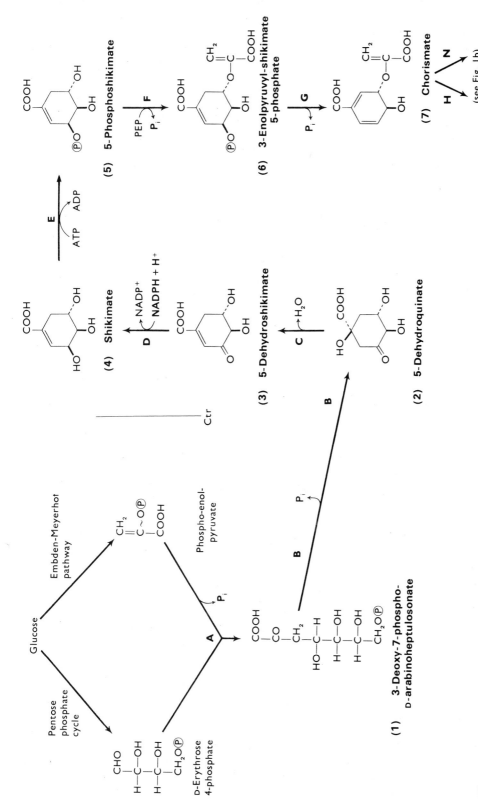

Figure 1a. The family of aromatic amino acids: enzymes **A**–**G**.

A Phospho-2-keto-3-deoxyheptonate aldolase
B 5-Dehydroquinate synthase
C 5-Dehydroquinate dehydratase
D Shikimate dehydrogenase
E Shikimate kinase
F 3-Enolpyruvyl-shikimate 5-phosphate aldolase
G Chorismate synthase

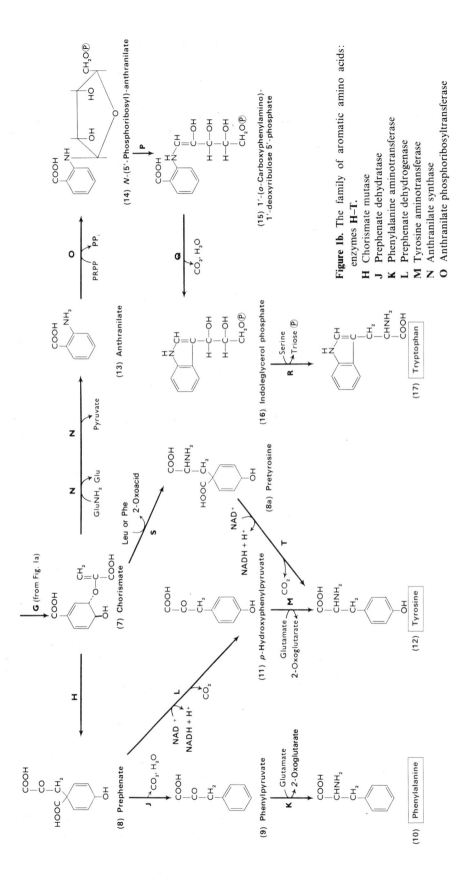

Figure 1b. The family of aromatic amino acids: enzymes **H–T**.

H Chorismate mutase
J Prephenate dehydratase
K Phenylalanine aminotransferase
L Prephenate dehydrogenase
M Tyrosine aminotransferase
N Anthranilate synthase
O Anthranilate phosphoribosyltransferase
P Phosphoribosyl-anthranilate isomerase
Q Indoleglycerol phosphate synthase
R Tryptophan synthase
S Pretyrosine aminotransferase
T Pretyrosine dehydrogenase

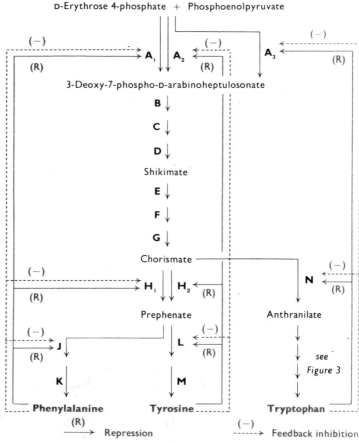

Figure 2. Control of aromatic amino acid biosynthesis in *Escherichia coli*.

→ Repression   ------→ Feedback inhibition

A_1, A_2 and A_3 are isoenzymes as are H_1 and H_2.

The genes specifying the sequence of enzymes N to R in Figure 1b (chorismate to tryptophan) constitute an operon (Figure 3) in most of the bacteria studied. Again variations in control patterns exist in different bacterial species, but *Escherichia coli* and *Salmonella typhimurium* have been the most extensively examined. Reactions P and Q in *E. coli* (but not in *Pseudomonas putida*) are catalysed by the same enzyme, which is a single polypeptide chain specified by the *trpC* gene. Tryptophan synthase (R) is composed of two kinds of polypeptide chain, one specified by the *trpB* gene (the B protein) the other by the *trpA* gene (the A protein). The polypeptide chain of the enzyme anthranilate phosphoribosyltransferase (O) specified by the *trpD* gene not only catalyses this reaction, but is also a component part of enzyme N (anthranilate synthase), the other peptide chain of which is specified by the *trpE* gene. Anthranilate synthase is feedback inhibited by tryptophan, and the site of tryptophan binding is to

the product of the *trpE* gene. All four enzymes of the operon are repressed co-ordinately by tryptophan. The close association of the enzymes formed (as well as of the genes of the operon) suggests that many of the intermediates recorded in Figures 1b and 3 do not occur in the free state. Tryptophan also contributes to cumulative feedback inhibition of glutamine synthetase (enzyme A, Figure 6a) which supplies one of the substrates of enzyme N (Figure 1b).

The aspartate family (Figures 4a and 4b)

Aspartate gives rise to five amino acids which are protein constituents, in addition to diaminopimelic acid, which is found in the cell wall peptidoglycans but not in protein.

Methionine is formed from homoserine via cystathionine (7) as shown in Figure 4b. In *Bacillus* species,

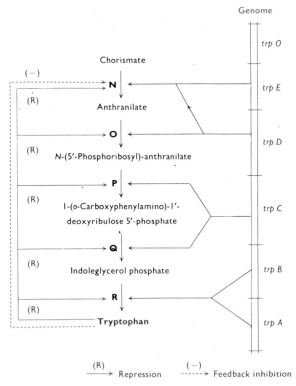

Genome

Chorismate

(−)

(R)

Anthranilate

(R)

N-(5′-Phosphoribosyl)-anthranilate

(R)

I-(o-Carboxyphenylamino)-I′-
deoxyribulose 5′-phosphate

(R)

Indoleglycerol phosphate

(R)

Tryptophan

(R)

trp O

trp E

trp D

trp C

trp B

trp A

N

O

P

Q

R

(R)
⎯⎯→ Repression

(−)
- - - - -→ Feedback inhibition

Figure 3. Control of tryptophan biosynthesis in *Escherichia coli.*

The genes form a cluster at min. 27 on the chromosome and the enzymes are co-ordinately repressed by tryptophan. Enzyme N is also feedback inhibited by tryptophan.

O-acetyl- rather than O-succinyl-homoserine (6) is involved. The final reaction of this sequence, the methylation of homocysteine to give methionine (9), is more complex. Two alternative pathways exist.

The other pathway diverging from homoserine is the formation of threonine (11). This takes place via the phosphorylation of homoserine by ATP in the presence of *homoserine kinase* (J), to give O-phosphohomoserine (10).

Both of the amino acids derived from aspartate which have more than four carbon atoms derive their remaining atoms from C-2 and C-3 of pyruvate. There is evidence for other biosynthetic routes to isoleucine involving citramalic acid, in a number of bacteria.

Control

Control of amino acid biosynthesis in the aspartate family is complicated by the large number of branches in the pathway. Accordingly a number of different

mechanisms have evolved in different species for control of essentially the same biosynthetic sequence.

In *Escherichia coli* K-12 (see Figure 5), the main feature is the occurrence of a multiplicity of aspartokinases (enzyme B, Figure 4a). There are three such enzymes, two of which (aspartokinases II and III) are under simple repressive control by their respective end-products methionine and lysine. Aspartokinase I, which is inhibited by threonine, is under multivalent repression by threonine plus isoleucine. Aspartokinase II is physically associated with homoserine dehydrogenase (enzyme D) activity, both being repressed by methionine. Aspartokinase I is similarly associated with homoserine dehydrogenase I, and both activities are inhibited by threonine. Aspartokinase III is repressed and inhibited by lysine; it is not associated with enzyme D activity, since homoserine dehydrogenase is not involved in lysine formation. Mutant studies have shown that the 4-phosphoaspartate (3) produced by any of the aspartokinases can be channelled to the synthesis of the other end-products, so that there is only a single pool of compound (3) in the cell.

Lysine represses the formation of enzymes R to W and methionine the formation of enzymes E to H. Additionally, homoserine O-succinyltransferase (E) is feedback inhibited by methionine and also by S-adenosylmethionine, and dihydrodipicolinate synthase (Q) is inhibited by lysine. In *Salmonella typhimurium* enzymes J and K, like the aspartokinase I-homoserine dehydrogenase I complex, are regulated by a multivalent repression mechanism involving both threonine and isoleucine. In contrast to this control pattern, *Bacillus polymyxa* and *Rhodopseudomonas capsulatus* each possesses a single aspartokinase which exhibits co-operative feedback inhibition. In *Rhodopseudomonas sphaeroides* and *Bacillus licheniformis* aspartokinase is inhibited by aspartate 4-semialdehyde (4) which may possibly indicate sequential feedback inhibition.

In *Lactobacillus arabinosus*, asparagine (2) both inhibits and represses enzyme A.

Threonine dehydratase (L), which catalyses the first step in isoleucine biosynthesis, is feedback inhibited by isoleucine. This inhibition can be relieved by valine, an example of compensatory antagonism. The control of this enzyme and later enzymes in the pathway (M to P) will be discussed later, when valine biosynthesis is considered (p. 173). Recent work has shown that the genes *thrA*, *thrB* and *thrC* constitute a small operon in the *E. coli* genome under the repressive control of threonine plus isoleucine. *ThrA* codes for aspartokinase I (enzyme B), and the homoserine

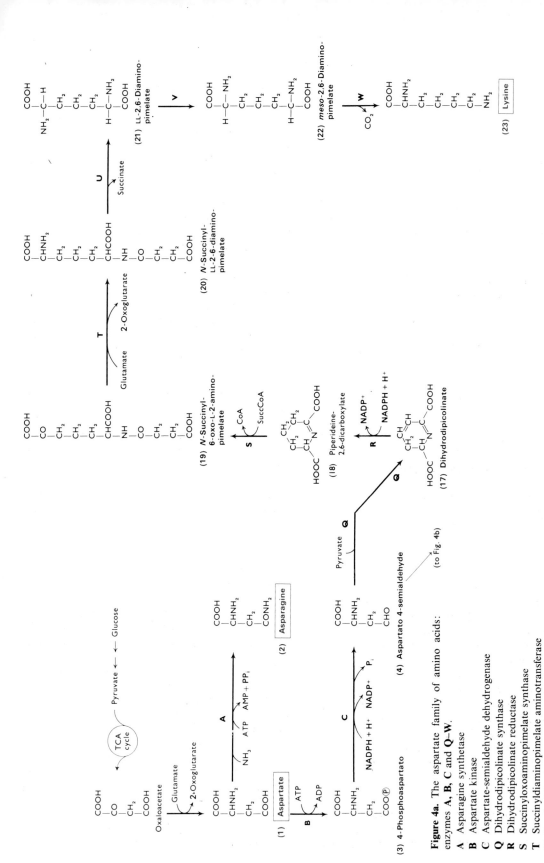

Figure 4a. The aspartate family of amino acids: enzymes **A, B, C** and **Q–W**.

A Asparagine synthetase
B Aspartate kinase
C Aspartate-semialdehyde dehydrogenase
Q Dihydrodipicolinate synthase
R Dihydrodipicolinate reductase
S Succinyloxoaminopimelate synthase
T Succinyldiaminopimelate aminotransferase
U Succinyldiaminopimelate desuccinylase
V Diaminopimelate epimerase
W meso-Diaminopimelate decarboxylase

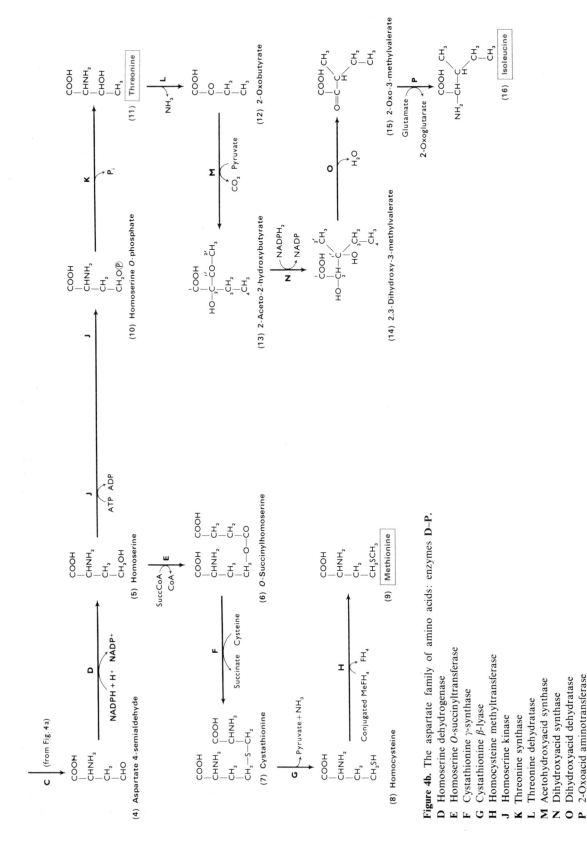

Figure 4b. The aspartate family of amino acids: enzymes **D–P.**

D Homoserine dehydrogenase
E Homoserine O-succinyltransferase
F Cystathionine γ-synthase
G Cystathionine β-lyase
H Homocysteine methyltransferase
J Homoserine kinase
K Threonine synthase
L Threonine dehydratase
M Acetohydroxyacid synthase
N Dihydroxyacid synthase
O Dihydroxyacid dehydratase
P 2-Oxoacid aminotransferase

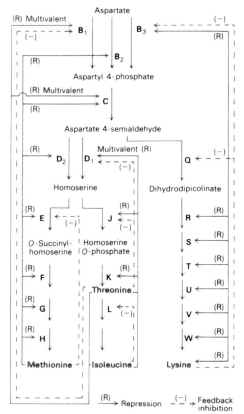

B₁, B₂ and B₃ are isoenzymes, aspartokinases I, II and III (see text); D₁ and D₂ are isoenzymes, homoserine dehydrogenases I and II.

Figure 5. Control of aspartate-derived amino acid biosynthesis in *Escherichia coli*.

dehydrogenase I (enzyme D_1) which are catalysed by a single polypeptide chain in *E. coli. ThrB, thrC* code for enzymes J and K in Figures 4 and 5.

The glutamate family (Figures 6a and 6b)

The amino acids glutamine, proline and arginine derive their carbon skeletons from glutamic acid. The additional carbon atom of arginine is derived from CO_2 and its two N atoms from ammonia and aspartic acid, respectively.

Glutamate (1) itself is formed from 2-oxoglutarate, which is an intermediate of the tricarboxylic acid cycle, either by transamination (2-oxoglutarate can act as amino acceptor for aminotransferases specific for almost all the L-amino acids) or else by direct reductive fixation of ammonia under the action of the NADH- or NADPH-dependent enzyme *glutamate dehydrogenase*. In most bacteria this enzyme appears to

be the principal, if not the only, primary reaction for the incorporation of free ammonia into amino acids.

Glutamine (2) is formed by reaction of glutamate with ammonia. The formation of the amide bond requires a molecule of ATP, but unlike the asparagine synthase reaction, the *glutamine synthetase* (A) reaction yields ADP and orthophosphate. Since this enzyme has a much higher affinity for ammonia than has glutamate dehydrogenase, under conditions of ammonia deficiency it can replace the latter enzyme as the primary port of entry of ammonia into amino acids. In these conditions, the enzyme *glutamate synthase* catalyses the reaction of glutamine with 2-oxoglutarate in the presence of NADPH to give two molecules of glutamate.

Control

Glutamine (2) is a branch-point compound with regard to its amide nitrogen atom which can be transferred to a variety of compounds in biosynthesis, e.g. histidine, tryptophan, glucosamine, carbamoyl phosphate, AMP, etc. In *Escherichia coli*, these metabolites exert a repressive effect on glutamine synthetase (A) and so does ammonia, since at high concentrations ammonia can replace glutamine in most of the reactions in which the latter acts as a nitrogen donor. Besides being subject to repression, the activity of pre-existing glutamine synthetase is also under the control of cumulative feedback inhibition. This controlling effect of feedback metabolites is modulated in *E. coli* by the occurrence of enzyme A in two forms (called respectively synthetases I and II), which show different susceptibilities to end-product inhibition. Glutamine synthetase II is more sensitive to feedback inhibition by CTP, AMP, histidine and tryptophan. Form I, which is less sensitive, is converted into form II by an adenylylation reaction requiring ATP and a specific activating enzyme. There is also a de-adenylylating enzyme which is inhibited by glutamine, and which reconverts form II to form I.

Proline (4) inhibits the formation of glutamate 5-semialdehyde (3) and also can repress enzymes B to E.

The control of the biosynthesis of arginine has been studied in much detail in *E. coli*. The genes controlling arginine biosynthesis do not constitute an operon, four are clustered and four occur at separate points on the genome. Nevertheless, these genes are all under the control of a single regulatory gene *argR*. In *E. coli* K-12 arginine represses all the enzymes of the biosynthetic pathway (F to N). In the B strain of *E. coli*, however, the apo-repressor protein produced by the *argR* gene does not combine so readily with arginine, so that the addition of exogenous arginine

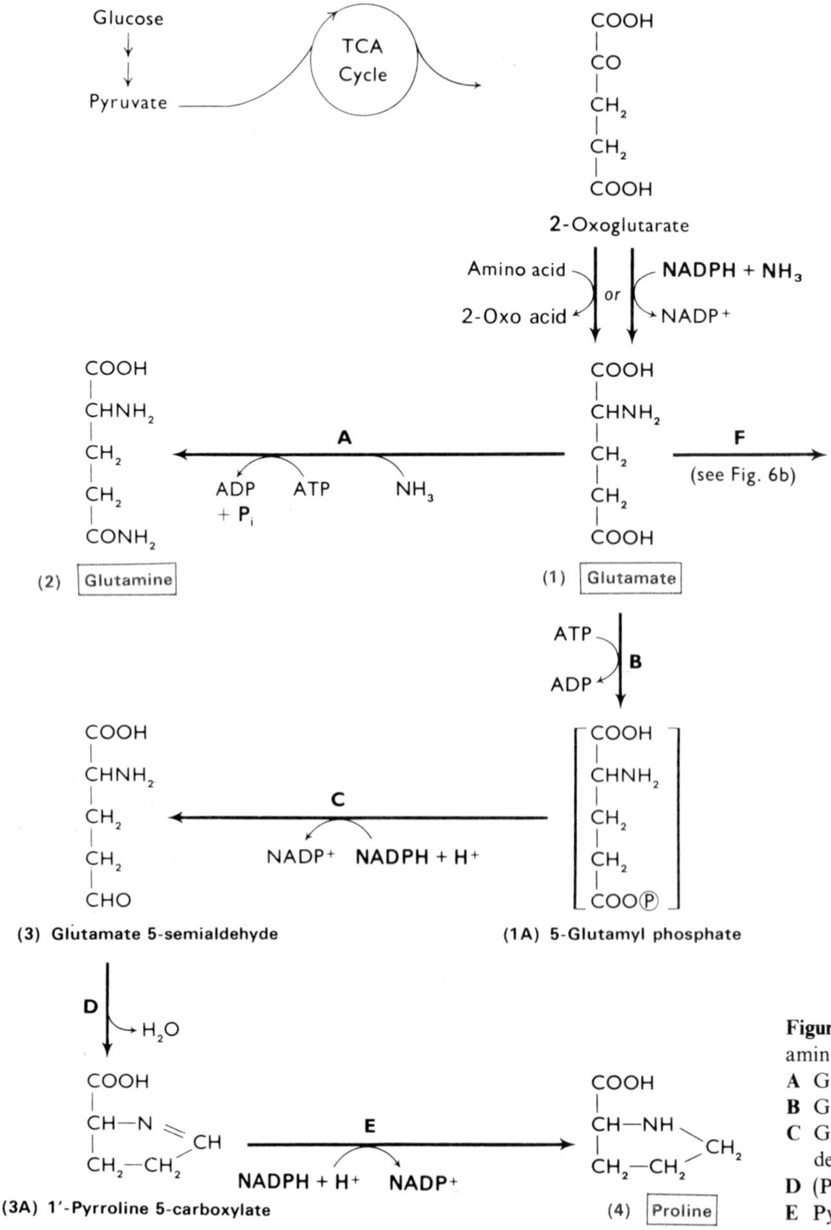

Glucose
↓
Pyruvate ──────→ TCA Cycle ──────→

COOH
|
CO
|
CH_2
|
CH_2
|
COOH

2-Oxoglutarate

Amino acid ⟍   ⟋ **NADPH + NH₃**
        or
2-Oxo acid ⟋   ⟍ NADP⁺

COOH   COOH
| |
$CHNH_2$ **A** $CHNH_2$ **F**
| ←───────────────────── | ─────────────────→
CH_2 ADP   ATP   NH₃ CH_2 (see Fig. 6b)
| + P_i |
CH_2 CH_2
| |
$CONH_2$ COOH

(2) Glutamine   (1) Glutamate

ATP ⟍
        B
ADP ⟋

COOH   ⌈ COOH ⌉
| | |
$CHNH_2$ **C** | $CHNH_2$ |
| ←───────────────────── | |
CH_2 NADP⁺   NADPH + H⁺ | CH_2 |
| | |
CH_2 | CH_2 |
| | |
CHO   ⌊ COO(P) ⌋

(3) Glutamate 5-semialdehyde   (1A) 5-Glutamyl phosphate

D ⟍ H_2O

COOH   COOH
| |
CH—N   CH—NH
| ⟍CH   **E** | ⟍CH_2
CH_2—CH_2 ─────────────────────→ CH_2—CH_2⟋
            NADPH + H⁺   NADP⁺

(3A) 1'-Pyrroline 5-carboxylate   (4) Proline

Figure 6a. The glutamate family of amino acids: enzymes **A–E**.
A Glutamine synthetase
B Glutamate 5-kinase
C Glutamate semialdehyde dehydrogenase
D (Possibly non-enzymic)
E Pyrroline 5-carboxylate reductase

has a paradoxical derepression effect. Arginine also inhibits the first biosynthetic enzyme, glutamate acetyltransferase (F) by feedback inhibition. Ornithine (9) is required to make *putrescine*, a precursor of polyamines; in the presence of excess arginine, which prevents ornithine biosynthesis, a new pathway for putrescine biosynthesis is induced involving the decarboxylation of arginine. In *Pseudomonas aeruginosa*,

arginine *induces* enzyme J, rather than repressing it. This is because the enzyme is involved (as an aminotransferase for ornithine) in arginine degradation as well as in biosynthesis.

The serine family (Figure 7)

There is some variation in the pathways by which serine is formed from 3-phosphoglycerate which is an

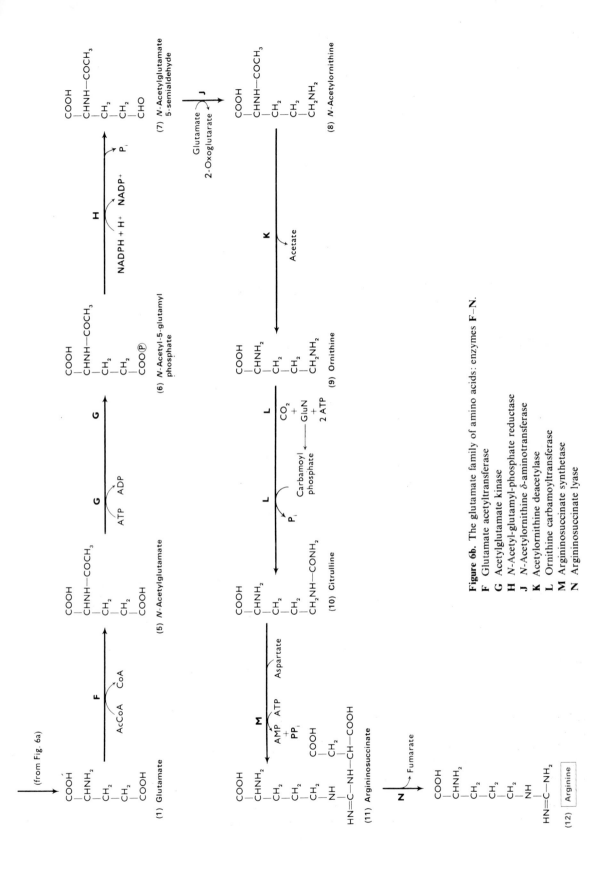

Figure 6b. The glutamate family of amino acids: enzymes **F**–**N**.

F Glutamate acetyltransferase
G Acetylglutamate kinase
H N-Acetyl-glutamyl-phosphate reductase
J N-Acetylornithine δ-aminotransferase
K Acetylornithine deacetylase
L Ornithine carbamoyltransferase
M Argininosuccinate synthetase
N Argininosuccinate lyase

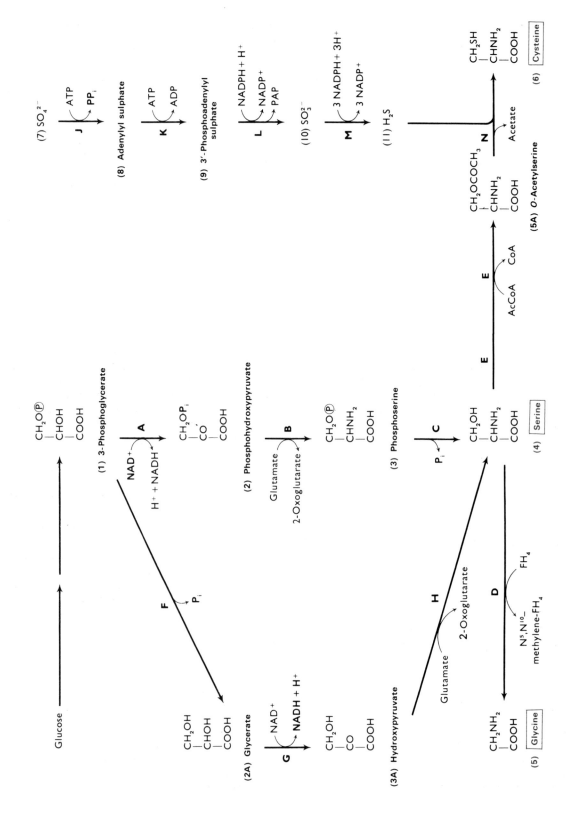

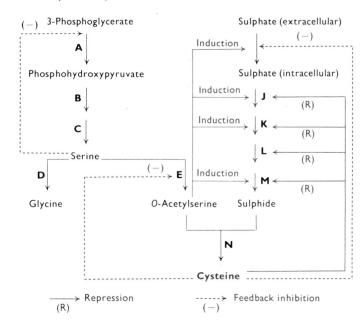

Figure 8. Control of cysteine biosynthesis in *Escherichia coli*.

———→ Repression (R)	------→ Feedback inhibition (−)

intermediate in glucose degradation. In some organisms, a non-phosphorylated pathway (involving glycerate and hydroxypyruvate and the enzymes *phosphatase* (F), *glycerate dehydrogenase* (G) and *serine aminotransferase* (H)) is thought to predominate, while in others, e.g. *Escherichia coli* and *Salmonella typhimurium*, a phosphorylated pathway shown as the main sequence in Figure 7 predominates.

Serine is the precursor of glycine and the sulphur-containing amino acid cysteine. Cysteine (6) is formed in bacteria via *O*-acetylserine (5A), which is then converted to cysteine by reaction with hydrogen sulphide (11) formed by reduction of sulphate by a series of ATP- and NADPH-dependent steps, as shown in Figure 7.

Control

The serine pathway can be divided into two routes, that supplying the C_3 carbon skeleton for cysteine formation (enzymes A to C and E) and that of sulphate reduction (enzymes J to N) (see Figure 8). In *Escherichia coli*, in the first route, the initial enzyme, phosphoglycerate dehydrogenase (A) is feedback inhibited by serine. Enzyme multiplicity in this pathway is absent. No repression of enzymes A to C by serine has been observed, but since serine is a precursor of a large number of compounds made from C_1 units (purines, thymine, methionine and histidine) as well as glycine and cysteine, repression by serine would not be likely to occur. Cysteine represses enzymes J to M in the sulphate-reduction pathway. Additionally, it feedback-inhibits serine *O*-acetyltransferase (E) and the permease catalysing uptake of sulphate by the cell. *O*-Acetylserine induces the sulphate permease and enzymes J, K and M. In the presence of cysteine both sulphate reduction and the formation of *O*-acetylserine are shut off. When the cysteine concentration falls the feedback inhibition of enzyme E is relieved and *O*-acetylserine is formed. This induces the enzymes of sulphate reduction which are also derepressed since cysteine is no longer present.

Figure 7. The serine family of amino acids: enzymes **A–N**.

A Phosphoglycerate dehydrogenase
B Phosphoserine aminotransferase
C Phosphoserine phosphatase
D Serine hydroxymethyltransferase
E Serine *O*-acetyltransferase
F Phosphoglycerate phosphatase

G Glycerate dehydrogenase
H Serine aminotransferase
J Sulphate adenylyltransferase
K Adenylyl sulphate kinase
L Phosphoadenylyl sulphate reductase
M Sulphite reductase
N Cysteine synthase

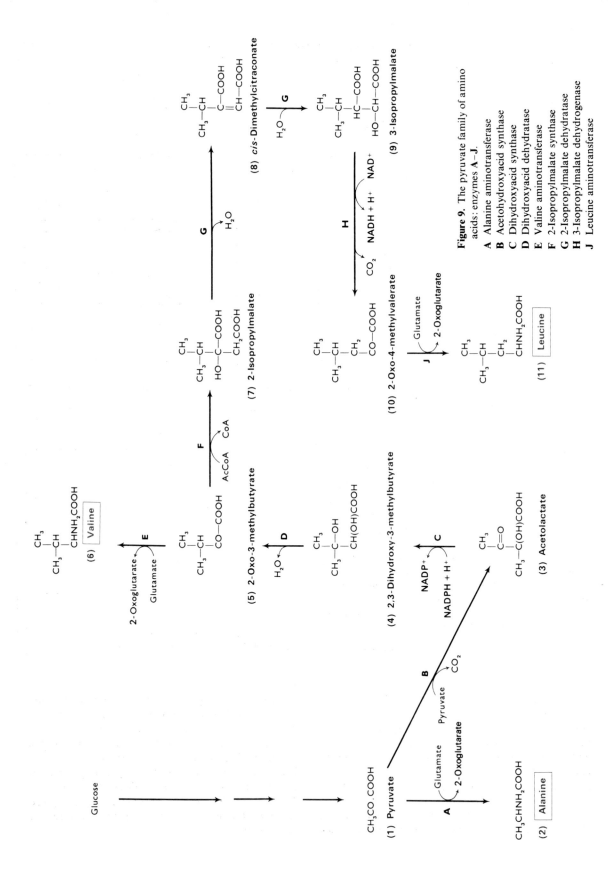

Figure 9. The pyruvate family of amino acids: enzymes A–J.

A Alanine aminotransferase
B Acetohydroxyacid synthase
C Dihydroxyacid synthase
D Dihydroxyacid dehydratase
E Valine aminotransferase
F 2-Isopropylmalate synthase
G 2-Isopropylmalate dehydratase
H 3-Isopropylmalate dehydrogenase
J Leucine aminotransferase

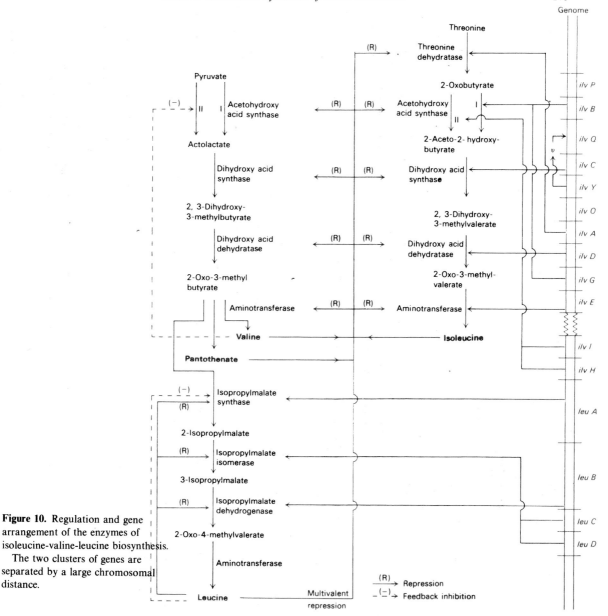

Figure 10. Regulation and gene arrangement of the enzymes of isoleucine-valine-leucine biosynthesis.

The two clusters of genes are separated by a large chromosomal distance.

The pyruvate family (Figure 9)

Alanine (2) is formed from pyruvate (1) by transamination with glutamate (and probably other amino acids) catalysed by *alanine aminotransferase* (A). In some bacteria alanine can be formed directly by the NADH-dependent amination of pyruvate catalysed by *alanine dehydrogenase*.

Pyruvate is also the precursor of valine and leucine. The first three steps in the formation of these two

amino acids from pyruvate are catalysed by enzymes B, C and D (Figure 9). These enzymes, *acetohydroxyacid synthase, dihydroxyacid synthase* and *dihydroxyacid dehydratase*, are the same as those involved in the corresponding three steps in the biosynthesis of isoleucine from 2-oxobutyrate and pyruvate (enzymes M, N and O in Figure 4b). Similarly the *aminotransferase* giving rise to valine (enzyme E, Figure 9) also catalyses the formation of isoleucine (Figure 4b, enzyme P).

Control

In *Escherichia coli* and *Salmonella typhimurium* the enzymes catalysing the conversion of compound 7 to compound 10 are under the control of four closely-linked genes, forming an operon (see Figure 10). The enzymes are co-ordinately repressed by leucine. Leucine also inhibits by feedback the first enzyme on its branch pathway (F).

The enzymes of the common isoleucine-valine biosynthetic pathway, acetohydroxy acid synthase (B in Figure 9, M in Figure 4b), dihydroxy acid synthase (C in Figure 9, N in Figure 4b), dihydroxy acid dehydratase (D in Figure 9, O in Figure 4b) and the aminotransferase (E in Figure 9, P in Figure 4b) are under the control of four closely-linked genes in the Enterobacteriaceae, along with the gene (*ilvA*) for threonine dehydratase (L in Figure 4b). Operator genes are also present, *ilvP* which controls structural gene *ilvB*, *ilvO* which controls genes *ilvA*, *ilvD* and *ilvE*, and possibly *ilvQ*, controlling the *ilvC* gene. *IlvQ* is thought to be the site of action of a protein called *upsilon*(*v*), coded for by the *ilvY* gene. In the presence of acetohydroxy acid, upsilon binds to *ilvQ* and allows the synthesis of dihydroxy acid synthase coded for by *ilvC*. The order of these and the reactions catalysed by the corresponding enzymes are shown in Figure 10.

In *Escherichia coli* K-12, the *ilvADE* cluster of genes (controlled by the *ilvO* operator) is multivalently repressed when isoleucine, valine, leucine and pantothenic acid, all of which arise by the common pathway (Figure 10), are simultaneously present in the medium. If any one of these compounds is not present in the medium, the enzymes specified by these three genes are derepressed. Enzyme multiplicity is also observed in *Escherichia* and *Salmonella*. Acetohydroxy acid synthase (enzyme B) occurs as two separately controlled isoenzymes. One of these (I), containing two types of polypeptide chain (specified respectively by the *ilvB* and *ilvG* genes) is not inhibited by valine, while the other (II) containing two types of polypeptide chain (specified by the *ilvH* and *ilvI* genes) is feedback-inhibited by valine. *IlvH* and *ilvI* are located near the *leu* genes on the chromosome, at some distance from the other *ilv* genes (Figure 10).

Histidine and its relations with purine biosynthesis
(Figure 11)

Extensive isotopic tracer studies have revealed that histidine derives its five-carbon backbone from ribose, its C-2 and N-3 from the same source as C-2 and N-1 of the purine ring, and its N-1 from the amide of glutamine.

Histidine and the purines arise from a common precursor, 5-phospho-α-D-ribosyl-diphosphate (2) which is formed from ribose 5-phosphate (1), an intermediate of the pentose phosphate cycle (Chapter 3).

Compound (7) has not yet been characterized. It is formed by transfer of the amide group of glutamine to (6) catalysed by an *amidotransferase* (F). This intermediate is then cleaved by a *cyclase* (G) to give two compounds: D-*erythro*-imidazoleglycerol phosphate (8), which is the precursor of histidine, and 5-amino-1-(5′-(phosphoribosyl))-imidazole-4-carboxamide (20). The latter compound, which is a precursor of the purine nucleus, can be formed from phosphoribosyl pyrophosphate by an independent pathway. Thus, if the purine precursor is reconverted to ATP, we have a cyclic system for histidine formation:

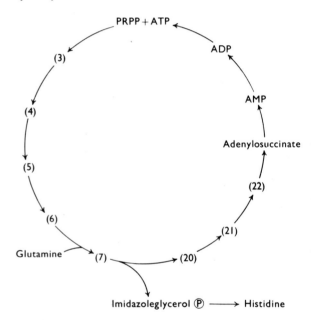

Control

The control of histidine biosynthesis has been studied in great detail in *Salmonella typhimurium*.

The first enzyme in the sequence, ATP phosphoribosyltransferase (B) is feedback-inhibited by histidine. 2-Thiazolealanine, an antagonist of histidine, inhibits growth by mimicking this feedback inhibition while being unable to replace histidine in proteins. Mutants which are resistant to 2-thiazolealanine have an altered enzyme B which is not sensitive to feedback inhibition by either histidine or 2-thiazolealanine. Enzyme B is also affected by the energy charge level of the cell.

The genes controlling the structure of the enzymes

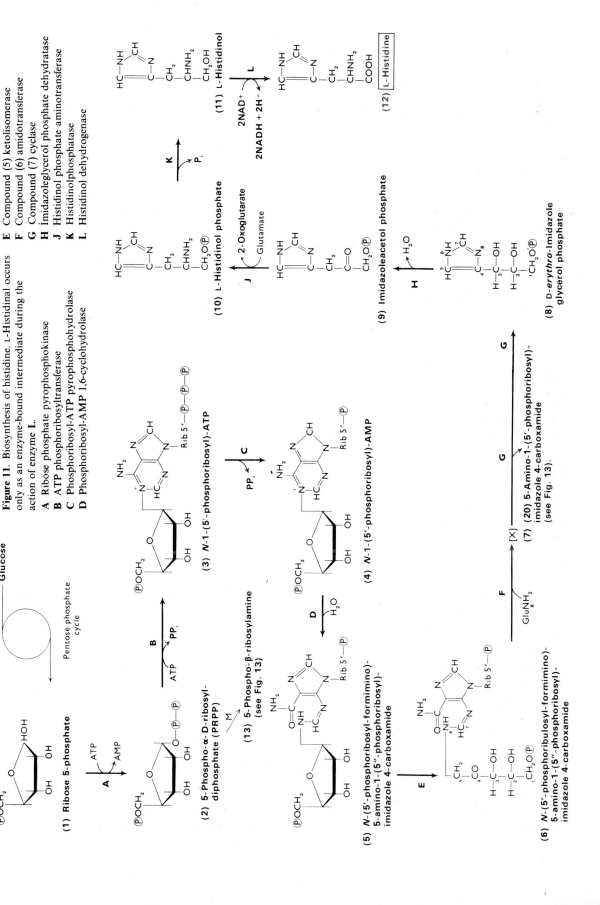

Figure 11. Biosynthesis of histidine. L-Histidinal occurs only as an enzyme-bound intermediate during the action of enzyme **L**.

A Ribose phosphate pyrophosphokinase
B ATP phosphoribosyltransferase
C Phosphoribosyl-ATP pyrophosphohydrolase
D Phosphoribosyl-AMP 1,6-cyclohydrolase
E Compound (5) ketolisomerase
F Compound (6) amidotransferase
G Compound (7) cyclase
H imidazoleglycerol phosphate dehydratase
J Histidinol phosphate aminotransferase
K Histidinolphosphatase
L Histidinol dehydrogenase

Glucose
Pentose phosphate cycle

(1) Ribose 5-phosphate
(2) 5-Phospho-α-D-ribosyl-diphosphate (PRPP)
(3) N-1-(5′-phosphoribosyl)-ATP
(4) N-1-(5′-phosphoribosyl)-AMP
(5) N-(5′-phosphoribosyl-formimino)-5-amino-1-(5″-phosphoribosyl)-imidazole 4-carboxamide
(6) N-(5′-phosphoribulosyl-formimino)-5-amino-1-(5″-phosphoribosyl)-imidazole 4-carboxamide
(7) [X]
(8) D-erythro-Imidazole glycerol phosphate
(9) Imidazoleacetol phosphate
(10) L-Histidinol phosphate
(11) L-Histidinol
(12) L-Histidine
(13) 5-Phospho-β-ribosylamine (see Fig. 13)
(20) 5-Amino-1-(5′-phosphoribosyl)-imidazole 4-carboxamide (see Fig. 13).

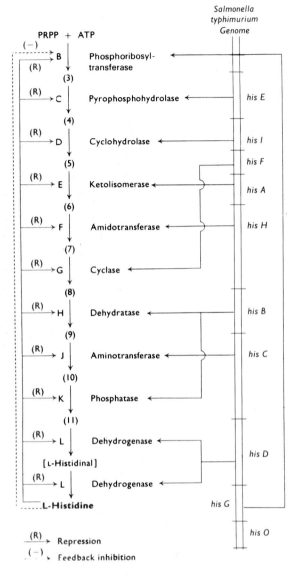

Figure 12. The histidine operon.

The operon has an operator gene *hisO* and nine structural genes (*A* to *I*) which are co-ordinately controlled and which determine the structures of the 10 enzymes (the dehydratase H and phosphatase K are different functions of the same protein). The order of the genes does not follow the biochemical reaction sequence. The numbers in brackets and enzyme letters refer to Figure 11. The whole operon maps at minute 65 on the *Salmonella* chromosome. Enzyme B is feedback inhibited by histidine.

of histidine biosynthesis map in a closely-linked operon (Figure 12). The letters used to denote genes should not be confused with those used to denote the enzymes. The enzymes show co-ordinate repression,

and this system was the first in which this phenomenon was observed. Mutations in the histidine operon may show *polarity*, i.e. a mutation in any gene on the map may cause not only a low level of activity of the enzyme coded for by that gene, but also a low level of all the enzymes coded for by genes lying beyond it (i.e. away from the O gene) in the genetic map. Those enzymes that are coded for by genes lying between the operator and the mutation site show normal levels of activity. Co-ordinate repression still occurs, but the ratio of activities of any two of the enzymes may differ from that in the wild type. The occurrence of polarity mutants is strong evidence for the idea of a *polycistronic messenger RNA* (Chapter 6). A number of other regulatory genes are important in histidine biosynthesis.

In *Bacillus subtilis*, histidinol phosphate aminotransferase (enzyme J) also catalyses the transamination step of phenylalanine and tyrosine biosynthesis (Figure 1b, reactions K and M), and its gene is one of a cluster concerned in aromatic biosynthesis.

BIOSYNTHESIS OF PURINE AND PYRIMIDINE NUCLEOTIDES

Purine nucleotides

This topic has already been mentioned in connection with histidine biosynthesis. The ultimate precursors of the individual atoms of the purine nucleotide ring are given in Figure 3 (p. 23) and see also Figure 13.

Control

Study of the control of purine biosynthesis is complicated because (a) histidine biosynthesis is closely inter-related; (b) nucleotides can be formed from pre-existent bases or nucleotides present in the medium (the 'salvage pathway') and (c) inter-conversion of nucleotides can occur when they are added to crude extracts. Considerations of space prevent discussion of the 'salvage pathway' here.

The first enzyme common to both purine and histidine biosynthesis, enzyme A in Figures 11 and 13, is inhibited in the Enterobacteriaceae by ADP, CTP, GTP and UTP and also by tryptophan. The inhibition caused by the latter compound is cumulative with that due to the nucleotides. It is possible that this enzyme may be controlled by the ATP/ADP ratio in the cell. Histidine does not affect enzyme A. Enzyme M (PRPP amidotransferase) is feedback inhibited by AMP (24) and IMP (22) in *Escherichia coli* and by GMP (26) and AMP (24) in *Klebsiella aerogenes*. There is some evidence to suggest that the latter two

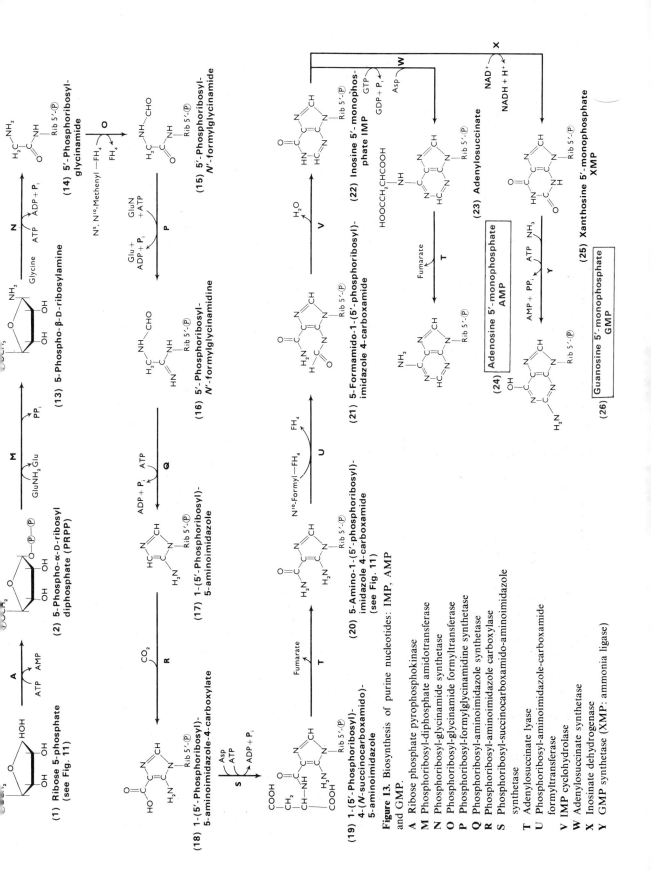

Figure 13. Biosynthesis of purine nucleotides: IMP, AMP and GMP.

A Ribose phosphate pyrophosphokinase
M Phosphoribosyl-diphosphate amidotransferase
N Phosphoribosyl-glycinamide synthetase
O Phosphoribosyl-glycinamide formyltransferase
P Phosphoribosyl-formylglycinamidine synthetase
Q Phosphoribosyl-aminoimidazole synthetase
R Phosphoribosyl-aminoimidazole carboxylase
S Phosphoribosyl-succinocarboxamido-aminoimidazole synthetase
T Adenylosuccinate lyase
U Phosphoribosyl-aminoimidazole-carboxamide formyltransferase
V IMP cyclohydrolase
W Adenylosuccinate synthetase
X Inosinate dehydrogenase
Y GMP synthetase (XMP: ammonia ligase)

(1) Ribose 5-phosphate (see Fig. 11)

(2) 5-Phospho-α-D-ribosyl diphosphate (PRPP)

(13) 5-Phospho-β-D-ribosylamine

(14) 5′-Phosphoribosyl-glycinamide

(15) 5′-Phosphoribosyl-N′-formylglycinamide

(16) 5′-Phosphoribosyl-N′-formylglycinamidine

(17) 1-(5′-Phosphoribosyl)-5-aminoimidazole

(18) 1-(5′-Phosphoribosyl)-5-aminoimidazole-4-carboxylate

(19) 1-(5′-Phosphoribosyl)-4-(N-succinocarboxamido)-5-aminoimidazole

(20) 5-Amino-1-(5′-phosphoribosyl)-imidazole 4-carboxamide (see Fig. 11)

(21) 5-Formamido-1-(5′-phosphoribosyl)-imidazole 4-carboxamide

(22) Inosine 5′-monophosphate IMP

(23) Adenylosuccinate

(24) Adenosine 5′-monophosphate AMP

(25) Xanthosine 5′-monophosphate XMP

(26) Guanosine 5′-monophosphate GMP

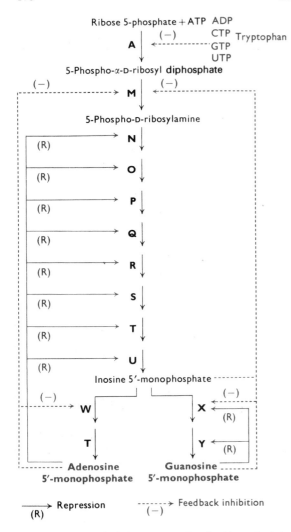

Ribose 5-phosphate + ATP ADP
A | (−) ←--------- CTP Tryptophan
                   GTP
                   UTP
5-Phospho-α-D-ribosyl diphosphate

(−) --------→ M ← (−)

5-Phospho-D-ribosylamine

(R) ----→ N
(R) ----→ O
(R) ----→ P
(R) ----→ Q
(R) ----→ R
(R) ----→ S
(R) ----→ T
(R) ----→ U

Inosine 5'-monophosphate

(−) --→ W   X ← (−)
                  ← (R)
T   Y ← (R)

Adenosine   Guanosine
5'-monophosphate   5'-monophosphate

——————→ Repression   -------→ Feedback inhibition
(R)   (−)

Figure 14. Control of the enzymes of purine biosynthesis in the Enterobacteriaceae.

inhibitors bind at different sites. Other enzymes between compound 13 and IMP (22) do not show feedback inhibition. Figure 14 shows these relationships.

Inosinate dehydrogenase (X) is feedback-inhibited by GMP but not by adenine nucleotides, while adenylosuccinate synthase (W) is inhibited by ADP.

Repression also occurs in this pathway. Most of the enzymes of the biosynthetic sequences shown in Figure 13 are either totally or partially repressed when purine bases, nucleosides or nucleotides are present in the medium. The genes controlling purine biosynthesis do not show an operon-type organization but are scattered over the chromosome and re-

pression is not co-ordinate. In the guanine branch of the pathway, however (Figure 13), the enzymes inosinate dehydrogenase (X) and GMP synthase (Y), specified by the genes *guaB* and *guaA* respectively, are co-ordinately repressed and the genes form a small operon.

Inhibition and repression by purine nucleotide derivatives extends to some of the enzymes of one-carbon metabolism such as methylene-tetrahydrofolate dehydrogenase which produces the formyl units required by enzymes O and U in Figure 13.

Pyrimidine nucleotides

The precursors of the pyrimidine nucleus have been shown in Section 4 (Figure 3, p. 23). Pyrimidines are synthesized (Figure 16) from aspartic acid and carbamoyl phosphate (1). This compound is also a precursor of arginine (see p. 169). Aspartate arises from the tricarboxylic acid cycle by transamination of oxaloacetate, while carbamoyl phosphate is formed by the action of a glutamine-dependent *carbamoyl phosphate synthase* (A).

Control

Some of the earliest observations on repression and feedback inhibition were those made on the biosynthesis of pyrimidine nucleotides in *Escherichia coli*, and aspartate carbamoyltransferase (B) is the enzyme on which much of the early work on allosteric proteins was done (Chapter 8). It is potently inhibited by CTP (Figure 15; see also Chapter 8, p. 326) and it is activated by ATP. Other enzymes subject to feedback inhibition are carbamoyl phosphate synthetase (A) and CTP synthetase (J). The latter is inhibited by CTP, while GTP and UTP activate it. Carbamoyl phosphate synthetase (A) is feedback-inhibited by UMP and this inhibition is antagonized by ornithine, an intermediate of arginine biosynthesis which also requires carbamoyl phosphate. In *Pseudomonas aeruginosa* the feedback inhibitor of enzyme B is UTP. In *Bacillus subtilis* and *Streptococcus faecalis* it is not subject to feedback control.

The enzymes C and F in *E. coli* are co-ordinately repressed by addition of uracil although only two genes *pyrC* and *pyrD* are at all close together on the chromosome. The pyrimidine nucleoside phosphates are probably the active repressing agents, as shown in Figure 15. Carbamoyl phosphate synthetase (A), which is also concerned with arginine biosynthesis (Figure 6b), is cumulatively repressed by arginine and a cytosine derivative, while enzyme B is non-co-ordinately repressed by both uridine and cytidine nucleotides. In

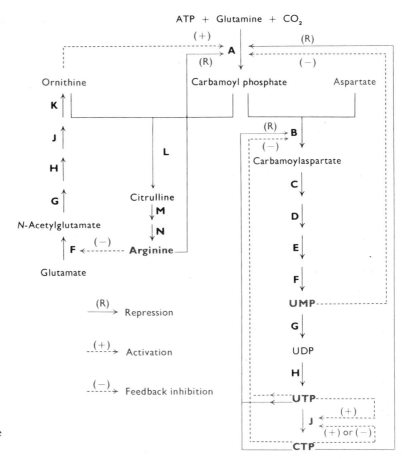

Figure 15. Control of pyrimidine biosynthesis in *Escherichia coli* and *Salmonella typhimurium* and its relation to arginine biosynthesis. Letters on the left-hand side of the diagram refer to enzymes in Figure 6b, while those on the right-hand side refer to enzymes in Figure 16.

contrast, in *P. aeruginosa*, not only is there no linkage between any of the genes, but the enzymes of pyrimidine biosynthesis are non-repressible ('constitutive').

Deoxyribonucleotides

The deoxyribonucleotides are formed from the corresponding ribonucleotides without cleavage of the bond between base and sugar but in coliform bacteria a major fraction of dUTP is derived by deamination from dCTP. The probable intermediates are indicated in Figure 17. Ribonucleotide reductase of *Escherichia coli* is under complex feedback inhibition and activation by a variety of deoxyribonucleotides so as to provide a balanced supply of precursors for DNA synthesis. The enzyme is also repressed by deoxyribonucleotides.

BIOSYNTHESIS OF LIPIDS

The lipid content of bacteria may vary from about 2–3% of the dry weight, most of which is associated with the lipoprotein of the cytoplasmic membrane, to more than 50% in those organisms which store lipid as a reserve material. The types of lipid differ according to the organism, and the main kinds, triglycerides and phospholipids, have been mentioned in Chapter 3.

We will first consider the synthesis from glucose of glycerol and long chain fatty acids. It has been found that *sn*-glycerol 3-phosphate and not free glycerol is involved in lipid synthesis. This compound is produced by reduction of dihydroxyacetone phosphate, itself a normal intermediate of glycolysis, by the action of *glycerol-3-phosphate dehydrogenase*, an NAD-dependent enzyme:

$$
\begin{array}{ccc}
CH_2OH & & CH_2OH \\
| & & | \\
C{=}O & + NADH + H^+ \rightleftharpoons & CHOH + NAD^+ \\
| & & | \\
CH_2O\textcircled{P} & & CH_2O\textcircled{P}
\end{array}
$$

The synthesis of long chain fatty acids is complex and is achieved by a series of reactions which have

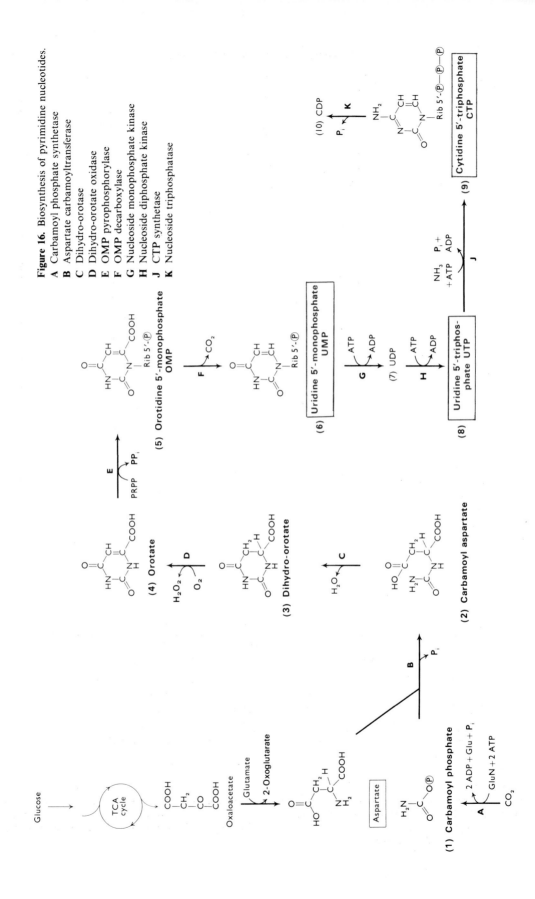

Figure 16. Biosynthesis of pyrimidine nucleotides.

A Carbamoyl phosphate synthetase
B Aspartate carbamoyltransferase
C Dihydro-orotase
D Dihydro-orotate oxidase
E OMP pyrophosphorylase
F OMP decarboxylase
G Nucleoside monophosphate kinase
H Nucleoside diphosphate kinase
J CTP synthetase
K Nucleoside triphosphatase

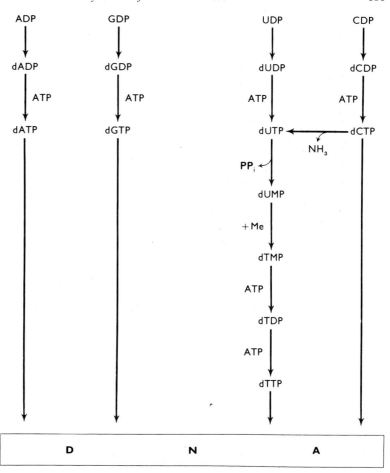

Figure 17. Formation of deoxyribonucleotides. Purine and pyrimidine ribonucleoside derivatives are reduced to the deoxy-compound without separation of the sugar from the base.

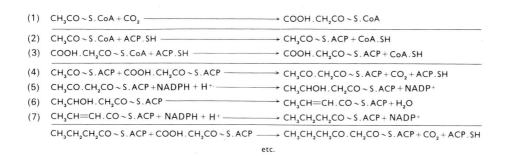

Figure 18. Biosynthesis of fatty acids.

Simplified representation of the reactions leading from acetyl-coenzyme A to long chain fatty acids and involving the acyl carrier protein (ACP.SH). The enzymes catalysing each step are as follows: (1) acetyl-CoA carboxylase; (2) ACP acetyltransferase; (3) ACP malonyl-transferase; (4) 3-oxoacyl-ACP synthase; (5) 3-oxoacyl-ACP reductase; (6) a family of 3-hydroxyacyl-ACP dehydratases; (7) enoyl-ACP reductases.

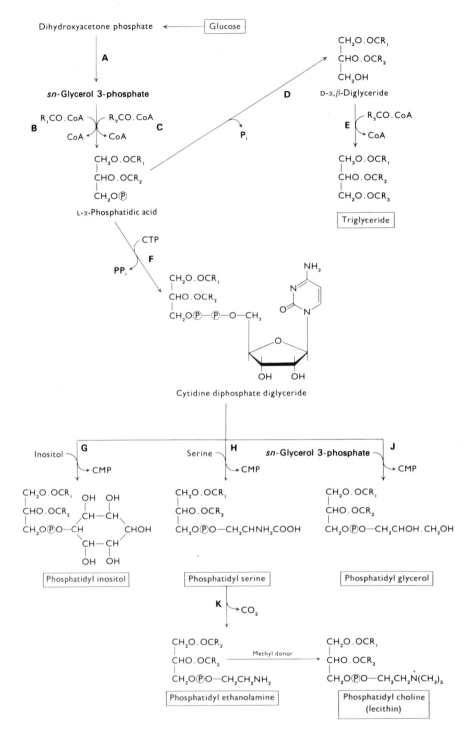

Figure 19. Biosynthesis of triglycerides and phospholipids. (Note that fatty acyl $\sim$ S.ACP rather than fatty acyl $\sim$ S.CoA functions as the fatty acyl donor in some micro-organisms.) The enzymes catalysing each step are as follows: **A**, *sn*-glycerol 3-phosphate dehydrogenase; **B**, **C**, glycerol 3-phosphate acyltransferases; **D**, phosphatidate phosphatase; **E**, diacylglycerol acyltransferase; **F**, phosphatidate cytidylyltransferase; **G**, CDP-diglyceride-inositol phosphatidyltransferase; **H**, CDP-diglyceride-serine *O*-phosphatidyltransferase; **J**, glycerol phosphate phosphatidyltransferase; **K**, phosphatidylserine decarboxylase.

acetyl coenzyme A as their starting point but, unlike fatty acid oxidation, do not have thioesters of coenzyme A as their intermediates. Instead, thioesters of a low-molecular weight protein, called *acyl carrier protein* (ACP) are involved in a manner to be described. A simplified outline of the series of reactions is given in Figure 18.

The cycle of Figure 18 is repeated until a fatty acid of the required chain length is obtained and then the fatty acid dissociates from the acyl carrier protein by the action of a hydrolytic deacylase. It will be noticed that acetyl $\sim$ S.CoA furnishes the two carbon atoms at the methyl end of the fatty acid chain and malonyl $\sim$ S.CoA contributes the rest of the carbon atoms. The overall reaction for the synthesis of palmitic acid (C_{16}) may be summarized as:

$$8CH_3CO \sim SCoA + 7ATP + 14NADPH + 14H^+$$
$$\longrightarrow CH_3(CH_2)_{14}COOH + 7ADP + 7P_i$$
$$+ 8CoASH + 14NADP^+$$

The reducing power required in the form of NADPH is principally generated by the oxidation of glucose 6-phosphate to ribulose 5-phosphate and CO_2 in the pentose cycle (p. 134).

In *Escherichia coli* extracts the major product of the fatty acid synthase system is *cis*-vaccenic acid, the Δ^{11} isomer of oleic acid:

$$CH_3(CH_2)_5 \quad (CH_2)_9COOH$$
$$C{=}C$$
$$H \qquad\quad H$$

cis-Vaccenic acid.

Evidence from experiments with isotopes using *Clostridium butyricum* indicates that while octanoate (C_8) and decanoate (C_{10}) are precursors of both saturated and unsaturated fatty acids, C_{12} and C_{14} saturated acids are incorporated only into saturated acids. These observations suggest that the pathways for synthesis of saturated and unsaturated fatty acids diverge at the C_{10} and possibly at the C_8 level. This has been confirmed by work with purified enzymes.

Although in animal cells the first enzyme of fatty acid biosynthesis, acetyl CoA carboxylase, is controlled by an allosteric activation with citrate, in *E. coli* fatty acid and phospholipid biosynthesis are under the control of ppGpp (guanosine 5'-diphosphate-3'-diphosphate, see p. 236). This compound inhibits acetyl-CoA carboxylase, preventing *de novo* fatty acid biosynthesis. It also inhibits *sn*-glycerol 3-phosphate acetyltransferase (when acyl-CoA is the substrate, but not when acyl-ACP is the substrate), thus preventing phospholipid biosynthesis from acyl-CoA (derived from fatty acids in the growth medium or from lipid turnover).

The biosynthesis of lipids and phospholipids is shown in Figure 19, with the enzymes involved.

FURTHER READING

1 Greenberg D. M. (1969) Biosynthesis of amino acids and related compounds, Part 1; Rodwell V. W. (1969) Biosynthesis of amino acids and related compounds, Part II. Both in *Metabolic Pathways*, 3rd edn, Volume 3 (ed. Greenberg D. M.). Academic Press, New York.

2 Hartman S. C. (1970) Purines and pyrimidines. In *Metabolic Pathways*, 3rd edn, Volume 4 (ed. Greenberg D. M.), pp. 1–68. Academic Press, New York.

3 Gots J. S. (1971) Regulation of purine and pyrimidine metabolism. In *Metabolic Pathways*, 3rd edn, Volume 5 (ed. Vogel H. J.) pp. 225–255. Academic Press, New York.

4 Bloch K. & Vance D. (1977) Control mechanisms in the synthesis of saturated fatty acids. *Ann. Rev. Biochem* **45**, 113.

5 Gatt S. & Barenholz Y. (1973) Enzymes of complex lipid metabolism. *Ann. Rev. Biochem.* **42**, 61.

6 White A., Handler P. and Smith E. L. (1973) *Principles of Biochemistry*, 6th edn. McGraw-Hill, New York.

7 Cohen G. N. (1967) *Biosynthesis of Small Molecules.* Harper and Row, New York, Evanston and London.

8 Lehninger A. L. (1975) *Biochemistry*, 2nd edn. Worth, New York.

9 Truffa-Bacchi P. & Cohen G. N. (1973) Amino acid metabolism. *Ann. Rev. Biochem.* **42**, 113.

Chapter 5
Class III Reactions: The Structure and Synthesis of Nucleic Acids

BASE-PAIRING

The structure, properties and synthesis of DNA and RNA depend fundamentally on the ability of adenine to pair specifically with thymine (or uracil) and of guanine to pair with cytosine. The pairing occurs through the formation of hydrogen bonds as shown in Figure 1. The bases themselves are flat molecules and the hydrogen bonding occurs in the same plane as the bases, so that the base-pairs as a whole are planar. In the adenine-thymine and adenine-uracil pairs two hydrogen bonds are formed, whereas in the guanine-cytosine pair there are three. This difference accounts at least in part for the GC pairing being rather more stable than the AT and AU pairings, as will be seen later. Although each base can assume different tautomeric forms (Figure 5, p. 284) only the forms shown in Figure 1 are important in the pairing mechanisms. The specific recognition between A and T or U and between G and C is referred to as *base complementarity*. In fact a considerable number of other types of base-pairing are stereochemically feasible, and some are believed to occur in the structure of tRNA (p. 195) and in codon-anticodon recognition (p. 203), but apart from these special instances there is little reason to believe that they play any wider role in living systems.

Two further features of the base-pairings represented in Figure 1 are worth noting. Firstly, the distance between the sugar moieties in the paired nucleotides is independent of the type of base-pair, and the same is true of the spatial orientation of the sugar residues with respect to each other and with respect to the plane of the paired bases. Secondly, the relative position of the sugar residues remains the same if the positions of the bases in a given type of nucleotide-pair are reversed. One can sum up these facts by saying that if a nucleotide is part of a polynucleotide chain and its complementary nucleotide is associated with it by hydrogen-bonding, then the position of the attached nucleotide in relation to the chain is fixed independently of the bases involved. This relationship is fundamental to the role played by nucleic acids in living systems. It explains why DNA molecules from different sources can show infinite variety in their base sequences while preserving a very stable, characteristic macromolecular structure. It also provides a sound basis for the replication and transcription of nucleotide sequences since the enzymic formation of the phosphodiester linkages in the products can proceed by the same type of mechanism in all cases.

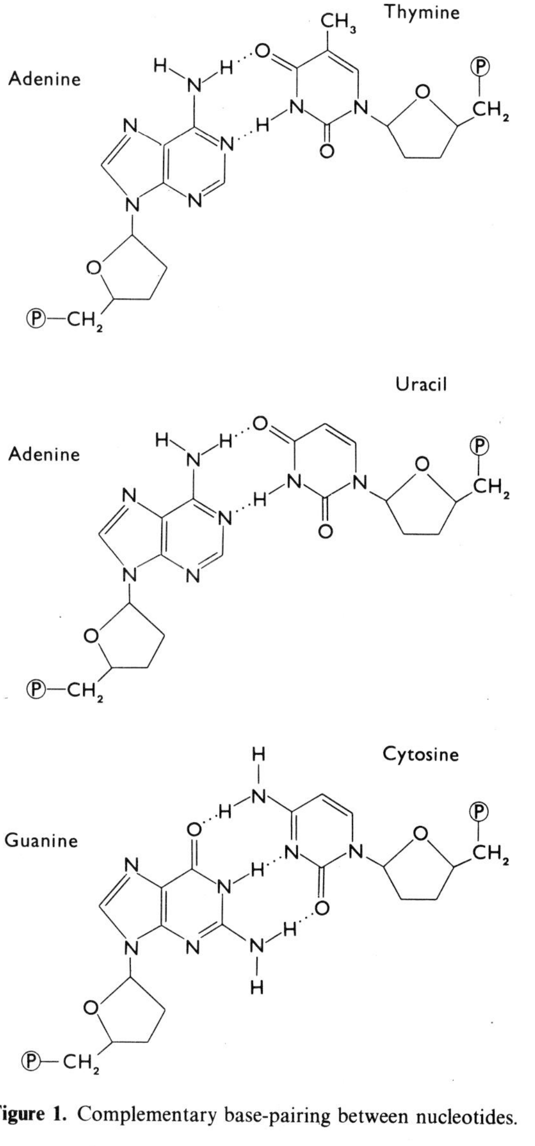

Figure 1. Complementary base-pairing between nucleotides.

STRUCTURE OF DNA

The primary structure of DNA consists of a linear unbranched chain of deoxyribonucleotides joined by phosphodiester linkages between the 3′ and 5′ positions of the 2′-deoxyribose units (Figure 3, p. 12). It may be considered as a continuous deoxyribose phosphate backbone from which the nitrogenous bases project in linear array. It is the precise sequence of the bases which endows the molecule with its biological significance; this sequence represents the encoding of the genetic information which characterizes the organism in which the DNA is contained. The molecular mechanisms by which this coded sequence is duplicated and expressed will be considered later, as will the techniques by which long tracts of nucleotide sequences in DNA (and RNA) can be determined in the laboratory (p. 195). At this stage it is appropriate to describe the form in which DNA exists in the cell, for the molecular organization of the DNA molecule is directly concerned with its unique biological functions.

The size of DNA molecules

In vivo, DNA (as well as RNA) does not exist as the 'naked' molecule but is involved in interactions with other substances, particularly proteins, many of which may be presumed to have regulatory functions. Much of the protein is basic, so that its positively charged groups help to neutralize the negatively charged phosphates of the nucleic acid. However, the organization of DNA-protein complexes in bacteria does not lead to the appearance of well-defined *nucleosomes* containing specific histone molecules in defined stoichiometric ratios, nor to the formation of higher-order chromosomal structures such as can be seen in the nuclei of eukaryotic organisms. In addition to its interaction with proteins the DNA of bacteria seems to be associated with the cell membrane; this is significant with respect to the mechanism and control of its replication (pp. 90 and 222).

By lysing cells extremely gently and analysing the products directly on a sucrose gradient the DNA can be released apparently intact in the form of a nucleoid or 'folded chromosome' (Figure 2). The overall dimensions of folded chromosomes appear to change little upon isolation, indicating that the structural integrity of the chromosomes *in vivo* has been preserved. Depending upon the temperature of cell lysis the folded chromosomes may be isolated attached to, or free from, membrane fragments. Both RNA and proteins are involved in maintaining the integrity of the folded structure; if either or both of these compounds are damaged extensive unfolding occurs. Membrane-free chromosomes contain 60% DNA, apparently arranged in about 50 independently supertwisted loops, together with 30% RNA and 10% protein, the majority of which seems to be core RNA polymerase (see p. 231).

Solutions of purified DNA are highly viscous, the viscosity of a particular preparation depending very much upon how carefully it was treated during the extraction and purification procedures. Although, as Cairns has shown, it is possible to liberate the total DNA of *Escherichia coli* as an apparently continuous molecule of molecular weight approximately $2 \cdot 5 \times 10^9$, most preparations of bacterial DNA have molecular weights in the region of 10^7 after they have been purified to remove protein and RNA. This is because the enormous size of DNA molecules renders them highly susceptible to hydrodynamic shear forces; the mechanical agitation with organic solvents necessary to remove other materials during DNA purification is quite sufficient to produce breakage into fragments. Even so, DNA samples isolated from bacteria are among the largest molecules which can be obtained in reasonably pure form from living materials. The enormous size of DNA can be used to advantage in purifying it because when cell extracts are treated with ethanol or *iso*-propanol the DNA precipitates in the form of fibrous threads which can be collected on a glass rod and thus selectively removed from smaller macromolecules.

Two techniques of visualizing DNA molecules have been developed so that their size and conformation may be revealed. In Cairns' method the DNA is first labelled *in vivo* with [^{3}H] precursors such as [^{3}H] thymidine; it is then very gently released from the cells and overlaid with a photographic emulsion. After exposure for several weeks the film is developed, revealing the positions of DNA molecules by near-continuous lines of silver grains produced by decay of the incorporated tritium atoms (see Figure 23, p. 212). Kleinschmidt's method does not require that the DNA be radioactively labelled; the DNA is simply allowed to spread over a film of basic protein molecules which act as a supporting medium, the film is caught on an electron microscope grid and shadowed with a heavy metal to outline the DNA molecules covered with basic protein, and the grid is then observed in the electron microscope. Not only have these experiments shown that the DNA of a bacterial nucleoid or of a virus can be extracted in one piece, but it has also been demonstrated that an unbroken molecule of *Escherichia coli* DNA appears to be circular, in agreement with the circular nature

Figure 2. Membrane-attached folded chromosome of *Escherichia coli*.
   Fast-sedimenting material from cells lysed very gently on top of a sucrose density gradient was prepared for electron microscopy by spreading on a monolayer of cytochrome *c*, staining with uranyl acetate, and shadowing with platinum. The scale bar represents 1 μm. Photograph kindly provided by Dr. A. Worcel, Dept of Biochemical Sciences, Princeton University.

of the bacterial genome as deduced from genetic studies (see Chapter 7). While such experiments provide impressive evidence for the size and apparent integrity of enormous DNA molecules the resolution that can be achieved is limited; consequently it cannot yet be concluded that the DNA of a bacterium has an uninterrupted phosphodiester backbone. There may exist breaks at certain points in the molecule, or even occasional non-nucleotide 'linkers', for instance amino acids or short peptide chains. Some viruses, however, have been found to yield truly circular DNA molecules, and here it can be stated with much greater certainty that the circles do not contain breaks or 'linkers'.

Secondary structure of DNA

The great viscosity of DNA solutions is due to the stiff rod-like character of the molecules, and this in turn is due to their highly organized secondary structure stabilized by base-pairing between strictly complementary nucleotide sequences. Although the gross base-composition of DNA from different bacteria varies between wide limits, there is always a close correspondence between the contents of adenine and thymine and of guanine and cytosine (Table 1). The significance of this relationship was not fully appreciated until the work of Watson and Crick on the secondary structure of DNA. These workers studied the X-ray diffraction patterns produced by oriented fibres of DNA and proposed the now-famous double-helical model for the structure of DNA. By suggesting that DNA consists of two polynucleotide chains wound into a helical form with hydrogen-bonding between complementary bases in the opposing strands, Watson and Crick not only accounted for the observed equivalence of the amounts

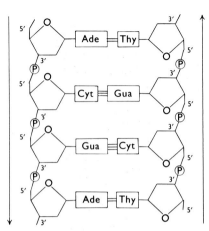

Figure 3. Sequence complementarity between opposite strands of DNA is determined by hydrogen bonding between bases.

Note the opposite polarity of the strands, indicated by the direction of the 3'–5' linkages and the position of the oxygen in the deoxyribose ring.

of A and T and of G and C, but they laid the foundation for a whole new understanding of the molecular events concerned in the storage, replication and expression of genetic information. The essential postulate in the Watson-Crick model is that the *sequence* of bases in one strand is exactly the complement of that in the other, so that every nucleotide participates in a sterically similar hydrogen-bonded association with a nucleotide in the opposite strand (Figure 3). Since, as pointed out earlier, the stereochemistry of the hydrogen-bonding is such that the relative orientation of the sugar-phosphate residues is not affected by the nature of the base-pairs, the duplex structure can take up the same conformation irrespective of the actual

Table 1. Base-composition of bacterial DNA.

	Proportion of base, moles %				GC content (% G + C)	$\frac{A+T}{G+C}$
	Adenine	Thymine	Guanine	Cytosine		
Clostridium perfringens	35	35	15	15	30	2·33
Proteus vulgaris	30	29	20	21	41	1·47
Escherichia coli	25	25	25	25	50	1·00
Aerobacter aerogenes	21	22	29	28	57	0·75
Micrococcus lysodeikticus	14	14	36	36	72	0·39

Values are given to the nearest integer.

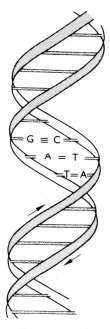

Figure 4. Diagrammatic representation of a DNA double helix.

The direction of the arrows corresponds to the direction of the 3'–5' phosphodiester linkages as represented in Figure 3. Note that the molecule appears to be traversed by two grooves of unequal widths. This feature arises because the polynucleotide chains are separated by a distance less than half that required for a complete turn of the helix (34 Å). The difference in size of the grooves has been exaggerated here; compare Figure 5.

base-sequences involved although it obviously requires the A = T and G = C equalities. The conformation assumed is that of a right-handed double helix having the sugar-phosphate backbones on the periphery and the hydrogen-bonded base-pairs stacked in the core of the molecule with their planes perpendicular to the helix axis (Figure 4).

The two polynucleotide chains are antiparallel, i.e. they run in opposite directions (see Figures 3 and 4), so that rotation through 180° does not alter the appearance of the molecule. There are ten base-pairs for every complete turn of the helix. A better impression of the structure can be gained from Figure 5 which shows a double helix constructed from space-filling atomic models.

Since each nucleotide is involved in hydrogen bonding the stability of the structure is very high. As well as the hydrogen bonding it is likely that other forces, arising from the ordered hydrophobic stacking of the base-pairs, also contribute to the stability of the DNA

double helix. Be that as it may, hydrogen bonding is surely the principal element which determines strict accuracy of base-pairing and is thus of paramount importance for the biological role of DNA.

In certain circumstances there exists a further level of organization of DNA molecules which can properly be referred to as *tertiary structure*. This occurs where the DNA molecule exists in a form where its ends are not free to rotate independently, as in a closed circular duplex, and is called *supercoiling*. It involves the imposition of a twisted or superhelical state upon the fundamental Watson–Crick double helix and may arise as a result of the binding of various proteins to the DNA as in the course of replication or transcription. The ability of helix-unwinding and destabilising proteins to modify the supercoiling of DNA is of paramount importance in the replication of the chromosomes of bacteria and their viruses (see p. 219). Supercoiling can only exist so long as the two strands of the double helix remain intact; a single 'nick' introduced into either strand (e.g. by endonuclease attack) provides a point of free rotation that allows one strand to rotate about the other and thus eliminates the superhelical state.

Denaturation

Disruption of the helix is a dramatic event which may be triggered by raising the temperature above a critical value or by exposing the DNA to extremes of pH. The change which takes place, called *denaturation*, is more or less irreversible and displays the characteristics of a cooperative process, i.e. the rupture of a critical number of hydrogen bonds facilitates the breakage of neighbouring bonds leading to the collapse of the entire double helix. Denaturation may be observed by measuring the ultraviolet absorbance of the DNA; when the helix breaks up there is an increase of some 40% in the absorbance at 260 nm (Figure 6). In addition to this *hyperchromic effect* there is a pronounced decrease in the viscosity of the solution as the strands separate and the buoyant density of the DNA in CsCl density gradients is increased. If the DNA carries genetic markers which can be studied in transformation experiments (Chapter 7) a considerable loss of transforming activity is observed.

The temperature at which denaturation occurs (often called the 'melting' temperature) depends upon the ionic strength of the medium. While the dimensions of the double helix are not significantly affected by the nature and concentration of salts present, the stability of the structure depends very much upon the

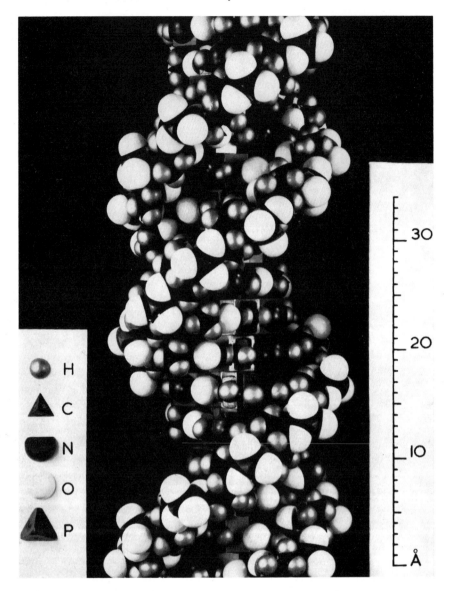

Figure 5. Scale model of the DNA molecule. Photograph kindly provided by Dr Watson Fuller.

ability of salts to neutralize the repulsive forces between the negatively-charged phosphate groups of
the two strands. Marmur and Doty have shown that
there is an approximately linear dependence of melting temperature on the logarithm of the cation
concentration. Furthermore, the melting temperature
under given ionic conditions increases in proportion
to the GC content of the DNA; this provides evidence
that the GC pairing is more stable than the AT pairing, as might be expected if the GC pair had three
hydrogen bonds while the AT pair had only two.

Once the strands of the helix have become
separated the resulting chains of denatured DNA are
free to assume whatever conformation is energetically
most favourable. Generally they simply collapse into
a tangled mass called a *random coil* but the nature of
the random coil structure they take up is strongly
dependent upon the salt concentration. At low ionic
strengths the neutralization of the repulsive charges
on the phosphate groups is poor and the coiled structure remains relatively open; at higher ionic strengths
the coil becomes more compact. These differences in

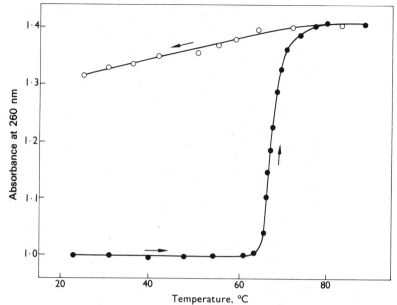

Figure 6. Thermal denaturation of double helical DNA.

A sample of DNA from bacteriophage T2 was heated to 90° and then cooled, as indicated by the arrows. The absorbance at the various temperatures is expressed relative to the starting value. In this experiment the 'melting' temperature is 67·2°. The relatively small, gradual decrease in absorbance on cooling represents the development of random-coil structure by the separated strands (Waring, unpublished data).

structure are evidenced by variations in the ultraviolet absorption of the DNA, which suggest that a certain amount of interaction occurs between the nucleotide bases in denatured DNA. Quite probably some hydrogen-bonding takes place between complementary bases which happen to come close together, but long helical regions like those of the native double helix are highly improbable. It should be emphasized that base interactions in denatured DNA must be virtually entirely *intra*-strand and that interactions between different strands probably play very little or no part in the formation of random-coil structures.

When the hydrogen bonds of a base-pair have been disrupted the amino groups of the bases become available for reaction with reagents such as formaldehyde or formamide. Once they have reacted in this way the modified bases are prevented from subsequently forming H-bond interactions with other bases, so that the denatured, single-stranded state of the polynucleotide chain is permanently 'frozen'. This fact has been exploited by a number of workers, especially Inman, to study denaturation. If DNA is treated with formaldehyde under conditions where denaturation is just beginning and then is prepared for electron microscopy by the Kleinschmidt technique, partially denatured states of the DNA molecules can be visualized (Figure 7). It can be seen that local separation of the strands occurs earlier in some regions of the molecules than in others. These 'early melting' regions are presumed to be regions

relatively rich in AT pairs, i.e. base-pairs with the lowest stability, while the regions which resist separation are thought to be relatively GC-rich. Denaturation maps can be constructed showing characteristic locations of early melting regions along the length of the molecules.

Single-stranded DNA is found naturally-occurring in certain bacteriophages (such as ϕX174) which infect *Escherichia coli*. The physical properties of DNA extracted from this phage clearly resemble those of the denatured DNA produced by heat or alkali treatment of ordinary double-helical DNA, and moreover ϕX174 DNA does not contain equivalent amounts of A and T or of G and C—a feature which has proved useful for showing the complementary relationship between template and product in reactions catalysed by the nucleic acid polymerases. DNA from ϕX174 is circular as well as being single-stranded. It is resistant to digestion by exonucleases (enzymes which progressively remove nucleotides from the ends of nucleic acid molecules), suggesting that it has no free ends, and it appears circular when shadowed preparations are viewed in the electron microscope. Immediately after infection of *E. coli* by ϕX174, the entering (+) strand serves as a template for the synthesis of a complementary (−) strand, resulting in a double-helical, circular *replicative form* (RF). This is then replicated to produce a number of RF molecules which serve as templates for the synthesis of progeny (+) strands; these are packaged

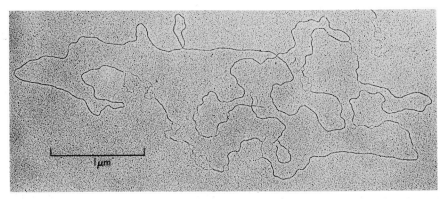

Figure 7. A partially denatured molecule of replicating bacteriophage lambda DNA.

    The DNA was extracted from cells of *Escherichia coli* infected with the phage; it shows an apparently circular double helix which has been partly replicated (compare the autoradiograph of the *E. coli* chromosome in Figure 23, p. 212). Alkali treatment was used to effect partial denaturation, and at several points it can be seen that the bold line of the intact double helix is interrupted by a region where the individual strands have become separated and are visible as two thinner lines. Photograph kindly provided by Dr R. B. Inman.

into newly-synthesized phage heads and finally the cell bursts releasing the progeny phages (see below Figures 33 and 34 and pp. 216–221).

Renaturation

Although complete denaturation of double-stranded DNA is effectively irreversible under most conditions, re-formation of a true double helix from separated single strands (*renaturation*) can be observed in special circumstances. The problem is to re-align the strands with their complementary base-sequences in register and this can only be achieved by repeated trial-and-error collisions between polynucleotide chains. Conditions must be adjusted so that when a pair of complementary nucleotide sequences happen to meet a 'nucleation' occurs and the helical structure then 'zippers' up along the remaining length of the strands. In practice, the heat-denatured DNA has to be cooled very slowly or held for a long time at a temperature some 25° below the melting temperature of native DNA; under these conditions the reformation of helical structure is energetically favoured once nucleation has occurred, and collisions due to thermal agitation give a reasonable frequency of nucleation. The renaturation reaction proceeds with second-order kinetics, as would be expected for a process requiring collision between two complementary species. Of course, the problem of getting complementary nucleotide sequences into register becomes greater as the size and complexity of the strands increases. Because of this a high degree of renaturation has only been observed with DNA from relatively simple organisms such as viruses and bacteria.

    In eukaryotic cells the genetic material seems to contain a substantial proportion of nucleotide sequences which are reiterated or repeated many times; sometimes thousands or millions of copies of a particular sequence may occur. Strands of denatured DNA containing these sequences engage in a form of 'renaturation' (better called reassociation) at easily measurable rates. Whether these repeated sequences reflect the existence of multiple copies of genes (in a strict sense) is doubtful. At all events, in bacteria this sort of sequence multiplicity or gene repetition does not seem to occur to any significant extent, except in respect of the limited numbers of copies of genes for the stable RNA species known to be present in the genomes of organisms such as *E. coli* (Table 1, p. 239).

STRUCTURE OF RNA

The primary structure of RNA is closely similar to that of DNA; it contains a linear arrangement of ribonucleotides linked through the 3′ and 5′ positions of the ribose residues (Figure 3, p. 12). As well as having a 2′ hydroxyl group in the sugar, RNA differs from DNA by having uracil in place of thymine. Certain types of RNA, particularly tRNA, also contain a significant proportion of unusual nucleotides such as pseudouridylic acid and ribothymidylic acid (Figure 8). As with DNA, no evidence has been found for the existence of branching in the molecule.

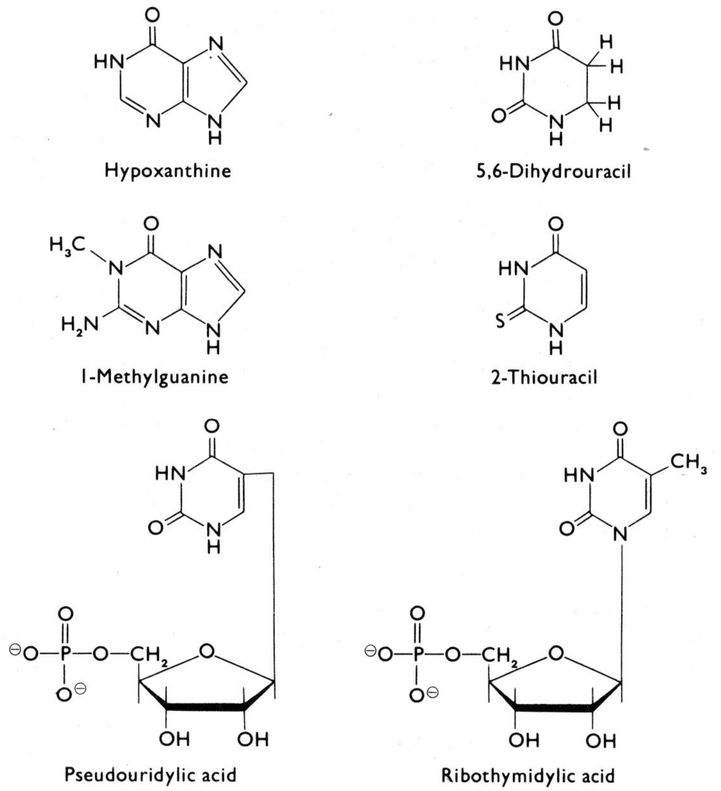

Figure 8. Some unusual bases and nucleotides found in RNA.

A more comprehensive list of 'minor' bases and their derivatives is given by Zachau (*Angewandte Chemie Internat. Ed.* **8**, 711 (1969)). The ribonucleoside and ribonucleotide containing hypoxanthine are known as *inosine* and *inosinic acid* respectively. Note that ribothymidylic acid is the analogue of thymidylic acid (as found in DNA) containing ribose in place of 2'-deoxyribose. The base in pseudouridylic acid is normal uracil joined to the ribose by C_5, instead of by N_1 as in uridylic acid.

In bacteria, as in other types of organism, three kinds of RNA can be distinguished; ribosomal (rRNA), amino acid transfer (tRNA) and messenger (mRNA). Their secondary structure is not nearly so well defined as that of DNA but rather resembles the structure assumed by denatured, single-stranded DNA, which is consistent with the belief that RNA molecules are single polynucleotide chains. Indeed, the simple fact that most forms of RNA do not have the A = T, G = C equality of base-composition shows that they could not be entirely double-helical molecules. However, it is now clear that rRNA and tRNA do adopt a partially ordered configuration in solution and that this molecular organization consists of short and possibly imperfect double-helical regions formed by looping of the chain, connected by non-helical stretches of the polynucleotide.

When solutions of RNA are heated the secondary structure of the molecules is disrupted and, as seen with DNA, there is a hyperchromic effect in the ultra-violet absorption. Unlike DNA, however, the increase in UV absorption occurs much more gradually and extends over a wide temperature range showing that the interactions between the bases vary widely in stability to heat. This would be expected if there existed helical regions which had different lengths or imperfections or, most likely, a mixture of both. The RNA 'melting' curve is more or less fully reversible, for the recovery of native-like secondary structure requires only the reformation of *intra*-strand interactions rather than collision between separate complementary strands in register—cf. renaturation of DNA. Only in special circumstances, e.g. when some tRNA's are heated in the absence of Mg^{2+}, has a sort of irreversible denaturation and loss of biological activity been observed (see below).

The probable nature of helical regions in natural cellular RNA's has been deduced mainly from X-ray diffraction studies on two unusual RNA's: first, the completely double-helical RNA which constitutes the genetic material of reovirus, an animal virus, and second, the synthetic double-helical complexes which are formed by mixing equimolar proportions of synthetic homopolymers such as polyriboadenylic acid and polyribouridylic acid. These materials are better suited to X-ray diffraction work because of their long-

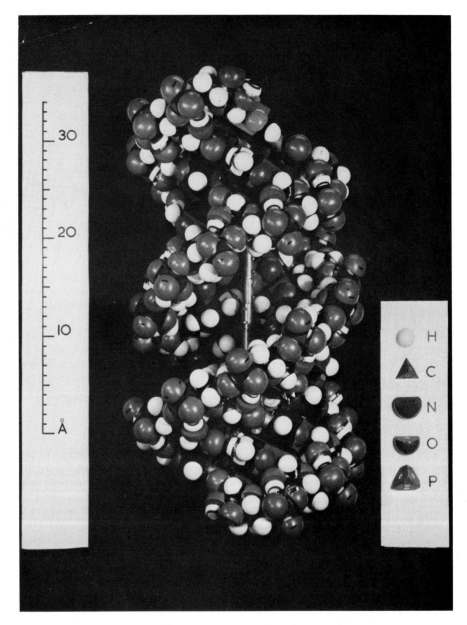

Figure 9. Space-filling model of a helical RNA molecule. Photograph kindly provided by Dr Watson Fuller.

range regularity of structure which makes them pack neatly into fibres. So far as it goes, the limited information from X-ray diffraction of fragments of natural rRNA yields similar results. It is found that RNA helices are *A-type*; they differ from the DNA helix (*B-type*) in that the base-pairs are tilted and further away from the helix axis, and that there are eleven or twelve base-pairs per turn (Figure 9).

These differences apparently result from the presence of the 2′ hydroxyl group on the ribose sugar, which imposes a constraint on the puckering of the sugar ring preventing the RNA ribose-phosphate backbone from adopting the same conformation as the deoxyribose-phosphate backbone of DNA. Significantly, the normal B-type DNA helix can 'click' into an A-type conformation resembling the

RNA helix—it does so in fibres of packed DNA molecules when the relative humidity is lowered—which may be important in the transcription of RNA from a DNA template (p. 230).

Sequencing

The pioneering work of Sanger, Holley and their colleagues in the early 1960s led to the initial development of methods for determining the nucleotide sequences of RNA molecules. Their approach was analogous to that which proved so successful in determining the amino acid sequences of proteins. Although superseded in recent years by more effective techniques, these early efforts provided most of the data currently available on tRNA molecules. Sanger's original methods involve enzymic cleavage of [^{32}P]-labelled RNA into small oligonucleotides which are separated, identified, and sequenced by end-group analysis, and reconstruction of the starting sequence by study of large fragments produced from the original molecule by limited cleavage with nucleases.

Transfer RNAs provided the best substrates for these studies because they are relatively small (about 80 nucleotides long) and their rare modified nucleotides greatly facilitate identification of overlapping sequences. Hundreds of tRNAs, from higher organisms as well as bacteria, have now been examined and a few representative sequences are shown in Figure 10. A striking feature is the substantial homology between different molecules: not only do all have the 3′ terminal sequence −pCpCpA,* but they also have a phosphate group at the 5′ end (usually attached to a terminal G but not in the special initiator species tRNA$_F^{Met}$) and they all possess the same sequence —pGpTpψpCpPu— at about the same position along the length of the molecule. Even more striking homologies become apparent when the sequences are written in the form of a *clover leaf* arrangement (Figure 10) which is generally believed to represent the secondary structure of all tRNA's. Written in this form, the tRNA molecule contains four helical segments: one terminates in the —pCpCpA sequence to which the amino acid becomes esterified (the amino acid arm), the next sustains a fairly large loop in which dihydrouridylic acid residues often occur (the dihydro-U or D loop), the next sustains a loop of seven nucleotides of which the middle three are believed to be the anticodon (the anticodon loop), and the fourth

sustains a seven-nucleotide loop starting with the universal —TpψC— sequence (the TψC loop). The region between the anticodon arm and the TψC arm is very variable between different tRNA's and may characterize each molecule for recognition, e.g. by the amino acid activating enzymes. Each helical region contains a characteristic number of base-pairs, though in some molecules the pairing seems to be imperfect and pairing between G and U (which is possible if small deviations from a perfectly regular helical structure are allowed) seems to be necessary. Structure-function relationships of tRNA in the context of its role in protein synthesis will be found in Chapter 6; evidence for its conformation is discussed below.

The sequence of the smallest ribosomal RNA, 5 S RNA, from *E. coli* has been elucidated (Figure 11). As yet the function of this molecule is obscure, but its structure reveals a curious feature in that there is considerable internal homology of nucleotide sequence which becomes evident if the chain is considered in two parts. This internal homology could mean that 5 S rRNA is the product of a gene which, in its primordial state, became duplicated and then the sequences of the two halves diverged during the course of evolution. The complete nucleotide sequence of the 16 S rRNA has also been determined and that of the 23 S rRNA is now known.

Numerous recent advances in molecular biological technique have combined to increase the power and capability of sequencing methods by orders of magnitude, and to extend their applicability to DNA. Chief among these are the techniques of molecular cloning and plasmid isolation (p. 319), the discovery and use of restriction enzymes to promote specific cleavage of DNA into fragments and the consequent facility for mapping of large genetic structures (p. 312), and, above all, the development of gel electrophoresis technology so that polynucleotides containing 100 residues or more can be resolved into series of discrete bands, each band differing from its neighbour by one extra nucleotide. Table 2 lists a selection of restriction enzymes which have proved particularly useful for sequence studies on DNA. It can be seen that the specific sequence of nucleotide pairs recognised by each enzyme in double-helical DNA is *palindromic*, i.e. is rotationally symmetrical about the mid-point. This seems to be a common feature of the action of these enzymes. For the purposes of the present discussion, however, the important point is that treatment with purified restriction enzymes causes cleavage of any double-helical DNA into a reproducible number of defined fragments, each of which is amenable to

---

* See p. 255 for explanation of this notation.

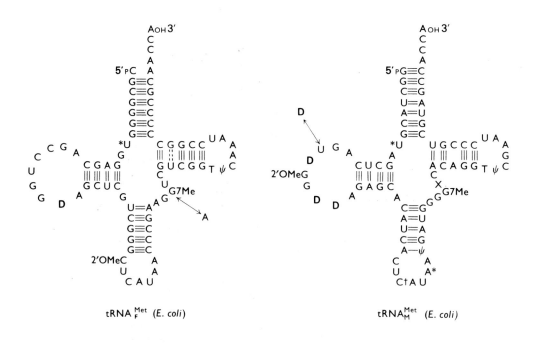

tRNA$_F^{Met}$ (E. coli)   tRNA$_M^{Met}$ (E. coli)

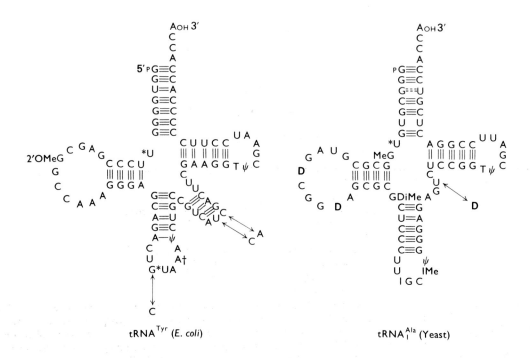

tRNA^{Tyr} (E. coli)   tRNA$_I^{Ala}$ (Yeast)

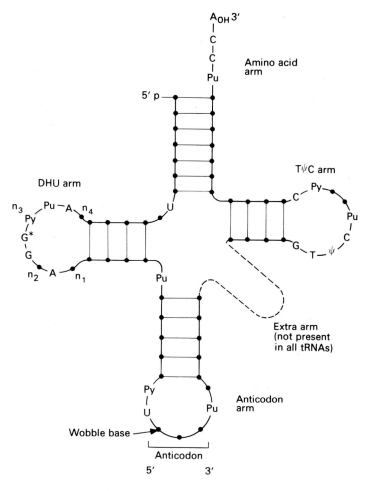

Generalised structure of tRNA

Figure 10. Nucleotide sequences of four transfer RNA's together with a generalized structure. Abbreviations for nucleotides:

A: adenylic acid	Pu: purine nucleotide
G: guanylic acid	Py: pyrimidine nucleotide
C: cytidylic acid	
U: uridylic acid	$n_1 = 0$ to 1
T: ribothymidylic acid	$n_2 = 1$ to 3 $\rbrace$ residues in
I: inosinic acid	$n_3 = 1$ to 3 $\rbrace$ DHU arm
D: 5,6-dihydrouridylic acid	$n_4 = 0$ to 2
ψ: pseudouridylic acid	

X represents an unknown nucleotide. Special modifications of nucleotides are represented by the substituent or by * or † where the nature of the modification is not certain. The 5′ ends of the sequences bear a phosphate group represented by p, and the 3′ ends are shown A_{OH}. In some cases variant or mutant sequences have been found; here the nucleotide replacements are indicated by a double-headed arrow. The $G^* \rightarrow C$ replacement in tRNA^{Tyr} alters its coding properties such that it acts as an *amber* (UAG) suppressor (see p. 258). After Waring, *Ann. Reports Chem. Soc.* **65B**, 551 (1968).

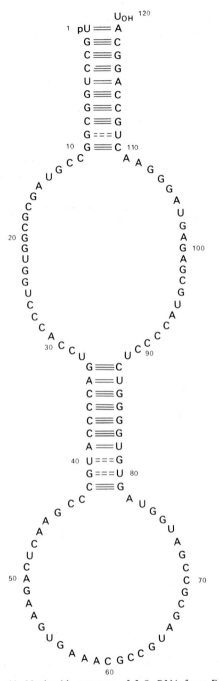

Figure 11. Nucleotide sequence of 5 S rRNA from *Escherichia coli*.

The first ten nucleotides from the 5′ end are almost perfectly complementary to ten nucleotides near the 3′ end, as are the tracts 33–42 and 79–88: although these sequences have been apposed to indicate the possible formation of base-paired helical regions the actual conformation of the molecule is unknown. After Brownlee *et al.*, *Nature* **215**, 735 (1967).

Table 2. Restriction enzymes.

The arrows indicate the points within the recognition sequence of 4 or 6 nucleotide-pairs where strand cleavage occurs. Note the rotational symmetry of each sequence, as a result of which the point of cleavage in each strand (written 5′ → 3′) is identical. For a more comprehensive listing of known restriction enzymes see Roberts (*Crit. Rev. Biochem.* **4**, 123–164 [1976]).

Producing organism	Enzyme	Specificity
Arthrobacter luteus	*Alu* I	↓ — A G C T — — T C G A — ↑
Bacillus amyloliquefaciens H	*Bam* I	↓ — G G A T C C — — C C T A G G — ↑
Escherichia coli RY13	*Eco* RI	↓ — G A A T T C — — C T T A A G — ↑
Haemophilus aegyptius	*Hae* III	↓ — G G C C — — C C G G — ↑
Haemophilus haemolyticus	*Hha* I	↓ — G C G C — — C G C G — ↑
Haemophilus influenzae Rd	*Hind* III	↓ — A A G C T T — — T T C G A A — ↑
Haemophilus parainfluenzae	*Hpa* I	↓ — G T T A A C — — C A A T T G — ↑
Haemophilus parainfluenzae	*Hpa* II	↓ — C C G G — — G G C C — ↑
Providencia stuartii 164	*Pst* I	↓ — C T G C A G — — G A C G T C — ↑
Streptomyces albus G	*Sal* I	↓ — G T C G A C — — C A G C T G — ↑

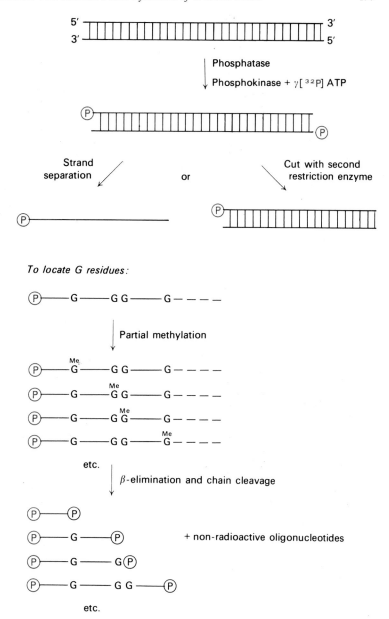

Figure 12. Schematic illustration of a method for sequencing DNA by chemical degradation to guanine-terminated fragments. Based on suggestions of Drs F. Sanger and G. G. Brownlee.

sequence determination by one or other of the 'gel read-off' methods.

One such method, devised by Maxam and Gilbert, is outlined in Figure 12. A chosen fragment is first treated with a phosphatase to remove any terminal phosphate groups and expose the 5' hydroxyl at the end of each strand. These termini are then labelled with radioactive phosphate groups transferred from $\gamma[^{32}P]$ ATP under the influence of a kinase. The next

step is to isolate material having only one strand labelled, either by separating the strands (e.g. in alkali) or by treatment with a second restriction enzyme to produce two sub-fragments. The sample is then treated under mild conditions with a reagent which specifically or preferentially attacks one of the four types of nucleotide bases. Conditions are so mild that, on average, only one of the potentially susceptible nucleotides is attacked per strand, and that at

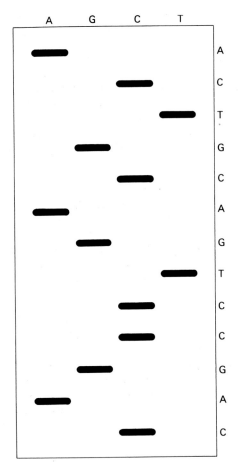

Figure 13. Hypothetical example of sequence-determination for a DNA fragment.

Four parallel vertical electrophoresis lanes are shown, each containing the products of reactions in which attack occurred specifically at A, G, C or T nucleotides as indicated by the letter above each lane. Each group of molecules travels a distance related to its length and appears as a dark band when the gel slab is subsequently allowed to expose a photographic plate (autoradiograph). The nucleotide sequence of the parent fragment (right) can be read off directly from the pattern of bands.

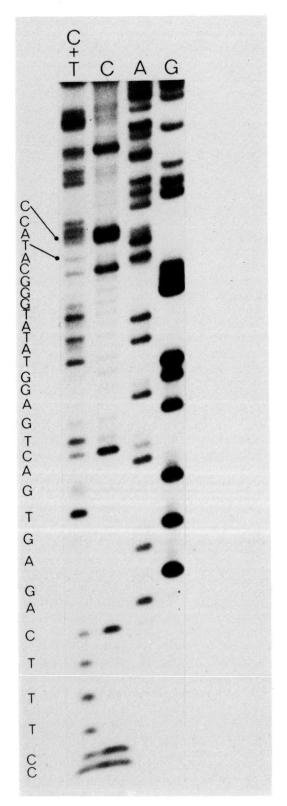

Figure 14. Autoradiograph of a sequencing gel produced according to the method of Maxam and Gilbert (*Proc. Nat. Acad. Sci. U.S.A.* **74**, 560–564 [1977]).

The labelled sample, a DNA restriction fragment, was cleaved in separate reactions specific for the nucleotides indicated above each track. The sequence of the last 33 nucleotides in the fragment can be read directly as indicated on the left of the diagram. Photograph kindly provided by Dr. G. G. Brownlee.

random. In the next step the damaged base is released and a β-elimination reaction ensues which cleaves the oligonucleotide chain at that point. The resulting mixture contains labelled oligonucleotides of different lengths, each length corresponding to the distance (i.e. number of residues) from one of the susceptible nucleotides to the labelled 5′ end. In Figure 12 the example given is for a guanine-specific reaction. Thus, electrophoresis of the products from that experiment should produce a series of bands, each of which reveals the distance of a G residue from the 5′ end. If parallel experiments are undertaken employing reactions aimed at each of the other three types of nucleotide in turn, and the fragment patterns from all tests are run in parallel tracks on the same gel electrophoresis slab, the complete nucleotide sequence can be read directly off an autoradiograph developed from the slab (Figure 13). The versatility of this approach is heightened by the fact that the degradative procedures need not be entirely specific for any single type of nucleotide base, so long as conditions can be found in which one type of base is substantially more susceptible than the others. This, in fact, is often the case in practice. For instance, much work has been done employing methylation using dimethyl sulphate as a purine-specific reagent under two different sorts of conditions. In the first, G residues are attacked more than A; in the second, the susceptibility is the reverse. Likewise, in using aqueous hydrazine as a pyrimidine-specific probe it has been usual to adopt conditions under which both C and T residues are attacked, and also those in which virtually only the C residues react. It is quite easy to locate the T residues 'by difference', i.e. by looking for fragments which appear under the former but not the latter conditions. An illustration of a gel constructed in this manner is shown in Figure 14.

Other ingenious methods have been devised to exploit the extraordinary resolving power of electrophoretic gels to allow direct read-off of a nucleotide sequence from patterns of labelled fragments of either DNA or RNA. Some involve the production of arrays of fragments, each terminating with a known nucleotide, by enzymic degradation of a labelled polynucleotide. Others involve the synthesis of fragments in specifically 'primed' reactions employing one of the various nucleic acid polymerases or reverse transcriptase (see p. 237). A particularly powerful version of the latter procedures employs dideoxy nucleoside triphosphates added to polymerase reaction mixtures to serve as chain terminators. These nucleotide analogues become incorporated at the end of a growing chain and, having no 3′ hydroxyl group to partici-

pate in internucleotide linkage, effectively prevent further chain growth and thus yield fragments terminating with a known base.

Examples of biologically important nucleotide sequences determined by these methods (viral genes, operators, promoters, terminators, ribosome-binding sites, etc.) will be found in other chapters, but the most impressive, involving the determination of sequences of thousands of nucleotides in each case, are the genomes of phages MS2 and φX174 illustrated in Chapter 6 (pp. 270 and 273).

The conformation of tRNA

Early evidence that tRNAs do indeed possess the helical regions defined by the clover leaf model came primarily from work on nucleolytic or chemical attack on the molecule. It was found that nucleotides which are supposed to be involved in the helical stems of the clover leaf were relatively resistant to the action of nucleases and various chemical reagents. This is in accord with the known specificity of these reagents: in general they work poorly if at all on nucleotides which are involved in base-pairing or are otherwise 'buried' in the secondary structure. However, it is also found that certain regions, such as the TψC loop, which one might expect to be readily attacked are not. For this and other reasons it is believed that tRNA possesses additional higher-order tertiary structure stabilised by interactions between different arms of the clover leaf. The existence of further structural forces, probably involving interaction with Mg^{++} ions as well as polyamines, could account for the phenomenon of 'irreversible' denaturation mentioned earlier.

Definitive evidence for a well-defined tertiary structure is available in one instance at least, that of the tRNA[Phe] of yeast which has been crystallised and subjected to X-ray diffraction analysis. The relation between the clover leaf and the folded molecule is illustrated in Figure 15. It is more easily visualised if the clover leaf is imagined to be twisted so that the TψC arm is on the left and the D arm on the right. Two major helical or pseudo-helical elements characterise the gross anatomy of the structure; one is formed by the stacking of the amino acid stem on the TψC stem; the other, lying roughly at right angles to the first, consists of the D stem (augmented by a few nucleotide-pairs at either end) leading on to the anticodon stem. Significantly, the site of amino acid esterification (adenosine 76) and the anticodon (Gm 34, A 35 and A 36) lie in exposed positions at opposite

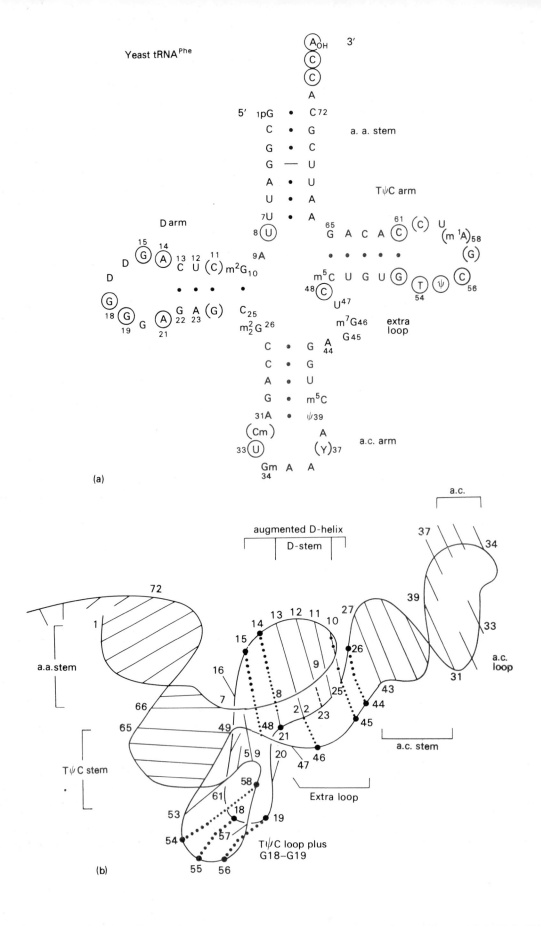

(a)

(b)

extremes of the molecule. In the centre of the structure where the two major helical sections meet the essential shape of the molecule is dictated by a number of important interactions involving portions of four discrete regions: the TψC loop, the D loop, the 'extra' loop, and nucleotides U8 and A9 which link the amino acid stem to the D arm. These interactions include highly specific base-stacking forces as well as an extensive network of hydrogen bonds, several of which arise from novel types of interactions between nucleotides. In particular, almost half of the hydrogen bonds involve 2'-hydroxyl groups of ribose moieties as acceptors, donors, or both. Other of the specific hydrogen bonds derive from interactions between three nucleotides (base 'triples') often joining nucleotides widely separated in the primary sequence e.g. A9-U12-A23 and C13-G22-m^{7}G46, triples which help to hold four chains together to augment the D-helix.

There is good reason to believe that the major features of the structure of tRNA^{Phe} as illustrated in Figure 15(b) are common to all tRNA's. Firstly, a generalized model of this sort would account satisfactorily for the observed chemical reactivity and enzymic susceptibility of nucleotides in other species of tRNA. Secondly, the peculiar interactions noted above which are specifically required to maintain the tertiary conformation of tRNA^{Phe} mostly involve nucleotides (encircled in Figure 15(a)) which are invariant in all tRNA's. Where the interactions involve the participation of semi-invariant (bracketed) bases, or bases which are variable between different species of tRNA, it seems that coordinated changes of sequence have occurred so that the tertiary interactions may be preserved. Lastly, some insight may be gained into the significance of certain of the apparently esoteric modifications of the four basic nucleotides in tRNA.

The orientation and conformation of the anticodon arm in the model deserve special consideration. The anticodon stem is not simply continued straight on from the D-helix as is the amino acid arm from the

TψC stem, but is tipped by about 20°. It may, in fact, be 'hinged' into place so that its conformation might possibly alter during protein synthesis. As it stands, the conformation of the anticodon loop bears a striking resemblance to a speculative model suggested years earlier (Figure 16). It has five bases stacked on the 3' side as if to continue one chain of the helical stem, and then makes a sharp turn so that the remaining two pyrimidines in the loop are loosely stacked on the 5' side. This places the three nucleotides of the anticodon in an exposed position at the extreme tip of the molecule, where they can interact *via* hydrogen bonding with a triplet of nucleotides in a mRNA strand as if that strand formed the opposing strand of a regular helix. The nucleotide at the top of the tRNA stack bears the first base of the anticodon, which is known to engage in unusual 'wobble' pairings (see Chapter 6, p. 263) with the third base of the mRNA codon yielding the characteristic pattern of degeneracy in the genetic code (p. 256). Being at the top of the stack, this first anticodon base enjoys special freedom of movement which can account very nicely for its ability to pair with any of two or even three different bases in the third place of the mRNA codon.

NUCLEIC ACID HYBRIDIZATION

The ability of two polynucleotide chains to interact and form a base-paired helical structure represents a sensitive test for complementarity in their base-sequences, and since the base-sequence of a polynucleotide is the chemical expression of its genetic information content the test becomes a measure of the 'genetic relatedness' of the two strands. The phenomenon of renaturation of DNA, which has already been described (p. 192), is an example of sequence matching between strands which presumably exhibit perfect complementarity; however, the test can also be applied in less extreme cases where sequence complementarity is only partial, the one criterion for successful reaction being that such helical regions as can form must be stable under the experimental conditions. Thus the possibility exists, at least in principle, of comparing the genetic information content of DNA from different sources. Moreover, sequence-matching is not restricted to DNA–DNA interactions but can also be observed between single-stranded DNA and RNA. In this case a DNA–RNA 'hybrid' results the structure of which resembles that of double-stranded DNA except that one strand is a polyribonucleotide (with uracil) and the other is a polydeoxyribonucleotide (with thymine). As discussed earlier, the substitution of uracil for thymine would not be expected to

---

Figure 15. Schematic diagram showing (a) the clover leaf form, and (b) the chain-folding and tertiary interactions between bases in yeast tRNA^{Phe}.

In (a) the nucleotides modified by methylation are indicated by m with a superscript number to define the site of substitution; Y represents a fluorescent substituted adenine derivative characteristic of this species of tRNA. In (b) long straight lines indicate base-pairs in the double-helical stems. Shorter lines represent unpaired bases. Dotted lines represent base-pairs outside the helices. After Jack, Ladner & Klug, *J. Mol. Biol.* 108, 619–649 (1976).

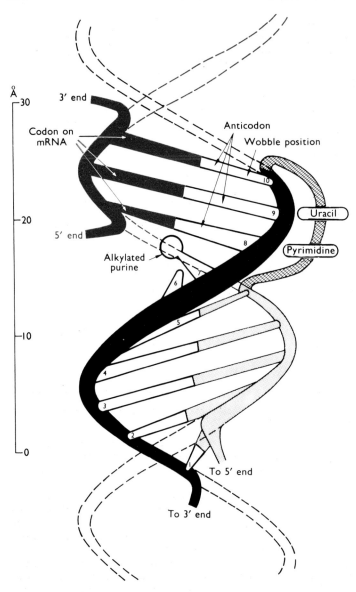

Figure 16. A molecular model for pairing between the anticodon loop of tRNA and a mRNA codon.

The base-pairs of the helical stem of the anticodon arm are arranged in the 11-fold A-type helical conformation and numbered 1–5. Bases 6–10 are stacked on top as if to continue this helix in a regular fashion. Dashed lines indicate the regular geometry of the extrapolated helix. The three bases of the anticodon, at the top of the stack, form regular base-pairs with a mRNA strand whose three-nucleotide codon is positioned as if it formed part of the complementary strand of the continued helix. After Fuller & Hodgson, *Nature* **215**, 817 (1967).

modify the stereochemistry of hydrogen-bonding, so the formation *in vitro* of DNA–RNA hybrids is not surprising on theoretical grounds. Interestingly, the form of a DNA–RNA hybrid helix has been found by X-ray diffraction to be the 11 or 12-fold A-type, which shows that the inability of the ribose sugar ring to assume the puckered conformation needed to give a 10-fold B-type helix dominates and determines the conformation of the deoxyribose-phosphate backbone in the DNA strand.

Analysis of relationships between nucleic acids using these tests of base-sequence complementarity has been immensely valuable in probing the problems concerned in the genetic control of growth and metabolism. Experimental approaches fall into two categories, depending on whether the process of sequence-matching occurs in free solution or with one of the reacting polynucleotides (always the DNA strand) immobilized on a support. The first category is exemplified by the density gradient method of Schildkraut, Marmur and Doty, where one of the reacting species of polynucleotide is obtained from organisms grown in the presence of a heavy isotope (often ^{15}N) so that its density is slightly higher than that of the normal polynucleotide. Normal ('light') and labelled ('heavy') nucleic acids can then be distinguished by centrifuging at high speed for a long time in concentrated solutions of caesium salts (about 7·7 molal). The centrifugal force tends to sediment the heavy Cs⁺ ions, but is opposed by the tendency of the salt to diffuse back and maintain uniformity of concentration throughout the system. The net result is that at equilibrium a concentration gradient exists, with the density of the solution increasing in proportion to the distance from the centre of rotation. The range of densities covered can be chosen so that each type of nucleic acid (containing ^{14}N or ^{15}N) comes to an equilibrium position in the gradient determined by its characteristic buoyant density. Thus to investigate possible sequence complementarity between two samples of DNA, one of the samples is obtained in a 'heavy' form. It is mixed with an equal quantity of the other 'light' sample, the mixture is heated to denature the native helical structures, the denatured material is incubated under conditions suitable for sequence-matching and 'renaturation', and finally the nature of the products is analysed in a CsCl density gradient. Pairing between complementary strands derived from different samples is demonstrated if a native-like hybrid fraction containing molecules with one ^{15}N strand and one ^{14}N strand is found (Figure 17). In the case of *complete* nucleotide sequence homology between the samples, three bands are observed corresponding to renatured ^{14}N-DNA, the hybrid, and renatured ^{15}N-DNA. They are found in the proportion 1 : 2 : 1, which is the proportion expected for random pairing of strands in a mixture containing equal quantities of complementary ^{14}N and ^{15}N strands. Less than 100% homology between the strands of the two samples results in the appearance of a much lower proportion of hybrid molecules.

Density-labelled DNA can also be obtained by growing organisms in a medium containing 5-bromodeoxyuridine (BUDR) which efficiently substitutes for thymidine in DNA synthesis, leading to the formation of 'heavy' molecules containing bromouracil (BU) in place of thymine. The density gradient technique is also applicable to the study of base-sequence complementarity between DNA and RNA. In this case density-labelling is not necessary because the buoyant density of RNA in CsCl is considerably higher than that of DNA, so the DNA–RNA hybrid is readily resolved. The samples of DNA and RNA for analysis are mixed, heated and 'annealed' to form the hybrid under much the same conditions as are used for renaturation of DNA, and then RNAase is added to digest away unhybridized RNA. RNAase does not attack RNA if it is hydrogen-bonded to another polynucleotide, whether DNA or RNA.

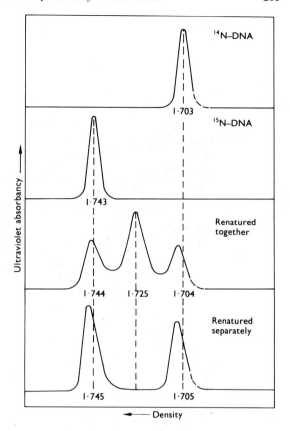

Figure 17. Renaturation between homologous samples of DNA, one density-labelled and the other unlabelled.

The top two curves show the characteristic banding patterns of double-helical DNA from *Bacillus subtilis*, unlabelled and density-labelled. The lowest curve shows the result obtained when samples of each type of DNA are renatured separately and mixed just prior to centrifugation. The curve next to the bottom shows the 1 : 2 : 1 banding pattern which results when the two DNA's are renatured together. Renatured samples must be treated before centrifugation with a phosphodiesterase which specifically attacks single-stranded DNA and thus removes unreacted denatured DNA. After Schildkraut *et al.*, *J. Mol. Biol.* **3**, 595 (1961).

DNA–DNA or DNA–RNA hybrids formed in free solution may also be separated by chromatography on hydroxyapatite columns. This method has the advantage that it is readily applicable on the preparative scale, so that relatively large quantities of hybrids may be obtained. In this way, hybrids between rRNA and DNA, or tRNA and DNA, have been fractionated and purified by several successive cycles of hybridization between the appropriate purified RNA and denatured DNA followed by hydroxyapatite chromatography. The products constitute essentially pure preparations of the bacterial DNA cistrons which code for rRNA's or for tRNA's.

A third technique, applicable to the detection and estimation of DNA–RNA hybrids, consists simply of filtering the solution containing the hybrids through nitrocellulose membrane filters. These filters have the curious property that they absorb denatured DNA and DNA–RNA hybrids, but not RNA or native (or renatured) DNA. Nowadays nitrocellulose filters are mainly used in a different way for hybridization experiments (see below), but recently they have proved invaluable for yet another remarkable property, namely that they also absorb the *lac* repressor (a protein) which binds to DNA containing the *lac* operon, so that labelled DNA which has formed a complex with the repressor is retained by the filter while free, uncomplexed native DNA passes through. This provides a rapid and simple means of measuring repressor—operator gene interaction (see p. 272).

A natural disadvantage of all free-solution hybridization reactions is that renaturation of the separated DNA strands competes with the hybridization reaction. For this reason, most hybridization work is performed using our second category of methods where the denatured DNA is trapped on some supporting material. The support can be an agar gel or, better, nitrocellulose filters. Denatured DNA can be baked on such filters which are then incubated with labelled RNA, treated with RNAase, and washed. It only remains to count the filters to assess the extent of hybrid formation. Only slight modification is needed to enable the same procedure to be used to measure hybridization with labelled denatured DNA. Generally, comparison between the nucleotide sequences of two species of polynucleotides is effected by a process of competition, e.g. by using reaction mixtures containing fixed amounts of denatured DNA on filters and of one species of RNA radioactively labelled, and including varying amounts of the other RNA in unlabelled form. To the extent that the nucleotide sequences of the 'cold' competitor RNA resemble those of the 'hot' RNA they will successfully compete with the 'hot' RNA for binding to the DNA and thus reduce the amount of radioactivity which becomes associated with the filters.

Nucleic acid hybridization has proved a powerful tool for investigating genetic relatedness among microbial species, relationships between bacteriophages and their hosts, and the mechanisms by which the bacterial chromosome is replicated and directs the activities of the cell through the synthesis of RNA. More will be said about these experiments in later sections of this chapter and in Chapter 7 and Appendix A.

It has also made possible two notable advances in the study of gene structure at the level of the DNA molecule: (1) direct mapping of deletion mutations along the length of the molecule, and (2) isolation of the *lac* operon DNA of *E. coli*.

(1) If DNA's from a deletion mutant and the corresponding wild-type organism are denatured and annealed together, the resulting hybrid molecules are visibly shorter than the wild-type DNA in the electron microscope and a 'loop' is visible at the position of the deletion (Figure 18). The 'loop' is formed by the sequence in the wild-type strand which is missing in the strand from the mutant.

(2) Beckwith and his colleagues isolated two specialized transducing phages which had incorporated the *E. coli lac* operon into their DNA's in opposite orientations. DNA from each phage was denatured and the complementary H and L strands separated by complexing with synthetic poly U,G (for the basis of this separation see p. 230). The H strands from each phage, which were complementary for the genes of the *lac* operon but not for the phage genes, were annealed to allow the *lac* sequences to interact and form a double helix, and the four non-complementary single-stranded tails of the helical section were finally digested away with a single-stranded DNA-specific nuclease (Figure 19). The remaining material was a duplex DNA containing most of the genes of the *E. coli lac* operon.

Synthetic polynucleotides

From what has already been said it will be appreciated that synthetic polynucleotides have contributed significantly to our understanding of the structure and function of natural nucleic acids. Outstanding, however, is the role they have played in the elucidation of the genetic code (p. 255). The first synthetic polynucleotides to be made were homopolymers—essentially RNA molecules containing only a single type of nucleotide. These are readily synthesized *in vitro* by polynucleotide phosphorylase, an enzyme which is abundant in many species of bacteria but whose role *in vivo* is obscure. It catalyses the following reaction:

$$nXDP \rightleftharpoons (XMP)_n + nPi$$

X can be a nucleoside containing adenine, guanine, uracil or cytosine—or indeed hypoxanthine or other

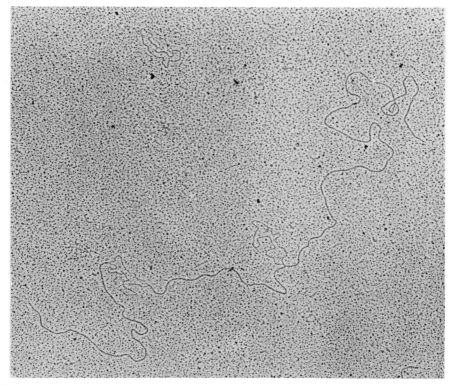

Figure 18. Electron micrograph of a heteroduplex (hybrid) DNA molecule formed by annealing denatured wild-type λ DNA with denatured DNA from a double mutant, λb2b5.

The b2 mutation is a deletion of slightly more than one-tenth of the total nucleotide sequence, revealed here as a loop located approximately half way along. The b5 mutation is a substitution in which a sequence present in the wild-type has been lost and replaced by a shorter, unrelated sequence: it is visible here as a non-helical region near the bottom left-hand end of the photograph. Note the difference in lengths of the mutant and wild-type single strands. At the top of the picture is a small circular single-stranded DNA molecule (φX174 DNA) included to act as a length calibration; its circumference is 1·9 μm. Photograph by Dr M. Wu kindly provided by Prof. N. Davidson.

bases. The products are polyadenylic acid, etc. (poly A, poly G, poly U and poly C) with IDP, the nucleoside diphosphate containing the base hypoxanthine, giving polyinosinic acid (poly I). Poly I is structurally analogous to poly G since hypoxanthine has a 6-hydroxy group although it lacks the 2-amino group of guanine. Because poly G is difficult to prepare, poly I is often used instead in model systems. When more than one nucleoside diphosphate is present the product of polynucleotide phosphorylase action is a *co-polymer* with a random sequence of nucleotides.

Double-stranded helical structures form when 'equimolar' solutions of poly A and poly U, or of poly I and poly C, are mixed. 'Equimolar' in this case means solutions which contain equal quantities of the corresponding nucleotides, rather than equal numbers of polymeric molecules which may be of different sizes. In effect, the complementary polynucleotides simply wrap round each other in a helical configuration. Here there is no question of aligning the molecules so that uniquely complementary sequences come into register, and the reaction proceeds very rapidly and efficiently at room temperature. The structures formed have been characterized as hydrogen-bonded double helices by the usual criteria, including X-ray diffraction patterns and thermal denaturation behaviour showing a sharp melting temperature at which a large hyperchromic effect occurs as the strands separate.

Later, small oligodeoxyribonucleotides were prepared by chemical synthesis, and it turned out that they acted as excellent templates for reiterative copy-

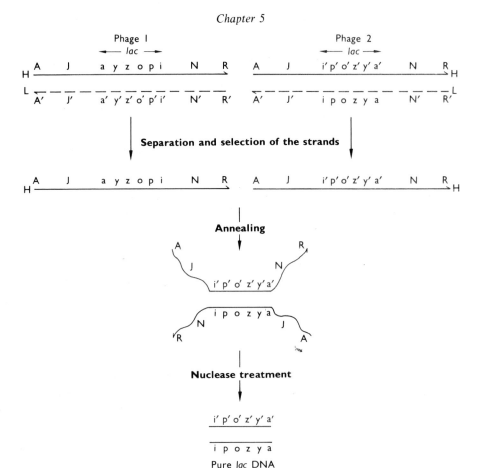

Figure 19. Isolation of *lac* operon DNA of *Escherichia coli*.

DNA molecules which constitute the genomes of two bacteriophages, A and B, are shown. Each contains the *lac* operon, but the orientation of the *lac* region is opposite in the two genomes as shown by the order of the genetic markers. A, J, N and R are markers on the phage chromosome. The *lac* markers are the repressor structural gene (*i*), the promoter (*p*), the operator (*o*), the β-galactosidase structural gene (*z*), the *lac* permease structural gene (*y*), and the galactoside transacetylase structural gene (*a*). Complementary sequences are indicated by primes. The 5′ end of each DNA strand is indicated by an arrowhead, and H and L designate the complementary strands. From Shapiro *et al.*, *Nature* **224**, 768 (1969).

ing by DNA and RNA polymerases (see below). By this means an impressive array of synthetic polynucleotides was prepared, including some which contain repeating di-, tri- or tetra-nucleotide sequences, both deoxyribo- and ribo-. These, too, proved invaluable in checking the genetic code (p. 256). A prodigious extension of this approach resulted in the total synthesis, in Khorana's laboratory, of the gene for a tRNA. This is a DNA double helix 77 base-pairs long whose nucleotide sequence corresponds precisely to that determined by Holley for the alanine tRNA of yeast (Figure 10). It was assembled (Figure 20) by synthesizing pairs of complementary oligonucleotides which formed small helical sections with protruding ends; each end was made complementary to that of the next section of the molecule such that the sections could associate by hydrogen bonding, and then the gaps in the chains were sealed by the polynucleotide ligase enzyme (see below). The availability of synthetic genes, as well as the isolation of specific operons (p. 317), opens up exciting possibilities for the study of gene transcription and translation *in vitro*.

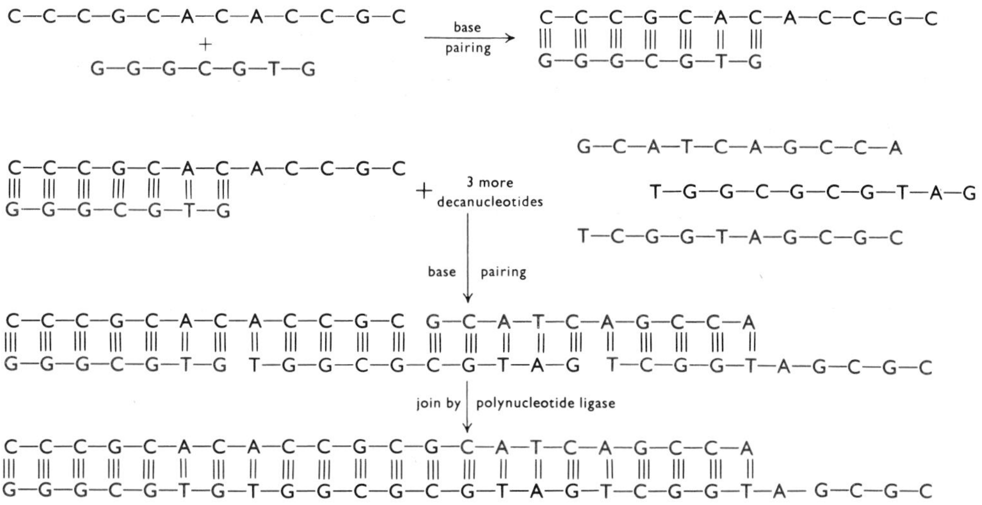

C—C—C—G—C—A—C—A—C—C—G—C
        +   → base pairing
G—G—G—C—G—T—G

C—C—C—G—C—A—C—A—C—C—G—C
||| ||| ||| ||| ||| || |||
G—G—G—C—G—T—G

C—C—C—G—C—A—C—A—C—C—G—C
||| ||| ||| ||| ||| || |||
G—G—G—C—G—T—G

      + 3 more decanucleotides

G—C—A—T—C—A—G—C—C—A

T—G—G—C—G—C—G—T—A—G

T—C—G—G—T—A—G—C—G—C

base | pairing

C—C—C—G—C—A—C—A—C—C—G—C G—C—A—T—C—A—G—C—C—A
||| ||| ||| ||| ||| || ||| || ||| ||| ||| ||| ||| || || ||| || ||| ||| ||| ||
G—G—G—C—G—T—G T—G—G—C—G—C—G—T—A—G T—C—G—G—T—A—G—C—G—C

join by | polynucleotide ligase

C—C—C—G—C—A—C—A—C—C—G—C—G—C—A—T—C—A—G—C—C—A
||| ||| ||| ||| ||| || ||| || ||| ||| ||| ||| ||| || || ||| || ||| ||| ||| ||
G—G—G—C—G—T—G—T—G—G—C—G—C—G—T—A—G—T—C—G—G—T—A— G—C—G—C

Figure 20. Assembly of a portion of the synthetic gene for yeast alanine tRNA. After Agarwal *et al.*, *Nature* **227**, 27 (1970).

SYNTHESIS OF DNA

The genetic material of a cell must have two properties. It must be able to be replicated exactly and it must be able to provide information for the synthesis of the determinants of all structure and function, i.e. the proteins. The physical basis for both of these properties is now believed to depend upon the capacity of uniquely ordered sequences of bases in DNA to act as templates for polynucleotide synthesis. We have already seen how the phenomenon of complementary base-pairing enables DNA molecules to assume a rigidly defined, highly stable, secondary structure and we can now consider how the same base-pairing mechanisms underlie the processes of replication and transcription of the coded nucleotide sequence.

The scheme of DNA replication outlined on p. 26 is now almost universally accepted. Because the two strands of the original double helix become separated and are subsequently incorporated into different daughter double helices, the mechanism is called *semiconservative* to distinguish it from a fully *conservative* mechanism where both strands of one duplex would be newly synthesized and the other duplex would consist of the old strands associated together in their original state. The principal evidence which demonstrates the *semiconservative* mode of DNA synthesis in growing bacteria comes from the classic experiment of Meselson and Stahl in 1958. They grew *Escherichia coli* in a medium containing $^{15}NH_4Cl$ as the sole source of nitrogen; after fourteen generations of growth in this medium, the nitrogen of the bases in the DNA of the organisms was virtually all ^{15}N so that this DNA was 'heavy' and could be separated from normal ^{14}N-containing ('light') DNA in a CsCl density gradient. The organisms were then abruptly changed to a medium containing ^{14}N and samples containing equal numbers of bacteria were removed at intervals over the next few generations. The bacteria in each sample were lysed by addition of detergent and the crude lysate was centrifuged in concentrated CsCl until the DNA had banded at its characteristic density (Figure 21). After a short period of growth in the new medium the band formed by the ^{15}N-DNA of the organisms began to diminish and a new band appeared at a density exactly halfway between that of ^{15}N-DNA and that of ^{14}N-DNA. This new band, formed by DNA molecules containing equal proportions of ^{15}N and ^{14}N, became stronger until after one complete generation, it was the only form of DNA present. During the next generation a band appeared corresponding to DNA containing ^{14}N only; this band increased but the hybrid band remained until at the end of the second generation equal quantities of the two types were present. During subsequent generations the ^{14}N-DNA band continued to increase in relation to the hybrid band.

These results were as expected for semiconservative replication, and the appearance of a band corresponding to hybrid molecules is strong evidence against a fully conservative mechanism. The nature of the hybrid band can be investigated by heating the DNA beyond the 'melting temperature' or by raising the

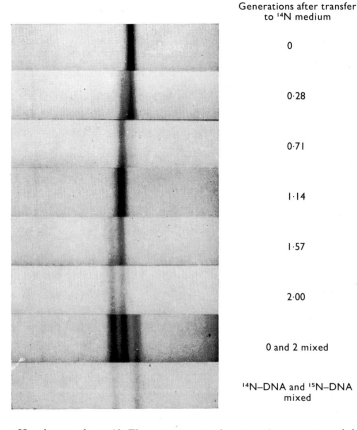

Generations after transfer
to ^{14}N medium

0

0·28

0·71

1·14

1·57

2·00

0 and 2 mixed

^{14}N–DNA and ^{15}N–DNA
mixed

Figure 21. CsCl density gradient analysis of *Escherichia coli* DNA after transfer of ^{15}N-labelled organisms to ^{14}N-medium.

The photographs, taken with ultra-violet light, reveal the positions of the DNA molecules in the ultracentrifuge cell as dark bands. The density increases from left to right. Photographs kindly provided by Dr M. Meselson.

pH value to about 13. These treatments denature the DNA and the resulting single-stranded material can be analysed in a CsCl density gradient to see if individual strands are hybrid. However, only two bands are observed, having the densities of denatured ^{15}N-DNA and denatured ^{14}N-DNA. The relative proportions of bands of different densities also indicate semiconservative replication (Figure 22) and show that the whole bacterial chromosome is reproduced in this fashion.

The mechanism of semiconservative replication

For the two strands of a parent double helix to appear as components of the daughter double helices it is necessary that the original strands become separated either before or during the replication process. Such separation requires expenditure of energy as is shown by the need for a high temperature to provide enough thermal energy to dissociate the strands of DNA *in vitro*. Moreover, because the double helix is a coiled structure, there must be physical rotation of the unreplicated part relative to the new daughter helices, a process which will be

opposed by a considerable viscous drag. Finally, it must be remembered that the DNA of *Escherichia coli* appears to be a single molecule 1·1–1·4 mm long, that it is circular, and that it is packed into a cell about 0·5 μm wide and 2 μm long. Thus, even more than twenty-five years after the announcement of the double-helical structure of DNA, the synthesis at a growing point which moves along the chromosome formed a Y-shaped structure still presents formidable problems.

Autoradiographic studies by Cairns in 1963 on the DNA of *Escherichia coli* first showed how a circular chromosome might be replicated from a single origin (Figure 23). A thymine-requiring strain was grown in [^{3}H] thymidine for about two generations; the labelled DNA was extracted very gently and embedded; it was allowed to expose a film as described earlier (p. 186). After the first generation the DNA had one strand radioactive and the other not; during the next cycle of growth these 'hybrid' duplexes were replicated in a semiconservative fashion to give one daughter helix with both strands labelled and the other with only one labelled. The difference in labelling could be distinguished because the density of

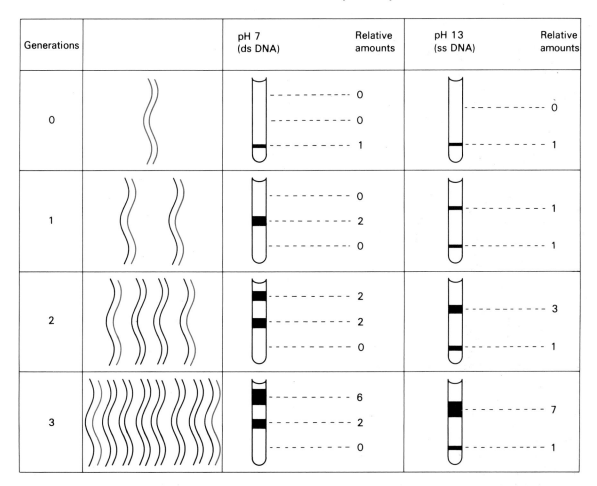

Generations		pH 7 (ds DNA)	Relative amounts	pH 13 (ss DNA)	Relative amounts
0			0 0 1		0 0 1
1			0 2 0		1 1
2			2 2 0		3 1
3			6 2 0		7 1

Figure 22. Semi-conservative replication of DNA (cf. Meselson & Stahl experiment, Figure 21).

silver grains produced by a given length of the former was double that of the latter. Visualization of the replicating intermediates of many circular DNA's, bacterial and viral, shows that a θ-like shape appears on initiation of the replicating bubble at the origin (Figure 24). That the replication is bidirectional in many systems has been shown by ingenious experiments in which, after a short pulse of low specific activity radioactive precursor, higher activity precursor is added (see Figure 25).

In Chapter 2 we saw that DNA replication in *Escherichia coli* takes about 40 min and is continuous from initiation to completion. If the doubling time for the cells is less than 40 min, then a second growing-point (origin) appears before completion of the existing round of chromosome replication. We can calculate the rate of growth *in vivo* of daughter chains at the replication fork. The non-replicating chromosomal DNA has $M_r = 2.5 \times 10^9$ (in agreement with the measured length of 1·1–1·4 mm) and this corresponds to 8×10^6 nucleotides. If the two forks traverse this in 40 min, the rate of incorporation of new nucleotides is 2×10^5 per min, or 3300 per sec. Since two new chains are growing at each fork, the rate of growth of each is about 825 nucleotides per second. Incidentally, assuming that the DNA helix has ten base-pairs per turn, the rate of unwinding of the helix at each fork can be calculated as about 5000 revolutions per minute.

Direction of synthesis of DNA

The complementary strands of DNA are anti-parallel, i.e. their 3' to 5' sugar-phosphate linkages run in opposite directions (see Figures 3 and 4). Consequently, simultaneous replication of both strands at a single fork implies that one daughter strand is being extended at its 3' end while the other is growing at its 5' end (Figure 26a). As will be seen, there is abundant

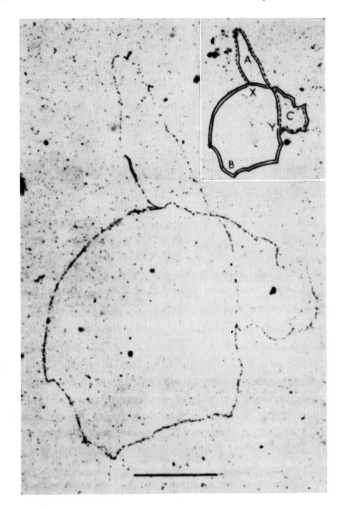

Figure 23. Autoradiograph of the chromosome of *Escherichia coli* K 12 Hfr labelled with ³H-thymidine for two generations. The scale shows 100 μm. *Inset*, the same structure represented diagrammatically (labelled strands ——, unlabelled strands - - - -) and divided into three sections (A, B and C) which arise at the two forks X and Y.

Photographs kindly provided by Dr J. Cairns.

evidence for elongation in the $5' \rightarrow 3'$ direction, both *in vivo* and with enzymic synthesis *in vivo*, but none for addition of nucleotides to the 5' end, giving elongation in a $3' \rightarrow 5'$ direction.

Okazaki fragments

The likely solution to this problem was provided by the discovery of Okazaki in 1964 that in cells infected with phage T4, much of the newly-synthesized DNA was in pieces 1000–2000 nucleotides long. These 'Okazaki fragments' are synthesized in the $5' \rightarrow 3'$ direction as can be shown by labelling with radioactivity for as little as 2 sec, isolating the nascent DNA as single strands by centrifugation in an alkaline sucrose gradient, and then treating them with exonucleases specific for the 3' and the 5' ends. Okazaki suggested that *discontinuous* synthesis in a $5' \rightarrow 3'$ direction, followed by joining of the fragments, could explain

the apparent $3' \rightarrow 5'$ synthesis. The 'leading' strand might be made continuously $(5' \rightarrow 3')$ and the 'lagging' strand discontinuously (Figure 26b) or both might be discontinuous (Figure 26c).

Similar Okazaki fragments have been found in other bacterial and eukaryotic systems and, although some pieces may result from excision of uracil misincorporated into DNA (see below, p. 225), the pool of fragments is believed by most workers to contain precursors for discontinuous synthesis of DNA.

Polynucleotide ligase

A mechanism for joining DNA pieces together is required not only for Okazaki fragments in replication, but also for processes involved in genetic recombination (see p. 289), the repair of DNA molecules (see p. 226), and a number of others such as interaction of the chromosome with prophages, sex factors, etc. (see

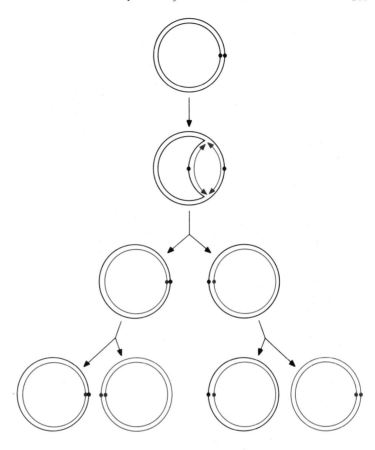

Figure 24. Representation of bidirectional replication of circular DNA.

Chapter 7). The appropriate enzymes, called *polynucleotide ligases*, having such properties, were found in 1967 in *Escherichia coli* and in T4-infected *E. coli*. They catalyse the formation of a 3′–5′ phosphodiester linkage between two DNA chains, one with a 5′-phosphate and the other with a free 3′-hydroxyl. The two pieces must be exactly adjacent and be H-bonded to a continuous complementary strand of DNA (Figure 27).

The involvement of polynucleotide ligase in DNA replication has been confirmed by the existence of temperature-sensitive mutants which produce enzyme active at 20° but not at 43–44°. At the latter temperature, large amounts of Okazaki fragments accumulate but on cooling to the lower temperature, they disappear. The ligase reaction is highly specific—it will not work with a 3′-phosphate and a 5′-hydroxyl—and a co-

Figure 25. Bidirectional replication of the *Escherichia coli* chromosome. After a short pulse of [³H] thymine (5 Ci per mmol), high specific activity [³H] thymidine (52 Ci per mmol) was added. After six minutes the DNA was isolated and autoradiographed. A high density of silver grains is seen on *both* sides of the origin, indicating bidirectional synthesis. After P. L. Kuempel *et al.* in 'DNA Synthesis *in vitro*' (R. Wells and R. Inman, editors) pp. 463–472. University Park Press, Baltimore. 1972.

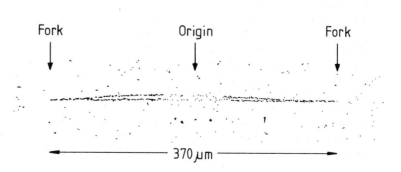

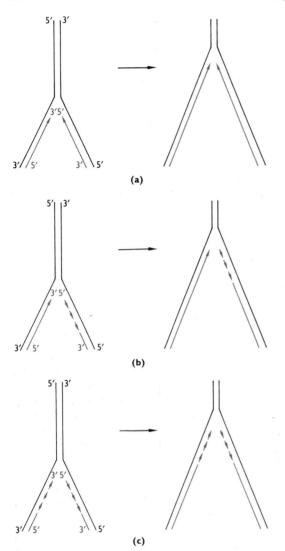

(a)

(b)

(c)

factor is needed: NAD for the *E. coli* enzyme and ATP for the enzyme induced by phage T4. In each case an enzyme-AMP intermediate is involved:

$$E + NAD \text{ (or ATP)} \rightleftharpoons E\text{-AMP} + NMN \text{ (or PPi)}$$

$$E\text{-AMP} + \text{(P)-5'DNA} \rightleftharpoons E[\text{AMP-(P)-5'DNA}]$$

$$DNA\text{-3'OH} + E[\text{AMP-(P)-5'DNA}] \rightleftharpoons$$
$$DNA\text{-3'O(P)-5'DNA} + AMP + E$$

$$DNA\text{-3'OH} + \text{(P)-5'DNA} + NAD \rightleftharpoons$$
$$\text{(or ATP)}$$

$$DNA\text{-3'O(P)-5'DNA} + AMP + NMN$$
$$\text{(or PPi)}$$

The substrate is, in effect, a 'nicked' double-stranded DNA, because a break or 'nick' like that shown in Figure 27 is produced by a single cleavage with an endonuclease. Such nicks can readily be repaired by the ligase.

DNA POLYMERASES

In 1956 Arthur Kornberg found an enzyme in *Escherichia coli* which appeared to be exactly what was necessary for the replication of DNA. *In vitro*, if given

Figure 26. Possible mechanisms of elongation of DNA chains at the replicating fork.

(a) By direct addition of nucleotides to the growing ends of both daughter strands. (b) By discontinuous synthesis of one strand by joining of fragments. (c) By discontinuous synthesis of both daughter strands.

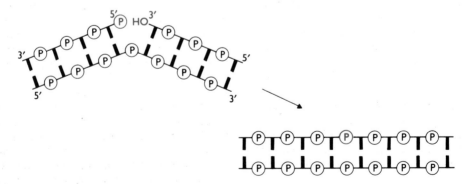

Figure 27. Joining of polydeoxyribonucleotide strands by the action of polynucleotide ligase.

the four deoxyribonucleotides* (dATP, dGTP, dCTP, and dTTP) together with Mg^{2+} and a preparation of DNA, it catalysed the synthesis of more DNA of the same composition:

$$dATP + dGTP + dCTP + dTTP \xrightarrow[Mg^{2+}]{DNA} \text{More DNA} + PPi$$

The other product of the reaction is inorganic pyrophosphate. If one of the dNTP's is labelled with ^{32}P in the α-position, the reaction can be followed by measuring the conversion of the radioactivity from acid-soluble to acid-precipitable form:

$$
\begin{array}{l}
n_1 \text{ dATP*} \\
n_2 \text{ dGTP} \\
n_3 \text{ dCTP} \\
n_4 \text{ dTTP}
\end{array}
\xrightarrow[Mg^{2+}]{DNA}
\begin{bmatrix}
n_1 \text{ dAMP*} \\
n_2 \text{ dGMP} \\
n_3 \text{ dCMP} \\
n_4 \text{ dTMP}
\end{bmatrix}
+ (n_1 + n_2 + n_3 + n_4)PPi
$$

(acid-soluble)   (acid-insoluble)
                              DNA

Omission of any one of the four dNTP's, or of the DNA, prevents the reaction although, after a long delay, the enzyme with dATP and dTTP alone will catalyse synthesis of a large poly (dA-dT) copolymer containing alternating residues of dAMP and dTMP. With dGTP and dCTP alone, it makes a mixture of homopolymeric poly (dG) and poly (dC). These artificial reactions do not have biological significance.

This enzyme was at first called DNA polymerase but is now known as DNA polymerase I, or Pol I, because there are others, both in *Escherichia coli* and other organisms. By using various DNA preparations much was discovered about its action and many of its properties are common to those of other DNA polymerases. These can be summarized as follows:

(a) *Direction*: all synthesize DNA from $5' \rightarrow 3'$, none from $3' \rightarrow 5'$.
(b) *Precursors*: all use the four dNTP's, eliminating inorganic pyrophosphate.
(c) *Primer*: all require an oligonucleotide primer, complementary to, H-bonded to, and anti-parallel to a DNA template; the primer must have a free 3'-OH and may be composed of ribonucleotides rather than deoxyribonucleotides.
(d) *Template*: all use a DNA strand as template to produce a complementary strand.
(e) *Product*: all synthesize DNA covalently linked to the 3'-OH end of the primer and complementary to, H-bonded to, and anti-parallel to the template:

† These will be collectively referred to as dNTP's.

(f) *Proof-reading*: prokaryotic DNA polymerases have $3' \rightarrow 5'$ exonuclease activity which allows excision of mispaired nucleotides incorporated in error:

(note that the eukaryotic enzymes do *not* have this nuclease activity.)

Accuracy (or 'fidelity') in copying DNA is essential and, while it has been argued from chemical considerations that the intrinsic mistake frequency in incorporating nucleotides is unlikely to be less than 1 in 10^5, by having a proof-reading step this can perhaps be reduced to 1 in 10^{10} which would approach a tolerable level of mutational load.

In chemical terms, the mechanism of chain growth is believed to be *via* nucleophilic attack by the 3'-OH of the terminal nucleotide of the growing chain on the 5'-pyrophosphate of the incoming triphosphate (Figure 28).

Kornberg's DNA polymerase (Pol I) seemed to have most of the appropriate properties for it to be the replicating enzyme but in 1969 de Lucia and Cairns reported the isolation of a mutant of *Escherichia coli*, extracts of which contained less than one percent of the normal level of this polymerase. However, the strain grew and replicated its DNA at normal rates. In the following year a second polymerase (Pol II) was described and in 1971, a third (Pol III). All three are present in wild-type *E. coli* and have the properties described above but they differ in other respects (Table 3). The most notable differences are the template-primer requirements which are less stringent for Pol I, and the existence of the $5' \rightarrow 3'$ exonuclease activity only in Pol I. Although this enzyme is a single polypeptide chain, it can be cleaved by trypsin into a larger piece ($Mr = 76\,000$) having polymerase and

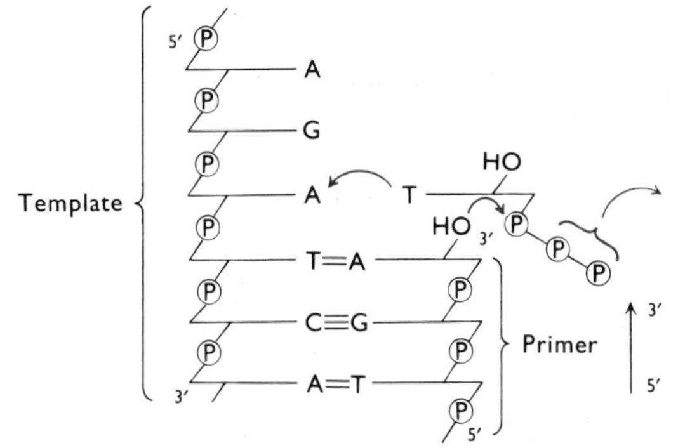

Figure 28. Mechanism of chain growth catalysed by DNA polymerase. After A. Kornberg, *Science* **163**, 1410 (1969).

$3' \rightarrow 5'$ exonuclease activities, and a smaller fragment ($Mr = 36\,000$) which has the $5' \rightarrow 3'$ exonuclease activity. The three activities have important roles in DNA replication (gap-filling, 'proof-reading', and excision of primer) and are illustrated in Figure 29.

It seems that five recognition sites are involved in the active centre(s) of Pol I (Figure 30). However, although this enzyme cannot be the major polymerase concerned in chromosome replication, nevertheless it was used by Kornberg in 1967 for the synthesis *in vitro* of biologically active, infective bacteriophage ϕX174 DNA. The procedure is outlined in Figure 31. The circular single-strands of DNA from the virus

(+ strands) were incubated with Pol I and ligase to produce complementary circles (− strands) which were density-labelled with bromouracil (see p. 205). A small oligonucleotide had to be added as primer as well as the circular template. The partly synthetic replicative form (double-circular RF) was nicked with an endonuclease to permit separation of the strands, and the synthetic (−) strands were purified on a CsCl gradient. Half of them were circular and half were linear, because the nuclease could nick either strand of the RF with equal probability. The density-labelled (−) strands were then used as templates for a second round of the same procedures leading to the formation of wholly synthetic, non-density-labelled (+) strands. These were infective for *Escherichia coli*. This remarkable experiment showed the impressive fidelity of complementary copying by the polymerase, for each strand contains 5377 nucleotides and the biologically active sequence was preserved through two cycles of synthesis.

DNA REPLICATION IN BACTERIOPHAGES

Since the first synthesis of biologically active, DNA, much has been learned from the study of temperature-sensitive mutants (*dna* mutants) and much use has

Table 3. DNA polymerases of *Escherichia coli*: Comparison of properties of Pol I, Pol II, and Pol III.

	Pol I	Pol II	Pol III
Molecular size (*Mr*)	109 000	120 000	140 000 + 40 000
Activity (nucleotides added per second)	16	1	250
Molecules per cell	400	40	10
Sensitivity to SH-blocking agents	+	−	−
Inhibition by antiserum to Pol I	+	−	−
Template/primer			
Nicked DNA	+	−	−
Gapped DNA	+	Short gap +	Short gap +
Long template with short primer	+	−	−
Polymerization $5' \rightarrow 3'$	+	+	+
Exonuclease $3' \rightarrow 5'$	+	+	+
Exonuclease $5' \rightarrow 3'$	+	−	−
Probable function	Excision of primer, filling of gaps between DNA fragments	?	Addition to primer to form fragments or continuous strand

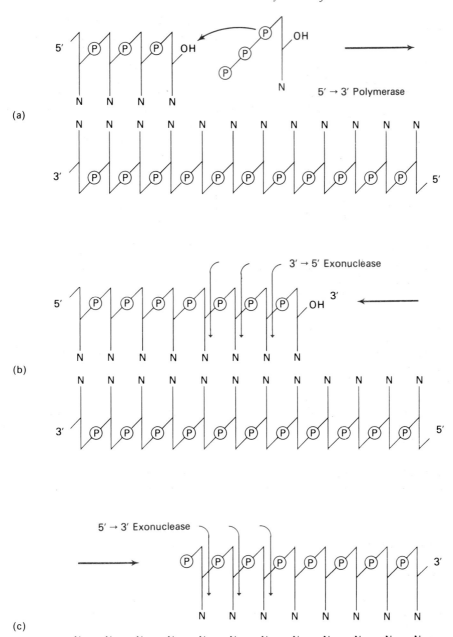

Figure 29. Polymerase and exonuclease activities of DNA Polymerase I (Pol I).
(a) Polymerase: addition of nucleotide residues $5' \rightarrow 3'$.
(b) $3' \rightarrow 5'$ Exonuclease: hydrolysis of successive nucleotides.
(c) $5' \rightarrow 3'$ Exonuclease: hydrolysis of successive nucleotides.

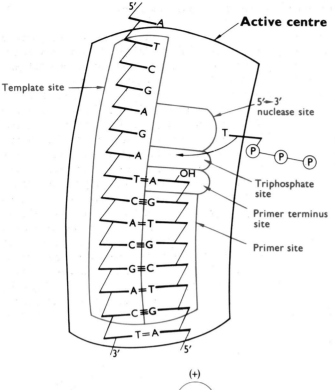

Figure 30. Kornberg's model to account for multiple catalytic sites within the active centre of DNA polymerase I. After A. Kornberg, *Science* **163**, 1410 (1969).

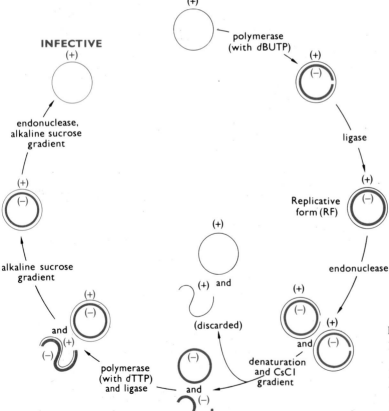

Figure 31. Synthesis of infective φX174 DNA *in vitro*.

*d*BUTP = bromodeoxyuridine triphosphate. Density-labelled strands are drawn with a thick line. Synthetic strands are shown in red. After M. Goulian *et al.*, *Proc. Nat. Acad. Sci. U.S.* **58**, 2321 (1967).

been made of bacteriophage systems in order to unravel the problems of the synthesis of DNA *in vivo*. It is now apparent that there are several different mechanisms, but all involve a number of proteins and a number of steps. These usually include (a) recognition of the initiation site (the 'origin'); (b) synthesis of primer RNA or exposure of single-stranded DNA to act as primer; (c) extension of primer by a DNA polymerase; (d) degradation of the primer if it is RNA; (e) filling of the gap thus left. They may also include (a') nicking of a circular DNA to prepare an initiation site and (f) sealing a nick to complete a circular molecule. Further, it seems that most DNA replication at some stage involves interaction of single-stranded DNA with a protein(s) variously called DNA-unwinding, DNA-binding (DBP), helix-destabilising (HDP), or single-strand binding (SSB) protein. In addition, several unwinding enzymes (*E. coli* unwinding enzyme I, *E. coli rep* protein, *E. coli dnaB* gene product) have been described which are DNA-dependent ATPases.

Perhaps the simplest examples of DNA replication are those in which some single-stranded DNA bacteriophages (such as G4, M13, fd) have their DNA converted to the duplex *replicative form*, RF, by the synthesis of a complementary strand on the viral strand. The recognition site is probably a region rich in G and C which is capable of intra-chain H-bonding. The single-strand circle combines with SSB protein and the primer is synthesized. This cannot be DNA since no known DNA polymerase can *initiate* a new strand: all merely extend an existing primer. On the other hand, RNA polymerases can begin *de novo*. Formation of phages M13 and fd is inhibited by rifampicin, an antibiotic which specifically prevents RNA initiation (see p. 235) and this has helped to prove that the primer is RNA—formed by the action of the host *Escherichia coli* RNA polymerase making a complementary copy of part of the DNA. In phage G4, the primer is synthesized by a different enzyme, the product of the *E. coli* gene *dnaG*, a protein of about

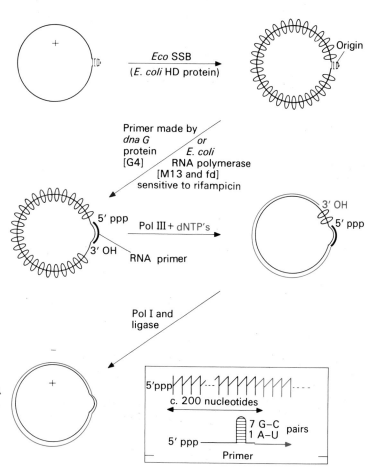

Figure 32. Formation of replicative form (RF) from single-stranded DNA bacteriophages such as G4, M13 and fd. The origin for RNA primer synthesis is thought to be rich in G and C and capable of intra-chain H-bonding (see insert).

65000, of which there are about ten molecules per cell. The holoenzyme DNA polymerase III which has four subunits including the catalytic 140 000 polypeptide, now uses the primer and completes the synthesis of the strand complementary to the template. DNA polymerase I (Pol I) then excises the RNA primer and fills the gap. Finally, DNA ligase seals the two ends together producing the double-helical circular RF (Figure 32).

The formation of ϕX 174 RF is much more complicated and involves not fewer than eleven separate proteins, at least six of which prepare the recognition site for the synthesis of RNA primer by the *dnaG* protein (Figure 33).

The formation of more viral strand DNA (as opposed to complementary strand) occurs by the so-called 'rolling circle' mechanism. This is illustrated for ϕX 174 in Figure 34. No primer synthesis (or degradation) is necessary as the existing viral strand (+) acts as the primer. The RFI DNA is negatively superhelical and a single nick is made by the ϕX-coded *cisA* protein which itself becomes cova-

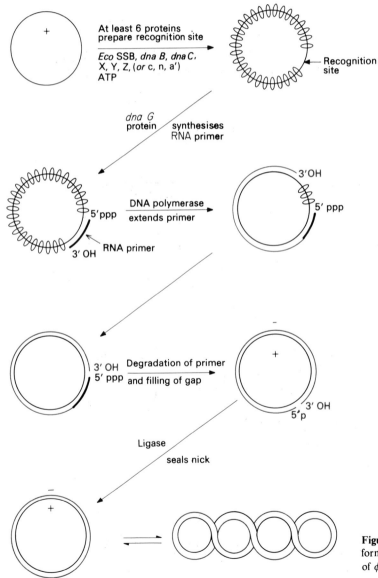

Figure 33. Formation of replicative form (RF) from single-stranded DNA of ϕX 174.

lently attached to the 5'-phosphate so exposed. The *E. coli rep* protein (with ATPase activity) causes unwinding of this strand from the still circular (−) strand and the (+) strand becomes coated with SSB protein which stabilizes it in the single-stranded form. Holoenzyme Pol III synthesizes the complementary copy using the (−) strand as template and adding to the 3'-OH of the (+) strand. When a length corresponding exactly to that of (+) strand has been completed, a cut is made by a specific endonuclease and the piece released has its ends joined, probably by the nick-closing activity associated with the *cisA* protein.

The bacterial viruses so far discussed contain single-stranded circular DNA, which is the (+) or sense strand. For reproduction (*in vivo* and *in vitro*) the complementary or (−) strand has first to be made, yielding the double-stranded circular replicative form which can either be multiplied as such or can be used in the formation of closed circular single (+) strands for inclusion in new phage particles. Other bacteriophages contain double-stranded linear (T1–T7 and λ phages of *Escherichia coli*) or double-stranded circular DNA (PM2 of *Pseudomonas spp.*).

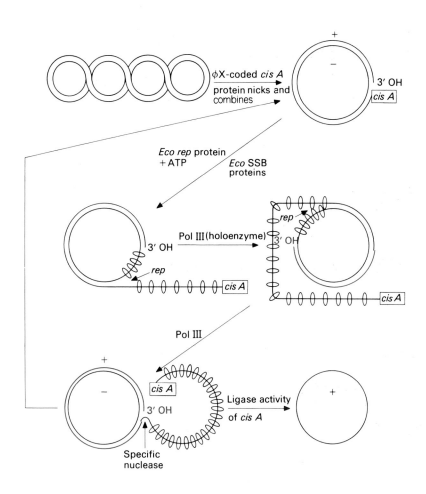

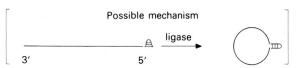

Figure 34. Formation of single-stranded φX 174 DNA by rolling circle mechanism from double-stranded RF form.

REPLICATION OF THE BACTERIAL CHROMOSOME

The problems of DNA replication in bacteria compound most of those found in viral systems and include others. The chromosome is an enormously long, double-stranded, closed circular molecule of DNA (see above p. 186). Replication begins at a unique, specific site (the origin) and involves local supercoiling, and unwinding, and separation of the two antiparallel strands. Synthesis by DNA polymerase cannot begin before formation of RNA primer and must proceed discontinuously on at least one of the two template strands. Primers must be excised and gaps filled before the 'Okazaki fragments' are joined to form continuous strands. Ultimately, the two progeny double helices have to be circularized.

At least fourteen *dna* genes are known to be involved in sustained DNA synthesis in *Escherichia coli*.

Of these, six encode proteins which have specific replication functions *in vitro*, but none of the mutations of the other eight has yet been associated with a defective replication protein (Table 4). Furthermore, at least six polypeptides shown to be essential in phage DNA replication have not yet been connected with any of these genes.

We can now elaborate a tentative model of chromosome duplication in *Escherichia coli*. The unique origin is probably related to a site of attachment of the DNA to the membrane. Initiation may only occur in a negatively supercoiled region introduced by the enzyme *DNA gyrase* which may also relieve any positive supertwist which occurs in front of the replication fork as the strands unwind during replication. The gyrase (also called *topoisomerase II*) is a tetramer of two pairs of chains (M_r 105 000 and 95 000), coded for by *nalA* (or *gyrA*) and *cou* (or *gyrA*) and is inhibited by nalidixic acid and by coumermycin (and the less

Table 4. Genes and proteins involved in DNA synthesis in *Escherichia coli*.

Locus	Protein
dnaB	c. 250 000 (48 000 subunit); c. 10 molecules per cell; ssDNA-dependent rNTPase (ATPase)
dnaC (D)	c. 25 000; interacts with *dnaB* protein
dnaE (polC)	180 000; c. 10 molecules per cell; DNA polymerase III
dnaG	c. 65 000; c. 10 molecules per cell; 'primase'; RNA polymerase; makes ribo-, deoxyribo-, and mixed oligonucleotides
dnaZ (H)	c. 125 000; functions with Pol III as part of holoenzyme
ssb-I	c. 80 000 (18 500 subunit); ssDNA-binding protein; helix-destabilizing protein, HDP; DNA-binding protein, DBP
dnaA *dnaI* *dnaJ* *dnaK* grpC *dnaL* *dnaM* *dnaP* *dnaQ*	Some of the proteins may be involved only in initiation
	Other proteins
lig	DNA ligase
polAex	5′→3′ exonuclease activity of DNA polymerase I
rpoB	RNA polymerase β subunit
nalA (gyrA)	105 000 subunit of DNA gyrase
cou (gyrB)	95 000 subunit of DNA gyrase
rep	c. 67 000; involved in replication of ϕX174 RFI; probably in *E. coli*; DNA-dependent ATPase; unwinding enzyme
?	DNA elongation factor I; β component of DNA polymerase III holoenzyme
?	DNA elongation factor III; c. 65 000; γ or δ component, forms complex with *dnaZ* protein
?	Replication factors X, Y, Z; factors i, n, a
dnaF	Ribonucleotide reductase—no longer considered to be a *dna* gene
dnaS	dUTPase—no longer considered to be a *dna* gene

active novobiocin). It probably becomes wrapped round a stretch of double-helical DNA, introducing compensatory positive and negative supercoiling. Then, by breaking and re-sealing one or both strands accompanied by ATP breakdown, it selectively removes the positive twist and dissociates from the duplex leaving it with a negative twist. The two antibiotics just mentioned and their analogues inhibit DNA replication and conditional lethal mutations have been mapped in the *nalA* and *cou* genes. The action of gyrase is dependent on the ATP content of the cell—and a high content at that, since the K_m of gyrase for ATP is about 300 μM. Unless adequate supplies of energy can be generated, the negative supercoiling of the DNA disappears. Thus, this may be a control point in DNA metabolism.

Gyrase may be required not only for host cell DNA replication but also recombination, transcription from certain promoters, and synthesis of some ds DNA phages, both circular ($\phi X174$) and linear (T7).

The unwinding, separation, and stabilization of the two strands of DNA at the fork involve a protein with ATPase activity, probably the *rep* protein (needed for ϕX RF replication) and also a number of molecules of the cooperative single-stranded DNA-binding protein (SSB or HD protein) which does not require ATP. Each of the separated strands can then act as a template for synthesis—discontinuous on at least one strand since, although anti-parallel, both are made more or less simultaneously and in the $5' \rightarrow 3'$ direction. For discontinuous synthesis, the primer, a short stretch of RNA, is made by the *dnaG* protein, an RNA polymerase. The DNA polymerase, Pol III, then adds deoxynucleotides to this in the $5' \rightarrow 3'$ direction, using the $3' \rightarrow 5'$ strand as template and forming the so-called Okazaki fragments. Note that the $3' \rightarrow 5'$ exonuclease activity of Pol III can 'proof-read' and excise misincorporated nucleotides (see above, p. 215). This continues until the 3'-OH of the newly synthesized chain approaches the 5' end of the previous RNA primer (Figure 35). The $5' \rightarrow 3'$ exonuclease of Pol I then successively excises the RNA nucleotides while its $5' \rightarrow 3'$ polymerase activity adds deoxynucleotides to the Okazaki fragment thus filling the gap. The enzyme DNA ligase finally seals the 'nick' between 3'-OH and 5'-phosphate. Thus although enzymic elongation of the 'lagging' strand occurs *away* from the fork, by making successive fragments and linking them together, the extension is effectively in the direction of movement of the fork.

The counter or 'leading' strand of DNA may also be made discontinuously or it may be made continuously in the $5' \rightarrow 3'$ direction after a single initial priming. Some work suggests that after a short pulse of [³H] thymidine in *Escherichia coli* or phage T 4, all the labelled DNA may be found as chains of 1000–2000 nucleotides (10 S) when sedimented in an alkaline sucrose gradient. However, the matter is not resolved conclusively and it has been shown that not all fragments labelled during a short pulse come from *de novo* priming. Sometimes dUTP behaves as an analogue of dTTP, leading to incorporation of uracil, and sometimes cytosine in DNA becomes spontaneously deaminated to yield uracil. Cells are believed to have a specific uracil excision-repair mechanism as shown in Figure 36. This could result in temporary breaks in DNA which could give rise to fragments like those described by Okazaki.

Ultimately, however, the products of replication consist of two double helices, identical to each other and to the pre-existing DNA. Further, they each contain one old and one newly synthesized strand. How they are circularized and separated as the progeny chromosomes is still the subject of speculation but just as the origin of replication is believed to be associated with the cytoplasmic membrane, so the separation may be due to extension of the cell membrane—the *replicon hypothesis*.

Repair of damaged DNA

Frequently damage to DNA is caused by ultraviolet irradiation which produces covalent linkages between adjacent thymine residues. These *thymine dimers* are genetically damaging but can be excised and replaced (Figure 37). A specific endonuclease makes a nick; a $5' \rightarrow 3'$ exonuclease excises 20–500 nucleotides including the dimer leaving a single-stranded gap; the gap is filled by a $5' \rightarrow 3'$ DNA polymerase using the single-stranded chain as template; and a ligase seals the nick. Possibly, Pol I acts as both exonuclease and polymerase.

SYNTHESIS OF RNA

The metabolism of RNA is intimately tied up with the process of protein synthesis, for the essential role of RNA *in vivo* seems to be that of acting as genetically coded molecules which participate directly in the protein-forming reactions. To be sure, the ultimate repository of the cell's hereditary information is its DNA but the DNA itself does not apparently take part directly in the chemical combination of amino

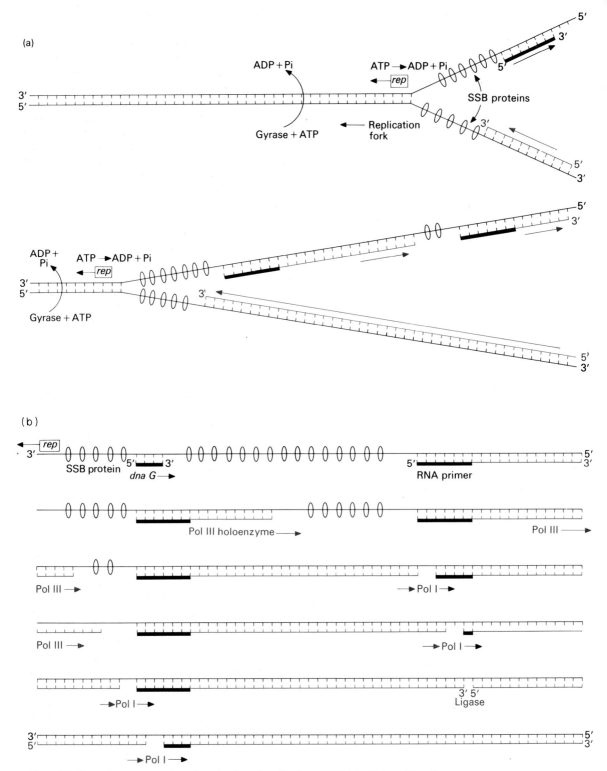

Figure 35. Tentative representation of mechanism of replication of DNA in *Escherichia coli*.
(a) Events at the replication fork assuming one strand is made discontinuously and the other continuously.
(b) Events on the lagging strand which must be made discontinuously.

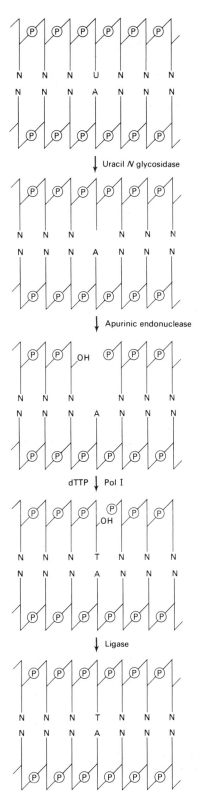

Figure 36. Excision of misincorporated Uracil and replacement by Thymine

acids to form polypeptide chains. The immediate control over the all-important amino acid sequence in nascent polypeptides is exerted by the different classes of RNA molecules, and it is only by directing the synthesis of RNA that the DNA determines the structure of the proteins whose properties are responsible for the phenotype of the organism. For clarity of presentation, therefore, the synthesis of RNA insofar as it relates to the synthesis of proteins is dealt with in the chapter on protein synthesis (p. 260 *et seq.*). We may concern ourselves here with the mechanisms by which the coded nucleotide sequence of DNA determines the nature of the RNA and in particular with the enzymic processes involved.

It is now well established that in bacteria the synthesis of all three classes of RNA is directed by the DNA of the cell. More explicitly, there are genes or regions of the DNA whose nucleotide sequences specifically determine the nucleotide sequences of the different kinds of RNA. This by itself does not mean that all RNA molecules must be synthesized by transcription from a DNA template for it is quite possible to imagine that once a few prototype molecules had been made they could themselves 'prime' further synthesis of similar molecules by RNA-dependent RNA-synthesizing reactions. However, there has never been conclusive evidence for RNA-directed RNA synthesis in bacteria, except in cells infected with RNA viruses, so we must look to DNA-directed RNA synthesis as the sole source of RNA in normal growing organisms.

SYNTHESIS OF rRNA AND tRNA

Genes which specify the nucleotide sequences of tRNA and the three classes of rRNA (23 *S*, 16 *S* and 5 *S*) were originally discovered by DNA–RNA hybridization studies using purified preparations of each class of RNA (Table 1 of Chapter 6). We have already seen how the formation of these hybrids has been exploited to enable isolation of DNA cistrons coding for the various RNA's (p. 271). rRNA's and tRNA are relatively rich in guanine. It may be calculated that the cistrons in double-helical DNA coding for the larger rRNA's must have a G + C content of 53–54%; thus in bacteria whose overall DNA base-composition differs substantially from this value it can be predicted that fragments of their DNA containing the rRNA cistrons will be significantly more or less dense than

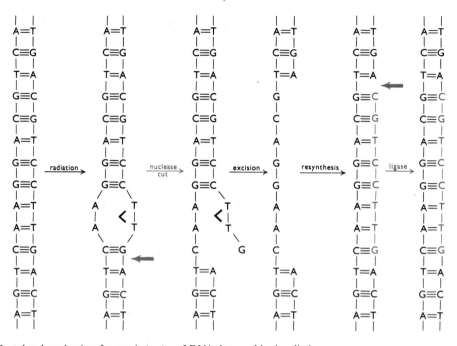

Figure 37. Postulated mechanism for repair *in vivo* of DNA damaged by irradiation.
     Radiation causes formation of a thymine dimer. The region containing this is excised and repaired. The figure is diagrammatic. The number of nucleotides excised is estimated to be 20–500, but their relation to the damaged region is uncertain.

the average in CsCl gradients and will have a correspondingly different 'melting' temperatures. These facts have proved useful in investigating the disposition of the 23 *S* and 16 *S* rRNA cistrons within the chromosome. In DNA of exponentially growing *E. coli* just under 0·4% of the sequences form hybrids with these rRNA's, from which it may be calculated that there are 6 or 7 copies of the cistrons for each of the larger rRNA's per chromosome. In other bacteria the number seems to be about the same. The cistrons are located in two discrete clusters on the genetic map. By studying hybridization between 23 *S* or 16 *S* rRNA and fragments of single-stranded DNA sheared to low molecular weight it was found that the 23 *S* and 16 *S* cistrons seem to be arranged in tandem, i.e. in pairs containing one cistron for each rRNA. Moreover, in *E. coli* (but probably not in all bacteria) each pair is associated with a cistron coding for 5 *S* rRNA as well. The cistrons complementary to the mature rRNA species are not contiguous but are separated by 'spacer' regions.

This arrangement of the rRNA genes is evidently important, for in bacteria as well as in higher organisms rRNA molecules are formed from larger precursors by a sequence of events known as *processing* or *post-transcriptional modification*. After exposure of cells to a short pulse of [^{32}P]-labelled phosphate, radioactivity is found in precursor RNA molecules (p23 and p16) which are larger, by about 150–200 nucleotides in each case, than the mature 23 *S* and 16 *S* rRNAs. The radioactive label in p23 and p16 RNA's can be 'chased' into the respective mature forms and the kinetics indicate a precursor-product relationship. Recently, the formation of an even earlier precursor has been observed, apparently a primary transcript of the entire assembly of genes plus spacers as well as extra 'leader' and terminal sequences. This primary transcript (designated p30 *S*) is believed to contain the sequences of the three rRNA molecules in the order (5' → 3') 16 *S*, 23 *S*, 5 *S*. It can be formed *in vitro* by incubating isolated nucleoids with nucleoside triphosphates and thus enabling polymerase molecules already engaged in synthesis to complete the product. It can also be isolated from a mutant strain of *E. coli* that is deficient in RNAase III. It is not normally found in wild-type cells which suggests that in these the p30 *S* RNA is subject to cleavage (presumably mediated by RNAase III) before completion of the whole transcript. Subsequent maturation steps, probably catalysed by additional specific nucleases, trim further nucleotides from both ends of the molecules to give products of the correct lengths (Figure 38).

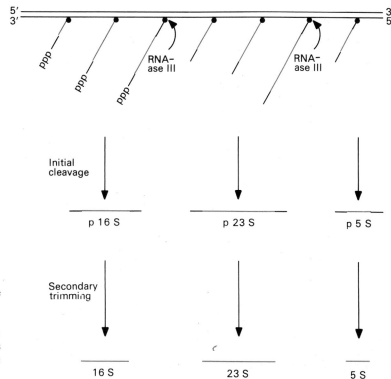

Figure 38. Schematic illustration of processing of rRNA precursors. The chromosomal DNA strand which codes for synthesis is shown with its 5′ and 3′ ends indicated; ribonucleotide sequences present in the mature rRNA products are drawn in red. The 5′ end of the primary transcript bears a triphosphate grouping as shown; this is removed, perhaps as part of an oligonucleotide, before p16 *S* rRNA is released. Sites of cleavage by RNAase III are indicated by arrows.

Some of the later steps seem to be dependent on protein synthesis and may involve association with nascent ribosomal proteins to form discrete pre-ribosomal particles. Although the significance of this complex pattern of maturation is not fully understood, it ensures that all three rRNA species will be synthesized proportionally and at a rate consistent with the cell's requirements.

Cistrons which code for tRNA also exist in clusters. About 0·04–0·05% of the DNA of *E. coli* hybridizes with tRNA, implying the existence of some 60 cistrons. Since there are about the same number of different species of tRNA, this suggests that each species of tRNA molecules is coded for by a single cistron. The clusters may contain as many as 5–7 tRNA sequences although some isolated single-copy genes occur. In all cases these are associated with 'leader' and 'trailer' sequences as well as with spacers which are initially transcribed with the RNA as large precursors. These precursors are readily isolated from temperature-sensitive mutants of *E. coli* which produce a defective form of RNAase P, which accordingly is identified as one of the enzymes involved in the processing. As in the case of rRNA's noted above, a complex and precise pattern of cleavage and trimming by specific

nucleases is required to generate the final products. Much interest attaches to the mechanisms by which these maturation enzymes recognize which appropriate internucleotide bonds to attack in the precursor molecules. Although much is known about the relevant sequences, the specificity of the cleavage sites is still obscure. In the case of RNAase P recognition cannot consist merely of interaction with a characteristic sequence surrounding the site of cleavage but must in addition involve recognition of conformational features related to the arrangement of stems and loops.

Post-transcriptional modification also involves formation of the methylated and otherwise altered nucleotides (Figure 8, p. 261) which are characteristic of rRNA and tRNA. The initial products of transcription of the rRNA and tRNA cistrons contain only the four normal nucleotides. During the process of maturation, sequence-specific enzymes are required to convert nucleotides at defined points in the molecule into the modified forms. The majority of the modifications, e.g. methylation, are simple, while others are more complicated, e.g. conversion of uridine to pseudouridine or dihydrouridine. Enzymes which catalyse some of these processes have been characterized.

The modifications can occur either on the precursor molecules or on the products of cleavage depending, presumably, upon the relative rates of reaction of the modifying enzymes with the intact or cleaved polynucleotides. Both types of reaction occur: for instance, the p30 S precursor of rRNA's contains all of the methylations found in 23 S rRNA but few, if any, of those found eventually in 16 S rRNA. Clearly, the sequences destined for the larger rRNA are methylated at an early stage of processing, whereas those destined for 16 S rRNA only undergo modification subsequent to cleavage by RNAase III. In fact, the latter reactions cannot occur in the absence of protein synthesis, the methylation of 16 S rRNA being effectively inhibited by chloramphenicol.

The 3′ terminal trinucleotide —CCA$_{OH}$ common to all tRNA's is an integral part of the initial transcript for many tRNA's and becomes the physical terminus of the molecule after cleavage by one of the maturation enzymes (RNAase Q). For others, this trinucleotide is wholly or partially missing from the initial transcript and must be added post-transcriptionally by the specific nucleotidyl transferase which serves to restore the sequence to tRNA molecules which have lost it during the course of normal metabolic turnover.

MESSENGER RNA

rRNA and tRNA cistrons together account for no more than one per cent of the bacterial chromosome. Of the remainder, a large fraction will hybridize efficiently with RNA synthesized in cells during a short exposure to a radioactive precursor. Much of this RNA must surely be mRNA transcribed from structural genes, but it is most unlikely that all of the 99% of the DNA is active in mRNA synthesis. Some is in the form of regulator genes rather than structural genes, and there may be regions with other genetic functions which need not necessarily involve the production of mRNA. Investigation of mRNA synthesis is hampered by the heterogeneity and metabolic instability of this class of RNA, in contrast to the rRNA's and tRNA's which are stable and easy to isolate because of their defined molecular size. In eukaryotic cells, mRNA's are the products of complex processing mechanisms which involve cleavage, splicing, and addition of nucleotides at both the 5′ and 3′ ends ('capping' and addition of a poly A tract). Prokaryotic mRNA's, on the other hand, appear to function largely as unmodified primary products of gene transcription. However, polycistronic mRNA's exist in prokaryotes and may undergo cleavage by specific

nucleases. A case in point is the 'early' T7 mRNA synthesized in *E. coli* infected with that bacteriophage (p. 372), which consists of molecules considerably smaller than the long transcripts produced by *E. coli* RNA polymerase acting on a T7 DNA template *in vitro*. These long transcripts can be cleaved to generate products essentially the same size as those found *in vivo* by incubation with RNAase III, the enzyme responsible for the initial processing of rRNA in *E. coli*.

Inhibition by drugs which bind to DNA

The evidence that all forms of RNA are synthesized by DNA-dependent mechanisms in normal bacteria came originally from studies on the action of the antibiotic actinomycin D. This substance completely blocks the synthesis of all forms of RNA in susceptible organisms without apparently affecting any other metabolic process to anything like the same degree. Furthermore, it forms a stable complex with native, double-helical DNA (but not with RNA) as a result of which the capacity of the DNA to act as a template for RNA polymerase is drastically reduced. Perhaps surprisingly, DNA can still act as a template for DNA polymerase activity in the presence of actinomycin, though this may be correlated with the fact that DNA polymerase I, at least, does not act on a truly double-helical template.

However, since binding to DNA appears to be the principal if not sole interaction between the antibiotic and a cell receptor, the fact that actinomycin blocks all *de novo* RNA synthesis in bacteria implies strongly that all RNA synthesis is directed by DNA. This interpretation of the action of actinomycin has found wide application: for instance, the antibiotic has been widely used as a tool for demonstrating the involvement (or non-involvement) of DNA-dependent RNA synthesis in a great variety of living processes. It has also been employed to investigate the breakdown of unstable species of RNA and to reveal the occurrence of RNA-dependent RNA synthesis in a number of virus-infected systems.

Some bacteria are not sensitive to actinomycin, most probably because they have a permeability barrier which prevents the antibiotic from entering the cell. *Escherichia coli* is insensitive for this reason. Nevertheless, the production of RNA by these organisms can be inhibited by another class of drugs, the acridines. One of the most widely used acridines is proflavine (3 : 6 diaminoacridine). Like actinomycin, this drug forms complexes with DNA and, when sufficient drug molecules are bound, the synthesis of

RNA is completely stopped. Compared with actinomycin, however, proflavine is a much less specific tool for interfering with DNA-dependent RNA synthesis; its interaction with DNA results in at least as powerful inhibition of DNA synthesis as RNA synthesis, and it also forms complexes with RNA. There are also differences between the complexes formed by the two kinds of drug with DNA: actinomycin binds specifically to sites containing deoxyguanosine residues and blocks the template activity of the DNA when relatively few sites have become occupied, whereas complex formation with proflavine shows no marked base-specificity and a large amount of drug must be bound before template activity is lost. The proflavine-DNA complex is of special interest for other reasons, however. The acridines are powerfully mutagenic drugs and it seems that their mutagenicity can be correlated with the nature of their interaction with DNA: being planar molecules they slot in or *intercalate* between the stacked base-pairs of the double helix in such a fashion that the affected base-pairs become separated by twice the normal distance (Figure 39). This mode of interaction may account for the striking capacity of acridines to generate frame shift mutations, i.e. mutations arising from the insertion or deletion of a base-pair in the DNA of the mutant. Acridine-induced frame shift mutations played an important part in establishing the triplet nature of the genetic code (p. 254).

Asymmetric transcription

In general, only one of the paired DNA strands acts as a template for RNA synthesis. The reason for this seems fairly clear; if both strands were to code for the synthesis of RNA there would be two kinds of RNA produced by each cistron in the DNA and this would demand a curious degeneracy in the genetic code if two different forms of mRNA were both to be translated at the level of the ribosome to give rise to the same type of polypeptide. There would also be a definite possibility of the two (complementary) RNA's from a single cistron finding each other and forming a perfectly base-paired double helix. This has never been observed in normal bacteria although such structures are known to occur in some RNA viruses. Of course, it is possible that both DNA strands could be transcribed and then one of the RNA products immediately degraded, but there is no evidence for a mechanism of this kind. However, it should be noted that the genomes of small DNA-containing bacteriophages may contain regions of nucleotide sequence which are transcribed as part of different

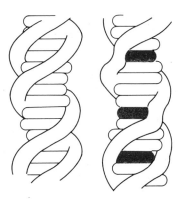

Figure 39. Sketches representing the secondary structures of normal DNA (left) and DNA containing intercalated proflavine molecules (right).
The helix is drawn as viewed from a remote point, so that the base-pairs and the intercalated proflavine appear only in edgewise projection, and the phosphate-deoxyribose backbones appear as smooth coils. From Waring, *Symp. Soc. Gen. Microbiol.* **16**, 235 (1966).

mRNA molecules. These subsequently function to produce different polypeptides (p. 271). This is exceptional, however, and no doubt arises from the demands of stringent economy in organization of their genetic material.

Three lines of evidence support the general conclusion that only one of the DNA strands codes for synthesis of RNA *in vivo*.

(1) In *E. coli* infected with bacteriophage ϕX174, phage-specific messenger RNA is only synthesized after the infecting (+) strand has been converted to the circular double-helical replicative form (RF). This RNA does not form hybrids with pure (+) strands of DNA isolated from phage particles, but it hybridizes readily with denatured RF–DNA. Evidently the template for its synthesis was the (–) strand of the RF.

(2) In certain instances it is possible to resolve the complementary DNA strands into separate peaks by centrifuging to equilibrium in CsCl gradients. This can be done with the RNA of several bacteriophages, notably α, which infects *Bacillus megaterium*, and SP8, which infects *Bacillus subtilis*. The basis for the separation resides in the fact that the A = T, G = C relationship required of the double helix does not necessarily fix the base-compositions of the individual strands—one may be relatively purine-rich and the other correspondingly pyrimidine-rich. This is the case for the two phages mentioned, and it is found that the phage-specific mRNA synthesized in infected

cells is complementary only to the pyrimidine-rich strand.

(3) A similar, though less complete, fractionation of the complementary strands can be effected with denatured pneumococcal transforming DNA in a CsCl gradient. When the two fractions, enriched for one or the other of the strands, were used to transform recipient cells it was found that one fraction caused the appearance of phenotypic changes one generation time before the other, suggesting that the 'slower' fraction had had to participate in a cycle of replication before its genetic information could be expressed.

The above examples, and others referred to earlier, show that the ability to isolate the complementary DNA strands has proved a valuable tool in the study of macromolecular synthesis. One of the most effective means of achieving complete resolution of the strands derives from an observation of Szybalski, who found that one strand frequently interacts preferentially with a synthetic polyribonucleotide, so that it bands at a substantially more dense position in a CsCl gradient than the other, less reactive strand. Guanine-rich ribopolymers have proved particularly effective in separating the strands of several phage DNA's, apparently by complexing with sequences rich in dC residues which are asymmetrically distributed between the complementary strands. In many cases the strand which preferentially binds poly G is the strand which is predominantly transcribed into RNA *in vivo*. The relation between these findings may not be fortuitous; it has been suggested that pyrimidine-rich clusters may help to identify the transcribing strand *in vivo* because they form part of recognition sites for the various proteins concerned with transcription.

The strands of phage T4 DNA are efficiently separated by complexing with the synthetic random copolymer poly (U,G). During the lytic infection cycle, at least three distinct classes of T4-specific RNA are synthesized: up to 2 min after infection only *immediate early* RNA is made; this is followed by *delayed early* RNA; and after 10–12 min *late* RNA species appear. Both species of early RNA are found to be complementary to the *C* strand of T4 DNA (the one which interacts preferentially with poly (U,G), whereas late RNA is complementary to the other strand (*W*). Thus, while the rule is still obeyed that only one strand of a given DNA sequence is transcribed into RNA, the transcribed sequences need not all be located on the same physical strand of the DNA. Evidently in T4-infected cells the synthesis of RNA directed by late cistrons involves a switch of the transcribing en-

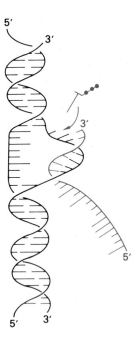

Figure 40. Synthesis of RNA by Watson–Crick base-pairing with a DNA template.

DNA strands are shown in black, with the double helix in the B-form having ten base-pairs per turn. The growing RNA strand is shown in red, with a ribonucleoside triphosphate about to be incorporated at the growing 3'-hydroxyl end. The RNA forms a transient hybrid helix with the template DNA strand, which is drawn with tilted base-pairs to indicate its likely A-form conformation having eleven base-pairs per turn. When the RNA is displaced from the hybrid the tenfold DNA helix re-forms as the region of local unwinding moves away.

zyme(s) from the *C* strand to the *W* strand. The mechanism by which this switch occurs presents an interesting problem which will be considered later and in Chapter 9.

We can picture the events during synthesis of RNA on a double-helical DNA template as follows (Figure 40). (1) The DNA helix opens up to allow the bases of one strand to interact by Watson–Crick base-pairing with incoming ribonucleoside triphosphates. (2) The growing RNA strand forms a transient, perhaps helical, DNA–RNA hybrid with the transcribing DNA strand. (3) An exchange of strands occurs, releasing the growing RNA strand and re-forming the DNA helix as the growing-point moves away. It is worth noting that this scheme involves two changes in helical involvement for the transcribing DNA strand. If the transient hybrid were to assume the 11-fold A-type helical conformation found with DNA–RNA hybrids *in vitro* (p. 204), the necessary conformational changes in the sugar-phosphate backbone of the DNA strand as it switched from B → A → B-type helical states could play a significant part in the overall process of synthesis. In particular, these conformational changes might play a major role in assuring the strand-exchange required to release the RNA free from its template (process (3) above).

VISUALIZATION OF BACTERIAL GENES IN ACTION

Electron microscopy on cell contents, released by osmotic lysis of fragile cells of *E. coli*, has provided striking confirmation of widely accepted ideas about gene action. Some photographs are shown in Figure 41. Long DNA fibres can be seen, to which are attached granular strings of varying lengths; the strings are believed to be polyribosomes, i.e. RNA molecules in process of synthesis covered with ribosomes which have already commenced translation of the nascent RNA strands into protein (Figure 41(a) and (b)). Important conclusions are: (1) Long stretches of the DNA appear to be genetically inactive, at least so far as RNA synthesis is concerned. (2) In the active regions, the lengths of the attached polyribosomes increase in a regular fashion along a portion of the chromosome, often with an irregularly shaped granule which could be an active RNA polymerase molecule at one end. (3) Free polyribosomes, not attached to DNA, are rarely seen. (4) Presumptive tandem 23 *S* and 16 *S* rRNA cistrons may be recognized by the much larger number of fibrils growing from them (Figure 41(c)), and these fibrils seem to be picking up protein rather than ribosomes, which agrees with evidence that ribosomal proteins become associated with the rRNA's as they are synthesized.

RNA POLYMERASE

Bacteria, unlike eukaryotic cells, are believed to have a single enzyme for the synthesis of all classes of RNA. This RNA polymerase has been isolated from a number of organisms. Its properties are in many ways analogous to those of DNA polymerase: the substrates are ribonucleoside 5'-triphosphates, a divalent metal ion is required (Mn^{++} is usually added as well as Mg^{++}), and the reaction is completely dependent upon the presence of a DNA template. The incorporation of a labelled nucleotide (indicated by *) into acid-insoluble material provides a convenient assay:

$$\begin{array}{c} n_1 \text{ ATP*} \\ n_2 \text{ GTP} \\ n_3 \text{ CTP} \\ n_4 \text{ UTP} \end{array} \xrightarrow{\text{DNA}} \left[\begin{array}{c} n_1 \text{ AMP*} \\ n_2 \text{ GMP} \\ n_3 \text{ CMP} \\ n_4 \text{ UMP} \end{array} \right] + (n_1 + n_2 + n_3 + n_4) \, PP_i$$

(acid-soluble)   RNA (acid-insoluble)

The RNA polymerase from *Escherichia coli* has been purified to homogeneity, and in what follows attention will be concentrated on this enzyme. The minimal active form, termed *core polymerase*, is a zinc-enzyme composed of four subunits: two α-chains, each of relative molecular mass 41 000, one β-chain (M_r 155 000) and one β'-chain (M_r 165 000). It can be designated $\alpha_2 \beta \beta'$. As isolated from cells in various physiological states the enzyme may contain one or more of a variety of additional subunits or polypeptides. Chief among these are the σ and σ' subunits (M_r 86 000 or 56 000 respectively) which give the enzyme the capacity for asymmetric transcription of a double-helical DNA template. In this form it is referred to as *holoenzyme*. More is known about the role of σ than σ' in regulating the activity of the polymerase, partly because σ was discovered earlier, but also because it binds to the core enzyme more tightly so that the $\alpha_2 \beta \beta' \sigma$ form of holoenzyme is the one normally isolated. Even less is known about various other polypeptides which seem able to bind to RNA polymerase. Some of them (e.g. the elongation and initiation factors for protein synthesis) undoubtedly influence the specificity of transcription by the enzyme.

Core polymerase retains the basic polymerizing activity and requirements for reaction noted above. In fact, most methods of purification yield a mixture of holoenzyme and core polymerase, and it is now clear that most of the general properties of the enzyme are primarily those of the core polymerase.

In vitro RNA polymerase can use either double-stranded or single-stranded DNA as a template, although core polymerase is weakly active on the former. This is probably because it has to rely upon the occurrence of random nicks or discontinuities (resulting e.g. from endonuclease action) to provide points for initiation. In general, the product is high molecular weight RNA which is released from the DNA template. With a single-stranded template the initial product is a base-paired DNA-RNA hybrid. The fact that the product is not covalently linked to pre-existing polynucleotide shows that there is no need for a primer. The role of the DNA as a template for RNA synthesis is shown by the usual

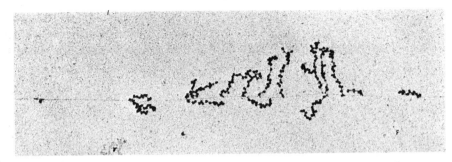

Figure 41. Visualization of bacterial genes in action.

(a) Polyribosomes attached to an active portion of the *E. coli* chromosome, an unidentified but undoubtedly large operon. The fainter granule at the extreme left is a putative RNA polymerase molecule presumably at or very near the initiation site for this operon. Polyribosomes exhibit imperfect gradients of increasing lengths as they become more distal to the initiation site. The shorter, most distal polyribosomes may have resulted from mRNA degradation. 83,000 × .

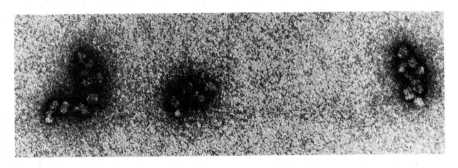

(b) An active segment of the *E. coli* genome at higher magnification showing polyribosomes which are attached to the chromosome by RNA polymerase molecules. 270,000 × .

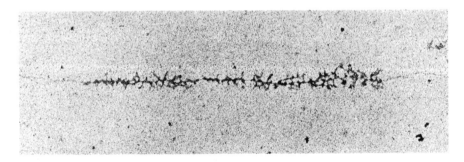

(c) A portion of an *E. coli* chromosome showing presumptive rRNA genes in action, with gradients of ribonucleoprotein fibrils on the 16 *S* and 23 *S* cistrons. 119,000 × .

Photographs kindly provided by Drs O. L. Miller, Jr., and B. A. Hamkalo [see *Science*, **169**, 392 (1970)].

criteria: (1) The base-composition of the product is strictly determined by that of the DNA—it is in fact complementary, as is evident from reactions employing synthetic polynucleotides as templates. The error rate, measured by determining the extent of incorporation of non-complementary nucleotides, is very low—typically less than 1 in 2000. (2) The nearest-neighbour nucleotide frequencies of natural DNA templates are faithfully reproduced in the product. (3) When denatured and annealed together, the template and product form DNA–RNA hybrids with high efficiency. (4) In appropriate circumstances the product may be shown to display biological activity.

However, it is clear from these observations that *both* strands of the DNA can serve as templates for RNA synthesis, and that the enzyme is capable of synthesizing RNA molecules complementary to *all* nucleotide sequences in the DNA presented to it. This is characteristic of the core polymerase and differs from the strand selectivity that occurs *in vivo*. It is thus apparent that additional controlling factors must be present in the organism. The role of σ and other factors in preventing the indiscriminate copying of all nucleotide sequences will be described later.

RNA polymerase synthesizes chains exclusively in the $5' \rightarrow 3'$ direction, the mechanism presumably involving nucleophilic attack by the 3'-hydroxyl of the terminal nucleotide of the growing chain on the 5'-pyrophosphate of the incoming nucleotide (cf. DNA polymerase). The direction of synthesis was established by changing the specific radioactivity of the nucleoside triphosphates during the course of a polymerase reaction, hydrolysing the product, and determining the specific activity of nucleotides derived from the 5' and 3' ends of the RNA chains. The end nucleotides are readily identifiable in alkaline hydrolysates of the product because the 5'-terminal nucleotide is released as a tetraphosphate, the 3'-terminal nucleotide as a nucleoside, and all intervening nucleotides as 2' or 3' nucleoside monophosphates (Figure 42). The specific activity of nucleotides derived from the 5'-ends was found to reflect the specific activity of the substrate nucleotides during the first stage of synthesis, while that of nucleotides from the 3'-ends reflected the specific activity of the substrates after it had been changed.

The direction of synthesis is confirmed by experiments using the antibiotic cordycepin (3'-deoxyadenosine) which acts as an inhibitor of RNA synthesis after its conversion *in vivo* to the 5'-triphosphate form. In this form, both *in vivo* and *in vitro*, it acts as an analogue of ATP and blocks the extension of RNA chains by becoming incorporated at their growing ends. The terminal cordycepin residue, lacking the 3'-hydroxyl group needed to react with the next nucleoside triphosphate, effectively prevents further polymerization.

It is interesting that the 5'-triphosphate grouping of the first nucleotide incorporated remains intact during subsequent growth of the chains: this provides a parallel with the genetic material of RNA bacteriophages, many of which have been shown to start with the

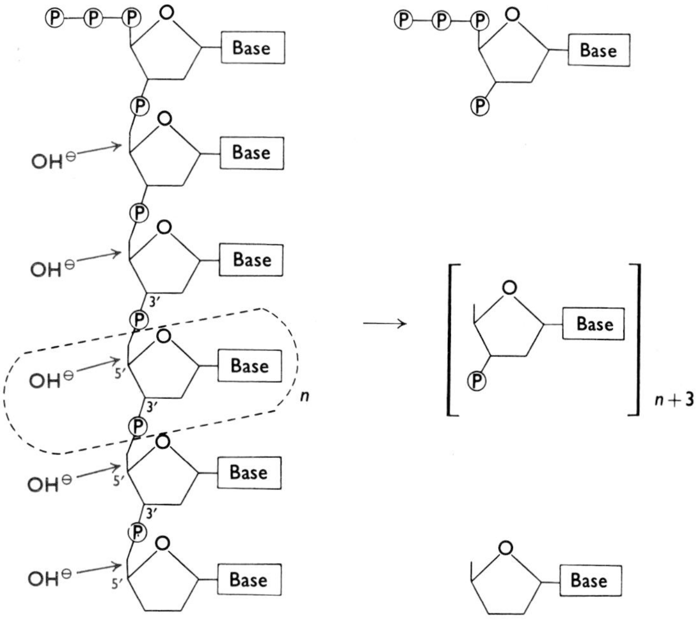

Figure 42. Alkaline hydrolysis of RNA formed in the DNA-dependent RNA polymerase reaction.

The 5' end of the chain yields a nucleoside tetraphosphate, the 3' end yields a nucleoside, and the residues in between give nucleoside monophosphates (an equilibrium mixture of 2' and 3').

sequence pppGp--- at their 5' ends. The parallel goes further, for RNA molecules synthesized by the polymerase *in vitro* are also initiated with a purine nucleotide. Rates of polymerization *in vitro* are rapid, from 40 to 100 nucleotides per second, and comparable with estimates of the rate of RNA chain growth *in vivo*.

Various workers have obtained evidence for the transient existence of a hybrid between the growing end of the RNA chain and one of the DNA strands during polymerase-catalysed synthesis on a double-helical template *in vitro*. This is consistent with a model for RNA polymerase action analogous to that represented in Figure 40, where synthesis occurs via base-pairing with the transcribed strand in a locally unwound region of the helix which progresses along the DNA as synthesis proceeds. According to this model the enzyme would contain at least three distinct sites: a DNA strand-separation site, a polymerization site, and a strand-exchange site where the growing RNA strand would be displaced from the hybrid with concomitant re-formation of the DNA helix. The inhibition of RNA synthesis by drugs such as actinomycin, which is reproduced with the polymerase reaction *in vitro*, is explained by their binding to the DNA helix and forming a more or less long-lived block preventing movement of the polymerase along the template, perhaps by making it more difficult to separate the DNA strands.

We must now explain the basis of asymmetric transcription *in vivo*, i.e. the mechanism which restricts RNA polymerase to copying only the correct strand of a DNA duplex. The σ factor present in *E. coli* holoenzyme plays an indispensable part in this process. The first evidence for its role in transcription was the finding of Burgess, Travers and their collaborators that the rather inefficient action of *E. coli* core polymerase with a T4 DNA template was markedly stimulated by addition of σ. It was subsequently found that, whereas core polymerase transcribed sequences more or less at random from both strands of T4 DNA, addition of σ factor restricted transcription to one strand only and the RNA produced behaved in hybridization tests like the 'early' RNA synthesized in T4-infected cells (p. 230). A similar result was obtained with the circular double-helical RF–DNA of phage fd (analogous to φX174 RF) as template: again, core polymerase copied both strands whereas addition of σ factor restricted the enzyme to transcribing only the (−) strand which is the one transcribed *in vivo*. In general, it appears that with virtually any phage DNA as template the core enzyme synthesizes RNA randomly from either strand, while in the presence of σ factor only those RNA species synthesized in infected cells immediately after infection are made.

Clearly the presence of σ in the *E. coli* holoenzyme enables it to recognize the correct DNA strand for transcription, and though most evidence comes from work with phage DNA's there is good reason to believe that σ exercises the same function in uninfected bacteria. The effect of σ is mediated at the level of initiation of RNA chains, by enabling the holoenzyme to recognize the nucleotide sequences in the DNA which act as signals for initiation—the *promoter* sites. RNA polymerase has a definite but relatively weak capacity for non-specific binding to DNA, but when it reaches a promoter site it is bound much more tightly. This initial interaction, termed *promoter recognition*, results in the formation of a so-called I complex which then undergoes a critical structural change to an activated state called an RS (rapid starting) complex which is able to initiate synthesis of an RNA chain very rapidly—within 0·2 sec. or so. This change, termed *promoter activation* or *opening*, involves a change in conformation of the promoter region and may be accompanied by conformational adjustments in the enzyme. The process of opening is highly co-operative and reversible; it exhibits a marked temperature dependence and the transition from the closed to the open form of a promoter can be characterized by a well-defined transition temperature (Figure 43). The resemblance between the plot in Figure 43 and the melting curve for a double-helical DNA (Figure 6) is so close that it suggests that activation involves the opening of the DNA double helix to expose a stretch of nucleotide bases ready to perform their template function. If this does occur, it must involve the disruption of some 6–8 base-pairs, which might represent the length of DNA which is transiently melted and hybridized to RNA during the copying reactions. Different promoters are characterized by different transition temperatures; those which have a low transition temperature are easily opened and probably behave as strong promoters (i.e. serve frequently as sites for initiation of RNA synthesis). Conversely, those with a high transition temperature probably behave as weak promoters because they are more difficult to convert to the active, open form. No doubt the transition temperature provides a measure of the energy required to open a promoter and is presumably related to the precise nucleotide sequence within it. Since the T_m of DNA (see Figure 6) rises with increasing GC content one can imagine that a higher proportion of GC base pairs in a promoter would render the strands more difficult to separate. This might provide a ready basis

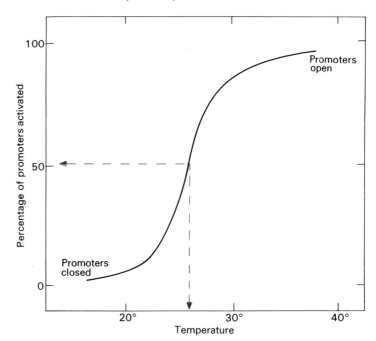

Figure 43. Promoter activation by RNA polymerase. In this example the transition temperature is 26°. Based on suggestions of Dr A. A. Travers.

on which to explain why different promoters have different transition temperatures. In any event, the energy required for promoter opening must come from the binding of RNA polymerase, i.e. promoter recognition, and this binding energy in turn will depend upon the recognition properties imparted to the enzyme by the σ subunit or other regulatory elements. Thus the effective transition temperature for a given promoter may be subject to variation under physiological conditions by factors which act *either* to stabilize or destabilize the structure of the promoter *or* to modify the recognition properties, and thus the binding affinity, of the polymerase holoenzyme (e.g. by allosteric interactions).

Regulation of transcription of specific genes can occur by binding of a repressor to a nucleotide sequence within or overlapping the polymerase recognition site. This may either impede recognition or prevent promoter opening. Positive regulatory factors such as the complex of cyclic AMP with its receptor protein can act in a similar fashion but binding leads to a lowering of the transition temperature for an otherwise inactive adjacent promoter sequence. This facilitates establishment of the open promoter complex by the holoenzyme. The balance of synthesis of the stable species of RNA (rRNA and tRNA) as opposed to mRNA involves delicate responses of polymerase molecules to the intracellular concentration of small-molecule effectors. Lastly, major changes

in the specificity of promoter selection by bacterial RNA polymerase occur during lytic bacteriophage development.

Promoter opening and the initiation reaction have been studied using the antibiotic rifampicin, an inhibitor of RNA synthesis which inhibits initiation but not propagation of RNA chains. The method involves 'challenging' the binary complex of RNA polymerase and DNA by simultaneous addition of rifampicin together with the four ribonucleoside triphosphate substrates. Polymerase already bound at initiation sites in the form of RS complexes (i.e. having open promoters) rapidly initiates synthesis of RNA chains and escapes the action of the inhibitor. Free polymerase molecules and those bound to the DNA either non-specifically or in the closed promoter (I complex) state are susceptible to inhibition. Thus, the fraction of enzyme-DNA complexes in the RS state, and the kinetics of promoter opening, can readily be determined. Rifampicin binds with high affinity to RNA polymerase from bacterial but not eukaryotic sources. Resistant mutants are fairly easy to isolate and it is found that their RNA polymerases, which are resistant to inhibition by rifampicin *in vitro*, have a much lower affinity for the antibiotic and contain an altered β subunit. Thus it, as well as the σ factor, may be implicated in promoter opening and initiation as well as elongation of chains.

The tight binding of holoenzyme at the open promoter complex results in the protection of a stretch of DNA about 40–45 base-pairs long from digestion by pancreatic DNAase. The actual initiation point for RNA synthesis lies near the middle of this fragment and, at least in some instances, forms part of a 'palindrome' or region of two-fold rotational symmetry. More striking, however, is the occurrence of a region of apparent sequence conservation preceding the starting point for transcription by 6 or 7 residues. It consists of seven nucleotide pairs closely related to the sequence:

$$(5') \quad -T-A-T-Pu-A-T-G- \quad (3')$$
$$(3') \quad -A-T-A-Py-T-A-C- \quad (5')$$

where Pu represents a purine nucleotide and Py a pyrimidine. There is good reason to believe that this 'conserved box' provides a common recognition region for promoters. It bears more than a superficial resemblance to poly (dA–dT), a polymer known to compete effectively with natural promoters for binding of RNA polymerase. Mutations in the heptanucleotide sequence of the conserved box are known to influence the efficiency of initiation at the *lac* promoter. However, other known promoter mutations fall outside the protected domain, lying in another region which displays sequence conservation about 35 base-pairs before the initiation point. This sequence may be required in the initial recognition step when the holoenzyme first binds to the closed promoter in the form of the I complex.

Once the open promoter RS complex has bound the initiating purine nucleoside triphosphate and formed the first internucleotide bond, the polymerization phase proceeds rapidly. It is catalysed by core polymerase, and the σ subunit dissociates from the complex to participate in renewed initiation with another core polymerase molecule. The ternary complex of polymerase, DNA template, and RNA product is quite stable and does not normally dissociate until chain termination occurs in response either to a specific signal in the template or to a termination factor. Although RNA polymerase holoenzyme is capable of synthesizing biologically active transcripts of discrete sizes from phage DNA templates *in vitro* without the aid of any accessory factors, the transcripts produced are often much longer than the corresponding natural mRNAs. It appears that proper recognition of termination signals in certain cases demands the activity of auxiliary factors, one of which, the rho factor of *E. coli*, is well characterized. It is normally isolated as a tetramer (M_r approximately 200 000) and in its presence the products synthesized from phage DNA templates such as lambda and T4

correspond to the natural mRNAs in both size and region of the genome transcribed. Rho does not exert its termination function by simple nucleolytic cleavage, but its mechanism of action seems to involve a curious RNA-dependent ATPase activity which is most effectively stimulated *in vitro* by poly C.

Recent evidence suggests that the expression of certain operons may be controlled by termination of transcription. One such operon is that which codes for the enzymes of tryptophan biosynthesis, where an *attenuator* located between the promoter site and the first structural gene of the operon apparently serves to limit 'read-through' by the polymerase into the cistrons which code for the enzymes of this pathway. Full expression of the structural genes demands the suppression of this attenuator by a factor sensitive to the intracellular concentration of tryptophan.

Transcription of rRNA cistrons

These cistrons constitute less than 1% of the bacterial chromosome, yet rRNA makes up the bulk of the RNA of the cell. When growth is rapid the rRNA cistrons must be transcribed at a high rate. Indeed, at the highest growth rates, 40–50% of all RNA polymerase molecules must be engaged in the synthesis of rRNA. Conversely, when cells are subjected to a sudden 'shift-down' to a nutritionally poorer medium the proportion of enzyme molecules synthesizing rRNA must drop at least ten-fold in favour of other transcriptional activities.

These changes are mediated by alterations in the pattern of promoter recognition as a response to the concentration of a 'signal' metabolite, guanosine-5'-diphosphate-3'-diphosphate or ppGpp. This nucleotide (and another one related to it, the 5'-triphosphate homologue pppGpp) is synthesized on the 70S ribosome in what has been termed an *idling reaction* which requires ribosomes, mRNA and uncharged tRNA. In effect, the uncharged tRNA serves to indicate that there is a shortage of amino acyl tRNA's, the immediate substrates needed for protein synthesis.

The classic experiment which reveals the regulatory role of ppGpp is the *stringent response* (see p. 110) seen when amino acid auxotrophs are starved for the required amino acid. The cells cease to make rRNA and tRNA, and the two unusual nucleotides accumulate. Mutants termed *relaxed*, fail to shut off stable RNA synthesis nor do they accumulate the guanosine nucleotides. Other strains of *E. coli* fail to accumulate pppGpp but still make ppGpp after amino acid starvation. Because they maintain the stringent control of rRNA synthesis it is concluded that ppGpp is the main mediator of the response.

The principal action of ppGpp is to alter the initiation specificity of RNA polymerase holoenzyme by inhibiting the formation of stable polymerase-rRNA promoter complexes. This is because of a substantial increase in the transition temperature for opening these promoters (from approx. 25°C to approx. 35–40°C measured *in vitro* at 75 mM KCl concentration). By contrast, if ppGpp is added after the polymerase has already become bound to rRNA promoters it has little or no effect on rRNA synthesis, showing that the effect of the nucleotide is purely on the initiation reactions, not on the rate of chain growth. Moreover, ppGpp has little effect on the transition temperature for opening of other promoters.

The K_i value for preferential inhibition of rRNA synthesis by ppGpp *in vitro* is 100–150 μM, a value which agrees well with the concentration required to effect half-maximal shut-off of rRNA production *in vivo*. Thus the evidence is strong that control of rRNA synthesis (and, by extension, of ribosome production) in bacteria can be attributed in large measure to fluctuations in the levels of one or two small molecule effectors which regulate the initiation specificity of transcription, most probably by an allosteric action on RNA polymerase. It has been suggested that RNA polymerase exists in more than one form—maybe a family of interconvertible forms—characterized by different patterns of promoter recognition and thus endowed with the capacity for selective initiation of different RNA species. According to this model, a variety of small molecule effectors (including ppGpp) might act by provoking a shift in the distribution of enzyme molecules between various functional forms. It is noteworthy that other components of the translation machinery, f-Met-tRNA$_F^{Met}$, IF-2 and EF-TuTs in particular, have been reported to modulate transcription by altering the initiation specificity of RNA polymerase. The elongation factors seem to act in opposition to ppGpp by decreasing the transition temperature for rRNA promoter opening. Thus, the complex structure of bacterial RNA polymerase (compared to the much simpler, often single-chain, enzymes which ostensibly perform the same catalytic function in mitochondria and virus-infected cells) may indicate a hierarchy of transcriptional control systems, each depending on a structural receptor on the enzyme for binding of an effector which alters the specificity.

VIRAL REPLICASES AND POLYMERASES

Although not strictly relevant to the processes of nucleic acid synthesis in bacteria, some mention must be made of the existence of two novel types of nucleic acid-synthesizing enzymes associated with viruses. The first type, known as *RNA replicase*, is essential for the growth and replication of RNA viruses and bacteriophages such as Qβ. The genetic material of RNA phages is single-stranded RNA. After entering the host cell, the (+) strand acts as a template for the synthesis of a complementary (−) strand, and the resulting double-stranded molecule then acts as a template for asymmetric synthesis of many more (+) strands to yield the progeny phages. Both these enzymic processes are mediated by a single replicase enzyme which is synthesized by the host cell ribosomes using the original infecting (+) strand as messenger RNA. At no stage is DNA involved in the polymerization reactions, which therefore proceed in a novel fashion inasmuch as the sequence of nucleotides in the growing RNA strand is directly determined by the sequence in an RNA template strand. Biochemically the reactions are perfectly analogous to the DNA-dependent polymerase reactions: the substrates are the four ribonucleoside triphosphates and the polymerization of growing RNA chains proceeds in the $5' \rightarrow 3'$ direction. Qβ replicase contains four polypeptide subunits, but of these only one is coded for by the RNA genome of the phage: the other three derive from the activity of host genes.

The second novel type of enzyme is an RNA-dependent DNA polymerase ('*reverse transcriptase*') associated with the virions of certain RNA tumour viruses which cause malignant transformation of mammalian cells, such as the Rous sarcoma virus and a number of leukaemia viruses. The transformed cells are genetically stable cancer cells, yet for generations they continue to yield small amounts of the transforming virus. The discovery of this remarkable enzyme was heralded by the work of Temin, who argued that the properties of the transformed cells could best be accounted for by postulating that the viral RNA served as a template for synthesis of a piece of DNA which then became stably integrated into the genetic material of the cells. His view is now vindicated. The reverse transcriptase employs the viral RNA strand as a template and converts it into a DNA–RNA hybrid, the DNA strand of which may then serve as a template for the synthesis of a truly double-helical DNA.

There is little reason to doubt that the reactions catalysed by RNA replicase and reverse transcriptase involve normal Watson–Crick base-pairing as do the better known DNA-dependent polymerase reactions. Lacking evidence to the contrary, it seems reasonable to conclude that the same base-pairing forces are the primary determinants for selection of nucleotides in *all* the nucleic acid polymerizing reactions. Thus the discovery of the two last mentioned enzymes serves to underline the striking versatility of base-pairing as a force dictating the structure and function of nucleic acids. Base-pair complementarity accounts well for the structures of DNA and RNA and for all four directions of flow of sequence information between them.

FURTHER READING

1 Alberts B. and Sternglanz R. (1977) Recent excitement in the DNA replication problem. *Nature* **269**, 655–661.

2 Brownlee G. G. (1979) Sequencing eukaryotic genes, or the anatomy of DNA. (Fourteenth Colworth Lecture) *Biochem. Soc. Trans.* **7**, 279–296.

3 Champoux J. J. (1978) Proteins that affect DNA conformation. *Ann. Rev. Biochem.* **47**, 449–479.

4 Denhardt D. T. (1979) DNA gyrase and DNA unwinding. *Nature* **280**, 196–198.

5 Fiddes J. C. (1977) The nucleotide sequence of a viral DNA. *Scientific American* **237**, 55–67.

6 Kornberg A. (1980) *DNA replication.* Freeman.

7 Kornberg A. (1979) Aspects of DNA replication. *Cold Spr. Harb. Symp. Quant. Biol.* **43**.

8 Krakow J. S. and Kumar S. A. (1977) Biosynthesis of RNA. In *Comprehensive Biochemistry* **24**, 105–184.

9 Lawrence E. (1978) Replication: twenty-five years after. *Nature* **274**, 210–212.

10 Maxam A. M. and Gilbert W. (1977) A new method for sequencing DNA. *Proc. Natl. Acad. Sci. USA* **74**, 560–564.

11 Nierlich D. P. (1978) Regulation of bacterial growth, RNA, and protein synthesis. *Ann. Rev. Microbiol.* **32**, 393–432.

12 Perry R. P. (1976) Processing of RNA. *Ann. Rev. Biochem.* **45**, 605–629.

13 Rich A. and RajBhandary U. L. (1976) Transfer RNA: molecular structure, sequence, and properties. *Ann. Rev. Biochem.* **45**, 805–860.

14 Travers A. A. (1977) Transcriptional control mechanisms. In *International Review of Biochemistry*, Volume 17 (ed. Clarke B. F. C.)

15 Waring M. J. (1981) Inhibitors of nucleic acid synthesis. In *The molecular basis of antibiotic action.* (by Gale E. F., Cundliffe E., Reynolds P. E., Richmond M. H. and Waring M. J.) 2nd edition. Wiley.

16 Watson J. D. (1976) *Molecular biology of the gene*, 3rd edition. Benjamin.

17 Wickman S. H. (1978) DNA replication proteins of *Escherichia coli. Ann. Rev. Biochem.* **47**, 1163–1191.

18 Wu R. (1978) DNA sequence analysis. *Ann. Rev. Biochem.* **47**, 607–634.

19 Ogawa T. and Okazaki T. (1980) Discontinuous DNA replication. *Ann. Rev. Biochem.* **49**, 421–457.

20 Tomizawa J. and Selzer G. (1979) Initiation of DNA synthesis in *Escherichia coli. Ann. Rev. Biochem.* **48**, 999–1034.

Chapter 6
Class III Reactions: Synthesis of Proteins

The outlines of the mechanism of protein synthesis have been sketched in Sections 5 and 6 (pp. 27–9) and it will be recalled that after activation the amino acids combine with specific transfer-RNA molecules and that these interact on a messenger-RNA template which is associated with ribonucleoprotein particles called ribosomes. There are hundreds of different kinds of proteins in each cell and each is made of polypeptides which are linear polymers of unique composition and sequence. In this chapter we will consider the components and reactions in more detail.

It is useful to have some idea of the distribution of the components in a bacterial cell and Tables 1 and 2 give this information in various ways— percentages, numbers of molecules, numbers of sub-units, total molecular masses per cell (molecular masses are referred to the atomic mass unit, u, which is one-twelfth the mass of the atom of ^{12}C; 6×10^{23} u are equivalent to about 1 g). The numbers in these tables are of necessity approximations but they do give an indication of the make-up of a bacterium and their significance will be increasingly apparent.

Table 1. Nucleic acid distribution in a bacterium

Volume of cell		10^{-12} ml
Dry matter per cell		2.5×10^{-13} g
molecular mass		1.5×10^{11} u
DNA per cell: % of dry mass		1.5–2.0% (say 1.67%)
molecular mass		2.5×10^9 u
no. of nucleotides		8×10^6
length		1100–1400 μm
Size of average structural gene		1250 nucleotides
molecular mass		4×10^5 u
Total no. of genes per cell		6400
Genes for tRNA		50 (0.02% of total DNA)
Genes for 16 S rRNA		6–10 (0.1% of total DNA)
Genes for 23 S rRNA		6–10 (0.2% of total DNA)
Genes for 5 S rRNA		6–10 (0.0075% of total DNA)
Genes for mRNA		thousands

	Sedimentation coefficient	Relative Molecular mass $\times 10^{-3}$	Nucleotides per molecule	Total no. molecules per cell	% total RNA	Total nucleotides $\times 10^{-6}$	Total mass u $\times 10^{-8}$
rRNA	5 S	40	120	12 000	80	64	200
	16 S	550	1600	12 000			
	23 S	1100	3200	12 000			
tRNA	4–5 S	25	73–93	150 000	15	12	36
mRNA	8–30 S	100–1500	300–4500	500–1000*	2–3	1.6–2.4	5–7.5
				Total RNA		80	250

Total RNA 10–20% of dry mass of cell, say 17%.
* Mean life of mRNA about 4% of the mean generation time of the bacteria. Hence number of molecules of mRNA synthesized per cell per generation = 12 500–25 000. The figures given are based on a variety of sources but may be about right for a bacterium of the size of *Escherichia coli*.

Table 2. Protein distribution in a bacterium

Total protein per cell: % of dry mass	60%
molecular mass	9×10^{10} u
amino acid residues	8×10^8
Size of average polypeptide: amino acid residues	200
molecular mass	22 500 u
No. of polypeptide chains per cell	4×10^6
No. of species of proteins	say 1000
No. of molecules of each species	$1-10^5$
70 S ribosomes per cell	12 000
protein/ribosome	40% (w/w)
molecular mass	$1 \cdot 1 \times 10^6$ u
amino acid residues	10^5
Total ribosomal protein per cell	
molecular mass	$1 \cdot 32 \times 10^{10}$ u
amino acid residues	$1 \cdot 2 \times 10^8$
molecules	$6 \cdot 6 \times 10^5$
% of total cell protein	14·5%
'Soluble' protein per cell: % of total	75%
molecular mass	$6 \cdot 7 \times 10^{10}$ u
amino acid residues	6×10^8
Size of average soluble polypeptide chain	
amino acid residues	230
molecular mass	25 000 u
No. of chains per cell	$2 \cdot 7 \times 10^6$

'Structural' protein (wall and membrane) say 10% of total cell protein

The figures are based on a variety of sources but may be about right for a bacterium of the size of *Escherichia coli.*

FRACTIONATION OF BACTERIAL CELLS

In order to analyse the components of an organism it is first necessary to take it to pieces. Bacterial cells can be disrupted by mechanical, chemical or enzymic means and the resulting *lysate* can then be fractionated in various ways. Low speed centrifugation gives a pellet consisting of unbroken cells, fragments of walls and membranes, perhaps some highly polymerized DNA, some ribosomes, and any large granules. The supernatant contains the rest of the ribosomes, all the soluble proteins, soluble RNA, and small molecular weight compounds such as salts, amino acids and nucleotides. If this supernatant is centrifuged at high speed (100 000 g) the ribosomes spin down but the supernatant still contains enzymes and RNA.

Much has been learned about cellular components and the mechanism of protein synthesis generally by the use of zone-centrifugation in sucrose density gradients and this technique is so useful that it will be described here. A mixture of particles of different sizes can be separated by centrifugation through a solution of graded density. The sample in a small volume is loaded on the surface of the solution in a swinging-bucket centrifuge tube and this is then spun at an appropriate speed. Particles of like sedimentation coefficient tend to form bands which move towards the bottom of the tube as centrifugation continues. After a suitable time the contents are removed.

This can be done by puncturing the bottom of the plastic centrifuge tube and collecting drops into a series of tubes or by pumping out the contents. The fractions or the continuous sample can be analyzed for absorbance at 260 nm (indicative of nucleic acids) and for radioactivity, etc (Figure 1).

If 0·1 ml of a whole cell lysate is layered on 5 ml of a 5% to 20% linear gradient of buffered sucrose and spun for 45 minutes at 37 000 r.p.m. in a centrifuge, a

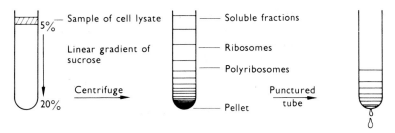

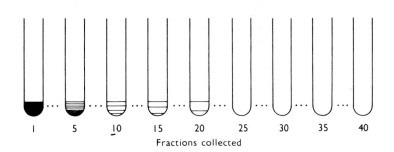

Figure 1. Zone-centrifugation.

This technique is useful for fractionation of components such as ribosomes and polyribosomes or of RNA's of different molecular mass. Note that centrifugation is *not* continued until equilibrium. Alternatively, after centrifugation the contents can be pumped out directly through a recording spectrophotometer.

graph like Figure 2 might result. Had the lysate first been treated with ribonuclease in media of various Mg²⁺ concentrations, results like those in Figure 3 might be obtained. The polyribosomes would have been degraded by ribonuclease to 70 *S* ribosomes, and low Mg²⁺ would have caused these to dissociate

(see below).

Thus it is possible by separating appropriate fractions and then dialysing or centrifuging them, to get various components in more or less pure form. In the study of protein synthesis, experiments can be carried

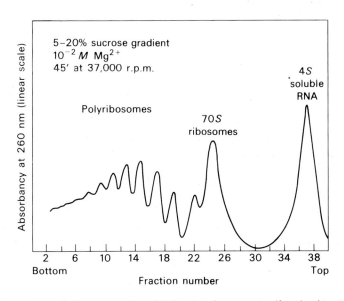

Figure 2. Separation of polyribosomes, ribosomes and soluble fractions by zone-centrifugation in a sucrose gradient.

The sample to be separated is loaded on the surface of a sucrose gradient. The steepness of the gradient and the duration of centrifugation determine the degree of separation of the components.

5%–20% gradient; 45 minutes at 37 000 r.p.m. Some of the larger polyribosomes have reached the bottom of the tube.

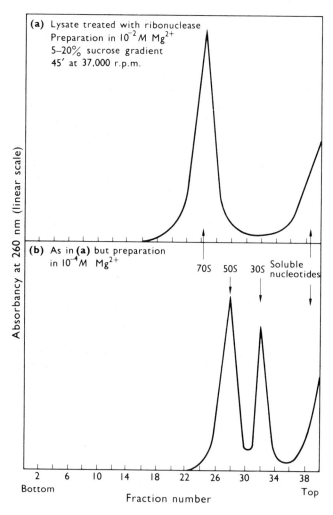

Figure 3. Effects of ribonuclease and reduced concentration of magnesium on ribosome pattern.

(a) Ribonuclease converts poly-ribosomes to 70 S ribosomes and soluble nucleotides. The soluble RNA (4 S) is also degraded. Preparation in $10^{-2}M$ Mg^{2+}.

(b) Same preparation in $10^{-4}M$ Mg^{2+}. The 70 S peak of ribosomes is replaced by two peaks consisting of 50 S and 30 S subunits respectively. This dissociation is reversible: $70\ S \rightleftharpoons 50\ S + 30\ S$. All of the 260 nm-absorbing material is precipitable by 5% TCA except the nucleotides.

out *in vivo* followed by fractionation to determine what has happened and where; alternatively, cell-free systems can be reconstructed from the fractions and then tested for activity. These *in vitro* experiments usually consist of incubating the components with a radioactive amino acid, adding trichloroacetic acid to precipitate protein, and measuring the radioactivity in the precipitate. The conversion of non-precipitable to precipitable counts is not, however, adequate proof that an amino acid has been incorporated by peptide linkage into polypeptide. There is an account of the criteria which have to be satisfied in an old but excellent review by Loftfield (see reference on p. 275).

RIBOSOMES

In Chapter 1 it was shown that the cytoplasm of bacteria appears to contain thousands of electron-dense particles about 10–20 nm in diameter. When cells are disrupted by mechanical, chemical or enzymic means these are released and can be isolated by differential centrifugation. Preparations can be purified by various techniques and the particles can be spun down at 100 000 g in one to two hours. They are found to consist of protein and ribonucleic acid and in 1958 they were given the name *ribosomes* by R. B. Roberts. Frequently they are characterized by their behaviour in an ultra-centrifuge and are, therefore, referred to by their sedimentation coefficient, e.g. 70 S ribosomes. Electron microscopy of isolated ribosomes and of sections of bacterial cells indicate that the predominant form is 70 S but that this is composed of two subunits—a 30 S and a 50 S piece. These can be separated (Figures 4a and 4b). The intact 70 S particle can be made to dissociate reversibly by altering the ionic composition of the medium in which it is suspended. In particular, reducing the Mg^{2+} concentration from $10^{-2} M$ to $10^{-4} M$ tends to cause dissociation into the subunits:

$$70 S \rightleftharpoons 30 S + 50 S$$

Bacterial ribosomes all have approximately the same chemical composition—just over 60% RNA and the rest protein—a total mass of about 2.7×10^6 u. Each 30 S subunit has a mass of 0.9×10^6 u and contains one molecule of 16 S rRNA of mass 0.55×10^6 u. Each 50 S particle is about 1.8×10^6 u, having rRNA of 23 S, mass 1.1×10^6 u but also a molecule of 5 S rRNA (see p. 198). 16 S and 23 S rRNA differ slightly in nucleotide composition and the latter is not a dimer of the former although many of its sequences occur twice. The sequences of 5 S and 16 S and 23 S are all known.

The proteins of ribosomes have been separated and the small subunit contains twenty-one different species (S1 to S21) and the larger subunit thirty-four (L1, L2, etc). Apart from L7/12, not more than one molecule of each protein occurs in each particle and only one (S20/L26) is common to both subunits and is presumably at the interface. L7 and L12 are identical except that N-terminal serine is acetylated in the former. Most of the proteins have a low (20–35%) α-helical content and are basic, containing much lysine. Their relative molecular masses range from 9600 to 65000 and nearly half have N-terminal methionine while many of the others have alanine (see below, p. 258). Most proteins occur once in every ribosome but some less frequently—these may be ones which join and leave during ribosome function. Many enzymic activities have also been found associated with ribosome preparations but may not be an integral part of the structure. Since ribosomes are involved in the formation of all kinds of proteins, one would expect to find traces of nascent enzymes on them.

The combination of proteins and RNA to form ribonucleoprotein particles requires cations, and Mg^{2+} and K^+ play this role *in vivo* with, perhaps, polyamines such as putrescine, spermidine and spermine also contributing to the neutralization of the many phosphate groups in the RNA. Suspension of ribosomes in media lacking Mg^{2+} or containing chelating agents such as ethylenediamine tetra-acetate (EDTA) can lead to disintegration of the particles. However, under controlled conditions it is possible to remove successively specific groups of proteins leaving 'core

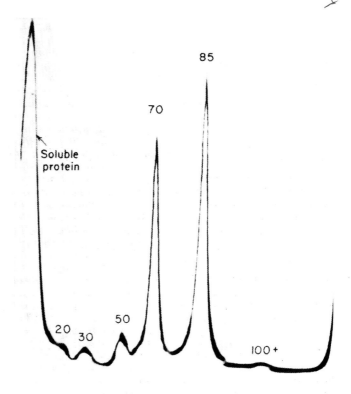

Figure 4a. Analytical ultracentrifugal analysis of ribosomes.

Pattern of ribosomes in a lysate from *Escherichia coli* in the exponential phase of growth. Preparation in medium containing $10^{-2}M$ Mg^{2+}. Schlieren plate taken 5 minutes after 50 740 r.p.m. was reached. The peaks are labelled with nominal sedimentation constants.

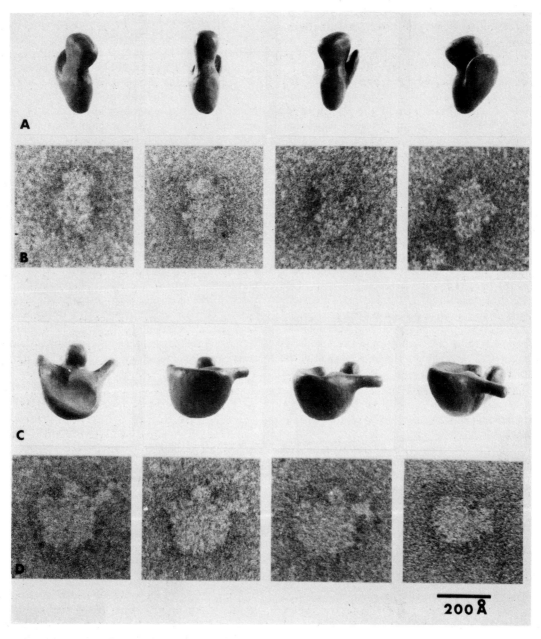

Figure 4b. *Escherichia coli* ribosomes.

Comparison of models (A and C) with electron micrographs (B and D) of 30 *S* (A and B) and 50 *S* (C and D) subunits in equivalent orientation.

From J. A. Lake, *J. Mol. Biol.* **105**, 131–159 (1976).

particles' and, eventually, only the RNA. Conversely, it is possible to reconstitute biologically active 30 S and 50 S subunits from the components but the process requires a particular sequence of addition and this may be related to assembly *in vivo* (see p. 365).

It is not possible by microscopy of living cells to see the form of ribosomes, and electron micrographs are of fixed and often sectioned preparations. However, it is generally thought that these particles do exist *in vivo* but in a hydrated form. The dimensions and shapes in solution are known with precision—hydrodynamic and X-ray scattering techniques suggest that 30 S subunits are oblate spheroids $22 \times 22 \times 5.5$ nm and that 50 S are $13 \times 17 \times 26$ nm; whereas electron microscopy shows 30 S as irregular prolate ellipsoids $8 \times 10 \times 19$ nm and 50 S as $16 \times 20 \times 23$ nm. Other techniques are being used to explore the topography of subunits, including neutron-scattering, fluorescence spectroscopy, cross-linking of neighbouring proteins with bifunctional reagents, photo-induced cross-linking of RNA with proteins, and, most excitingly, immune electron microscopy. This last involves using antibodies against individual proteins to 'stain' the ribosome. Some of the resulting models are shown in Figure 5.

Until recently it seemed that in order to understand the mechanism of protein synthesis, the 'function' of each of the $21 + 34$ proteins would have to be identified. It is now apparent that this view was naive and that functional regions are likely to include RNA as well as several interacting proteins. For example, in re-assembly of 30 S subunits, the association of S9, S10, S13, S14, and S19 is dependent on the presence of S7. All six proteins are near neighbours as shown by cross-linking and by immune electron microscopy. Further, all six can associate with nucleotide sequences towards the 3′ terminus of 16 S rRNA. Such groups of proteins together with the RNA sequence have been called 'assembly clusters'. They may well also be 'functional clusters'.

A large number of protein–protein neighbours have been discovered and most of the 21 proteins of 30 S subunits can be associated with specific regions of the 16 S rRNA. Less is known about the topography of 50 S subunits. However both subunits have properties in common:

1 Proteins vary in size; some are globular, many are quite elongated (e.g. S4, S7, S15 and S18, the last two stretching the whole length of the 30 S subunit; and L6, L18 and L25).

2 Much of the rRNA has secondary structure, there being, perhaps, dozens of regions of intra-chain base-pairing.

3 Protein—RNA—protein interactions abound.

4 Neither rRNA nor protein forms a 'coat' round the subunit.

5 The proteins are accessible to chemical reagents, enzymes, and antibodies.

6 The rRNA's are accessible to dyes and some enzymes.

POLYRIBOSOMES

When very gentle methods of cell disruption are used (e.g. lysis of protoplasts or spheroplasts—see pp. 59–61 in Chapter 1) most of the ribosomes are released as clusters joined together by a tenuous thread. It has been estimated that 80% or more of the ribonucleo-protein particles occur as these *polyribosomes*, also called polysomes or ergosomes. There is evidence that the material linking the particles is RNA—probably messenger-RNA. If a growing culture of bacteria is treated for a few seconds with a radioactive precursor of RNA (e.g. uridine) radioactivity is found in the polyribosomes but not in the ribosomes which can be derived from them by treatment with ribonuclease (Figure 6). In other words, there is RNA in polyribosomes which is not the ribosomal rRNA itself. Moreover, much of this labelled RNA forms hybrids with homologous DNA in a manner consistent with its being mRNA (see pp. 203–206, Chapter 5). Moreover, when further synthesis of RNA is inhibited by adding the antibiotic, actinomycin D (see p. 228, Chapter 5) much of the rapidly-labelled RNA is degraded during the next few minutes. Concomitantly the polyribosomes break down to yield single ribosomes. These findings all support the belief that poly-ribosomes consist of ribosomes strung together on a strand of mRNA.

Whereas prokaryotic organisms (bacteria and blue-green algae) contain 70 S ribosomes, eukaryotes (such as yeasts, protozoa, plants and animals) have particles of about 80 S, composed of subunits of 40 S and 60 S with correspondingly larger rRNA components. The ratio of protein to RNA is nearer 50/50 than 40/60. Eukaryotes, however, may have ribosomes resembling those of prokaryotes in such organelles as mitochondria and chloroplasts.

The site of protein synthesis *in vivo*

Just as radioactive precursors to RNA can be used, so amino acids can be added to label proteins. If they are added to growing cultures, radioactive protein will be

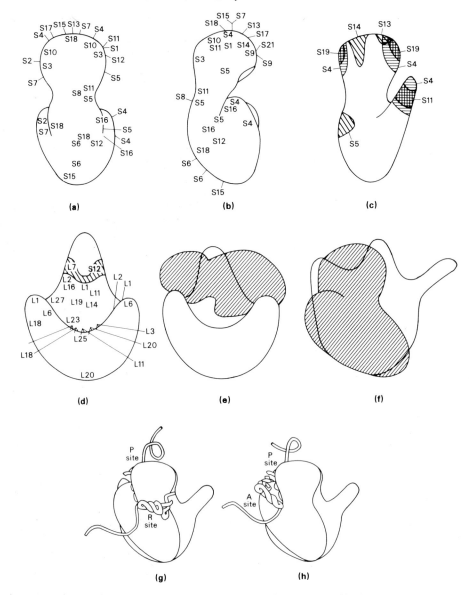

Figure 5. Models of ribosomes of *Escherichia coli.*

(a), (b), and (c) 30 *S* subunit showing localisation of proteins.

(d) 50 *S* subunit.

(e) and (f) 70 *S* couple.

(g) and (h) 70 *S* couple showing tRNA binding sites.

(a), (b), (d), and (e) from G. Stöffler and H. G. Wittmann.

(c) and (f) from J. A. Lake, J. Mol. Biol. **105**, 131–159 (1976).

(g) and (h) from J. A. Lake, Proc. Nat. Acad. Sci. USA **74**, 1903–1907 (1977).

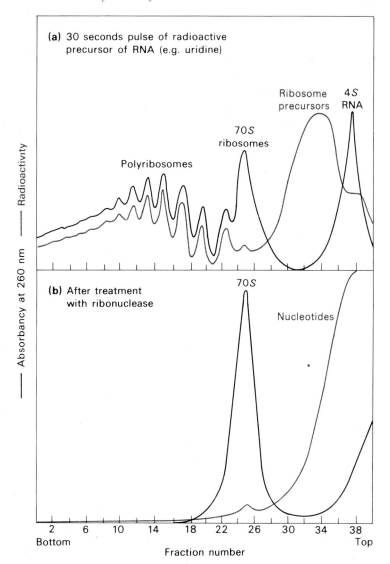

Figure 6. Presence in polyribosomes of an RNA component which is absent from 70 S ribosomes.

(a) A short pulse of a radioactive precursor to RNA (e.g. uridine) labels polyribosomes much more than 70 S ribosomes. There is also a slower sedimenting peak of radioactivity which is known to contain ribosome precursors.

(b) Ribonuclease converts the polyribosomes to 70 S ribosomes but their radioactivity is not in this peak; it is found in the TCA-soluble nucleotides.

found a few seconds later. The experiment can be done by squirting a solution of radioactive amino acid(s) into a vigorously stirred culture and then, after a few seconds, pouring the whole culture on crushed ice to reduce the temperature very rapidly and halt further metabolism. The cells are then broken and fractionated. Some species (e.g. *Bacillus megaterium*) can be lysed by adding the enzyme lysozyme a few seconds before cooling so that by the time the organisms have been centrifuged out of suspension (in the cold) their walls will have been digested and only fragile protoplasts will remain. These are highly sensitive to detergents such as Triton X100 which disrupts the cytoplasmic membrane releasing the contents.

The kinetics of transfer of radioactivity from amino acids through intermediates to the end-product protein can be studied by taking samples at intervals and examining them by this kind of technique. If this is done it is found that radioactive peptides occur first in association with ribosomes and polyribosomes. The activity here increases to a maximum in a few seconds and subsequent 'chasing' by adding an excess of non-radioactive amino acids causes replacement of labelled material on the ribosomes and appearance of radio-activity in the soluble protein fraction of the cell. This is illustrated in Figure 7. It seems, therefore, that nascent protein has a transient existence on ribosomes and determination of the flux through this

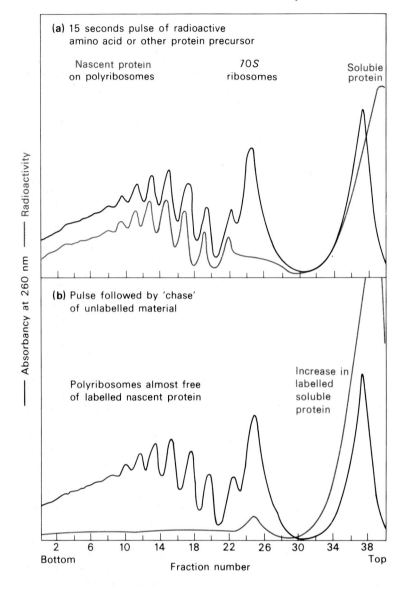

Figure 7. Appearance of nascent protein first in association with polyribosomes.

(a) Addition of radioactive amino acid to a culture growing exponentially. In a few seconds some soluble protein has become labelled but the bulk of the radioactivity which has been converted to peptide form is associated with the ribosomes and polyribosomes.

(b) If a short pulse of radioactive amino acid is followed by a large unlabelled 'chase' of the same amino acid, the radioactivity is swept on out of the polyribosome region and appears as soluble protein near the top of the gradient.

The nascent protein is transiently attached to the ribosomes; the soluble protein is the end product.

stage suggests that all proteins are assembled in this manner before being released and going to their functional sites.

AMINO ACID ACTIVATION

Although particulate material is involved in the making of proteins, soluble components are also essential as can be demonstrated by reconstruction experiments. Amino acids become incorporated into polypeptide only if both particulate and soluble fractions are present. As has been mentioned (p. 240) the soluble fraction contains both enzymes and soluble RNA. The latter is seen as a 4 S peak of material absorbing at 260 nm near the top of the sucrose gradient (see Figure 2, p. 241). Both it and enzymes are implicated in the early stages of protein synthesis.

Soluble enzymes catalyse a reaction of amino acids with ATP as follows:

$$NH_2\overset{|}{\underset{R}{CH}}.COOH + ATP$$

$$\longrightarrow NH_2\overset{|}{\underset{R}{CH}}.CO \sim AMP + PP_i$$

The products are inorganic pyrophosphate and an amino acyl-AMP in which there is a high-energy bond between the carboxyl of the amino acid and the phosphate of the AMP:

Amino acyl ~ AMP

Amino acyl-AMP remains bound to the enzyme but this complex will react with hydroxylamine to give a coloured amino acid hydroxamate which can be estimated:

$$Enz (NH_2\overset{|}{\underset{R}{CH}}.CO \sim AMP) + NH_2OH$$

$$\longrightarrow NH_2\overset{|}{\underset{R}{CH}}.CO.NHOH + AMP + Enz$$

There is a specific activating enzyme for each of the twenty amino acids and most have been obtained pure (M_r 50 000 to 200 000). These enzymes are present in the soluble fraction of cells and can be precipitated at pH 5 so they were referred to as the 'pH 5 precipitable fraction' or as 'pH 5 enzymes'.

AMINO ACID TRANSFER RNA (tRNA)

When unpurified soluble fractions are used as sources of amino acid activation, the reaction goes further and the amino acids are transferred to specific soluble RNA acceptors:

$$\alpha\alpha + ATP + Enz \longrightarrow Enz (\alpha\alpha \sim AMP) + PP_i$$

$$\overset{tRNA}{\longrightarrow} \alpha\alpha\text{-tRNA} + AMP + Enz$$

These RNA's are variously called soluble-RNA, sRNA, acceptor-RNA, amino acid transfer-RNA, tRNA. We shall use the latter terms. Each tRNA is specific for a single amino acid. The transfer RNA's can be purified from the soluble fraction but are *not* precipitated at pH 5. It is thus possible very easily to separate the activating enzymes from them. It turns out that the same enzyme activates an amino acid by forming the amino acyl-AMP and then transfers the amino acid residue to the tRNA, i.e. catalyses the whole of the reaction:

$$\alpha\alpha \longrightarrow \alpha\alpha \sim AMP \longrightarrow \alpha\alpha\text{-tRNA}$$

Sometimes tRNA must be present for the enzyme to catalyse a reaction between the amino acid and ATP—perhaps the sequence of binding of substrates to the enzyme is important.

In most bacterial cells the bulk of the ribonucleic acid is rRNA in ribosomes which can be sedimented. The soluble RNA is largely tRNA but there may also be ribosome precursors and traces of mRNA. The term transfer-RNA (tRNA) is a functional description whereas soluble-RNA (sRNA) is an operational definition. All transfer-RNA's are similar in many properties. They have sedimentation coefficients of about 4 S; their molecular masses are about 25 000; they are 73–93 nucleotides in length and contain pseudo-uridylic acid residues and methylated purines and pyrimidines (see Chapter 5, p. 196); and they

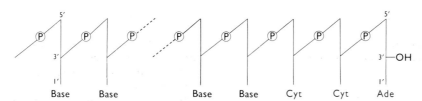

Figure 8. Diagrammatic representation of transfer-RNA.
The amino acid transfer-RNA's are about 80 nucleotides in length and have the sequence -cytidylyl-cytidylyl-adenosine at one end. There is at least one specific tRNA for each of the 20 amino acids. The vertical lines represent ribose residues which are joined by phosphate through position 3' and 5'. They have a purine or pyrimidine base attached to position 1'.

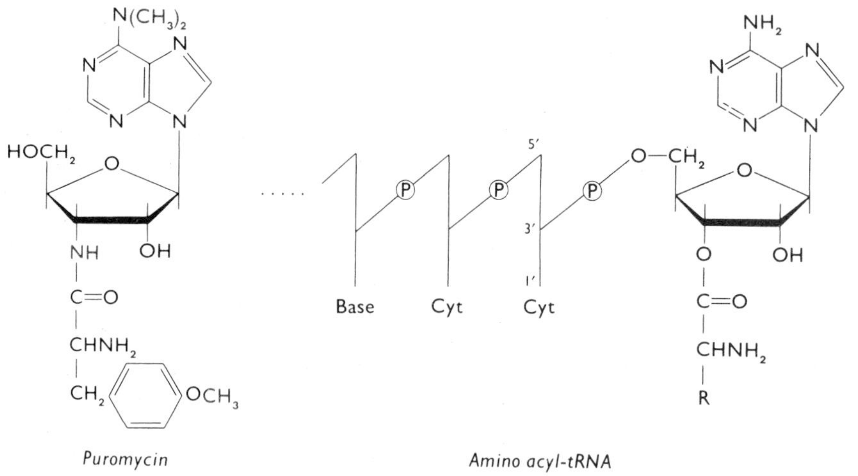

Figure 9. Comparison of structures of puromycin and the terminal sequence of amino acyl-tRNA.
Puromycin is an antibiotic which interferes with the synthesis of proteins by acting as an analogue of amino acyl-tRNA.

have considerable secondary structure (see Figure 10, p. 196). Finally, they all appear to have a sequence of -cytidylic acid, cytidylic acid, and adenosine at one end of the molecule (Figure 8). The amino acid is linked to the ribose of the terminal adenosine, being esterified to the 3'-OH after migration from the 2'-OH. Some antibiotics which inhibit protein synthesis are analogues of amino acyl-adenosine, e.g. puromycin (Figure 9). The ester bond in $\alpha\alpha$-tRNA has not the high energy of the mixed acid anhydride bond in $\alpha\alpha \sim AMP$ but is of higher energy than most simple esters because of the properties of the 2' and 3' hydroxyls of the furanose ring.

There exists at least one tRNA for each amino acid and for some several have been demonstrated. However, only one activating enzyme for each amino acid may occur. These enzymes are highly specific for each of the two stages but more so for the second. Thus the isoleucine enzyme activates isoleucine and valine and the valine enzyme activates valine and threonine but neither of the 'wrong' amino acids gets transferred to tRNA. Sometimes 'alien' amino acid does become incorporated into protein—7-azatryptophan and 2-azatryptophan (tryptazan) are incorporated in place of tryptophan having been activated by its enzyme. However, 5-methyltryptophan is neither activated nor incorporated but it does inhibit synthesis of protein. In general the fidelity of synthesis is very high indeed. The elucidation of primary, secondary and tertiary structures of tRNA's (see Chapter 5, pp. 195–203) helps us to understand why. Significant differences reside elsewhere apart from the anti-codon triplet (see p. 262).

Which species of tRNA occur in a bacterial cell depends on the stage and conditions of growth, and infection by bacteriophage may result in synthesis of phage-specific tRNA's and inactivation of pre-existing host species.

The anti-codon triplet is *not* the site recognized by the activating enzyme and, for instance, the pure *Escherichia coli* serine enzyme charges tRNA^{Ser} whether it has U C A or A G U as the anti-codon. Enzymes and tRNA's from heterologous species may also interact—an *Escherichia coli* enzyme will charge a yeast tRNA.

The tRNA's are precipitated by cold 5% trichloracetic acid (TCA) and the amino acyl residue is not removed by such treatment. It follows, therefore, that a radioactive amino acid will become TCA-precipitable if combined with tRNA and that precipitability is not an adequate indication of incorporation into polypeptide. However, there are procedures which leave proteins intact but which degrade amino acyl-tRNA complexes rendering the amino acid moiety soluble in TCA. Treatment with ribonuclease or with dilute alkali has this effect, as has hot TCA (95°C for a few minutes). It is thus possible to distinguish between the earlier stage of complex formation between the amino acid and ribonucleic acid, and the later stage of incorporation of amino acid into protein. The first part requires only the amino acid, the activating enzyme, ATP and the appropriate tRNA. No particulate fraction is needed. Indeed, the supernatant from high speed centrifugation (100 000 g for 1–2 hr) is an adequate source of the enzymes and tRNA's. The latter can be discharged of their amino acids by raising the pH value to 8·8 and incubating for 45 min before re-neutralizing. They can be subsequently recharged with labelled amino acids by incubating them in the presence of ATP. The tRNA's complexed with radioactive amino acids can then be isolated by extraction with phenol for use in the next stages.

INCORPORATION OF AMINO ACIDS INTO POLYPEPTIDE

Fractionation of disrupted organisms showed that for amino acids to become incorporated into polypeptides, the particulate fraction and the soluble fraction and also a supply of ATP are all necessary. If the soluble fraction was dialyzed to remove small molecules, then GTP was also needed and all the amino acids had to be added. When it became possible to prepare tRNA loaded with radioactive amino acids it could be shown that these complexes were obligatory intermediates. Even the presence of a large excess of unlabelled free amino acids did not reduce the conversion of labelled amino acyl-tRNA to polypeptide. The complex is not, therefore, just acting as a source of free amino acids and the sequence must be as follows:

$$\alpha\alpha \longrightarrow \alpha\alpha \sim AMP \longrightarrow \alpha\alpha\text{-tRNA}$$
$$\longrightarrow \text{Polypeptide}$$

That this is so has been shown unequivocally in cell-free studies but has not been convincingly demonstrated *in vivo*. The complexes cannot enter whole cells and the rate at which amino acyl-tRNA is formed and broken down is such as to have defied efforts to demonstrate its position as an obligatory intermediate. Molecules of amino acids have to be converted to protein at the rate of 8×10^8 per cell-generation, about $8 \times 10^8/50 \times 60$ molecules per cell per second. If there are $1·5 \times 10^5$ molecules of tRNA per cell, this implies that each molecule of tRNA would become charged with an amino acid and discharged again about twice per second (see Tables 1 and 2, pp. 239 and 240). It is believed that this does

happen *in vivo* but the difficulty of demonstrating it will be evident.

It has been established that polypeptides are built up from the *N*-terminal end by successive addition of amino acids carried by their tRNA's. Thus at any intermediate stage a peptide will carry a tRNA attached to the carboxyl of the most recently added amino acid and the next amino acyl-tRNA will react as in Figure 10. This process, catalysed by the ribosomal enzyme *peptidyl transferase*, would normally continue until the polypeptide was complete but the antibiotic, puromycin, acts as an analogue of amino acyl-tRNA and causes premature release of an incomplete polypeptide carrying the drug as its terminal residue (Figure 11). Such puromycin-terminated peptides have been found after addition of radioactively-labelled antibiotic to reconstructed cell-free systems. A useful application of puromycin in experimental systems is as a means of discharging nascent polypeptides from ribosomes. For it is in association with ribosomes that proteins are made and the newly-formed peptides can be separated still attached to these particles whether incorporation of amino acids has occurred *in vivo* or in a cell-free system.

The ingredients of cell-free systems for making polypeptides which we have established so far are the amino acyl-tRNA's, GTP, inorganic ions (particularly Mg^{2+} and K^+), 70 S ribosomes and soluble enzymes. ATP is necessary for activating amino acids but is probably not needed if amino acyl-tRNA's are provided. Soluble proteins are not fully characterized but include at least two elongation factors which have been called T and G factors. The former is involved in binding an incoming $\alpha\alpha$-tRNA to the ribosome while the latter has GTP-ase activity and is involved in the translocation step (see p. 266).

MESSENGER-RNA (mRNA)

For some years it was thought that the ingredients mentioned in the last paragraph were adequate and, indeed, such reconstructed systems do convert amino

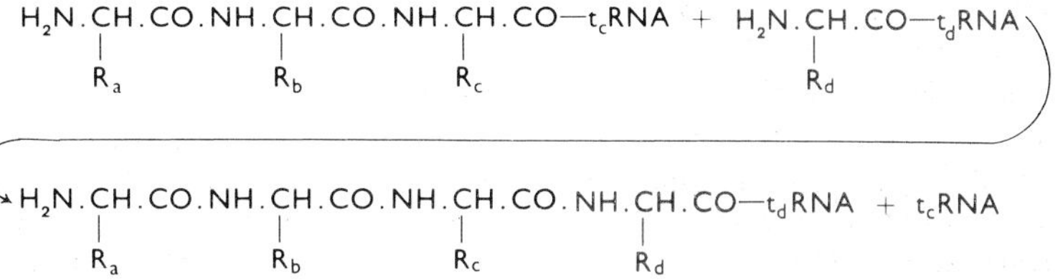

Figure 10. Elongation of a peptide chain.

The peptide combined via its carboxyl group to one tRNA reacts with an amino acyl derivative of another. The first is eliminated yielding an extended peptidyl-tRNA. The enzyme involved has been called *peptidyl transferase*.

$$H_2N.CH.CO.NH.CH.CO.NH.CH.CO-t_cRNA \; + \; H_2N.CH.CO-t_dRNA$$
$$\overset{|}{R_a} \qquad\qquad \overset{|}{R_b} \qquad\qquad \overset{|}{R_c} \qquad\qquad\qquad\qquad \overset{|}{R_d}$$

$$H_2N.CH.CO.NH.CH.CO.NH.CH.CO.NH.CH.CO-t_dRNA \; + \; t_cRNA$$
$$\overset{|}{R_a} \qquad\qquad \overset{|}{R_b} \qquad\qquad \overset{|}{R_c} \qquad\qquad \overset{|}{R_d}$$

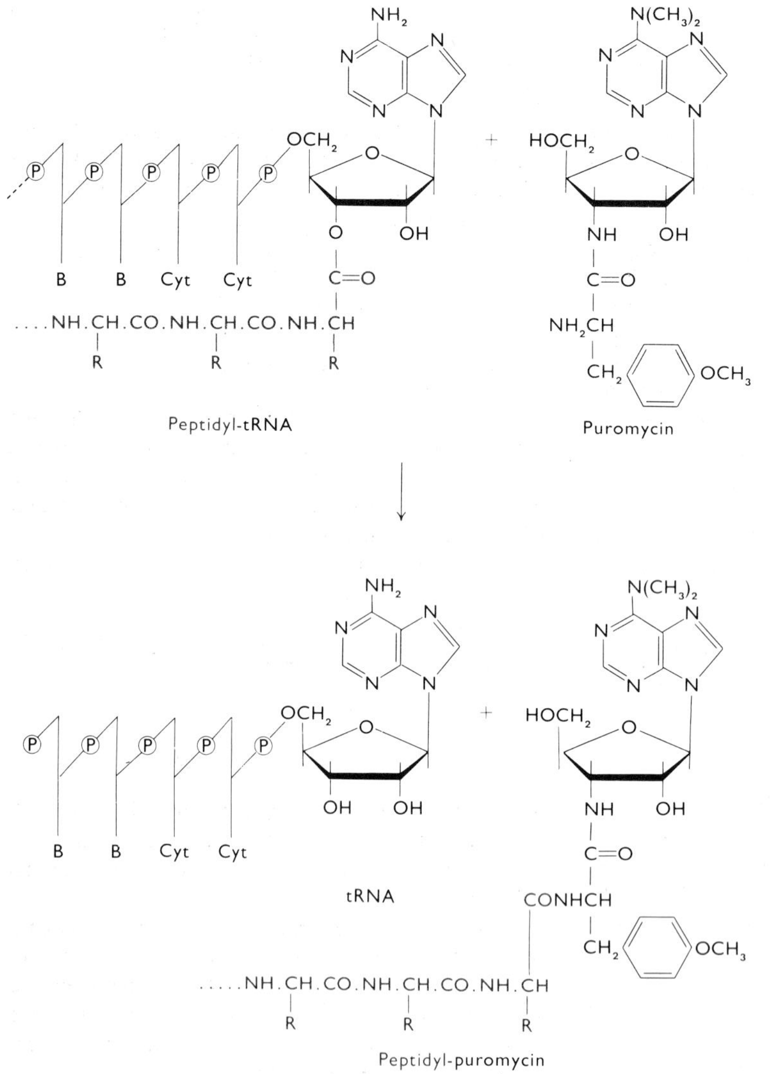

Figure 11. Premature termination of peptide chains by puromycin.

The antibiotic is an analogue of the amino acyl-adenosine end of an amino acyl tRNA and can react with a peptidyl-tRNA to yield peptidyl-puromycin and thus prematurely terminate the lengthening of the polypeptide chain.

acids to polypeptides. But it is difficult to see how a specific sequence of subunits can be assembled into a specific protein unless the 'programme' for this is present in the ribosome. For a time it was tacitly assumed that this was so and that the ribosomal-RNA contained the 'Message'. There are theoretical objections to this assumption and experimentally it has been shown to be wrong.

Cell-free incorporation systems prepared from bacteria (usually from *Escherichia coli*) had trivially low activity compared with intact cells which might be many orders of magnitude more active in polymerizing amino acids. Attempts were made by many people to increase *in vitro* performance by adding additional components to the usual amino acids, tRNA's, ribosomes, soluble enzymes, ATP, GTP, etc. Nirenberg and Matthaei in 1961 tried adding preparations of RNA of various kinds and they included polyuridylic acid (poly U) which had been synthesized enzymically from uridine diphosphate by polynucleotide phosphorylase (Figure 12). The addition of some natural RNA preparations increased incorporation but the most dramatic effects were found with poly U. It caused enormous stimulation of incorporation *but only of one amino acid*, namely, phenylalanine. The product was shown to be poly-phenylalanine. Later many other polynucleotides of known base sequence were found to enhance conversion of particular

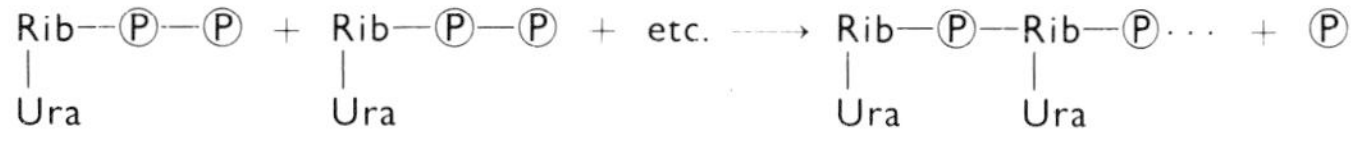

Figure 12. Formation of poly U from
UDP using polynucleotide phosphorylase.

Rib—(P)—(P) + Rib—(P)—(P) + etc. ——→ Rib—(P)—Rib—(P)··· + (P)
| | | |
Ura   Ura   Ura   Ura

amino acids to polypeptides and the relationship of
the nucleic acid bases (purines and pyrimidines) to
the amino acids is well established (see below,
p. 256). It has also been shown that a natural RNA
may direct the formation of a specific polypeptide in a
reconstructed system. However, RNA extracted from
purified ribosomes does not have this property and it
is now known that rRNA does not act as messenger.
It is possible that nascent rRNA has activity which is
lost subsequently on 'maturation'—precursor rRNA's
are of greater molecular weight than those found in
mature ribosomes and some of the bases are modified
by, for example, methylation *after* the polynucleotide
is formed. However, the 16 S rRNA precursor could
be messenger for only three to five of the 30 S ribosomal proteins and the 23 S precursor for only six to ten
of the 50 S ribosomal proteins. Thus at most about
one-quarter of the 50–60 proteins could be specified
by rRNA. The fact that ribosomes without addition of
mRNA are active in promoting amino acid incorporation in cell-free systems is probably due to pieces of
natural mRNA adhering to the particles.

We have seen above (p. 245) that ribosomes can
exist as clusters held together by a strand of RNA and
that it is on these polyribosomes that protein synthesis occurs *in vivo*. This can also be demonstrated *in
vitro* and polyribosomes are often more active in promoting amino acid incorporation than are single ribosomes. Moreover, it is possible to aggregate single
70 S ribosomes into polyribosomes by adding suitable
preparations of RNA, including polynucleotides such
as poly (U). An *in vitro* system appears, therefore, to
require mRNA in addition to ribosomes and tRNA's.

THE GENETIC CODE

The information for determining the sequence of
amino acids in a polypeptide chain resides ultimately
in the DNA of a structural gene. The nucleotide sequence of this is *transcribed* into a complementary sequence in messenger-RNA by the mechanism described
in the previous chapter (p. 223). This mRNA nucleotide sequence has to be *translated* into an amino acid
sequence. There are several ways in which this 'code'
has been deciphered. We shall first make some general
statements which will be amplified later.

1 There are four species of nucleotide (dA, dT, dG,
dC) in DNA and four (A, U, G, C) in mRNA. Both
kinds of nucleic acid are linear polymers and the transcription process is as follows:

$$dA \longrightarrow U$$
$$dT \longrightarrow A$$
$$dG \longrightarrow C$$
$$dC \longrightarrow G$$

2 There are twenty different amino acids (see
Figure 2, p. 11). Other amino acids found in proteins
may be modifications occurring after polypeptide
formation, e.g. hydroxylation of proline, methylation
of lysine to ε-N-methyl lysine.
3 The code is triplet, non-overlapping, and is read
sequentially.

dA dT dG   dC dA dC   dT dT dA   dG dC dA   dT dG dA

might correspond to

$\alpha\alpha_1$ $\alpha\alpha_2$ $\alpha\alpha_3$ $\alpha\alpha_4$ $\alpha\alpha_5$

4 The triplets do not recognize amino acids as such
but, rather, the tRNA to which they are attached.
5 The code is universal, i.e. is the same for all
organisms.
6 Of the 64 possible triplets (4^3) 61 are meaningful
for specific amino acids, i.e. the code is 'degenerate' in
the sense that more than one triplet may represent the
same amino acid. The remaining triplets indicate
'chain termination'. Points 1 and 2 require no elabor-
ation but we must consider the others.

Sections 5 and 6 (p. 28) mentioned the likelihood
that the code is triplet in character since a doublet
code only provides 4^2 different pairs of nucleotides
whereas there are 20 amino acids. Conclusive proof
that the code is triplet, non-overlapping, and is read
sequentially has now been obtained but the same
sequence may be read more than one register (see
Figure 26). Let us first consider the implications of point
3. Assuming that a DNA sequence of nucleotide is:

dA dT dG   dC dA dC   dT dT dA   dG dC dA   dT dG dA

and that the left-hand end is the beginning of the
message in which triplets of nucleotides code for
amino acids, we can predict the consequences of certain changes. For instance, removal or addition of a

single nucleotide would radically alter the message from that point onwards:

dA dT d$\bar{\text{G}}$ dC dA dC dT d$\overset{+}{\text{T}}$ dA dG dC dA dT dG dA

A deletion of the first dG would change the whole sequence:

dA dT dC dA dC dT dT dA dG dC dA dT dG dA ...

An addition of dX between the pair of dT's would cause alteration from that triplet onwards:

dA dT dG dC dA dC dT dX dT dA dG dC

dA dT dG dA ...

However, the deletion together with the addition would only alter a small part of the sequence (the first three triplets in this instance):

dA dT dC dA dC dT dX dT dA dG dC dA dT dG dA

It is thought that some mutagenic chemicals such as acridine orange (see pp. 228 and 283) are effective because they cause the addition or deletion of a nucleotide pair from the DNA. If this is so and if the DNA is coding for protein, then three deletions at the places indicated:

dA dT dG d$\bar{\text{C}}$ dA dC dT d$\bar{\text{T}}$ dA d$\bar{\text{G}}$ dC dA dT dG dA

would give:

dA dT dG dA dC dT dA dC dA dT dG dA ...

Three additions at the same sites would give:

dA dT dG dX dC dA dC dT dY dT dA dZ dG dC dA

dT dG dA ...

These triple mutants would not only have a small number of altered amino acids but would also have one fewer or one more amino acid than the wild-type. It has been found experimentally with acridine-induced mutants of the bacteriophage T4 that single mutants were unable to grow, some double mutants could and some could not, and some triple mutants were semi-normal. Although there is no certainty as to which are additions and which are deletions, these findings were interpreted to mean that an addition plus a nearby deletion could restore the bulk of the message to normal and that three neighbouring additions (or deletions) could do the same. It was suggested also that the observations made it likely that the code was triplet or a multiple of that. As we shall see there are even firmer bases for the belief in the validity of point 3.

The genetic information we have been discussing must be transcribed as follows:

DNA sequence

dA dT dG dC dA dC dT dT dA dG dC dA dT dG dA

mRNA sequence

U A C   G U G   A A U   C G U   A C U

There is no evidence that a particular ribonucleotide triplet has any affinity for any particular amino acid. Rather, the recognition is of the transfer-RNA to which the amino acid is esterified. This has been demonstrated very beautifully in the cell-free system from reticulocytes (immature red blood cells). This preparation can synthesize the polypeptide chains of haemoglobin and can be supplied with the amino acids attached to their tRNA's. What was done was to modify the amino acid chemically *after* it had been attached to tRNA and to show that it was still incorporated into the peptide as if it had not been so altered. Cysteine attached to tRNA^{Cys} was treated with Raney nickel which converted the amino acid to an alanyl residue (Figure 13). When the product was added to the *in vitro* reticulocyte system it was found that the polypeptide produced contained alanine at some sites where cysteine was expected. Similar experi-

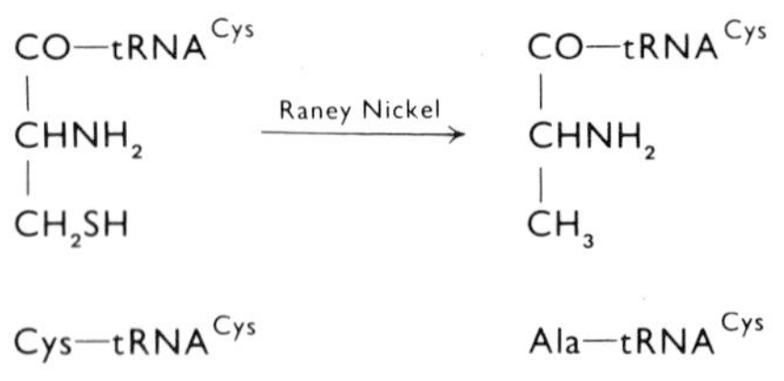

$$
\begin{array}{ccc}
\text{CO—tRNA}^{\text{Cys}} & & \text{CO—tRNA}^{\text{Cys}} \\
| & \xrightarrow{\text{Raney Nickel}} & | \\
\text{CHNH}_2 & & \text{CHNH}_2 \\
| & & | \\
\text{CH}_2\text{SH} & & \text{CH}_3 \\
\\
\text{Cys—tRNA}^{\text{Cys}} & & \text{Ala—tRNA}^{\text{Cys}}
\end{array}
$$

Figure 13. Conversion of the cysteinyl residue to the alanyl residue while still attached to the transfer-RNA specific for cysteine.

When the altered amino acyl-tRNA was used in a cell-free system making haemoglobin polypeptide chains, some alanine was found in positions where cysteine was expected. This confirms that the tRNA (normally) recognizes the mRNA and so delivers the appropriate amino acid to the correct site.

ments have been carried out using synthetic poly-nucleotides as mRNA and demonstrating that the product was determined by the tRNA rather than by the amino acid attached to it. This is excellent justification of point 4 above.

The universality of the code (point 5) is suggested by the results of *in vitro* experiments in which amino acids linked to tRNA from one species of organism are functional with ribosomes and mRNA of another, e.g. *Escherichia coli* and mammalian reticulocytes, respectively. There is an obvious evolutionary significance in this. However, mitochondria are exceptional in having a few codons which are read differently.

Deciphering the genetic code

The allocation to the 64 ribonucleotide triplets of specific functions is now complete. Table 3 lists these and it should be noted that the sequence of the three bases is important and that they are recorded in a conventional order. Thus A U G has phosphate diester links from the 3' of the adenine nucleoside to the 5' of the uracil nucleoside and from the 3' of this to the 5' of the guanine nucleoside whereas G U A would be the converse (Figure 14).

We shall now discuss methods which have been used to decipher the code. The most rigorous would be to establish the complete nucleotide sequence of a DNA structural gene or the corresponding mRNA and then to relate this to the amino acid sequence of the appropriate polypeptide. Techniques for determining polypeptide sequences have been available for some time and those for nucleic acids have now been developed. Many sequences of proteins and nucleic acids are known and in several instances the correspondence between DNA, mRNA, and polypeptide has been established (e.g. regions of the *lac* operon and β-galactosidase (p. 272) and the complete sequence of 5386 nucleotides of the DNA of the *E. coli*

phage φX174 together with much of the amino acid sequence of the nine proteins it encodes). Further, the RNA of phages such as R17, Qβ and MS2 has been much studied to relate the more than 3000 nucleotides to the 1000 amino acids in the three proteins specified (see p. 270).

An early method used in decipherment was to relate amino acid incorporation by cell-free systems to the base sequence of synthetic polyribonucleotides acting as artificial messengers. Thus poly (dA) acts as a template for RNA polymerase to synthesize poly (U) and this acts as a code *in vitro* for the synthesis of poly-phenylalanine. It was concluded that the triplet U U U stands for phenylalanine. Similarly, poly (A) and poly (C) yield polymers of lysine and proline, respectively. Mixed polymers can also be synthesized by adding the enzyme polynucleotide phosphorylase to mixtures of ribonucleoside disphosphates. From the relative amounts of nucleotides in the product and on the assumption that the sequence is random, the frequency of all possible triplets can be calculated. If such a polynucleotide is used as an artificial messenger the amino acids incorporated will depend in kind and amount on the nature of the code and the frequency of occurrence of the various triplets. Thus if poly (A, U) is a random sequence of A's and U's, all the following triplets will be present:

U U U, U U A, U A U, A U U, U A A, A A U,

A U A and A A A.

The relative abundances of these can be predicted from the composition and it can be determined experimentally which amino acids become incorporated when this polynucleotide is used as messenger. It should be noted, however, that the three triplets U_2A will have the same frequencies as each other as will the three UA_2. In other words the sequence of the three nucleotides will not be apparent. From Table 3

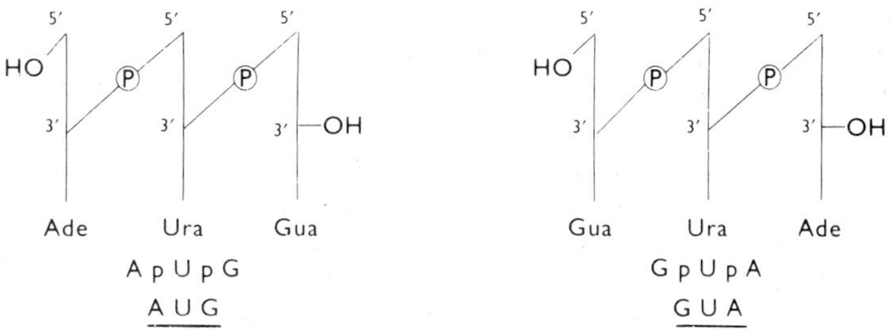

Figure 14. Conventional abbreviations for oligonucleotides.
The form A U G implies a phosphodiester link between the 3' and 5' positions of the adenine and uracil nucleosides respectively. Similarly there is a link between the 3' position of the uridine and the 5' position of guanosine. A p U p G likewise means that the 5' hydroxyl of the ribose attached to adenine and the 3' hydroxyl of that attached to guanine are unsubstituted.

Table 3. The Triplet Code

5'-OH Terminal base	Middle base				3'-OH Terminal base
	U	C	A	G	
U	U U U $\Big\}$ Phe U U C U U A $\Big\}$ Leu U U G	U C U $\Big\}$ U C C $\Big\}$ Ser U C A U C G	U A U $\Big\}$ Tyr U A C U A A Ochre U A G Amber	U G U $\Big\}$ Cys U G C U G A Opal U G G Trp	U C A G
C	C U U $\Big\}$ C U C $\Big\}$ Leu C U A C U G	C C U $\Big\}$ C C C $\Big\}$ Pro C C A C C G	C A U $\Big\}$ His C A C C A A $\Big\}$ Gln C A G	C G U $\Big\}$ C G C $\Big\}$ Arg C G A C G G	U C A G
A	A U U $\Big\}$ A U C $\Big\}$ Ile A U A	A C U $\Big\}$ A C C $\Big\}$ Thr A C A A C G	A A U $\Big\}$ Asn A A C A A A $\Big\}$ Lys A A G	A G U $\Big\}$ Ser A G C A G A $\Big\}$ Arg A G G	U C A G
G	G U U $\Big\}$ G U C $\Big\}$ Val G U A G U G	G C U $\Big\}$ G C C $\Big\}$ Ala G C A G C G	G A U $\Big\}$ Asp G A C G A A $\Big\}$ Glu G A G	G G U $\Big\}$ G G C $\Big\}$ Gly G G A G G G	U C A G

Most of the 64 triplets of nucleotides are allocated to specific amino acids. A few have special functions. Thus A U G represents the chain-initiating N-formyl-methionine. U A A, U A G and U G A indicate chain termination.

it can be seen that the eight triplets will code for the amino acids Phe, Leu, Tyr, Ile, Asn and Lys. Another preparation of poly (A, U) of different composition would have different abundances and the altered relative amino acid incorporations can help to indicate which triplet corresponds to which amino acid.

Polynucleotides of the form X U U U U U U ... were also prepared and used to investigate from which end the message is read and also to find out what X U U stands for. The resulting polypeptide should be either:

$$NH_2CH.CO—Phe—Phe—Phe...Phe.COOH$$
$$\Big|$$
$$R$$

or

$$NH_2Phe—Phe—Phe—Phe...PheCO.NH.CH.COOH$$
$$\Big|$$
$$R$$

The evidence indicated that translation begins from the 5' end of the polynucleotide and that this corresponds with the N-terminal end of the polypeptide.

Polynucleotides of regular alternating sequence were also made. Poly (X-Y) programmed formation of a polypeptide composed of two alternating amino acids. Thus poly (U-C) yielded poly (seryl-leucyl):

$$\underline{U\ C\ U}\ \ \ \underline{C\ U\ C}\ \ \ \underline{U\ C\ U}\ \ \ \underline{C\ U\ C}\ \ \ U\ C...$$

  Ser   Leu   Ser   Leu ...

Even more striking was that a polynucleotide composed of repeating units of A A G acted as a messenger for three homopolymers—poly-lysine, poly-arginine and poly-glutamic acid. This is consistent with the following allocations of triplets:

AAG AAG AAG AAG ...
Lys   Lys   Lys   Lys

A A G codes for lysine

AGA AGA AGA AGA ...
Arg   Arg   Arg   Arg

A G A codes for arginine

GAA GAA GAA GAA ...
Glu   Glu   Glu   Glu

G A A codes for glutamic acid

Since for the series A A G A A G A A G A A G A A ... only the three triplets mentioned do occur and since no other amino acids are incorporated and since only homopolymers are formed, this is further evidence that the code is a non-overlapping, triplet one in which contiguous triplets are read sequentially without bypassing a single nucleotide.

A few polymers having repeated tetra-nucleotides were made and found to yield polymers of four amino acids in sequence; for instance poly (UAUC) codes for tyrosine, leucine, serine and isoleucine:

UAU CUA UCU AUC UAU CUA UCU
Tyr   Leu   Ser   Ile   Tyr   Leu   Ser

A second method of approach for deciphering the code is one which is related more directly to natural systems. It is the study of amino acid replacements in proteins. A mutational change involving the *alteration* (not deletion or addition) of only a single nucleotide may result in the replacement of a single amino acid by another at a particular place in a polypeptide. Thus the DNA triplet dA dA dA is transcribed to the mRNA triplet U U U and this is translated as phenylalanine. A genetic alteration to dA dA dT would result in the RNA triplet U U A which represents leucine. Hence leucine would replace phenylalanine at *one* specific place in the appropriate protein. There are, of course, many possibilities:

DNA Triplet	mRNA Triplet	Amino Acid
dA dA dA	U U U	Phenylalanine
dA dA dT	U U A	Leucine
dA dT dA	U A U	Tyrosine
dT dA dA	A U U	Isoleucine

etc.

But with hindsight we can see that a *single change* should only produce a limited number of alternative replacements. This has been found to be so in naturally occurring mutants of this kind and the translation of the code must be consistent with these observations. Artificial changes of a similar kind can be induced by chemical modifications, e.g. nitrous acid may convert purine and pyrimidine amino groups to hydroxyls and thus alter the bases in DNA (see Chapter 7, p. 285).

The third method to be described depends on a somewhat different principle. We have seen that mRNA can be attached to ribosomes and that amino acyl-tRNA appears to recognize an appropriate nucleotide region in the mRNA. The whole mRNA is not necessary for this association—a triplet of nucleotides is enough. Thus ribosomes plus the triplet U U U form a complex with phenylalanyl-tRNA but not with any other amino acyl-tRNA nor with any free amino acid. All of the 64 trinucleoside diphosphates (XpYpZ) have now been tested individually with all the amino acyl-tRNA's. In very many instances the results are clear-cut and indicate a specific relationship between a particular amino acid and one or more triplets. An advantage of this method is that it gives unequivocal information about the sequence of the three bases as well as the overall composition of the triplet.

Finally, the complete nucleotide sequences of various nucleic acids can be related to the amino acid sequences of the proteins for which they code (see p. 272). It is gratifying that the results of *in vitro* incorporation studies, of amino acid replacements, of investigations on binding, and of translation of natural mRNA, confirmed each other so that the translations given in Table 3 are established. It should be noted that only three of the 64 triplets do not code for an amino acid (see below) and also that certain generalizations can be made about those which do. In every instance XpYpU and XpYpC represent the same amino acid, i.e. either pyrimidine can be in the third position. Usually this is also true for the purines, XpYpA and XpYpG being alternatives. Indeed for eight amino acids the code is essentially doublet rather than triplet in that the third position can be occupied by any one of the four bases. (This does not imply, however, that these alternatives may not have different functions *in vivo* and we shall return to this shortly.) Finally, the assortment of triplets does not seem to be random—those amino acids which are related structurally or metabolically often have related codes. A consequence of this which may be significant is that a mutation leading to replacement may cause substitution of a related amino acid rather than one which is quite dissimilar. This might well minimize the adverse effects.

The beginning and ending of polypeptide chains

Termination codons

When polypeptides are synthesized *in vitro* in a recon-structed, cell-free system it seems that the polynucleo-tide used as 'artificial' messenger can be read from various starting points. Thus the sequence A A G-A A G A A G which we mentioned above can ap-parently be read as a sequence of A A G or A G A or G A A. Manifestly this would be undesirable *in vivo* with a natural messenger-RNA since it would involve the synthesis of a great deal of useless polypeptides. Messages must have a beginning and an ending. There is evidence that the three triplets which have not been assigned to amino acids (sometimes called 'nonsense') represent termination codons and are normally present at the end of each message. They are U A A, U A G and U G A. Mutants have also been found in which a single base change converts an amino acid codon to one of these three and this re-sults in premature termination of polypeptide syn-thesis and release of incomplete peptide. The site of the mutation determines the amino acid at which the peptide ends and it has been established that the next triplet in the mRNA is U A A or U A G or U G A. These three termination codons do not have corre-sponding tRNA's but are recognized by specific ter-mination proteins or *Release Factors* of which RFI interacts with U A A or U A G and RF2 with U A A or U G A (see below, p. 267). However, an additional *suppressor* mutation (see p. 281) at another locus may over-ride the effect of a mutation such as we have been considering. The consequence of this is to cause the formation of a modified tRNA with an anti-codon complementary to the termination codon produced by the first mutation. Thus an amino acid is inserted

and complete polypeptide is made. An example in-volving tyrosine is shown in Figure 15. Codons for several other amino acids can mutate to termination codons and suppression does not always result in in-sertion of the same amino acid as in the wild-type protein. If different, the protein formed may have altered biological activity.

Some natural mRNA's carry information for more than one polypeptide chain, so-called *polycistronic messengers* (see also Chapter 8, p. 331) and have intra-cistronic as well as terminal stop codons.

Initiation codons

There are some curious features about the beginnings of polypeptide chains. We mentioned that the riboso-mal proteins of *Escherichia coli* nearly all begin with methionine or alanine (p. 243). It is also found that the soluble proteins (the bulk of the total) of the same organism have N-terminal methionine (45%) or alan-ine (30%) or serine (15%). As there are of the order of a thousand different species of proteins it is certainly surprising to find so few of the twenty possible amino acids at the N-terminal ends of polypeptides.

There are two different transfer-RNA's for methion-ine and they have been called $tRNA_f^{Met}$ and $tRNA_m^{Met}$. When methionine is bound to the former it can be enzymically formylated both *in vitro* and *in vivo*, yielding N-formyl-methionyl-$tRNA_f$. The anal-ogous methionyl-$tRNA_m$ cannot be so formylated. Moreover, the former complex is precursor to N-terminal methionine whereas the latter delivers methionine to other positions in the polypeptide chain. Both tRNA's have the same anti-codon triplet and interact with the codon A U G ($tRNA_f$ can also recognize the codon G U G which normally codes for valine) but there are differences in structure elsewhere. As far as is known all polypeptides are synthesized beginning with N-formyl-methionine and nearly half have the formyl group removed enzymically to yield proteins with N-terminal methionine. Others have the methionine removed also—by a specific aminopeptidase—yielding a protein with N-terminal alanine or serine or, less frequently, some other amino acid. Thus the RNA of the small phage f_2 acts, *in vitro*, as mRNA for the synthesis of a protein beginning

N-formyl-Met—Ala—Ser—Asn—Phe—Ser— ⋯

whereas the phage protein made *in vivo* begins

Met—Ala—Ser—Asn—Phe—Ser— ⋯

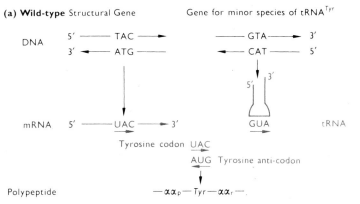

(a) Wild-type Structural Gene   Gene for minor species of tRNA^{Tyr}

(b) Mutation in Structural gene ('Amber')

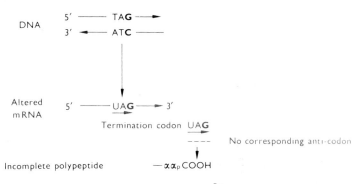

(c) Mutation in gene for minor species of tRNA^{Tyr}

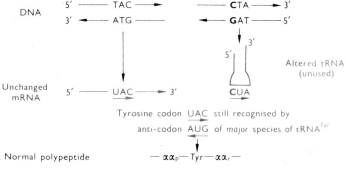

Figure 15. Mutation in a structural gene and its suppression by mutation of a gene for a minor species of tRNA.

Mutation converting a tyrosine codon to UAG, a termination codon, yields an 'amber mutant' producing incomplete polypeptides terminating just before the position where a tyrosine residue occurs in the wild-type (b).

Mutation leading to an altered minor species of tRNA^{Tyr} has no effect since other species of tRNA^{Tyr} interact with the unchanged tyrosine codon (c).

The effects of the structural gene mutation giving UAG can be suppressed by the second mutation changing the tRNA. This results in the UAG being translated into tyrosine rather than acting as a terminator (d).

(d) Mutation in Structural gene *and* **Suppressor mutation** in gene for minor species of tRNA^{Tyr}

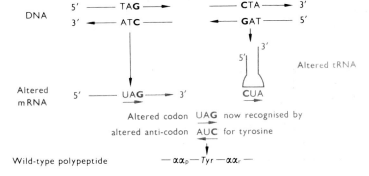

259

Synthetic mRNA						Product
A U G	A A A	A A A	A A A	...	A A A	(f)Met-(Lys)$_n$ on ribosomes
A U G	U U A	A A A	A A A	...	A A A	(f)Met-Leu-(Lys)$_n$ on ribosomes
A U G	U U U	A A A	A A A	...	A A A	(f)Met-Phe-(Lys)$_n$ on ribosomes
A U G	U U U	U A A	A A A	...	A A A	f-Met-Phe released from ribosomes

Figure 16. Initiation and termination codons (see p. 256).

Further, in model systems with synthetic mRNA's the results shown in Figure 16 were obtained.

Synthetic polyribonucleotides

$$AUG(A)_n, \qquad AUGU_2(A)_n,$$
$$AUGU_3(A)_n \text{ and } AUGU_4(A)_n$$

were used as artificial messengers in the presence of labelled amino acids, ribosomes and appropriate supplements.

With the first three, a peptide containing a labelled methionine (presumably formylated) remained attached to the ribosome. In addition it contained many lysine residues and, as indicated, no other residue, or one leucine, or one phenylalanine.

Only with $AUGU_4(A)_n$, which contains the termination codon U A A, was a peptide released from the ribosome—the dipeptide *N*-formyl-methionyl-phenylalanine.

This experiment proves that the code is triplet, non-overlapping, sequential, read from 5′ to 3′, that A U G initiates and U A A terminates, and that the peptide is made N → C.

NATURAL MESSENGER-RNA

It should now be abundantly clear that there is a component of bacterial cells to which the name messenger-RNA must be given. But it was an elusive substance, the subject of much controversy. The existence of such a class of RNA was predicted before it was demonstrated in any convincing way. It was postulated that mRNA should have a composition like that of DNA since it was to transmit genetic information to the protein-forming systems; that it should be polydisperse in size to account for the many different sizes of protein molecules; and that it should be capable of being used perhaps only a few times before being degraded—this because of the rapidity with which the production of particular proteins could be switched on and off in bacteria.

The kinetics of enzyme synthesis after induction are discussed in Chapter 8 and it is shown that the maximum rate can be achieved in a minute or two and that synthesis can cease equally rapidly. Moreover,

under certain conditions up to 5% of the total protein production may be turned over to the making of one species. These properties seem to demand great flexibility in formation, utilization and discard of mRNA.

The concept of mRNA is of necessity a functional one—mRNA is that RNA which carries a message, i.e. the information for the formation of a polypeptide. This has led to a number of misconceptions of which some are widely current. In fact mRNA does not have to have the composition of DNA although it has been referred to as DNA-like or D-RNA. It should, rather, be complementary in nucleotide composition to *one strand of DNA*—and therefore be like the other strand if the DNA is of the usual double-helical type. Much work has been carried out with *Escherichia coli* and it just happens that in this organism the composition of each strand is like that of the other as well as being complementary to it and, in addition, that for this DNA, dA = dC = dG = dT. Secondly, one should not expect the composition of mRNA to resemble the total DNA of a cell, i.e. the thousands of genes of all kinds. Rather, a specific molecule of mRNA should be complementary in composition to one strand of the corresponding structural gene. If much of the genome (chromosome) is not composed of structural genes or if the relative abundances of the various mRNA's were different from those of the corresponding genes, one would not necessarily expect overall relationships in composition between total DNA and total mRNA.

It is also misleading to suggest that all the RNA labelled during a short period of growth in the presence of a radioactive precursor (so-called pulse-labelled RNA or rapidly labelled RNA) is necessarily all mRNA or that all mRNA is obliged to have a short life. At any instant the fraction which can be identified as being the mRNA is only one or a few hundredths of the total cellular RNA. However, whereas ribosomal-RNA and amino acid transfer-RNA are very long-lived, there is a fraction of RNA which does not survive very long—perhaps only a few minutes. This means that pulse-labelling will give a higher fraction of radioactivity in this labile RNA but even a pulse as short as 30 seconds may result in more than half of the radioactivity being in rRNA with a little in tRNA. Whether or not all the remainder is in mRNA is a matter for conjecture. If, after a labelling period of 30 seconds, excess of non-radioactive precursors are added to 'chase' the radioactivity out of the transient substances, it is found that rRNA's and tRNA's are stable end-products whereas the rest of the radioactivity is in a labile component. This is shown in Figure 17 where the total RNA extracted by phenol has been centrifuged through a sucrose gradient. The positions of the 23 *S* and 16 *S* rRNA's are seen as is the 4 *S* peak of tRNA's. The radioactivity after 30 seconds is distributed rather diffusely over a range of sizes. After 'chasing', only the rRNA and tRNA are labelled. It is highly probable that the transient component is mRNA as it can be shown to have a composition unlike that of rRNA or tRNA and to be hybridizable with DNA.

In general it is difficult to reduce the specific activity of the nucleotide pool by adding unlabelled substances and so the 'chase' has to be relatively long. There is, however, another way in which the disappearance of label from some types of RNA but not from others can be shown. This is by adding the antibiotic actinomycin D which prevents further synthesis of all kinds of RNA. Under these conditions breakdown of the labile RNA is roughly exponential with time and so it is possible to determine the mean half-life by studying the kinetics of loss of radioactivity from the unstable RNA fraction (Figure 18). The addition of the drug also brings about a progressive impairment in the rate of protein synthesis and this reduction is also exponential with time. Finally, there is an exponential decrease in the content of polyribo-

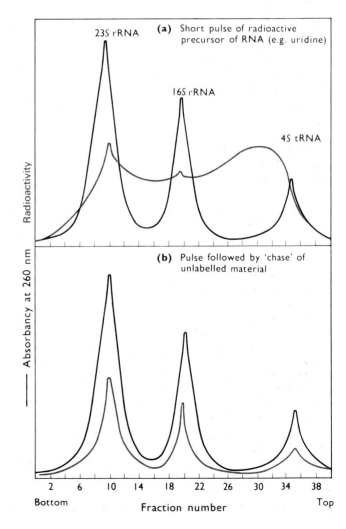

Figure 17. Sucrose gradient analysis of nucleic acids.

RNA can be extracted from bacteria with phenol and appears as three kinds—23 *S* and 16 *S* ribosomal RNA (rRNA) and 4 *S* soluble RNA (mainly tRNA). A short pulse of radioactive precursor shows a different distribution since the mRNA fraction, although heterogeneous in size and very small in amount, turns over very rapidly and accounts for about ⅓ of the label in short times. After 'chasing' the pulse with unlabelled precursor the radioactivity is retained only in the tRNA and rRNA components.

somes and a corresponding increase in the number of 70 S ribosomes. It would be remarkable if these phenomena were not interrelated and the simplest explanation is that they are consequent upon the breakdown of mRNA which holds 70 S ribosomes together as

polyribosomes and which is necessary for synthesis of protein.

It is technically difficult to isolate natural mRNA from bacterial cells but the nucleic acids of some RNA viruses can readily be extracted and studied *in vitro*. These natural messengers have been shown to programme the formation of characteristic viral proteins. Such experiments are much more convincing than those using poly U as mRNA.

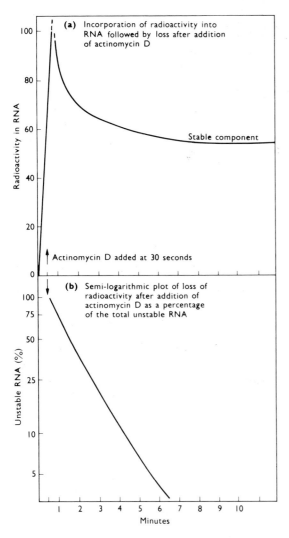

INTERACTION OF mRNA, tRNA AND RIBOSOMES

We have seen that ribosomes may be aggregated to form polyribosomes by addition of poly U and that natural polyribosomes are probably held together by mRNA. The binding *in vitro* can be shown to be to the 30 S subunit of the ribosome and the initial interaction seems to involve base-pairing between part of the pyrimidine-rich, 3' terminal dodecanucleotide of 16 S rRNA and a purine-rich sequence preceding the initiation triplet of the mRNA. The lengths of the complementary sequences vary between three and nine nucleotides, the average of those so far recognised being 4–5 (Table 4). Transfer-RNA, on the other hand, binds mainly to the 50 S subunit and there is evidence that there are two such sites on the particle. These correspond to one for the tRNA at the end of the nascent polypeptide and the other for the next incoming amino acyl-tRNA (Figure 19).

There must also be interaction between the mRNA and the tRNA if the latter is to place an amino acid correctly at the behest of the former. Since a triplet of nucleotides is established as the determinant for an amino acid and since the only chemical explanation available for the recognition is that of base-pairing, it is not surprising that the tRNA has a triplet which is complementary to that of the mRNA sequence. The mRNA triplet, the codon, is complementary to the tRNA triplet, the anti-codon. The interactions are between anti-parallel strands just as they are between the two strands of DNA (see Chapter 5, p. 188). Relationship could be thus:

Figure 18. Breakdown of labile RNA.

(a) shows the rapid incorporation of label from radioactive precursors into RNA. This can be stopped almost instantly by addition of actinomycin D which prevents synthesis. Subsequently about half of the label becomes soluble as RNA breaks down. This labile component is only a few percent of the total RNA and is probably largely mRNA.

(b) After subtracting the amount of the stable component (the asymptote in (a)) the decay of the labile component can be plotted. For a pulse of 30 seconds, the time for 50% decay is about 60 seconds and that for 80% is about 150 seconds.

DNA	dA dG dA ⟶ dT dC dT ⟵	dT dT dG ⟶ dA dA dC ⟵	dC dT dT ⟶ dG dA dA ⟵
mRNA codon	A G A ⟶	U U G ⟶	C U U ⟶
tRNA anti-codon	U C U ⟵	A A C ⟵	G A A ⟵
Amino acid	Arginine	Leucine	Leucine

Table 4. Possible sequences for ribosome-binding sites

mRNA: Purine-rich sequences preceding initiation triplets AUG and GUG

Escherichia coli		
lac Z	5′ . .	. A A U U U C A C A C A G G A A A C A G C U **A U G** . . .
lac I	. .	. A G U C A A U U C A G G G U G G U G A A U **G U G** . . .
gal E	. .	. A U A A G C C U A A U G G A G C G A A U U **A U G** . . .
gal T	. .	. U A U C C C G A U U A A G G A A C G A C C **A U G** . . .
trp A	. .	. G A A A G C A C G A G G G G A A A U C U G **A U G** . . .
trp E	. .	. G A A C A A A A U U A G A G A A U A A C A **A U G** . . .
Phage MS2		
A-protein	. . .	. G A U U C C U A G G A G G U U U G A C C U **G U G** . . .
coat	. . . .	. C C U C A A C C G A G G U U U G A A G C **A U G** . . .
replicase	. . . . . .	. A A A C A U G A G G A U U A C C C **A U G** . . .
Phage ϕX174		
A	. . . . .	. A A A U C U U G G A G G C U U U U U U **A U G** . . .
B	. . .	. U A A A G G U C U A G G A G C U A A A G A **A U G** . . .
C	. . .	. G A A G U G G A C U G C U G G C G G A A A **A U G** . . .
D	. . .	. A C C A C U A A U A G G U A A G A A A U C **A U G** . . .
E	. . . . .	. U G C G U U G A G G C U U G C G U U U **A U G** . . .
J	. . .	. A C G U G C G G A A G G A G U G A U G U A **A U G** . . .
F	. . .	. C C C C U U A C U U G A G G A U A A A U U **A U G** . . .
G	. . .	. U U U C U G C U U A G G A G U U U A A U C **A U G** . . .
H	. . .	. G C C A C U U A A G U G A G G U G A U U U **A U G** . . .

16 S rRNA: Pyrimidine-rich sequences at 3′ termini

Escherichia coli	3′ HO A U U C C U C C A C U A G . . . .
Pseudomonas aeruginosa	HO A U U C C U C U C PyX X G . . . .
Bac. stearothermophilus	HO A U C U U U C C U C C A C . . . .
Caulobacter crescentus	HO U C U U U C C U PyX X X G . . . . . .

mRNA sequences which are complementary to *E. coli* 16 S rRNA are in colour. (G-U pairs also occur but are not indicated.) The sequence in *E. coli* 16 S rRNA which is complementary to the probable site for ribosome-binding on the mRNA for ϕX174 protein H is also coloured.

where two of the six leucine codons are represented, the arrows indicating the 5′ → 3′ direction.

When purified tRNA's with known anti-codons, became available it was possible to test their interactions with trinucleotides and it was found that some tRNA's could recognize several different codons. Moreover, many tRNA's contained the base inosine at the 5′ position in their anti-codon. These observations led to the 'wobble' hypothesis which suggests that base pairings other than A—U and G—C can occur between the base at the 3′-end of the mRNA codon and that at the 5′-end of the tRNA anti-codon. The possible combinations are shown in Table 5.

From the hypothesis it can be predicted that 31 is the minimum number of tRNA's which could react with all 61 amino acid codons and, more specifically, that three tRNA's should exist for the six serine codons (U C U, U C C, U C A, U C G, A G U and A G C).

Table 5. The 'wobble' hypothesis

Base in 5′-position of anti-codon	Base in 3′-position of codon
G	C or U
C	G
A	U
U	A or G
I	A, U or C

The base in the 5′-position of the anti-codon of tRNA may be able to form H-bonds with more than one base at the 3′-position of the codon of mRNA. Thus guanine in the 'wobble' position can pair with cytosine or uracil; uracil with adenine or guanine; inosine with adenine, uracil or cytosine. In the other two positions of the anti-codon only the usual C—G and A—U pairs are possible.

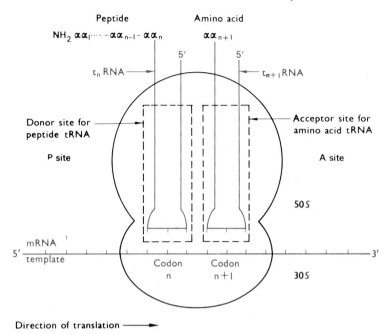

Figure 19. Representation of binding sites for transfer-RNA on 70 S ribosomes.

The mRNA is bound to the 30 S component. There are two sites for tRNA and these may involve both 30 S and 50 S subunits. One of them is an *acceptor site* for amino acyl tRNA (sometimes called the 'amino acid site'); the other is the *donor site* for the peptidyl tRNA (sometimes called the 'peptide site').

The antibiotic puromycin acts as if it could occupy part of the acceptor site and receive a peptide from the donor.

The latter prediction has been confirmed and it seems that organisms indeed have more than twenty species of tRNA and may be able to translate any of the 61 codons.

As well as interacting with mRNA, the transfer-RNA has to have a site specific for the amino acid activating enzyme and a region which has affinity for some part of the 50 S ribosomal subunit. These properties may account for the size of these molecules (about 80 nucleotides in length) and for their having regions of similar structure (see Chapter 5, p. 197). There is much speculation as to how these interactions occur. A chain of 200 amino acid residues (M_r *c*. 22 500) requires an mRNA of 600 nucleotides, i.e. about 200 nm in length if there is one nucleotide every 0·34 nm, or double that if the mRNA is in extended form. Since a 70 S ribosome has a maximum dimension of about 250 nm, it is clear that there must be relative movement between it and the mRNA so that the message can be read sequentially from beginning to end. A stretch of 15–20 nucleotides is protected by association with a ribosome against degradation by ribonuclease and it seems likely that ribosomes are closely packed along the length of the mRNA, perhaps one to every 10–20 nm. There is cumulative evidence that ribosomes begin translating mRNA while it is still being transcribed from DNA

and that degradation from the 5'-end may also begin before transcription is complete but only after some dozens of ribosomes have sequentially begun translation. Such a process might be represented diagrammatically as in Figure 20.

INITIATION, ELONGATION AND TERMINATION PROTEIN FACTORS

Many protein factors have been described which appear to be involved in the formation of polypeptides from amino acyl-tRNA's by complexes of ribosomes and mRNA. The peptidyl transferase itself is an integral part of the 50 S subunit but at least eight other proteins have been found in the high speed supernatant fraction of cytoplasm or can be washed off ribosomes by solutions of high ionic strength (e.g. 2 M NH_4Cl). Some of them are concerned in chain initiation, some in elongation and some in termination.

Initiation factors (IF's)

Three factors have been separated from various prokaryotes and shown to be transiently involved in forming the initiation complex according to the following:

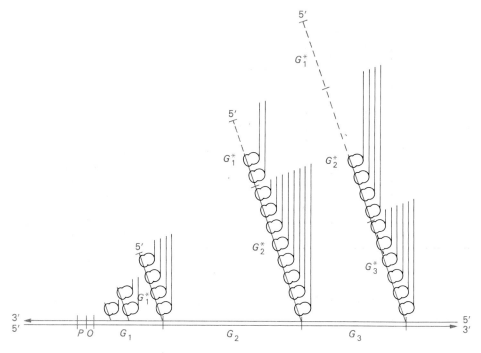

Figure 20. Diagrammatic representation of transcription and translation and degradation of mRNA.

Three structural genes, G_1, G_2 and G_3 form an operon with promoter, P, and operator, O. Transcription of one strand of DNA yields a polycistronic mRNA, the 5′-end of which is made first. Ribosomes become attached here and sequentially translate the three cistrons, G_1^*, G_2^*, and G_3^*. Degradation from the 5′-end begins before transcription is complete and no further ribosomes can be added. Those already attached complete translation.

$$70\ S \rightleftharpoons 50\ S + 30\ S$$

$$30\ S + \text{fMet-tRNA}_f + \text{mRNA} + \text{GTP}$$

$$\longrightarrow\ [30\ S, \text{fMet-tRNA}_f, \text{mRNA}, \text{GTP}]$$

$$[30\ S, \text{fMet-tRNA}_f, \text{mRNA}, \text{GTP}] + 50\ S$$

$$\longrightarrow\ [70\ S, \text{fMet-tRNA}_f, \text{mRNA}] + \text{GDP} + \text{Pi}$$

IF1 is a basic protein, M_r 9 500; IF2 has M_r between 95 000 and 117 000; and IF3 is a thermostable protein, M_r 21 000. Various modifications of these three have been described but their functions seem now well established. (Note that *seven* IF's have been identified in a eukaryote and have had specific functions ascribed to them.)

As we have seen, 70 S ribosomes can dissociate into 50 S plus 30 S but at physiological Mg^{2+} concentrations (4–5 mM) they tend to stay associated. Initiation, however, involves subunits and whereas IF1 increases the rate constant (k_1 and k_2) for association/dissociation, IF3 binds to 30 S subunits decreasing k_2 and preventing reassociation:

$$70\ S \underset{k_2}{\overset{k_1}{\rightleftharpoons}} 50\ S + 30\ S$$

$$30\ S + \text{IF1} + \text{IF3} \longrightarrow [30\ S, \text{IF1}, \text{IF3}]$$

This provides the necessary 30 S subunit which then reacts with a complex of N-formyl-methionyl-tRNA$_f$ and IF2 and GTP, expelling IF3:

$$[30\ S, \text{IF1}, \text{IF3}] + [\text{fMet-tRNA}_f, \text{GTP}, \text{IF2}]$$

$$\longrightarrow\ [30\ S, \text{IF1}, \text{IF2}, \text{fMet-tRNA}_f, \text{GTP}] + \text{IF3}$$

Now part of the 16 S rRNA (see p. 263 and Table 4) binds to a complementary ribosome-binding site on the mRNA. This is on the 5'-side of the first initiation codon and this AUG (or GUG) interacts with the anti-codon of the tRNA$_f$:

[30 S, IF1, IF2, fMet-tRNA$_f$, GTP] + mRNA

$\longrightarrow$ [mRNA, 30 S, IF1, IF2, fMet-tRNA$_f$, GTP]

The next steps in these initial reactions are attachment of a 50 S subunit (to form the complete 70 S) and expulsion of IF1:

[mRNA, 30 S, IF1, IF2, fMet-tRNA$_f$, GTP] + 50 S

$\longrightarrow$ [mRNA, 70 S, IF2, fMet-tRNA$_f$, GTP] + IF1

As yet there has been no breakdown of GTP and it can be replaced by its analogue (GMP-PCP) the hydrolysis of which is *not* catalysed by GTP-ase:

GTP

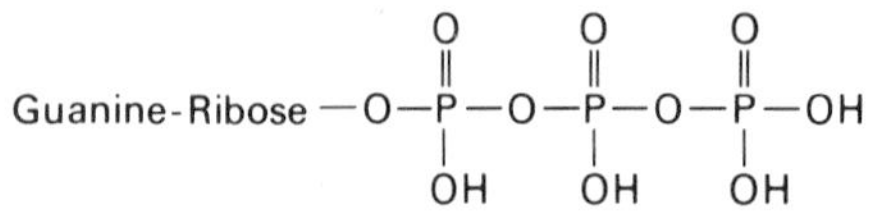

$$\text{Guanine-Ribose}-O-\overset{\overset{O}{\|}}{\underset{\underset{OH}{|}}{P}}-O-\overset{\overset{O}{\|}}{\underset{\underset{OH}{|}}{P}}-O-\overset{\overset{O}{\|}}{\underset{\underset{OH}{|}}{P}}-OH$$

GMP-PCP

$$\text{Guanine-Ribose}-O-\overset{\overset{O}{\|}}{\underset{\underset{OH}{|}}{P}}-O-\overset{\overset{O}{\|}}{\underset{\underset{OH}{|}}{P}}-CH_2-\overset{\overset{O}{\|}}{\underset{\underset{OH}{|}}{P}}-OH$$

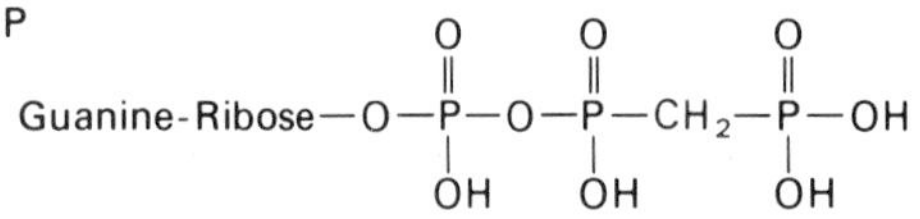

However, IF2 has GTP-ase activity, particularly when associated with the ribosome, and the following reaction occurs:

[mRNA, 70 S, IF2, fMet-tRNA$_f$, GTP]

$\longrightarrow$ [mRNA, 70 S, fMet-tRNA$_f$] + IF2 + GDP + Pi

The result is that N-formyl-methionyl-tRNA$_f$ occupies the P-site—which is analagous to having a peptidyl-tRNA there. The subsequent set of reactions occurs whichever is there and is repeated at each successive step in chain elongation. The hydrolysis of GTP appears to enable the rapid release of IF2 from the ribosome so that the elongation phase of protein synthesis can proceed.

Elongation factors (EF's)
Again three factors have been identified. EF-Tu is a single polypeptide, M_r 44000 and comprises about 5% of the soluble protein of *E. coli*, a similar molar quantity to that of tRNA but ten times that of EF-Ts (see below) and ribosomes. Free EF-Tu is extremely unstable (hence the 'u') but is protected by complex formation with GTP, GDP or EF-Ts. Replacement of GDP by GTP increases the affinity of the binary complex for $\alpha\alpha$-tRNA by at least 10^5.

EF-Ts is also a single chain, M_r 19000, but is much more thermostable (hence the 's'). It can catalyse the exchange of nucleotides attached to EF-Tu:

[EF-Tu, GDP] + EF-Ts $\rightleftharpoons$ [EF-Tu, EF-Ts] + GDP

[EF-Tu, EF-Ts] + GTP $\rightleftharpoons$ [EF-Tu, GTP] + EF-Ts

A ternary complex can then form by addition of amino acyl-tRNA:

[EF-Tu, GTP] + $\alpha\alpha$-tRNA

$\rightleftharpoons$ [EF-Tu, GTP, $\alpha\alpha$-tRNA]

and this can interact with a ribosome, the appropriate tRNA being selected by the next codon on the mRNA, so that it is bound to the A site and has its anti-codon associated with the mRNA codon. In the course of this binding reaction, the GTP is hydrolysed by the EF-Tu (which like IF2 has GTP-ase activity):

[EF-Tu, GTP, $\alpha\alpha$-tRNA] + [mRNA, 70 S]

$\longrightarrow$ [mRNA, 70 S, $\alpha\alpha$-tRNA, EF-Tu, GTP]

$\longrightarrow$ [mRNA, 70 S, $\alpha\alpha$-tRNA] + EF-Tu + GDP + Pi

The ribosome-mRNA complex now has the amino acyl-tRNA in the acceptor site (A site) and already has peptidyl-tRNA (or fMet-tRNA) in the donor site (P site). Transpeptidation occurs, the peptide (or N-formyl-methionine) being joined *via* its carboxyl group to the amino group of the amino acyl-tRNA, thus lengthening the peptide chain by one residue. A protein of the 50 S subunit, peptidyl transferase, catalyses this step (the same enzyme will condense the peptide (or fMet) with puromycin if it is present—see Figure 9, p. 250):

[Pep-tRNA, mRNA, 70 S, $\alpha\alpha$-tRNA]

$\longrightarrow$ [tRNA, mRNA, 70 S, Pep-$\alpha\alpha$-tRNA]

The discharged tRNA is ejected from the P site and the elongated peptidyl-tRNA is 'translocated' from the A to the P site while the mRNA moves by one codon relative to the ribosome. These movements require the third elongation factor, EF-G, which is a single polypeptide, M_r 80000. Like IF2 and EF-Tu, it can bind GTP and has GTP-ase activity, particularly in association with the ribosome. The energy of hydrolysis again is probably involved in the release of the factor from the ribosome:

[tRNA, mRNA, 70 S, Pep-$\alpha\alpha$-tRNA] + EF-G + GTP

$\longrightarrow$ [tRNA, mRNA, 70 S, Pep-$\alpha\alpha$-tRNA, EF-G, GTP]

$\longrightarrow$ [Pep-$\alpha\alpha$-tRNA, mRNA, 70 S]

+ tRNA + EF-G + GDP + Pi

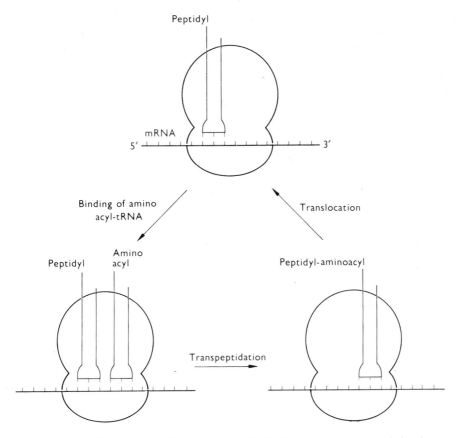

Figure 21. The three successive steps involved in the formation of each peptide bond.

The result of these steps is an empty A site, a P site occupied by a peptidyl-tRNA, and release of tRNA, GDP and phosphate.

Binding of αα-tRNA, transpeptidation, and translocation recur at each subsequent elongation and two molecules of GTP are cleaved (Figure 21). At some stage before completion of the chain, the formyl group is removed by hydrolysis and the *N*-terminal methionine may also be eliminated.

Release factors (RF's)
Eventually the polypeptide is completed but is still attached to the tRNA corresponding to its C-terminal amino acid. The subsequent codon in mRNA is a termination codon, U A A or U A G or U G A, for which there is no complementary tRNA but which is recognised by a protein termination factor, or *release factor*. One of these is RF1, a single polypeptide chain,

M_r 44 000, which recognises U A A or U A G; the other, RF2 is M_r 47 000, and recognises U A A or U G A. In the absence of RF1 and RF2 the polypeptidyl-tRNA remains attached to the ribosome; in the presence of the appropriate factor, hydrolysis occurs and the polypeptide is released. Normally there is about one molecule of each of RF1 and RF2 per fifty ribosomes. They bind to the A site and modify the specificity of the transferase so that the peptide is transferred to H_2O rather than to the $-NH_2$ of an αα-tRNA:

[Pep-tRNA, mRNA, 70 *S*]

$$\xrightarrow{\text{RF1 or RF2}} \text{Peptide} + \text{tRNA} + \text{mRNA} + 70\ S$$

A third factor RF3 has been said to stimulate RF1 and RF2 and a further factor, ribosome release factor (RRF), has been suggested to function with EF-G and

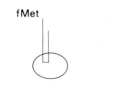

A 30 S subunit, stabilised by IF1 (and possibly IF3) binds a complex of N-formyl-methionyl-tRNA$_f$, GTP, and IF2.

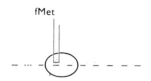

mRNA binds via a sequence complementary to a stretch of 16 S rRNA and so that an initiation codon AUG (or GUG) is base-paired to the anticodon of tRNA$_f$.

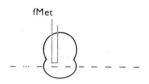

A 50 S subunit combines to yield a complete 70 S ribosome and IF1 is released.
GTP is split by IF2 to release GDP, Pi, and IF2.
fMet-tRNA$_f$ is now in the P site.

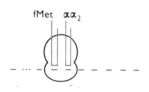

The second amino acid ($\alpha\alpha_2$) carried by its tRNA$_2$ and complexed with EF-Tu and GTP, is bound to the A site by codon-anticodon interaction.
GTP is split by EF-Tu to release GDP, Pi, and EF-Tu.

The carboxyl of N-formyl-methionine is joined to the NH$_2$ of $\alpha\alpha_2$ by peptidyl transferase, a protein of the 50 S subunit.

The discharged tRNA$_f$ is ejected from the P site.

The dipeptidyl-tRNA is translocated from A to P site while mRNA moves one codon relative to the ribosome. GTP is split by EF-G to release GDP, Pi, and EF-G.

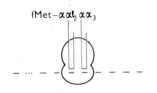

The third amino acid, $\alpha\alpha_3$, attached to tRNA$_3$ enters the A site.

The peptide is elongated by $\alpha\alpha_3$ and tRNA$_2$ is eliminated.

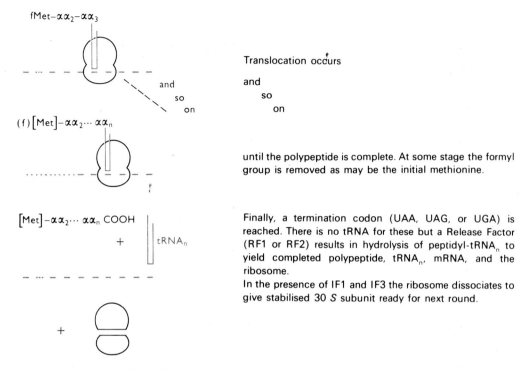

fMet–αα$_2$–αα$_3$

and
so
on

(f)[Met]–αα$_2$··· αα$_n$

[Met]–αα$_2$··· αα$_n$ COOH

+ tRNA$_n$

+

Translocation occurs

and
so
on

until the polypeptide is complete. At some stage the formyl group is removed as may be the initial methionine.

Finally, a termination codon (UAA, UAG, or UGA) is reached. There is no tRNA for these but a Release Factor (RF1 or RF2) results in hydrolysis of peptidyl-tRNA$_n$ to yield completed polypeptide, tRNA$_n$, mRNA, and the ribosome.

In the presence of IF1 and IF3 the ribosome dissociates to give stabilised 30 *S* subunit ready for next round.

Figure 22. Representation of steps in formation of polypeptide.

GTP after RF1 or RF2 has released the completed polypeptide:

[Pep-tRNA, mRNA, 70 *S*]

$$\xrightarrow{\text{RF1 or RF2}} \text{Peptide} + [\text{tRNA, mRNA, 70 } S]$$

$$\xrightarrow{\text{RRF + EF-G + GTP}} \text{tRNA} + \text{mRNA} + 70 \ S$$

If the message is polycistronic (see p. 258) ribosomes will re-initiate and begin reading the next cistron at the next initiation codon but, ultimately, when the whole message has been translated, the ribosome will reach the last termination codon. It then separates from the mRNA, becoming dissociated into subunits under the influence of IF1 and IF3. The [30 *S*, IF1, IF3] complex is then ready to begin another round of translation.

Figure 22 summarizes the steps in the biosynthesis of polypeptide chains so far as they are at present understood.

Location of components on the ribosome
In recent years there has been considerable increase in understanding of the structure and role of the ribosome and its components (and of the mode of action of antibiotics which interfere with protein synthesis, very often at the level of one molecule per ribosome—see p. 271).

Initiation factors, tRNA$_f$, and mRNA have been shown to be associated with various proteins of the 30 *S* subunit. IF1, 2 and 3, and mRNA all involve S1 and S12 among other proteins, and also the 3'-region of 16 *S* rRNA. The aminoglycoside antibiotics (e.g. streptomycin) bind to the 30 *S* subunit and some cause mis-reading of codons. Mutants resistant to (and even dependent on) these drugs may have an altered protein or 16 *S* rRNA of the smaller subunit.

The three factors with GTP-ase activity (IF2, EF-Tu, and EF-G) bind in a mutually exclusive manner to ribosomes and each involves protein L7/12 which may be near the interface between subunits.

The sites for tRNA binding have been explored in detail: the P site seems to involve proteins L2/L27 and L11/L18 whereas the A site implicates L16. This latter, and L11, are concerned in peptidyl transferase activity. Thus L16 binds chloramphenicol (CAP), an antibiotic which inhibits this enzyme. Likewise, the tetracyclines (TC's) prevent binding of αα-tRNA to the A site at L16. Further, thiostrepton is also an A site inhibitor and requires the presence of L11 for binding to ribosomes. Antibodies to L11 and L16 also inhibit the peptidyl transferase.

TRANSLATION

It will have become apparent that synthesis of protein is not just a matter of linking together amino acids in the right order. In a living organism the process is vastly more complicated because of the operation of intricate systems of control (see Chapters 4 and 8). A message is not just a polyribonucleotide beginning and ending with initiation and termination codons. Many are polycistronic—sometimes they may be 'alien' as when phage DNA is transcribed or when an RNA phage infects a cell.

Protein factors ensure that an appropriate 30 S subunit becomes attached at the specific oligonucleotide region of the mRNA so that the initiation codon of the first cistron to be translated is positioned correctly. With polycistronic mRNA from bacteria this is probably the cistron nearest the 5'-end but even so a mechanism must exist to prevent random initiation at A U G or G U G triplets other than those at the beginnings of messages. Studies of mRNA from the small RNA phages suggest that untranslated regions of the RNA and secondary structure may play important roles in these effects.

The RNA of phage MS2 consists of 3659 nucleotides and codes for three proteins (Figure 23). Starting at the 5'-end the 'leader sequence' of 129 nucleotides which is not translated, is followed by the A-protein cistron which begins with a G U G initiation codon and ends with U A G. There follow 26 intercistronic nucleotides, the coat protein cistron, 36 intracistronic nucleotides, the replicase cistron, and finally an untranslated 174 nucleotide segment at the 3'-terminus.

Each virus particle contains one A-protein molecule and about 180 coat protein molecules but none of the replicase. This last is subunit β, one of four polypeptides which make up the enzyme complex which replicates the virus RNA. Of the other three, α is the 30 S ribosomal subunit protein S1, γ and δ are EF-Tu and EF-Ts!

The nucleotide sequences of all three cistrons have been established by Fiers and his colleagues and the corresponding amino acid sequences can be deduced. A model for the secondary structure of the cistrons has been suggested (Figure 24).

The tricistronic messenger of the bacteriophage may not be exactly typical of bacterial messengers and some of the untranslated portions may be sites for attachment of RNA-primed RNA replicase. Such would not be necessary for bacterial RNA's which are only transcribed from DNA. However, it is probable that there are considerable similarities. There is not so much known about bacterial cistrons as there is about the simple RNA phages but the *lac* operon of *Escherichia coli* has been studied in some detail. Transcription of DNA to RNA has been dealt with in Chapter 5 (pp. 228) and the control mechanisms of transcription and translation will be discussed in Chapter 8 (pp. 328) but here we will summarise what is known of the molecular biology of β-galactosidase formation (Figure 25).

The regulator gene *I* is transcribed and translated to yield repressor protein which can be inactivated by combining with inducer (a derivative of lactose) or, in the absence thereof, can bind to the operator region of the DNA which precedes the first cistron of the *lac* operon, the *Z* gene for β-galactosidase. Between the end of the *I* gene and the beginning of the operator lies the promoter region. This consists of a sequence with *dyad symmetry* (loosely referred to as 'palindromic' see Chapter 5, p. 195) which can bind the complex of cAMP and the dimeric cAMP receptor protein (CRP). This enhances the binding of RNA polymerase at the adjacent initial entry site on the DNA which is rich in A-T pairs and is flanked by two G-C rich regions. The core enzyme together with sigma (σ) (see p. 231) then binds firmly to a sequence which does *not* have dyad symmetry and in which, therefore, the two chains can be distinguished, allowing strand selection for transcription. A similar sequence has been found in all promoter regions analysed and one strand can be represented T A T Pu A T Pu with minor variations. (This stretch of seven base-pairs has itself sometimes been referred to as the 'promoter'.)

Figure 23. Bacteriophage MS2.

The RNA codes for three proteins (shown in blocks). Untranslated regions are present at each end and between the cistrons. The number of nucleotides in each region is shown on top and the sequence from 5' to 3' is numbered below.

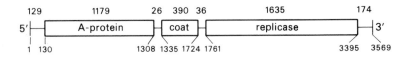

If active tetrameric repressor is present, it binds to the operator region (which also has dyad symmetry) and prevents progress of the polymerase. If repressor is absent or is combined with inducer, then transcription of one strand of the DNA begins at this locus and, after dissociation of sigma, the core enzyme continues with production of polycistronic mRNA. The first 38 nucleotide residues from the 5′ end of the mRNA are not translated but contain the ribosome-binding sequence which is complementary to a pyrimidine-rich sequence near the 3′ end of 16 *S* rRNA. Residues 39, 40, and 41 are A U G, the initiation codon for fMet. This amino acid is not found in completed β-galactosidase which has *N*-terminal threonine.

The *lac* operon has three cistrons, one each for β-galactosidase, galactoside permease, and galactoside transacetylase. However, the *his* operon of *Salmonella typhimurium* has nine structural genes which are transcribed as a single polycistronic mRNA (see Chapter 4, p. 176) from which at least several cistrons are translated with the same frequency.

Perhaps the most remarkable organisational arrangement so far uncovered has been that of the single-stranded, circular DNA of the bacteriophage ϕX174. It consists of approximately 5375 nucleotides and codes for nine known proteins, A, B, C, D, E, J, F, G, H, in order determined genetically. The nucleotide sequence has been established by Sanger and his colleagues using the 'plus and minus' method (see p. 195) and various mRNA's and protein sequences are also known. Remarkably, some stretches of the DNA have multiple functions—cistron B is entirely within A, and E is within D, but in each case with different reading frames. Moreover, an A U G initiation codon twice overlaps a U G A termination codon and in another instance, a U A A terminator overlaps an A U G initiator. Parts of the linear sequences are shown in Figure 26. As with MS2, it is likely that internal H-bonding gives rise to secondary structure with base-paired loops.

ANTIBIOTICS WHICH INTERFERE WITH SYNTHESIS OF PROTEIN

The complexity of the process of making a protein is indicated by the fact that more than 120 species of macromolecules (proteins and nucleic acids) are known to be involved. It is all the more surprising, therefore, that so seldom under physiological conditions does the 'wrong' amino acid become incorporated or does the process go awry in other ways.

However, there are substances, including many antibiotics, which interfere specifically with one stage or another. Compounds which inhibit or modify nucleic acid replication and transcription are dealt with in Chapter 5 but even with a correct message it is possible to get faulty translation. Inhibitors may act on either the smaller or larger subunit of the ribosome or on one of the protein factors involved.

Aurintricarboxylic acid (a triphenylmethane dye related to crystal violet) prevents the binding of mRNA to 30 *S* subunits *in vitro* and colicin E3 (a protein, 60 000 u, produced by some strains of *E. coli*) can split off a fragment of 49 nucleotides from the 3′-end of 16 *S* rRNA, again interfering with mRNA binding. Kasugamycin attaches to 16 *S* rRNA of sensitive species, preventing binding of mRNA; some resistant strains have two adjacent adenines whereas the sensitive ones have dimethyl-adenines. Other amino-glycosides such as streptomycin, kanamycin, and neomycin, can cause mis-reading—whereas poly (U) normally promotes formation of poly (Phe) *in vitro*, addition of streptomycin reduces incorporation of phenylalanine but stimulates that of isoleucine. Resistant and dependent strains have altered S12 but binding is believed to involve S3, 4 and 5. Pactamycin, an anti-tumour agent, acts on prokaryotes as well as eukaryotes (many substances are specific for one or other class of organism) again reacting with the smaller subunit.

The tetracyclines (including aureomycin and terramycin) act by competing with αα-tRNA for attachment to the A site on ribosomes and bottromycin probably does the same. Puromycin (see p. 250) mimics the amino acyl terminus of αα-tRNA and causes premature ending of elongation. Chloramphenicol and, probably, sparsomycin and other drugs, inhibit the 50 *S* peptidyl transferase activity—they also prevent its catalysing the formation of peptidyl-puromycin (see p. 252).

Spectinomycin (an aminoglycoside) and erythromycin (a macrolide) both effect translocation. The former may involve S5, an interface protein between the subunits and near to L7/12 which is known to be concerned in EF-G function. Erythromycin acts at the P site, preventing translocation and keeping all the nascent peptidyl-tRNA in the A site where it cannot be released by puromycin.

Fusidic acid, a steroid, also blocks translocation but by sequestering EF-G as a ternary complex (FUS, EF-G, GDP) on the ribosome. It has no action on IF2 or on EF-Tu but the latter is affected by kirromycin with which it forms a complex having GTP-ase activity in the absence of ribosomes. So kirromycin

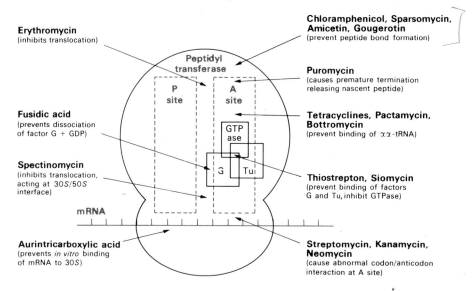

Figure 27. Probable sites of action of some antibiotics which interfere with synthesis of proteins.

prevents formation of [EF-Tu, GTP, αα-tRNA] which is necessary for binding αα-tRNA to the A site. It does not, however, affect EF-G or IF2, the other factors with GTP-ase activity.

Figure 27 suggests probable sites of action of some of these substances which interfere with synthesis of proteins in prokaryotes.

ASSEMBLY OF NATIVE PROTEINS

Many proteins consist of more than one polypeptide chain and some enzymes are very large molecules containing a number of subunits which may be identical so that the enzyme consists of an aggregate with several active centres. Thus β-galactosidase of *Escherichia coli* has a sedimentation coefficient of 16 S and a molecular weight of about 500 000. It can be dissociated by urea (which breaks H-bonds) into four subunits but is probably composed of 12 monomers of 40 000 molecular weight. The functional enzyme may be put together before release from the ribosomal site of synthesis and β-galactosidase activity can be found associated with 70 S ribosomes.

Tryptophan synthetase (see p. 163) is composed of two separable proteins components A and B and in *Escherichia coli* which is making the enzyme, some ribosomes have A protein and some B protein and some have both associated with them.

There is not a great deal which can be said about the mechanics of assembly of polypeptides into native functional proteins except that much may be in a sense 'spontaneous'. The ultimate secondary, tertiary and quaternary structures (see pp. 10–13) are largely determined by the amino acid sequence or primary structure of the constituent parts. Even elaborate cross-linking disulphide bridges may have an inevitability because some proteins which have had their native structure destroyed have been shown to be capable of renaturation. For instance, the ribonuclease molecule is a single chain of 124 amino acid residues and is stabilized by the S—S bonds of four cystine bridges. These can be broken and the enzyme denatured by reduction to cysteine residues in 8 M urea and all enzymic activity is lost. Re-oxidation restores activity almost completely and yields a product indistinguishable from the native protein. This is despite the fact that the reduced ribonuclease had eight sulphydryl groups which could have been combined in pairs in 105 different ways ($7 \times 5 \times 3$). Almost more surprising is the finding that heat-denatured β-galactosidase can be restored to partial activity.

It is now generally believed that there do not exist three-dimensional templates which force polypeptides to assume a particular tertiary structure but rather that chains once synthesized take up their configurations spontaneously since these are the ones which satisfy the stereochemical requirements of the amino acid sequence and energy considerations.

FURTHER READING

1 Loftfield R. B. (1957) The biosynthesis of proteins. *Prog. Biophysic. Biophys. Chem.* **8**, 347. (An excellent account of the position at that time and of the general implications.)

2 Nomura M., Tissières A. and Lengyel P. (eds.) (1974) *Ribosomes.* Cold Spring Harbor Laboratory.

3 Kurland C. G. (1977) Structure and function of the bacterial ribosome. *Ann. Rev. Biochem.* **46**, 173.

4 Weissbach H. and Pestka S. (eds.) (1977) *Protein synthesis.* Academic Press, London. (Includes a good account by Stoffler G. and Wittmann H. G. (1977) Primary structure and three-dimensional arrangement of proteins within the *E. coli* ribosome. pp. 117–202.)

5 Brimacombe R. (1978) The structure of the bacterial ribosome. In *Relations between structure and function in the prokaryotic cell.* (Eds. Rogers H. J., Stanier R. Y. and Ward J. B.) Twenty-eighth Symposium of the Society for General Microbiology, pp. 1–26. Cambridge University Press.

6 Grunberg-Manago M., Buckingham R. H., Cooperman B. S. and Hershey J. W. B. (1978) Structure and function of the translation machinery. In *Relations between structure and function in the prokaryotic cell.* (Eds. Rogers H. J., Stanier R. Y. and Ward J. B.) Twenty-eighth Symposium of the Society for General Microbiology, pp. 27–110. Cambridge University Press.

7 Brimacombe R., Stöffler G. and Wittman H. G. (1978) Ribosome structure. *Ann. Rev. Biochem.* **47**, 217–249).

8 Sanger F., Coulson A. R., Friedmann T., Air G. M., Barrell B. G., Brown N. L., Fiddes J. C., Hutchinson III C. A., Slocombe P. M. and Smith M. (1978) The nucleotide sequence of bacteriophage ϕX174. *J. molec. Biol.* **125**, 225–246.

9 Fiers W., Contreras R., Duerinck F., Haegeman G., Iserentant D., Merregaert J., Min Jou W., Molemans F., Raeymaekers A., Van den Berghe A., Volckaert G., and Ysebaert M. (1976) Complete nucleotide sequence of bacteriophage MS2 RNA: primary and secondary structure of the replicase gene. *Nature* **260**, 500–507,

10 Chambliss G., Craven G. R., Davies J., Davis K., Kahan L. and Nomura M. (1980) *Ribosomes: structure, function and genetics.* University Park Press, Baltimore

11 Cundliffe E. (1981) Antibiotic inhibitors of ribosome function. In Gale E. F., Cundliffe E., Reynolds P. E., Richardson M. H. and Waring M. J. *The molecular basis of antibiotic action*, 2nd edn. Wiley, New York.

Chapter 7
Genetics

Genetics is concerned with the *genotype*, or genetic constitution, and the way in which it determines the *phenotype*, or observable characteristics of an organism. At the population level, genetics deals with the changes which occur in the relative frequencies of different genotypes in populations as a result of mutation and selection. At the level of the individual cell, which is the aspect with which we are more particularly concerned in this book, it deals with the nature of the genetic material, its structure and mode of self-replication, and the way in which it carries out its directing function.

THE CHROMOSOME

All the genetic information needed for the growth and functioning of a bacterium is contained in a single chromosome, although individual cells may contain more than one copy depending on the organism and the conditions of growth. The chromosome consists of a single continuous molecule of DNA (about 4×10^6 base pairs in *E. coli*) (Chapter 5, p. 186). Under the microscope the DNA can be seen in stained cells as condensed structures called *nucleoids*. These structures are not enclosed in a membrane as they would be in the eukaryotic nucleus, but they do contain, in addition to DNA, proteins and RNA of possible structural importance. They are often found attached to the cell membrane and it is believed that this attachment is required for chromosome replication (Chapter 5).

The bacteria studied to date have a common important genetic characteristic: they are normally *haploid*. The nature of the haploid condition is illustrated in Figure 1a. It shows a cell containing two chromosomes. The chromosomes are identical except for a single change in one of them causing one of its many functions to be altered. This chromosome has just acquired a *mutation*, whilst its neighbour retains the *wild-type* version of that function. Before the bacterium divides it replicates its chromosomes (Stages 1–2). The bacterium then begins to divide, distributing the new chromosomes to the daughter cells (Stages 2–3). As a result two new cells are formed which like the parent cell contain two chromosomes. However,

the two daughters differ both from the parent and one another, since in one, both chromosomes carry the mutation while in the other, both have the wild-type function. The progeny of each chromosome remain together for one division, each new pair separating into daughter cells. Thus each chromosome becomes separated (or *segregates*) from its neighbour in the space of a single cell cycle. The conserved unit of inheritance is therefore a single copy of the chromosome. This is the defining characteristic of the haploid cell. A *diploid* cell conserves two copies of each chromosome (Figure 1b). That bacteria are haploid can be seen very clearly if cells are allowed to form colonies on indicator agar immediately after acquiring a mutation. The colonies formed show two separate *sectors*, one containing the progeny of the mutant bacterium that divided off shortly after plating, the other containing wild-type bacteria. An illustration of such sectors can be found in Figure 2.

Secondary genetic elements of bacteria: plasmids and temperate phages

Many bacteria contain additional and dispensable genetic information in DNA which is either separate or separable from the chromosome. Some of these molecules are able to replicate independently of the chromosome. They are called *plasmids*. Others are integrated into the chromosome and replicate as part of the host structure. The DNA of most *temperate bacteriophages* can be perpetuated in this way. Those particles which can replicate independently but which can also be integrated into the chromosome are called *episomes*.

Plasmids

Some plasmids are of interest because they code for mechanisms that promote mating between bacteria. These plasmids, the sex or *fertility factors* or *conjugative plasmids*, are probably important in nature and certainly useful to the geneticist and the genetic engineer. In addition, some plasmids carry, and transmit between bacteria, genes determining resistance to antibiotics. They are called *R-factors* and are of increasing importance in medical and veterinary science

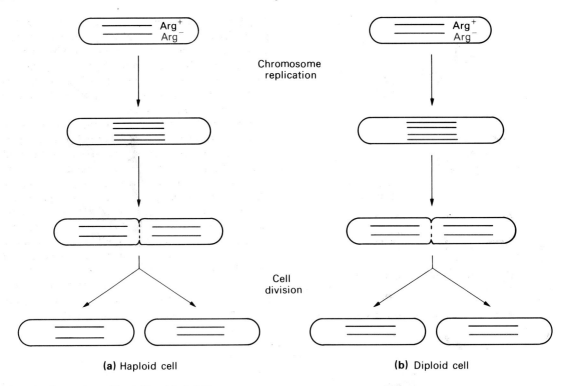

Figure 1. Comparison of haploidy with diploidy.

(a) In conditions of slow growth a haploid cell contains one chromosome after cell division but during fast growth there may be two. The figure illustrates what happens when a mutation occurs in one of them producing an inability to synthesize arginine (arg⁻). After chromosome replication and cell division the mutated chromosomes segregate from the wild-type. The conserved unit of inheritance is a single chromosome; the daughter cells and the parent cell all have different genotypes.

(b) In a diploid cell the conserved unit of inheritance is a pair of chromosomes and the cell has a mechanism for ensuring that both daughters inherit the same, mixed (hybrid) genotype as the parent.

(Table 1). Other plasmid-coded functions include restriction endonucleases, DNA repair enzymes, bacteriocin production, haemolysin synthesis and aromatic hydrocarbon degradation. *Agrobacterium tumefaciens* has a plasmid (Ti) which enables it to form tumours in certain plants.

One of the most important plasmids in bacterial genetics is the F factor of *E. coli* characterized by Lederberg, Hayes and Jacob. It promotes its own transfer by causing the bacteria to mate or *conjugate*. In addition, it can insert itself reversibly into the chromosome, i.e. it is an episome.

The F factor represents a group of plasmids which replicate in step with the chromosome. Each cell contains a low number of F factors (Table 1). Indeed, for the greater part of each cell division cycle there is probably only one F factor for each replicating chromosome. We infer that such plasmids have a precise mechanism ensuring that each cell inherits a new F at division. In fact, contemporary strands of the chromosome and of F are jointly inherited at division. It seems likely that they are both bound to a structure, probably a part of the membrane, which acts like the mitotic spindle of eukaryotes. Consistent with this view is the finding that the major part of plasmid DNA can be associated with the nucleoid on extraction from the cell.

Temperate bacteriophages

The other major class of secondary genetic elements contains the temperate bacteriophages. First discovered by Lwoff in *Bacillus megaterium* these viruses are now known in very many bacterial species. Perhaps the best known in microbial genetics is the bacteriophage λ of *E. coli*; it has basic features in common with phages of widely different species from

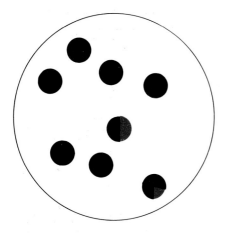

Figure 2. Sectoring of colonies produced by segregation of mutations.

Bacteria, treated with a mutagen, are plated directly on agar containing growth medium supplemented with lactose and an indicator dye. Most of the colonies are wild type, Lac⁺, (shown in black) but a few show sectors consisting of Lac⁻ cells (shown in red).

Pseudomonas to Streptomyces. Temperate bacteriophages have the ability on infecting their host to enter a passive state known as lysogeny. The virus, termed prophage in this state, survives by becoming inserted into the bacterial chromosome. It switches off its own multiplication by means of a repressor which prevents expression of most phage genes.

In nature, temperate phages in the prophage state are known to confer properties of considerable importance on their host. For example, somatic antigens of *Salmonella*, a common enteric pathogen, are synthesised under the direction of temperate phages, while the disease diphtheria is caused by a prophage-determined toxin made in *Corynebacterium diphtheriae*. In the laboratory the temperate phages are powerful genetic tools since, like plasmids, they can be used as vehicles both of genetic transfer and of partial diploidy. They provide model systems for recombination studies and since their DNA can readily be extracted and manipulated *in vitro* and *in vivo* they are widely used as vectors of foreign DNA in genetic engineering.

Table 1. The range of plasmids and their properties.

Name	Host	Number per cell	Size (M_r)	Phenotype conferred on host	Conjugative
F	*E. coli*	1–2	63 × 10⁶		yes (constitutive)
ColEl	*E. coli*	20 (1000–2000 in presence of chloramphenicol)	4.2× 10⁶	colicin synthesis	no
R1	*E. coli* *Salmonella paratyphi* *Shigella flexneri* *Proteus spp.*	1–2	62 × 10⁶	drug resistance (Ap Cm Km Sm Sp Su)* Restriction endonuclease (*Eco*RI) and modification enzyme	yes (repressive)
RP1	*Pseudomonas* and many other Gram-negative species	1–2	38 × 10⁶	drug resistance	yes, transmissible to a broad range of hosts
TOL	*Pseudomonas putida*		78 × 10⁶	toluene degradation	yes
Ti	*Agrobacterium tumefaciens*		~100 × 10⁶	plant tumours	yes

*Ap: ampicillin; Cm: chloramphenicol; Km: kanamycin; Sm: streptomycin; Sp: spectinomycin; Su: sulphonamide.

GENES AND MUTATIONS

We may now examine the major genetic elements of the bacterium in more detail. Their information is packaged in *genes*, each of which determines the synthesis of a single polypeptide or RNA molecule as the final product. The genes lie in a linear sequence on the chromosome. The continuous thread of DNA of which they are composed contains nucleotide sequences which act as start and stop signals for transcription and translation and which demarcate one gene from another (Chapters 5 and 6).

In genetics, genes are distinguished from each other not by their chemical structures but by their effects on the phenotype and by their distinctive patterns of transmission from cell to cell. There are two essential stages in classical genetic analysis. First of all the gene to be studied has to be made distinguishable by a mutation. It is axiomatic that if a gene is present in identical form throughout the species one cannot tell, by the methods of genetics, that it is there at all. There has to be a *difference* residing in the gene before its inheritance can be followed. So the induction and isolation of mutants is commonly the first step in a genetic investigation. A piece of the genome carrying a mutation (i.e. a heritable change) with a readily observable phenotypic effect is thereby *marked*, and the mutation in question is often referred to as a *marker*. The second stage of genetic analysis is the determination of the relative positions of a series of markers by their modes of transmission. We will deal first with some practical and theoretical aspects of mutation.

Mutation

The frequency of spontaneous mutation detectable in any given gene is very low, generally of the order of one per 10^5 to 10^7 cells per cell division. But taking all genes into account, mutation is not such a rare event, and in a population of say 10^{11} cells, such as can easily be obtained in bacteria, mutants of practically any desired type will be present and can be isolated given a suitable selective technique. Frequencies of mutation can, furthermore, be greatly increased by the use of mutagenic treatments. We shall return to mechanisms of mutation as such later on in this chapter, but to begin with it seems more appropriate to consider the different types of mutant which can be obtained and the methods used for their detection and isolation.

On pp. 288 *et seq*. the enzymic mechanism for the transcription of DNA base sequence into RNA base sequence was described. So far as we know, the primary effect of any mutation in the DNA will be to cause a complementary change in the base sequence of the RNA transcribed from it. This RNA is of three different kinds: ribosomal RNA, amino acid-transfer RNA and messenger RNA. The first two kinds of RNA, though they account for a high proportion of the total RNA, consist of relatively few kinds of molecule, and are transcribed from a correspondingly rather small fraction of the DNA. The mutations which we know most about are those which cause structural changes—generally simple amino acid substitutions—in proteins, and each of these is presumed to have its primary effect through a change in the base sequence of one or other of the hundreds or thousands of different kinds of messenger RNA molecules. Mutations of this class are usually detected through their effects on specific enzymes.

Auxotrophic mutants: mutants with defective or altered Class II enzymes

The most useful and intensively studied kind of mutant in bacteria is the *auxotroph*. An auxotrophic mutant has an additional nutritional requirement, for example for an amino acid, a vitamin or a purine or pyrimidine base or nucleoside, over and above the *minimal* nutritional requirements of the wild-type organism. A medium containing only the minimum nutrients for the growth of the wild-type is called minimal medium, and in the case of *Escherichia coli* or *Salmonella* species, it contains only essential inorganic salts and a simple carbon and energy source such as glucose. A typical auxotrophic mutant fails to grow on minimal medium but grows as well as wild-type if the specific substance which it requires is added. More fundamentally, an auxotroph has usually lost the ability to make one of the Class II enzymes essential for the biosynthesis of the substance which it requires.

The selection of auxotrophic mutants might seem a difficult task in as much as it involves seeking out a few cells which cannot grow on minimal medium from among a large number of cells which can. The difficulty was overcome by making use of the fact that penicillin kills only growing bacterial cells. In the *penicillin enrichment* method, cells of *E. coli*, usually after some sort of mutagenic treatment, are suspended in minimal medium plus penicillin. During a subsequent period of incubation the nutritionally normal, or *prototrophic*, cells start to grow and are killed by the penicillin. The auxotrophs, on the other hand, are prevented from growing by their nutritional deficiencies and tend to survive. They are subsequently recovered by washing the cell suspension free of penicillin and plating on supplemented medium. If only one class of auxotroph is sought, a master plate is prepared containing minimal medium supplemented only with the specific substance required by that class. Of the colonies which grow up some are usually wild-type which have, for some reason, survived the penicillin, but a good proportion of them are auxotrophs of the desired type. This method is very rapid and efficient and has been used more than any other for the isolation of auxotrophs, not only in *E. coli* but in several other species as well. The final isolation of auxotrophs is quickly achieved by the technique of *replica plating*.

In this method representative cells from the whole ensemble of colonies on the master plate are simultaneously transferred to a plate of minimal medium on the surface of a replicator which is usually a disc of sterile velveteen material supported on a round wooden block. The replicator is lightly pressed first on the surface of the master plate and then on the replica plate containing fresh medium. A pattern of colonies should grow on the replica plate identical to that on the master plate, provided that all the colonies are capable of growth on the medium in the replica plate. Since the replica plate contains minimal medium, auxotrophic mutants from the master plate will not grow. In this way all the colonies on the master plate, numbering perhaps a hundred or more, can be tested in one simple operation (Figure 3).

On the other hand, if auxotrophs of a range of types are sought, the master plate should be supplemented with a wide variety of substances: amino acids, purines, pyrimidines, vitamins. Replicas from the master plate are then made on a series of replica plates each containing only one supplementary substance.

By special techniques it is possible to obtain mutants in which an enzyme is altered in its properties rather than merely absent. For instance, among temperature-sensitive auxotrophs—able to grow on minimal medium at a lower temperature (say 27°) but not at a higher temperature (say 37°)—one can find many which have unusually thermolabile Class II enzymes. Again, many growth-inhibitory metabolite analogues exert their effects through acting as 'false feedback inhibitors' of enzymes subject to allosteric regulation (see Chapter 8). A mutant selected as resistant to one of these drugs may have an altered enzyme no longer subject to end-product regulation of its activity. This mechanism of drug resistance should be clearly distinguished from resistance to false feedback *repression*; this latter type of resistance is due to a breakdown in the normal control of *rate of synthesis* of one or more enzymes and does not involve any change in enzyme structure. Further consideration of non-repressible, or constitutive, mutants is deferred until the following chapter.

Mutants defective in Class I enzymes

A second class of mutants includes those which are unable to catabolize some substance which can be utilized by the wild-type. For instance, whereas wild-type *E. coli* can use not only glucose but also lactose, maltose, arabinose or mannitol (among other substances) as sole carbon and energy source, mutants can be obtained which are unable to grow on any one of these less usual carbon sources. In principle it is possible to adapt the penicillin-screening procedure to the isolation of such mutants, using a minimal medium with the substance in question as sole source of carbon. However, a more commonly employed procedure is to pick out the mutant colonies by eye after plating on a complex organic (broth) agar

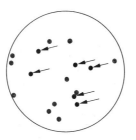

Master plate supplemented with tryptophan—colonies grown from survivors of penicillin treatment in unsupplemented medium

Replication

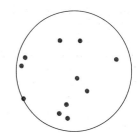

Replica plate without tryptophan—tryptophan auxotrophs indicated by arrows in master plate fail to grow

Figure 3. Identification of tryptophan auxotrophs by replica plating.

medium supplemented with the carbohydrate in question and with the dyes eosin and methylene blue. Such a medium containing say, lactose is commonly referred to as EMB-lactose medium. Cells which can ferment the sugar give colonies of a deep purple colour on this medium. Cells not able to ferment the sugar grow on the other carbon sources in the broth, but they give white colonies which show up in striking contrast with the wild-type ones. Even though such mutant colonies occur only at the rate of one in several thousand they can easily be picked out. The EMB technique is also useful in that it permits the identification of sectored colonies in which only one-half or one-quarter of the colony is mutant (white) while the remainder remains wild-type (purple). Such sectoring indicates that the cell which gave rise to the colony was genetically mixed, containing both mutant and wild-type genes. Some circumstances in which this can occur were mentioned above (cf. p. 275).

Mutants unable to utilize a given carbon source have most commonly lost the ability to make one of the Class I enzymes necessary for the catabolic pathway in question. For example lactose mutants may have lost β-galactosidase, while galactose mutants may have lost any one of the three enzymes specifically concerned in the conversion of galactose to glucose. In some cases, however, failure to utilize a carbon source may be due to an inability to concentrate it from the medium, such an inability being due to loss of a component of a specific transport system ('permease'). Since permease activity depends on specific proteins in the cell membrane it should be just as liable to mutational loss or damage as are enzymes. We may note here in passing that permease deficiency is responsible for some cases of drug resistance. For example the loss of an amino acid permease may confer resistance to a normally inhibitory amino acid analogue.

Sometimes loss of an enzyme by mutation may be detected by a direct test of the enzyme activity itself. For example a large number of colonies can be tested simultaneously for the presence of alkaline phosphatase. This is done by spraying the surface of the plate with a phenol phosphate and then with a mixture of sulphanilamide and nitrous acid which couple with any phenol released by enzyme action to form a bright red azo-dye. Colonies deficient in the enzyme remain white and stand out strikingly among the red, wild-type colonies. Alternatively, colonies can be sprayed with dinitrophenol phosphate which, on hydrolysis by the enzyme, colours the wild-type yellow with the dinitrophenol released. Large numbers

of mutants of *E. coli* deficient in alkaline phosphatase have been isolated by these two methods.

Mutants defective in Class III enzymes and other components of the protein-synthesizing system

Some of the most interesting and potentially valuable mutants are those defective in Class III enzymes—enzymes concerned with macromolecular synthesis. Such mutants are, however, among the most difficult to obtain. If the Class III enzyme function is seriously depressed the mutant will usually not grow on any medium since the product of the enzyme activity will be a large complex molecule which may be unstable and also unable to enter the cell. The most generally fruitful approach to this problem is through temperature-sensitive mutants. Temperature-sensitive mutants usually have specific enzymes altered in such a way that, although functional at the lower temperature, they become inactivated as the temperature is raised to the upper part of the range compatible with wild-type growth.

There is, in principle, no restriction on the kinds of enzymes which can be affected in this way. It is easy enough to identify and collect many temperature-sensitive mutants by replicating large numbers of colonies to plates which are then incubated at the higher temperature. Among those temperature-sensitive mutants whose defects at the higher temperature cannot be repaired by supplying metabolites of low molecular weight are many affected in enzymes involved in nucleic acid or protein synthesis. It has been possible, for instance, to devise methods for picking out those which are deficient in DNA synthesis. Such mutants have been extremely useful in analysing the role which DNA synthesis plays in various cell processes. Other temperature-sensitive mutants have proved to have labile aminoacyl-tRNA synthetases, and the analysis of such mutants in *E. coli* has led to the identification of the genes coding for several of these enzymes.

Another approach to the isolation of mutants altered in macromolecular synthesis is through the isolation of drug-resistant mutants. We have already seen that resistance to metabolite analogues may be due to derepression of enzyme synthesis or to loss of a permease or to the alteration of a Class II enzyme to make it resistant to feedback inhibition. Another possible reason for resistance to an amino acid analogue is that the corresponding aminoacyl-tRNA synthetase has been altered so that it no longer activates the analogue, though continuing to deal adequately with the normal amino acid. To the extent that the inhibitory effect is due to the incorporation of the analogue into protein, such an alteration will confer resistance.

For example a mutant of *E. coli* resistant to *p*-fluoro-phenylalanine was shown to have an altered phenyl-alanyl-tRNA synthetase.

Other drugs, such as streptomycin, spectinomycin and erythromycin, are inhibitory because of their high affinity for the ribosomes. Mutants resistant to these drugs are thought to have altered ribosomal proteins. For example, in *E. coli*, a specific ribosomal protein of the 50s subunit which is altered in an erythromycin-resistant mutant has been identified. The genes coding for this and other ribosomal proteins appear to be closely linked to each other and to the well-known gene governing streptomycin resistance.

Through the study of temperature-sensitive and drug-resistant mutants and by other techniques, many of the genes for other ribosomal proteins and enzymes concerned in polypeptide chain initiation and elongation have been identified.

Turning now from proteins involved in nucleic acid and protein synthesis to genes for the nucleic acid components of the protein-synthesizing system, we might hope to be able to identify genes both for ribosomal RNA (rRNA) and for transfer RNA (tRNA). We have, in fact, direct evidence that genes for rRNA exist, since it can be shown that an appreciable fraction of bacterial DNA (0·3% in *E. coli*) will, after denaturation to single strands, hybridise with purified ribosomal RNA, both 16 *S* and 23 *S*. It may be assumed that the rRNA is transcribed from this part of the genome. The amount of this DNA is 10-fold greater than would be required for just one gene for 16 *S* and another for 23 *S* RNA, and it seems that there must be about 7 copies of each gene. These copies are scattered along the length of the bacterial chromosome rather than clustered together at one location. No mutations in these genes have as yet been identified. The multiplicity of copies would no doubt make a change in just one gene difficult to detect.

Genes for tRNA species, on the other hand, have been identified by mutation. The mutational effect which has been identified is that of genetic *suppression*. As will be explained in more detail later in this chapter, many mutants which lack enzyme activities do so because the mutation has created a chain-terminating codon—UAG, UAA or UGA—in place of what, in the wild-type, is a codon for an amino acid. Selection for revertants from such enzyme-deficient mutants leads to the isolation of some in which the original mutation has been truly reversed—or at least an acceptable amino acid codon substituted for the chain-terminating one—but also others in which the original mutation is still present together with a second (*suppressor*) mutation at a quite separate position which has the effect of overcoming, at least partially, the chain-terminating effect. The general way in which these suppressors of chain termination work is through the modification of the anticodon of a species of transfer RNA so as to make it recognize the chain-terminating triplet. In the first case of this kind to be fully analysed it was shown that a modified tyrosine-specific tRNA was responsible for the suppression (see Chapter 6, p. 259). The normal anticodon, GUA, which recognizes the tyrosine codons UAU and UAC, had been changed to CUA, which now recognizes the chain-terminating codon UAG. Thus this class of suppressor mutation identifies a gene for tyrosine-specific tRNA. Actually the tRNA involved here is only one of two or three tyrosine-specific varieties controlled by different genes, and it contributes only a minor fraction of the total tyrosine-tRNA activity. Were a mutation of this kind to occur in a unique tRNA the effect would normally be lethal, and it is only the fact that *E. coli* has a 'spare' tyrosine tRNA which permits the ready isolation of this class of suppressor. Nevertheless, mutant suppressor versions of normally unique tRNA's have been identified in special stocks in which the gene concerned is present in two copies.

Resistance to bacteriophage

Mutants resistant to virulent bacteriophages can be readily selected by exposing large populations to enough virus to kill all non-mutant cells. Such mutants have been of great importance in the development of bacterial genetics. Resistance to bacteriophage is usually due to the loss or alteration of the specific component of the bacterial cell surface to which the virus normally attaches. Different bacteriophages attach to different specific chemical groupings, since mutation to resistance to one phage type does not generally confer resistance to unrelated types.

Spontaneous mutation frequencies

Mutations arise constantly in cell populations and there is no known way of preventing them or of predicting whether a particular cell will suffer a particular kind of mutation at any specified time. This is what is meant by calling experimentally unprovoked mutations *spontaneous*, but the term should not be taken as implying that the process is necessarily uncontrollable or, still less, without definite chemical causes. It is difficult to give meaningful values for typical spontaneous mutation frequencies since everything depends on the kind of mutation being counted. Since a gene has of the order of a thousand nucleotide pairs, the mutation frequency per gene may be expected to be about a thousand times the frequency per nucleotide pair. But not all changes of single nucleotide pairs will have observable effects on the phenotype, and the proportion which do will be very different in different experimental systems. If one is looking for mutations leading to a loss of a given gene function it is likely that a change in any one of several hundred nucleotide pairs within the gene will have this effect. If, on the other hand, one is counting reversions from a non-functional mutant form of the gene to a functional one then only a further mutation in the originally mutated codon, and perhaps 'suppressor' mutations at relatively few other sites, will contribute to the score. Typical orders of magnitude might be 10^{-8} for spontaneous mutation frequencies per nucleotide pair per cell division, and 10^{-5} per gene, but the proportion of mutations in the DNA which will be actually observed will depend on the sensitivity to genetic change of the phenotypic character under observation.

The chemical nature of mutations

Since the genetic material is DNA we would expect mutations to be either changes in single DNA nucleotide pairs or more complex changes involving sequences of nucleotide pairs. Various lines of evidence, some of which will be considered below, suggest that there are three principal kinds of mutation. Firstly, there are what Freese has termed *transitions*. These are changes in single nucleotide pairs such that a purine is substituted for a purine and a pyrimidine for a pyrimidine. Thus starting with an adenine-thymine (A-T) base pair a transition mutation would give a guanine-cytosine (G-C) base pair or *vice-versa*. Secondly, again using Freese's terminology, there are *transversions*, in which a purine is substituted for a pyrimidine and a pyrimidine for a purine. Thus A-T could mutate by transversion to either T-A or C-G, and G-C to either C-G or T-A. The third important class of mutations includes those involving changes in the total number of nucleotide pairs—that is either *additions* or *deletions* of nucleotide pairs or sequences of nucleotide pairs. When these additions or deletions involve single base pairs or short sequences of nucleotide pairs within genes they are likely to act as *frameshift* mutations, for reasons which are discussed below. On a larger scale one can occasionally find deletions of large parts of genes or of blocks of genes. Still rarer are transpositions of blocks of genes from one chromosomal position to another, or *inversions* of parts of the gene sequence. These grosser structural

changes have been very important in the analysis of gene position in relation to function in *E. coli*, but it would take us too far afield to discuss them in detail here.

How may one distinguish experimentally between different kinds of mutation in the DNA? In the past the only possible approach to this problem was through the use of mutagenic chemicals with some degree of specificity in their action on the genetic material. More recently, notably in the *lacI* system of *E. coli*, it has become possible to identify mutations directly. The *lacI* gene nucleotide sequence and the amino acid sequence of its protein product are known, as are the sequences of many of its spontaneous and induced mutants. Studies of this kind reveal the nature of both spontaneous and induced changes causing mutations and suggest ways in which different types of mutation may occur.

Mutations are readily caused either by chemical treatment or by irradiation. Both procedures result in some killing of the treated cells as well as in the induction of viable mutants, and in practice one always has to choose a dose of the mutagen which is high enough to give a useful yield of mutations among the survivors without being so high as to leave too few cells alive. More emphasis will be given here to chemical mutagens, since their action is more easily interpretable in chemical terms.

A great variety of chemicals were shown to be mutagenic even before very much was understood in detail about their probable mechanisms of action. With our present knowledge of the chemical properties of DNA we can now at least make good guesses about how each of the main classes of mutagens acts.

Two basically different kinds of chemical action on the DNA of living cells can be distinguished. Firstly, there are mutagens which act only at DNA replication. These include the base analogues, which can become incorporated into DNA in place of normal bases at replication, and drugs of the acridine class, which can become interpolated into the stack of paired bases in a DNA double helix and may throw out of phase the pairing of template and newly-synthesized strand during replication. The action of these compounds depends on growth of the cells, with replication of the DNA, during the treatment. The second class of mutagens act in a direct chemical fashion to modify the bases of non-replicating DNA. We will deal first with the base analogues.

Mutation induced by base analogues

The two base analogues which have been used more than any others are 5-bromouracil (or its nucleoside

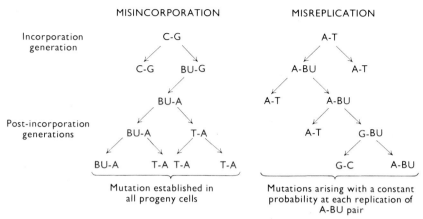

Figure 4. The different consequences of two different mechanisms of mutation induced by 5-bromouracil (BU) an analogue of thymine.

5-bromodeoxyuridine) and 2-aminopurine. The former will become incorporated into DNA in place of thymine, especially when the organism's own synthesis of thymine is blocked by genetic mutation or inhibited by the drug aminopterin. Under favourable conditions 5-bromouracil (BU) can replace thymine in DNA almost completely. 2-Aminopurine (AP) is an adenine analogue which can apparently be incorporated into DNA in place of adenine, although its incorporation is small in comparison with that of BU. Once incorporated, these analogues act most of the time as if they were the corresponding normal bases, but it seems likely that they make 'mistakes' in pairing with an unusually high frequency. BU is thought sometimes to pair with guanine instead of with adenine at DNA replication, while AP is probably apt to pair with cytosine instead of with thymine (Figure 4). This ambiguity of hydrogen-bonded pairing properties can be plausibly explained on the basis of the structures and chemical properties of the analogues (see Figure 5), but the detailed explanations need not detain us here. It is enough to point out that the mistakes made probably do not violate the rule that a purine always pairs with a pyrimidine and *vice versa*, and thus can only lead to transition mutations and never to transversions. There are, in principle, two ways in which a base analogue can cause a mutation, and the direction of the induced transition will be different in the two cases. The mistake may be made in the incorporation of the analogue (*misincorporation*) or in the base-pairing one or more cell generations after the analogue has been 'correctly' incorporated (*misreplication*). For example, if BU is correctly incorporated instead of thymine and then misreplicates it will have the effect of bringing in guanine where adenine should be. After a further

round of replication the transition A-T → G-C will have been completed, the intermediate stages being A-BU and G-BU. If, on the other hand, BU is misincorporated opposite guanine in place of cytosine, a further round of replication will establish the transition C-G → T-A, BU-G and BU-A being intermediate stages (Figure 4). By a similar argument, AP is expected to induce G-C to A-T transitions by misincorporation and the reverse change by misreplication. Since mutants induced by either analogue are usually capable of being reverted to wild-type by both it seems that both analogues can act in both ways. It is, however, observed in experiments on T4 bacteriophage that mutants induced in bacteriophage by hydroxylamine (which is probably specific for G-C to A-T transitions—see below) are much less strongly revertible by BU than by AP and that most mutants induced by BU are not revertible at all by hydroxylamine. From this evidence it is concluded that BU acts preferentially by misincorporation to induce G-C to A-T transitions.

Acridine compounds as frameshift mutagens

The first acridine drug which was shown to be mutagenic was proflavine, and it was used to obtain the first frameshift mutations in bacteriophage T4. Subsequently a variety of acridine compounds have been used for inducing frameshifts. One of the most widely used in recent years has been the acridine half-mustard, code-named ICR-191, the formula of which is as follows:

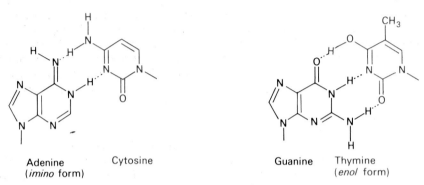

NORMAL BASE PAIRS

Adenine   Thymine

Guanine   Cytosine

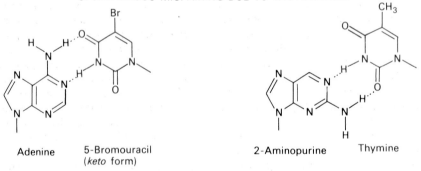

Adenine   Cytosine
(*imino* form)

Guanine   Thymine
(*enol* form)

(Similarly *imino*-cytosine might pair with adenine and *enol*-guanine with thymine)

SPONTANEOUS MISPAIRING DUE TO TAUTOMERISM

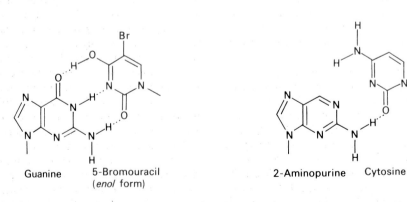

Adenine   5-Bromouracil
(*keto* form)

2-Aminopurine   Thymine

Guanine   5-Bromouracil
(*enol* form)

2-Aminopurine   Cytosine

AMBIGUOUS PAIRING PROPERTIES OF BASE ANALOGUES

Figure 5. Some probable mechanisms of mutagenesis involving alterations in base-pairing properties (see text).

While the function of the 'mustard' part of the molecule in mutagenesis is not completely clear, it seems likely that the reactive chlorine enables the compound to combine with the DNA so as to increase the chance that the acridine part of the molecule will become inserted in the stack of base pairs.

It is not safe to assume that all mutations induced by a reactive compound like ICR-191 are of the frameshift type, but many are. In practice a mutant is thought likely to be a frameshift if it cannot be reverted to wild-type except by further treatment with the same or another acridine mutagen. The interpretation is strengthened if the mutation exerts a polar effect on translation within an operon (see p. 338).

Mutagens which directly modify DNA

Mutagens of this class do not, at least in principle, depend for their action on DNA replication. The most interpretable effects are due to hydroxylamine, nitrous acid, and alkylating agents. Nitrous acid seems to react with DNA in various ways but its chief effect is probably to convert primary amino groups to hydroxyl groups and it thus tends to convert adenine to hypoxanthine and cytosine to uracil. Both these changes are expected to lead to transition mutations (in different directions) since uracil has the hydrogen-bonding properties of thymine rather than of cytosine, while hypoxanthine in its most stable tautomeric state would form hydrogen bonds more readily with cytosine than with thymine (Figure 5). These expectations about the action of nitrous acid are borne out by the finding that most mutants induced by nitrous acid are revertible by base analogues and *vice versa*. Hydroxylamine, which has a strongly mutagenic effect on bacteriophage, attacks only cytosine among the DNA bases so far as is known. It too induces transition mutations, as judged by the test of revertibility by base analogues, but probably only in the direction G-C to A-T.

Amongst the most carefully studied mutagenic alkylating agents are the ethylating agents diethylsulphate, ethylethanesulphonate (EES) and ethylmethanesulphonate (EMS), and the methylating agents methylmethanesulphonate (MMS) and *N*-methyl-*N*'-nitro-*N*-nitrosoguanidine (NG). These are potent mutagens, particularly NG, and cause mutation at doses which do not greatly affect cell survival; they are therefore the most commonly used mutagens. These compounds alkylate bases, in a number of positions, and while N^7-alkylguanine is formed in greatest amounts, it is probably the O^6-alkylguanine (constituting about 7% of the total alkylated bases) which leads to mutation by causing mispairing. Mutations induced by alkylating agents may also arise from mistakes occurring during the *repair* of this type of damage. Cells have enzyme systems for removing alkylated bases from one strand of the DNA and resynthesizing the missing sequence. Some of these repair systems are subject to errors (see p. 223). NG induces, for example, not only transitions, but transversions and frameshifts.

NG preferentially mutates DNA which is replicating, and often a series of closely-linked mutations is generated in a cell. This may be due to an action of NG on the process of replication (which it inhibits), or it may be due to misrepair occurring only in the region of the replication fork, other repair systems operating on the remaining alkylated DNA may be less prone to error.

Mutagenesis by irradiation

Ultraviolet and X-ray irradiation induce mutations in bacteria. *Escherichia coli* has several ways of repairing lesions introduced into the DNA by irradiation. The best understood mechanisms are those that evolved to reverse the effect of UV light. This agent induces dimers to form between adjacent pyrimidines on the same strand of DNA. The dimers block replication. Treated bacteria have three ways of removing such blocks: by splitting the dimer (photoreactivation), by precisely excising the dimer and repairing the gap (excision repair) and by postreplicative repair. Replication is arrested at the dimers probably because the DNA polymerase repeatedly excises any base as soon as it is inserted opposite the dimer. (This editing function of DNA polymerases is discussed elsewhere, p. 215.) However, replication can be reinitiated beyond the lesion leaving a gap which must be filled to ensure the integrity of the daughter strand. Repair of these gaps is called post-replicative repair and is at least in part directed by the *SOS system*, which is put into operation when DNA is damaged or replication stalls. The mechanism of repair by the SOS system is not known, but it does lead to mutation. Perhaps it suppresses the editing function of the polymerase thus allowing it to read past the dimer, but as a consequence lowering the fidelity of replication. A scheme of this kind is illustrated in Figure 6.

Since most species of bacteria are likely to suffer exposure to sunlight it is not surprising that they have evolved an elaborate insurance against UV irradiation. It might also function in other circumstances and be a source of spontaneous missense, and perhaps insertion and deletion mutations.

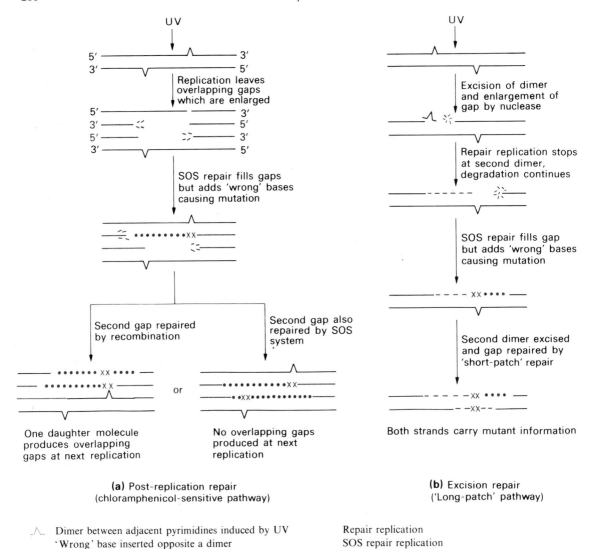

_A_   Dimer between adjacent pyrimidines induced by UV   Repair replication
      'Wrong' base inserted opposite a dimer   SOS repair replication
      New strands of DNA

Figure 6. Possible pathways of mutagenic (error-prone) repair of damage to DNA caused by UV radiation in *Escherichia coli.*
  (a) *Post-replication repair* involves DNA replication producing daughter strands with overlapping gaps. A repair system (known as SOS) is induced and polymerises nucleotides past the dimer, often inserting 'wrong' bases opposite the dimer.
  (b) *Excision repair* requires the activity of a nuclease (e.g. that associated with DNA polymerase I) excising nucleotides from one strand. If excision continues past a dimer, the SOS repair system is assumed to be induced, leading to replication past the non-coding pyrimidine dimers with a high probability of error. (See E. Witkin (1976) *Bacteriol. Rev.* **40**, 869–907.)

Mechanism of spontaneous mutation

We have said nothing so far about the probable chemical mechanism of spontaneous mutations. Watson and Crick, at the same time as they proposed the double-stranded structure and semi-conservative mechanism of replication of DNA, also suggested how spontaneous mutations might occur. They pointed out that the hydrogen atoms involved in the hydrogen-bonding of the paired bases of DNA might shift their positions by tautomerism. In fact, tautomerism is possible in each of the DNA bases, and to the extent that the tautomeric forms exist there is the possibility of some variation in pairing properties. For instance, the rare *imino* tautomeric variant of adenine, with a hydrogen atom in the N1 position and an imino group at C6, instead of no hydrogen at N1 and an amino group at C6, would pair with cytosine instead of with thymine (Figure 5). If this were the sole reason for spontaneous mutation spontaneous mutants should always be of the transition type and should thus be revertible by base analogues or nitrous acid. That the tautomerism hypothesis is not the whole explanation is indicated by the finding that, at least in bacteriophage, most spontaneous mutations are *not* revertible by base analogues. Many spontaneous mutants may thus be transversions and some are certainly frameshifts. There is evidence that the latter occur through some kind of slippage of one DNA strand relative to the other in regions of fortuitously reiterated sequence.

In theory, mispairing could give rise to transversion mutations. Topal and Fresco have suggested that mispairs may occur between two purines, provided one of them is in the rarer *syn* isomeric form with the base rotated 180° at its glycosidic bond. Such mispairs would of course lead to transversions. This proposal is supported by the observation that transversions are not induced by 2AP since if mutation is by mispairing, *anti*/*syn* purine combinations with 2AP would not be expected to give satisfactory hydrogen bonding, so that the appropriate pairs would not form.

This mispairing model of Topal and Fresco can be used to predict the frequency of *spontaneous* transitions and transversions. If each step in elongation of a DNA chain is quickly followed by an editing step to remove errors (such as that determined by DNA polymerase I) and if newly incorporated bases are free to tautomerize or isomerize before editing, then a cytosine in the rare *imino* form would pair with adenine at incorporation but would then return to the *amino* form and, very probably, be excised. From this model it is predicted that the rate of transition at a given base pair *in vivo* should be equal to the *square* of the appropriate tautomeric equilibrium constant. Similar arguments can be used to predict the rate of transversion *in vivo*. Tested in the *trpA* gene of *Escherichia coli*, these predictions are confirmed. There is therefore a good experimental basis for supposing that the majority of spontaneous base-substitution mutations are caused by mispairing during normal chromosome replication.

We should mention here that spontaneous insertion and deletion mutations are also caused by movable DNA elements now believed to be widespread in bacteria. These elements can insert themselves into genes. Their excision is commonly not exact and can cause deletions (see p. 316).

Spontaneous mutation frequency, or at least the frequency of spontaneous transition mutations, is itself subject to genetic control. Several genes have been found in *E. coli* which increase mutability; one of these, that discovered by Treffers, promotes A-T to G-C transitions specifically. The mutation frequency in T4 bacteriophage has been shown to be controlled by the nature of the DNA polymerase produced by this virus. Mutant forms of the polymerase have been identified which determine both higher and lower frequencies than are found in the standard type. Presumably there is natural selection for a frequency of errors in DNA replication which strikes a balance between excessive inaccuracy and excessive evolutionary conservatism.

The expression of mutations

When a mutation arises in a bacterial cell it may have an immediate effect on the observable properties (i.e. the phenotype) of the cell or, more likely, it may not be expressed for a few cell divisions. There are a number of possible reasons for delay in expression. In the first place, many rod-shaped bacteria, including the most thoroughly studied species *Escherichia coli* and *Salmonella typhimurium*, tend to have two or more chromosomes per cell, the number depending on the growth rate and the stage of the cell division cycle (Chapter 2). If a mutation occurs in one strand of one chromosome it would be expected that it would take two to three cell divisions before a cell was formed with all its DNA descended from the mutant strand. It very generally happens that the mutation causes a *loss* of function of a protein—in other words it leads to a *negative* phenotype—and that the positive phenotype continues to be expressed so long as at least one non-mutant DNA molecule remains in the cell. Delay in the expression of a mutation due to this cause may be called *segregation lag*. A second reason for delay in the expression of a loss of function may be that the functional protein characteristic of the non-mutant cell may persist and function for some time after the genotype has become wholly mutant. This may be termed *phenotypic lag* and, if it occurs, it will be superimposed on the segregational lag. Finally, and perhaps most interestingly, the mode of replication and transcription of the DNA may itself result in a lag quite independent of the nature of the protein affected by the mutation. As was explained (Chapter 6), each messenger is transcribed from only one of the two strands of the DNA, at least *in vivo*, and the strand used for transcription is always the same one. A mutation in the transcribed strand can have an *immediate* effect in the production of a new

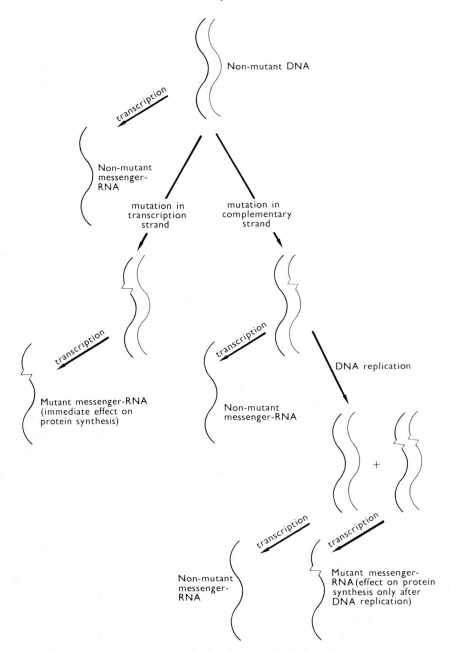

Figure 7. The steps intervening between a mutation (base substitution) and its expression in the phenotype.

kind of messenger which, if it causes a *positive* change in the enzyme complement of the cell, can have an immediate effect on the phenotype. A mutation in the other strand, however, cannot produce any phenotypic effect until the DNA has replicated. Figure 7 illustrates the principle.

It should be mentioned that there are strong indications that a completely new base *pair* may sometimes be substituted by mutation in a DNA molecule, without the necessity for a further round of DNA replication. Following a change in one base of a hydrogen-bonded pair, its now mismatched partner may become vulnerable to excision and replacement by the action of DNA polymerase I even in the absence of extensive DNA synthesis.

We should not leave this discussion of the expres-

Table 2. Properties of different classes of bacteriophage T4 transition mutants.*

Class	Revertibility by			Presumed base pair at mutant site	Phenotypic suppression by FU	Presumed base at mutant site on mRNA
	AP	BU	NH₂OH			
I	+	+ +	+ +	GC	0	G or C
2	+	±	0	AT	0	A
3	+	±	0	AT	+	U

* Based on data of S. P. Champe & S. Benzer (*Proc. Nat. Acad. Sci. Wash.*, **48**, 532).
Abbreviations: AP = 2-aminopurine
                BU = 5-bromouracil or 5-bromodeoxyuridine
                FU = 5-fluorouracil.

sion of mutations without mentioning the very important demonstration that the *expression* of certain mutations can be masked in the presence of 5-fluorouracil (FU), a uracil analogue which seems to interfere with transcription but which is not itself a mutagen. For instance, in *E. coli* it has been shown that FU will partially restore the ability to make alkaline phosphatase to certain mutants which are normally deficient in this enzyme. Suppression of a mutant phenotype by FU was first demonstrated among rII mutants of bacteriophage T4. The important property of such mutants is their inability to grow on the K12 (λ) strain of *E. coli*. A large number of rII mutants were identified as base-pair transitions on the basis of their revertibility with the base analogue 2-aminopurine (cf. p. 283). Within the transition mutant class a proportion were reverted strongly by 5-bromodeoxyuridine (BDU) and by hydroxylamine. For reasons which we discussed on p. 283 these mutants are likely to have had G-C at the mutant site. A second sub-class, roughly equal in number, were not reverted by hydroxylamine and were also insensitive, or relatively so, to BDU. They were presumed to have A-T at the mutant site. Within this second sub-class of mutants almost half were enabled to grow to a significant extent in K12 (λ) when FU was added to the infected bacteria. With two doubtful exceptions none of the first sub-class responded to FU in this way (see Table 2). The interpretation given to these results is as follows. FU is probably incorporated into the messenger RNA in place of U, but once in the messenger it is occasionally translated as if it were C. It will tend to 'cure' any transition mutant with A-T at the mutant site when the A is on the transcribed strand. This is because, in such a mutant, the mutant base (A) will be transcribed in the messenger as U or FU, and FU will be sometimes translated as if it were C, the base that would have been

transcribed from the wild-type base G. It must be emphasized that the effect of FU is only on the *phenotype*; it modifies the effect of DNA mutations without altering the mutations themselves. Table 2 summarizes the interpretation.

RECOMBINATION

The linear array of genes in the chromosome can be plotted on a genetic map which orders genes and measures the relative distance between them. Genetic maps can be constructed in various ways. Most methods depend on crossing mutants which carry different mutations. In a cross the DNA molecules of the two parent bacteria interact and a mechanism called *recombination* permits the chromosomes to exchange material. A map can be constructed on the assumption that the more distant two mutations are the more probable are exchanges between them; the rarer a class of recombinants the more closely linked are the two mutations. Thus a genetic *map* or *linkage group* is a catalogue of *recombination frequencies* taken to represent the underlying physical structure of the chromosome. The physical distances between genes on purified DNA can now be measured directly under the electron microscope or inferred from molecular mass measurements of defined fragments. These results largely vindicate the assumptions of classical mapping derived from recombination studies.

The mechanism of recombination

Although chromosome recombination is the major route of genetic analysis, it is still not understood at the molecular level. However, it is possible to outline the major features of the recombination process and it

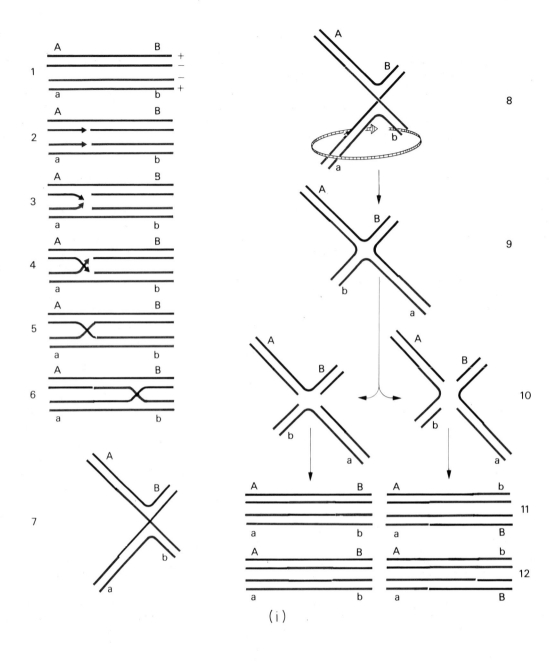

(i)

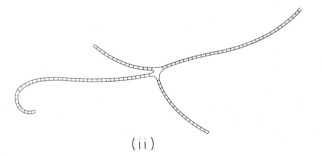

(ii)

seems that a model of recombination proposed by Holliday to account for the genetic behaviour of eukaryotes may also explain the structure of isolated prokaryotic DNA molecules undergoing genetic recombination. Recombination must occur in several steps: a model incorporating the main features is outlined in Figure 8. First, the two participating DNA molecules must be correctly aligned. Homology between the two molecules is no doubt important in this interaction. We know, for example, that donor DNA fragments transferred by conjugation are very efficiently integrated at the homologous position in the recipient chromosome. At the second stage the two molecules break. Current evidence favours the view that initially the breaks occur in one strand of each duplex generating two interstitial ends within each molecule. The breaks need not be at identical positions in the two molecules. At a later stage the ends reunite between the molecules to give at least one recombinant strand which during subsequent processing and replication will generate a recombinant chromosome. In eukaryotic systems both pairs of ends appear to reunite to give reciprocal recombinant progeny. In prokaryotes reciprocal recombination does occur but may not be a necessary aspect of recombination chemistry.

Three different experimental approaches have been used to study recombination. In the first, the products of individual and pooled crossovers are analysed. In the second recombination-deficient mutants have been found and used to uncover the steps of recombination biochemistry. More recently putatively recombinant DNA molecules have been examined by electron microscopy.

Genetic analysis has made two major contributions to our understanding of recombination. Firstly, a number of elegant experiments combining genetic analysis with DNA labelling have shown that both in phage and bacteria the products of a genetic exchange contain material from both parents. They are thus formed by physical breakage and reunion of the participating DNA molecules. Secondly, genetic analysis shows that the primary product of an exchange between two DNA molecules has a hybrid (*heterozygous*) joint in the region of the crossover; the recombinant molecules carry information from both parents in the joint region. Heterozygotic DNA molecules are formed by several different phages and their existence has been inferred in fungal systems. This suggests that the breaks in each participating molecule are staggered and are patched together during reunion.

Recombination-deficient (rec) mutants

Mutant studies have been pursued largely in bacterial and phage systems. In *E. coli* several different *rec* genes have been identified by this approach and for some of them at least, a special function is known. Mutants in *recA* lack virtually all recombination ability (at a frequency <0·01% of that seen in wild-type populations). Surprisingly, this gene codes for an enzyme with proteolytic activity and it remains to be discovered how this enzyme promotes recombination. The genes *recB* and *recC* code for different subunits of an ATP-dependent exonuclease acting on single and double-stranded DNA. It can also cut single-stranded DNA endonucleolytically. The variety of reactions catalysed *in vitro* by this enzyme, exonuclease V, makes it difficult to infer the precise function of this enzyme in recombination. Nevertheless studies on the *recB* and *recC* mutants have shown that it probably acts after the primary breakage and reunion step. Mated *recB⁻* or *recC⁻* recipients can generate from two non-complementing *lacZ* mutants, DNA molecules, presumably recombinants, that will code for wild-type β-galactosidase function. However, this activity is only transiently observed in the initial zygotes which do not go on to make Lac⁺ colonies. We infer that the enzyme participates in the maturation rather than the formation of the recombinant molecules.

Mutants are now known which identify several genes (*recF, K, J and L and sbcA and B*) whose products may play a part in the recombination process. The interactions of mutations in these genes suggest that bacteria, at least, have more than one potential pathway of recombination. This is part of the reason why mutant characterization alone has not revealed the nature of recombination. The precise role of these genes should become clear when a cell-free system for recombination studies is developed.

Figure 8. A model for genetic recombination.

(i) The prototype Holliday model. Breakage and reunion of strands of the same polarity in two homologous DNA molecules (1 to 5) gives an intermediate the crossover of which can migrate along the paired duplexes (6) or unwind (7–8) to give a chi-shaped (χ) structure (9). Subsequent nicks in the recombinant or non-recombinant strands allow separation of the two recombinant duplex molecules (10–11). Ligation of the recombinant strands gives intact molecules (12).

(ii) A diagram based on electron micrography indicating the way in which the four arms of a chi-form are covalently interconnected in the region of the crossover.

Electron microscopy of recombination intermediates

A first step in this direction has now been made. Complex DNA molecules which have the properties of recombination intermediates can be extracted from appropriate bacterial cells and processed for electron microscopy. The Col E1 plasmid is a small circular DNA duplex whose replication is not inhibited by chloramphenicol. Chloramphenicol-arrested cultures containing Col E1 can accumulate up to 1000 copies instead of the usual 20 or so per cell. Some of these molecules have a 'figure 8' structure and prove to have properties expected of an intermediate formed during recombination between two plasmids. Their major features appear in preparations treated with the restriction endonuclease, *Eco* RI. The plasmid has one site at which *Eco* RI cuts it. Each circle of the new structure is broken at a single site by the endonuclease. This treatment generates molecules termed *chi* (χ) structures, with two pairs of arms equal in length. These dimensions imply that the structure is composed of two plasmid molecules joined at a single region of homology. The contact region is short in any individual *chi* structure but may be anywhere in the plasmid molecule. Interestingly, these structures cannot be recovered from *recA* strains. They can, however, be found in normal amounts in a *recB-C* strain.

Perhaps the most striking property of these intermediates is the nature of the single-strand connections which can be seen under the appropriate conditions (Figure 8b). The two duplexes are covalently bound together by single strands exchanged between the molecules. This structure is predicted by models originally proposed by Holliday and illustrated in Figure 8a. According to this view of recombination, paired, homologous, duplex molecules break a single strand in a given region. Each new end dissociates from its complementary strand and re-anneals with that of its homologous partner. Formation of covalent bonds between the recombined strands, generates an intermediate in which the crossover itself can move by branch migration. The important point is that rotation of the arms of this structure gives a molecule with the properties seen in Figure 8b. Breaks in the strands of this structure lead to the formation of two separate recombinant molecules with the heterozygous regions discussed above. Perhaps *E. coli* exonuclease V is involved in this maturation stage. Finally, the recombined strands are re-sealed (perhaps after some local strand digestion and re-synthesis) to generate covalently linked final products. The discovery of *chi* structures provides the first direct physical evidence for the crucial molecular intermediate proposed in this model of recombination. This approach might serve to test other aspects of the model.

A comparison of genetic maps

The techniques of map construction will be discussed in the next section, but it will be profitable at this stage to take a brief look at the maps they generate. These have a number of important common features.

Maps have been constructed for a number of widely different bacteria. These include *Rhizobium*, *Pseudomonas spp*, *Salmonella*, *Vibrio*, *Proteus*, *Streptomyces*, *Bacillus subtilis*, *Erwinia*, *Shigella flexneri* as well as *E. coli*.

The first important generalization we can make is that all the bacteria studied to date appear to have a single linkage group. We can be quite confident that this is so in the case of species whose maps are well defined, i.e. where many mutations have been mapped. Even among those with fewer mapped loci there is no positive evidence for more than one linkage group, i.e. the genetic information appears to be in a single chromosome.

Secondly, the linkage maps are circular. Positive proof of circularity first established in *Escherichia coli* has now been extended to bacteria as different as *Pseudomonas aeruginosa* and *Streptomyces coelicolor* but not yet to all the other species listed above. Nevertheless, considering the broader aspects of DNA synthesis and its control we should not be surprised to find that most, if not all, bacterial linkage maps are circular.

Relatedness of genetic maps

Bacteria which have been regarded as closely related for taxonomic reasons turn out to have similar genetic maps. The best documented examples are from the enterobacteriaceae amongst which *E. coli* and *Salmonella typhimurium* have almost identical maps.

Conversely, unrelated bacteria, e.g. *E. coli* and *B. subtilis*, have quite different genetic maps. Therefore the structure of the chromosome is not unusually highly conserved in evolution. In fact, it can be argued that bacterial chromosomes which do survive gross rearrangements become fixed in separate species. The maps of *Escherichia coli* and *Salmonella typhimurium* which are closely related, show stably inherited inversions and deletions of blocks of genes.

Gene clusters

The genetic map of *E. coli* shows a marked tendency for genes of related function to cluster together. In *E. coli* these clusters are expressed and regulated as single units called operons (Chapter 8, p. 333). Such clustering appears in some other species (e.g. *Salmonella*) but it is rare in *Streptomyces* and rarer still in *Pseudomonas*. In fact there are indications as new maps begin to appear that clustering of related genes may not be a common feature of bacterial chromosomes.

Silent regions of the genetic map

The latest map of *E. coli* (Figure 9) contains close to 500 known genes. These are not distributed at all randomly, in fact some segments are virtually empty. For example the whole region between 30 minutes and 35 minutes contains only 2 known genes. Other segments are equally empty, e.g. 22 minutes–24 minutes and 6

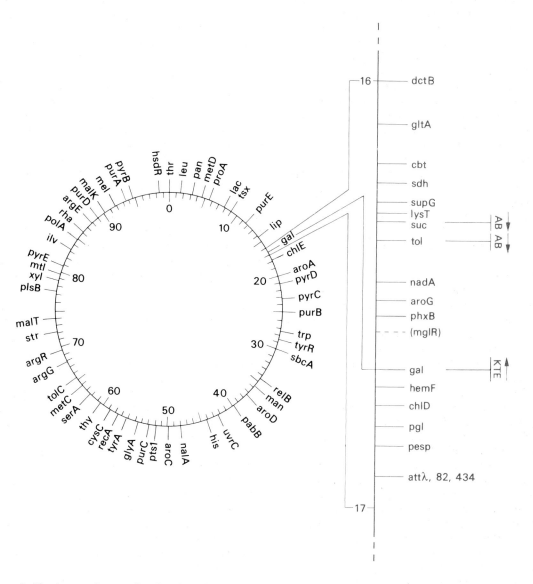

Figure 9. Circular genetic map of *Escherichia coli* K 12.

The map is divided into 100 minutes, beginning at the *thr* locus, and representing the time of entry of markers in mating experiments. About fifty loci are shown in the lefthand diagram. The positions of fifteen genes which occur between minutes 16 and 17 are shown in red on the right.

(See B.J. Bachmann and K.B. Low, (1980) *Bacteriol. Rev.* **44**, 1–56)

minutes–8 minutes. These silent regions can be explained in various ways. They could be artefacts of the mapping techniques, or the map distances could be genuine but represent DNA whose function is unusual. For instance they could contain genes for essential functions whose products cannot readily tolerate conditional lethal mutations such as those coding for ribosomal proteins. Another explanation is that part of the DNA could have a function not related to coding. For example, the genetically silent regions could serve a structural purpose, participating directly in the packing of the nucleoid or its binding to the membrane. In this regard it should be pointed out that the largest gap in the map (between 30 and 35 minutes) contains the segment where replication is believed to terminate and which therefore may have a special interaction with the membrane.

Genetic exchange between bacteria

The period 1944–49 was important for microbial genetics. During those years mutation was established as the source of variation in bacteria. This discovery attested to the eligibility of bacteria as tools for studying the basic processes of life. Its sequel, the discovery of recombination between bacteria, showed that they might be used to investigate the nature of the hereditary material, which by this time was already believed by many to be DNA.

Originally Lederberg and Tatum employed strains of *Escherichia coli* K12 from which a number of different auxotrophic mutants had been obtained. The parental strains participating in these crosses were multiple mutants. The first pair, for example, were *bio met* and *pro thr* double auxotrophs.* Mixtures of the two strains of bacteria were washed and spread on unsupplemented minimal plates. They gave rise to prototrophs which were rare (about 1 in 10^6 parents) but nevertheless much more common than the expected revertants (1 in 10^{14}). As we shall see the prototrophs were formed by mating between intact cells. This was the first example discovered of genetic transfer by *conjugation*.

Genetic transfer by *transformation*, in which recipient bacteria take up pure DNA, had already been demonstrated in *Pneumococcus* two years before. The third mode of transfer, discovered by Zinder and the Lederbergs in the early fifties, exploits bacteriophages which can package host DNA and inject it into recipient cells by *transduction*.

---

* These auxotrophs require biotin and methionine, or proline and threonine.

Conjugation, transduction and transformation remain the major routes (barring mutation) by which the genetic constitution of a bacterium can be radically changed in the laboratory, and probably also in Nature. As we shall see, each of them has a different place in genetic analysis.

Conjugation

One-way transfer and time of entry experiments

The key to the understanding of mating in *E. coli* lay in the recognition that different strains of the organism are of different mating types. Cell conjugation, in which pairs of cells are temporarily joined, depends on at least one of the partners possessing a *fertility factor*, called *F*. The ability to conjugate appears to be due to the presence of special filamentous cell wall appendages—sex-pili (see p. 78)—for the formation of which F is necessary.

When F$^+$ and F$^-$ populations, differentiated by genetic markers, are mixed, one can demonstrate the formation of recombinants by plating the mixture, on a selective medium, after an hour or so of incubation. For example, if the F$^-$ strain requires threonine and leucine (symbolized by *thr$^-$ leu$^-$*) and is streptomycin-resistant (symbolized by *str^r*), while the F$^+$ strain does not require the amino acids and is streptomycin-sensitive (*thr$^+$ leu$^+$ str^s*), plating of the mixture on minimal agar plus streptomycin will select for recombinants of type *thr$^+$ leu$^+$ str^r*, and colonies of this type are regularly found. If the mixed strains also differ with respect to markers which make no difference to growth on the selective medium (*unselective markers*), these markers have a strong tendency to be inherited by the recombinant colonies from the F$^-$ parent. This is explained by the finding that the F$^+$ strain acts as the genetic donor, and normally donates only relatively small pieces of the genome; the selected recombinants therefore tend not to inherit unselective markers from the F$^+$ strain unless the markers happen to be quite closely linked to the selected ones.

The frequency of recombinants given by an ordinary F$^+$ × F$^-$ mating is low—often of the order of 10^{-6}. It is now well estimated that this low fertility is the result of there being a great majority of F$^+$ cells which transfer no chromosomal markers at all to the F$^-$ cells and a small minority which do so with a high efficiency. Cells of this latter type, known as *high-frequency-recombination* or *Hfr*, are to be found in all F$^+$ populations and arise from F$^+$ cells by a kind of mutational event the nature of which will be made clear in our further discussion. When Hfr cells are isolated by plating they give rise to colonies all the cells of which are of the Hfr type. In this way reasonably stable Hfr strains can be established and it is such strains, rather than ordinary F$^+$ stocks, which are always used nowadays for genetic analysis.

Isolation of individual cells resulting from the first few divisions after conjugation has shown that, in an Hfr × F$^-$ mating, all the recombinants arise by subsequent division of the F$^-$ cells and none from their Hfr partners. This shows that there is a one-way transfer of genetic markers from the Hfr ('male') to F$^-$ ('female'). The Hfr markers are transferred in a definite time sequence as can be shown by the *interrupted mating* type of experiment. The principle of such an experiment is best illustrated by reference to one of the classical crosses. The F$^-$ parent strain carried mutations resulting in resistance to streptomycin (*str*^{r}), azide (*azi*^{r}) and bacteriophage T1 (*ton*^{r}), nutritional requirements for threonine and leucine (*thr*$^-$ *leu*$^-$) and inability to use galactose (*gal*$^-$) or lactose (*lac*$^-$). The Hfr strain had the contrasting wild-type characters (*str*^{s} *azi*^{s} *ton*^{s} *thr*$^+$ *leu*$^+$ *gal*$^+$ *lac*$^+$). Cells of the two strains were mixed in a suitable

liquid nutrient medium and, at different times after mixing, samples of the mixture were violently agitated in a mechanical blendor to interrupt mating and plated on minimal (glucose) medium containing streptomycin. On this medium only *thr*$^+$ *leu*$^+$ *str*^{r} recombinants were able to grow to form colonies, the *azi*, *ton*, *lac* and *gal* markers being unselective. The mechanical agitation separated the conjugating pairs of cells so that only those F$^-$ cells which had received *thr*$^+$ and *leu*$^+$ from the Hfr conjugant *before* the time of blending were able to form recombinants of the selected type.

The markers *thr* and *leu* are, in fact, quite closely linked and could therefore be regarded to a first approximation as a single unit. It was found that when mating was interrupted after less than 8 min no *thr*$^+$ *leu*$^+$ *str*^{r} recombinants were recovered. Starting at about 8 min, however, such recombinants began to appear and increased in frequency with time. The recombinants formed after the earliest possible interruption tended to contain no Hfr markers other than *thr*$^+$ *leu*$^+$ but after about 9 min of uninterrupted mating *azi*^{r} began to make its appearance among the recombinants followed in sequence by *ton*^{r}, *lac*$^+$ and

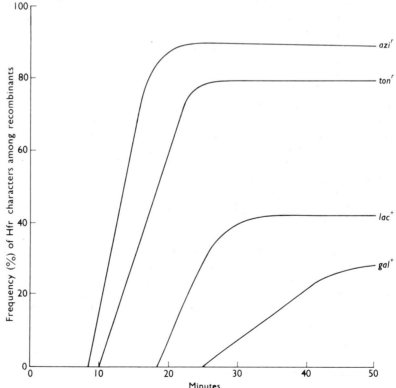

Figure 10. Interrupted mating in an *Escherichia coli* cross.

F$^-$ *thr*$^-$ *leu*$^-$ *azi*^{s} *ton*^{s} *lac*$^-$ *gal*$^-$ *str*^{r} × Hfr *thr*$^+$ *leu*$^+$ *azi*^{r} *ton*^{r} *lac*$^+$ *gal*$^+$ *str*^{s}. Conjugation was interrupted at various times after mixing; the cells were plated on streptomycin medium devoid of amino acid to select *thr*$^+$ *leu*$^+$ *str*^{r} recombinants; the frequencies of the unselected markers among these recombinants were determined.

gal⁺, starting at 11, 18 and 25 min respectively. The result is summarized in Figure 10, and can most reasonably be interpreted as meaning that the Hfr markers enter the F⁻ cell in a time sequence corresponding to their order on a linear chromosome.

The length of the sequence which is transferred depends on how long the cell pairs remain in conjugation. When the cells are free in liquid suspension, conjugation is usually broken off after only a fraction of the chromosome has entered the F⁻ partner. The lower maximum frequencies among recombinants shown by the later-entering markers in Figure 10 is mainly due to the lower probability of entry the longer the time required. On the other hand, if the cells are held immobile on a filter the whole chromosome may be transferred.

The sequence of markers *thr-leu-azi-ton-lac-gal* is characteristic of one particular Hfr strain. Other Hfr strains have the same linear order of genes, but the transfer may start at different points in the order and may proceed in either direction. The comparison of the sequences of markers transferred by a number of different Hfr's shows that they often overlap. The overlaps can be used to build up a continuous map of the whole genome. The feature which caused most surprise when this map was first made was that when all the different sequences were put together they formed a closed loop. Circular linkage maps were at that time completely novel. It is now known that the *E. coli* genome consists of a closed loop of DNA (cf. p. 186). Figure 9 shows the linkage map of *E. coli* as determined by time of entry experiments. The distances between the markers are expressed in minutes. The entire map is 100 minutes in circumference, this being the time necessary for entry of the complete Hfr chromosome under standard conditions.

Conjugation in other bacteria

Mating is not restricted to strains of *Escherichia coli* as was once thought, and there are now a wide range of conjugative plasmids which can promote Hfr-like transfer in quite a few Gram-negative species (see Table 1). The plasmid RP1 and RP4 are particularly interesting since they promote transfer between different Gram-negative genera, including *Pseudomonas*, *Escherichia*, the Proteus group, *Shigella*, *Myxococcus*, *Rhizobium* and *Agrobacterium* (see p. 315). These plasmids are useful tools for the genetic study of certain species for which no other, or less extensive, genetic manipulations are available. For example, the nitrogen fixation genes in *Klebsiella* form a closely-linked group which have been transferred into *E. coli* in a cross mediated by the plasmid.

Properties of the F factor—episomes

The fertility factor F has been studied in detail because of its importance in the conjugation of *E. coli*. When an ordinary F⁺ strain (i.e. not an Hfr) is mixed with an F⁻ strain, F⁺ and F⁻ cells conjugate and there is a very efficient transfer of the F factor to F⁻ cells but transfer of chromosomal genes occurs in only a small minority of cell pairs. Thus, F behaves as an infective factor transmitted independently of the chromosome and the whole population tends to become F⁺ within a few cell divisions. It is now known that the free F factor is a small closed loop of DNA, 94·5 kilobases (kb) long (i.e. about 25 μm, with molecular mass 5×10^7 u) which could code for about 50 to 80 genes. The genes determining the main features of conjugation have been located on the map of F; most are called transfer genes (*tra*) and these fall into a large cluster. The genes *tra A, L, E, K, B, C, F, H* and *G* are necessary for the formation of the F pilus by *E. coli* cells, and they are expressed as a single operon whose transcription requires the *tra J* gene product. In ordinary F⁺ cells the F factor is quite separate from the main chromosome and is an example of the class of extrachromosomal genetic elements known as *plasmids* (see p. 275).

Transfer of F requires one or both strands of the covalently-closed molecule to be cut at a specific site, designated *oriT*, which may be a nucleotide sequence recognized by an F-specific endonuclease. The cut permits transfer of a single DNA strand to the recipient, the 5′ end leading, and at the same time mating triggers DNA synthesis in the donor. These features can be accommodated in the rolling circle model for DNA synthesis as shown in Figure 11. The model proposes that a single strand is cut and peeled off its complementary, still circular, strand for transfer. At the same time nucleotides are added to the 3′ end of the open strand to replace the sequences transferred. This type of replication is illustrated in Figure 11 (for an Hfr × F⁻ mating). This suggests that transfer may continue beyond a single revolution of the donor circle, and it is in fact found that DNA transfer is normally coupled with synthesis in the donor and recipient, and that sex factor DNA reisolated from recipient cells is larger than unit length F. Once transferred, the single-stranded F molecule is converted to the double-stranded circular form in which it continues to be propagated in the autonomous state, thus establishing a new line of donor cells.

The behaviour of Hfr cells with regard to the transmission of fertility is quite different. Almost all the recombinants from Hfr × F⁻ matings are F⁻. It is as if two simultaneous changes have occurred in the

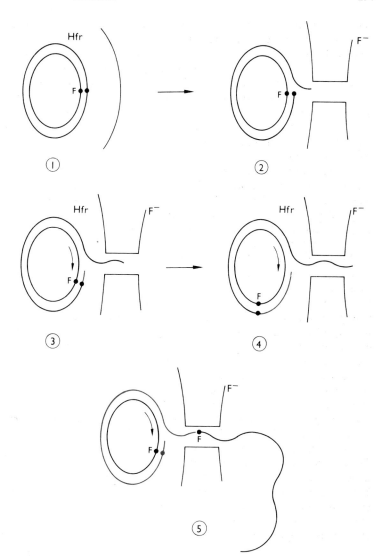

Figure 11. Hypothetical representation of DNA transfer during replication in an Hfr × F⁻ mating.

The F factor is thought to provide the point where new synthesis is initiated with generation of a free 5′ end. The F factor is replicated and transferred last. If conjugation continues long enough, a second copy of the chromosome can begin to be transferred as in (5). Newly synthesized strands are in red. The two DNA strands should be shown as helically wound but are shown here as parallel lines the space between which is highly exaggerated.

derivation of the Hfr from the F⁺: the chromosome has become transferable and the fertility factor has become non-infective. The explanation is suggested by the finding that those few F⁻ recombinants which have inherited markers from the extreme trailing end of the Hfr chromosome—those in other words, which have received the entire chromosome from the Hfr parent—are often Hfr in type. The capacity for transfer is evidently determined by something at the trailing end of the chromosome and this is believed to be the F factor, *integrated into the chromosome* and therefore no longer independently infective. The integration of the F factor has occurred at different points in different Hfr strains—always at the point which is transmitted to the F⁻ cell last.

The integration of the F plasmid into the chromosome depends on relatively short base sequences in common with various segments of the bacterial chromosome. Reciprocal exchanges between these short homologous regions lead to integration of the small loop into the large one as shown in Figure 12. The junctions between F and chromosomal DNA in several Hfrs have been studied in some detail by heteroduplex mapping techniques (p. 310) which show that the sites in F that interact most commonly with the chromosome are special *insertion sequences* (IS). These sequences are transposable elements generally shorter than 2 kb long. The best characterized of them are IS1, IS2, and IS3; each of these is represented in the chromosome of *E. coli* and other bac-

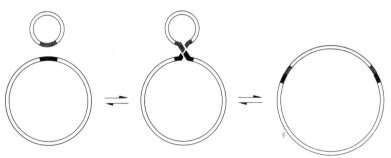

Figure 12. Suggested mechanism for integration of a plasmid into the chromosome by recombination.

The smaller circle (red) represents a plasmid, such as F factor or a prophage; the larger circle (black) represents the bacterial chromosome. The shaded areas represent either regions of genetic homology or DNA sequences which may recombine under the influence of specific enzymes.

teria examined. The F factor has two copies of IS2 and two of IS3, while the R factor, R100, has two of IS1 and two of IS3. Some Hfrs are formed by recombination between the homologous IS sequences of F and the chromosome. In such cases the IS sequence is repeated at each joint between plasmid and chromosomal DNA. This kind of integration should be reversible, and it is indeed known that Hfr strains tend to revert to the F$^+$ condition as well as to arise from it. In its capacity to exist either as part of the chromosome or as an independently transmitted extra-chromosomal element the F plasmid displays the classical properties of an *episome*. Other examples of episomes are the genomes of certain temperate bacteriophages, such as *lambda*, which will be mentioned later.

The capacity of the integrated F factor to promote chromosome transfer, like its capacity to replicate autonomously, is probably connected with its capacity to initiate DNA replication. Though there has been some dispute on the question in the past, the evidence is now strong that the Hfr chromosome is replicated during its transfer to the F$^-$ cell. This *transfer replication* may be induced by conjugation and starts at the point where the F factor is integrated. It is analogous to the transfer of the autonomous F factor discussed above (p. 296). Instead of the chromosome loop remaining closed, as in ordinary vegetative replication (cf. Chapter 5, p. 213), separation of single DNA strands next to the F factor is thought to result in the formation of a free 5′ end, which is then 'pushed' through the conjugation tube (which may be the sex pilus itself), by new strand synthesis at the fork in the duplex structure. Figure 11 illustrates the idea. There is some evidence that the donor DNA is transferred in single-stranded form and that only later, perhaps at the time of its integration into the recipient chromosome is a complementary strand synthesized. An alternative view, which is still taken seriously, is that the new complementary strand is synthesized simultaneously with transfer.

F prime factors (F′)

When integrated F is excised to return to the autonomous state, F$^+$ cells are usually regenerated in which both F and the chromosome return to their previous state, see Fig. 12. Sometimes, however, a rare excision event results from cuts at new sites. If one or both of these cuts are in the chromosome, *F prime factors* (F′) which carry a fragment of the bacterial chromosome are generated. The successful formation of an F′ factor requires the cut ends to be rejoined to circularize the excised structure. Thus F′ factors are formed by recombination events. In some cases these result from *rec*-directed exchanges between nearby or internal IS sequences.

The F′ factors are valuable genetic tools. They provide a means to make bacteria diploid for the bacterial genes they carry, and can therefore be used to analyse gene expression and its control (see p. 334). Since there is now a large collection of different F′ factors in which the entire chromosome is represented by segments, these elements can be used in low-resolution mapping. They have been purified intact and analysed by heteroduplex mapping against F itself, other F′ factors, and other structures.

F′ factors are also capable of chromosome mobilization. They are able to integrate by recombination between the chromosomal segment they carry and the homologous region on the chromosome. In an F′ culture the episome remains autonomous in most cells but about 10% of them have the F′ factor integrated at the same (ancestral) site. For this reason the culture as a whole gives oriented transfer.

Analogous structures (R primes) can be generated by other conjugative plasmids, notably RP4 and R68-45, which are both able to promote mating between widely different bacterial species (see pp. 315–317). In addition some Col factors, of the Col VB type and some F-like R factors, e.g. R1-drdl9 have been found to generate Hfr strains.

Mapping by transduction

General transduction

While chromosome transfer of the F-mediated type is known only in a few species, transfer of small pieces of the chromosome in bacteriophage particles (*transduction*) is known to occur in a wide range of species. The phenomenon of transduction was first described in *Salmonella typhimurium*, the bacteriophage agent in this case being PLT22 (P22 for short). The *Salmonella* phage P22 and the *Escherichia coli* phage P1 are examples of *general* transducing phages in that they are able to transduce any marker in the bacterial chromosome.

The procedure in a transduction experiment is to infect cells of one strain of the bacterium with bacteriophage, allow the cells to lyse, kill any cells which may have survived the infection with chloroform, and use the lysate to infect a second strain of bacteria carrying different genetic markers. Most of the infected cells of the second strain will probably be lysed, but among those which survive the infection some are found with markers acquired from the bacteria on which the infecting phage had been grown. Since the frequency of transduction with respect to any one marker is always low one has to use a selective method for the isolation of the transduced cells. In the most usual experimental design the second, or *recipient*, strain carries an auxotrophic marker which is not present in the strain (the *donor*) on which the bacteriophage was grown. By plating the surviving cells on medium lacking the supplement required by the original auxotrophic strain, one automatically selects those cells which have inherited the prototrophic character from the donor.

By choosing donor and recipient strains which are differentiated by several markers it is easy to show that, when transduction is mediated by P1 or P22, donor markers are generally transduced one at a time—it is quite uncommon to find that a recombinant cell has inherited more than one marker from the donor. This means that the donor chromosome is transferred to the recipient in small fragments, which is not surprising in view of strong evidence that the transduced material is carried within the very small particles of the transducing phage. A tendency of two markers to be transduced simultaneously (cotransduced) is evidence of their very close linkage. This point may be illustrated by reference to *E. coli* and the transducing phage P1. Using the phage grown on *thr⁺ leu⁺* cells to infect *thr⁻ leu⁻* cells, *thr⁺* transductants can be selected for by plating the recipient bacteria on medium containing leucine but devoid of

threonine. In one experiment only about 4% of the selected transductant colonies were found to have received *leu⁺* as well as *thr⁺*. When *leu⁺*. transductants were selected by plating on medium containing threonine but lacking leucine it was found that 98% of them carried *thr⁻* and only 2% *thr⁺*. Thus *thr⁺* and *leu⁺* show only about 2 to 4% cotransduction and the linkage of these two markers is very weak as judged by transduction tests. We saw, however, that *thr* and *leu* appear closely linked in interrupted mating experiments, differing in time of entry by less than a minute. This will give some idea of the resolving power of transduction analysis and, at the same time, its unsuitability for mapping more than very short genetic segments.

The situation is somewhat different with the *Bacillus subtilis* phage PBS1. The protein coat of this virus is capable of carrying considerably more bacterial DNA than either P1 or P22, and it has been calculated that as much as 8% of the *B. subtilis* genome can be transduced at one time. This permits linkage between much more widely spaced markers to be detected by transduction analysis in *B. subtilis* than in *E. coli* or *S. typhimurium*.

Transduction really comes into its own as a tool for the mapping of genetic fine structure, especially when one has a number of independently isolated and apparently similar auxotrophic mutants. The classical example of this type of analysis is the very extensive work on histidine-requiring mutants of *S. typhimurium*. Almost all these mutants turned out to map at different sites, as shown by the fact that almost any two of them, when one was used as the donor and the other the recipient in a transduction 'cross' with phage P22, gave some histidine-independent recombinants. The appearance of wild-type recombinants means, of course, that the donor is able to contribute the wild-type equivalent of the recipient mutant site, and *vice versa*. The frequency of prototrophs was usually considerably less than would have been obtained had the same recipient been crossed with a wild-type donor. This indicated that the mutant sites in the two mutants must be closely linked, so that the donor mutant site was cotransduced with the donor wild-type site in a high proportion of transductants (Figure 13). The frequency of prototrophic recombinants evidently depended on the frequency with which the two donor sites became separated, either by fragmentation of the donor chromosome during the formation of the transducing phage or by recombination after transfer to the recipient cell. Whichever way the separation occurred it was expected to be more frequent the greater the physical separation of the sites

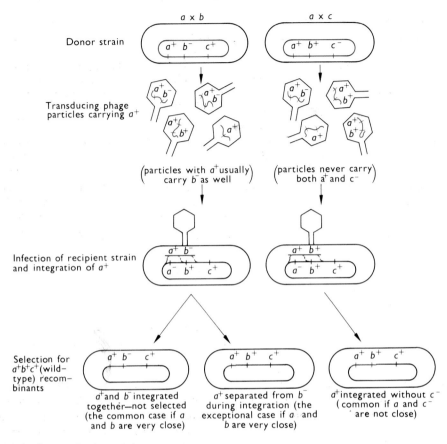

Figure 13. Mapping by transduction.

If *a* and *b* are closely placed and *c* and *d* are relatively distant, the cross *a* × *b* will yield few wild-type recombinants compared with *a* × *c*. The donor chromosome is shown in red and the recipient chromosome in black. Dashed lines indicate alternative possible positions for exchanges.

and so recombination frequency could be taken as a measure of genetic map distance. By comparing the recombination frequencies given by crosses of numerous pairs of mutants it was possible to build up a linear map showing the approximate order of the sites of the *histidine* (*his*) mutations. All the sites fell within one comparatively short region of the chromosome. One source of uncertainty in this kind of mapping is that different transducing phage preparations may have different transducing efficiencies. This difficulty could be overcome to some extent by determining for each cross the frequency of transduction of a standard donor marker outside the *his* region and expressing recombination within *his* as a percentage of this frequency. Fortunately, also, a considerable number of overlapping deletion mutants turned up in the *his* series and these were used to confirm and correct the conclusions drawn from crosses between

point mutants. The principles of deletion mapping are outlined on p. 310 and apply just as much to transduction as to any other way of performing genetic crosses. Some of the results of the fine structure mapping of the *Salmonella his* region are shown in Figure 12, p. 176.

Another general method of establishing an unequivocal order of sites in fine structure mapping is through the use of *three point crosses*, in which the donor and recipient strains are distinguished by three closely linked markers see Figure 19, p. 311.

Generalized transducing phages

The *generalized transducing phages* as they are called, have in common a number of properties which help us to understand how they transmit host DNA from donor to recipient. In the ensuing discussion we shall

Table 3. The range of transducing phages and their properties.

Name	Host	Phage genome size (M_r)	Location	Normal mode of transduction	Size of transduced DNA (M_r)	Ends of DNA	Cuts for packaging
λ	*E. coli*	36×10^6	Integrated	Specialised	$\leqslant 20 \times 10^6$	Unique, cohesive	Site specific
P1	*E. coli*	60×10^6	Not integrated	Generalised	60×10^6	Permuted terminally redundant	Headful
P22	*Salmonella typhimurium*	26×10^6	Integrated	Generalised and specialised	26×10^6	Permuted terminally redundant	Headful
PBS1	*Bacillus subtilis*	190×10^6	Probably not integrated	Generalised	190×10^6	—	—
F116	*Pseudomonas aeruginosa*	36×10^6	—	Generalised	36×10^6	—	—
011	*Staphylococcus aureus*	33×10^6	—	Generalised and specialised	—	—	—

have in mind phages P1 of *E. coli* and P22 of *Salmonella typhimurium*. They share features with other important transducing phages whose properties are summarized in Table 3.

These phages typically yield linear DNA molecules which are anatomically quite different from those found in specialized transducing phage such as λ. First, the terminal sequences are double-stranded. Secondly the ends are not unique. Each molecule contains all the phage genes in an order which is any one of a set of circular permutations, so an individual molecule could carry any phage gene in a terminal position. Finally, the termini are redundant: the same sequence is found at both ends of the molecule:

ABC   ABC

These remarkable features can be understood in the light of the phage maturation process. There is much evidence to show that the DNA of these phages is replicated into a long concatemer containing many repeats of the genome. A population of mature phage DNA molecules such as we have discussed can easily be generated from such concatemers. The trick is for the excision mechanism to cut lengths rather greater than the sum of the gene lengths. Thus in our example the first cut will generate a phage ABC – – – XYZABC,

the second DEF ... XYZABCDEF and so on. The importance of this mechanism is that it measures appropriate DNA lengths (phage headfuls) without regard to DNA sequence. It also generates phages which on infection can circularize following a reciprocal crossover between their redundant ends.

This mechanism, developed to handle maturing phage DNA, can go astray and proceed instead to cut and package fragments of the host chromosome. The resultant particles differ from specialized transducing particles in three important respects: they contain only bacterial DNA, they cannot replicate themselves and, finally, they contain segments from all parts of the bacterial chromosome. Thus it is not possible to obtain high-frequency transducing lysates from these systems, since the transducing particles are produced by the chance packaging of bacterial DNA fragments in about 0·5 to 1% of the total phage particles. Each particle has a capacity for transducing DNA which varies, according to the phage, from about 80 kb for P1 to 281 kb in *Bacillus subtilis* phage PBS1 (about 8% of the *B. subtilis* genome).

Incorporation of the DNA of these particles can produce two types of transductant, at least in the case of P1 and P22. *Complete transductants* arise which have substituted donor genes for their own. They are therefore haploid and stable. By contrast, others, called *abortive transductants*, carry the donor DNA in a form which is neither autonomously replicated nor integrated into the chromosomes. Since it is inherited unilinearly, that is by only one of the two daughter cells at each division, the proportion of diploids falls exponentially. Recent work suggests that abortive transductants carry the donor DNA in a circular form. Since it was packaged as a linear molecule the DNA must have circularized after infection. The same study also indicates that these circles are not covalently closed. Instead their ends are held together probably by a protein molecule, since protein denaturation opens the circles. The conditions of these experiments exclude the possibility that this protein is synthesized after infection by 'helper' phages. We must conclude that it is either of recipient origin, or is included in the phage head.

Bacterial genes are not included randomly for transduction. Recent genetic and physical analysis shows that P22 and P1 start to package host or phage DNA headfuls from defined sites. Thus a population of transducing phages contains host DNA fragments which are not cut at random from the chromosome. In fact there are mutants of P22 which transduce more efficiently than normal and show an altered pattern of joint transduction for adjacent markers. They affect a gene involved in phage DNA maturation; the gene probably codes for a nuclease which acts preferentially at certain DNA sites. Thus the mutant nuclease may cleave additional sites in the chromosome thereby changing the probability that adjacent markers will be picked up in the same particle. At the same time studies on *Escherichia coli* have revealed a remarkable correlation between the transduction frequency of a gene and its position on the map. It will be recalled that many of the known genes of *Escherichia coli* tend to form clusters on the linkage map. It turns out that markers within these clusters, particularly a group close to the origin of replication, are up to 10-fold more readily transduced than those in or near the 'silent' regions. One suggestion is that such variation could be due to the gross structure of the chromosome. If its DNA were composed of a series of loops it is quite possible that some parts of the molecule would be more accessible for incorporation into the phage or for recombination in the recipient bacterium.

Specialized transduction
The specialized transducing phages of *Escherichia coli* include λ, ϕ80, 82, 434 and ϕ170. The genomes of such phages can attach to a specific site on the bacterial chromosome. They normally pick up genes close to that site. To understand how this is done we must

first consider the physical interactions between phage and host chromosomes.

Lysogenization
On infecting a sensitive cell, temperate phages may multiply, lysing the cell and liberating about 100 progeny. Alternatively an infecting phage genome may code for a repressor which blocks further phage gene expression and, therefore, phage development. This permits the infected cell to become *lysogenic*. Typically, lysogenic bacteria carry the phage as *prophage* permanently integrated into a specific site on the chromosome. However they can be *induced* to liberate phage by treatments which inactivate or circumvent the repressor. Induction causes prophage excision and multiplication.

The phage λ will serve as a model for prophage attachment (Figure 14). It infects the *E. coli* cell as a linear molecule, 49 kb in length, bearing at each end complementary single-strand sequences twelve bases long. Pairing of these *cohesive ends* generates a circular molecule which is covalently closed by the action of ligase. This circular molecule is integrated into the chromosome by a single reciprocal crossover as is the sex factor F. The circular λ molecule has a sequence termed *att* located midway between the cohesive ends. It is this sequence which crosses over with the chromosome to permit integration. The *att* site on the phage has its counterpart in the bacterial chromosome (*att*λ). Each element has three components. There is a core region (O) of fifteen base-pairs and two flanking sequences called on the left P and on the right P' in the phage and B and B' in the bacterium. The integration of λ follows an exact, site-specific exchange in this core to give two recombinant *att* sites.

$$\text{POP'} + \text{BOB'} \rightarrow \text{POB'} + \text{BOP'}$$

Sequencing studies of these elements have shown that the O sequence is identical in all the above combinations. The four flanking segments, which are essential for integration, prove to have different sequences. Each of these is conserved after integration. Thus the substrate and product sequences have all the features expected if the site-specific exchange occurred by identical cuts in the phage and bacterial core sequences to give ends which joined reciprocally.

It is possible that the strand cuts in each are staggered to give cohesive ends which join by reannealing and ligation but this attractive hypothesis has yet to be proved.

What function do the flanking sequences serve? Undoubtedly they provide binding sites for the machinery of site-specific recombination. So far two

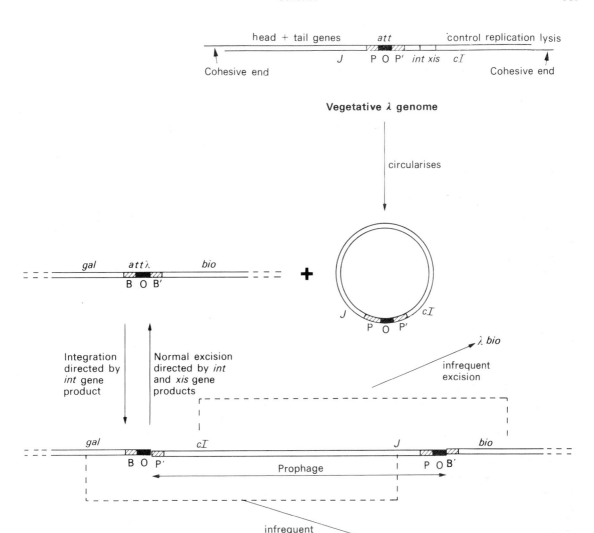

Figure 14. Integration and excision of λ and the formation of transducing phages.

The λ genome is indicated in red, showing the attachment site with its core region, O, and flanking sequences, P and P′. The *Escherichia coli* chromosome (in black) has the same core region, O, but different flanking sequences, B and B′. Site-specific exchange between the phage and the bacterial genomes, in the core sequences, leads to integration of λ to form prophage. Excision normally occurs at the same site but, infrequently, abnormal cuts occur outside P O B′ and B O P′ to generate transducing phages carrying either *gal* or *bio* from *Escherichia coli*.

proteins are known to be required. Both are phage determined. One is an 'integrase', the product of the *int* gene. It alone is required to promote the forward exchange in the reaction: POP′ + BOB′ ⇌ BOP′ + POB′. The reverse exchange occurs when prophage λ is induced. This exchange requires the product of a second gene, *xis*, in addition to the *int* product. The mechanism by which these proteins catalyse specific genetic exchanges remains relatively uncharted territory for the present.

Specialized transducing particles

Very rarely a prophage can be excised abortively by cuts outside BOP′ and POB′. They remove a part or the whole of the phage together with a segment of the adjacent bacterial chromosome. This structure can circularize and, provided certain basic requirements are satisfied, it can be propagated and packaged like a normal phage. The specialized transducing phage thus formed can infect and lysogenize a new cell. As a consequence, the cell stably inherits those bacterial genes incorporated by the new phage just as it maintains the normal prophage.

The picking up of bacterial DNA by specialized transducing phages is commonly associated with loss of phage DNA (see Figure 14). As a result a transducing phage may be defective. However, such phages can grow in bacteria jointly infected with normal phage (helper).

A question that arises is which parts of the DNA of the phage are essential? First, the cohesive ends cannot be deleted, otherwise the circular transducing phage prototype could not be cut and packaged, nor could such a phage recircularize on infection. Second, the region containing the phage's own replication origin must remain. Third, there is a minimum quantity of DNA which must be packaged to make a viable phage. This, of course, depends both on how much phage DNA is lost and how much bacterial DNA is picked up. The capacity of the phage coat imposes an upper physical limit (about 55 kb) on the amount of DNA which can be incorporated in a transducing phage. That is why only those genes close to the prophage attachment site can normally be transduced by λ-type phages. Phage λ itself attaches close to 17 minutes on the chromosome (Figure 9) and can transduce the *bio* genes lying clockwise and the *gal* genes anticlockwise each less than ~ 20 kb from the prophage. The phage $\phi 80$ (att $\phi 80$ about 27′) transduces markers from *cysB* to *tyrT* which are again about 40 kb apart.

How is a transductant formed? The transducing phage infects and is processed like a normal plaque former. It circularizes and, if it lysogenizes the cell, is integrated by an *int*-mediated exchange between its own *att* site and the host's. Interestingly these phages retain the recombinant *att* sites of the ancestral prophage, BOP′ in λ *gal* and POB′ in λ *bio* because they were created by abnormal excision events. Nevertheless the *int* and *xis* proteins can catalyse exchanges between them and the BOB′ of the host. Alternatively, a crossover between the chromosome and the bacterial DNA in the phage will permit prophage insertion.

Of course a normal lysate contains very few transducing particles—about one phage-forming particle (pfp) per 10^6 cells—and for that reason is called an LFT lysate (low frequency of transduction). By contrast the induction of a transductant yields a lysate containing many copies of the transducing particle.

Those particles which are defective require helper phage in order to grow. In these cases helper will also be present in the lysate. Such lysates are termed HFT (high frequency of transduction). They typically contain equal quantities of transducing and helper particles.

Once established, the transductant is stable and carries its acquired bacterial genes as part of the prophage. Since the host's original copies of these genes remain, the transductants become partial diploids.

The technique of specialized transduction by λ can now be exploited to pick up genes from all parts of the chromosome. Very rarely prophage λ can become inserted at abnormal sites on the chromosome. The normal integration (*int*) machinery drives these exchanges which use POP′ of the phage and the *int* protein. Since such events are much rarer than normal, the *int* protein probably selects chromosomal sites with sequences similar to, but not identical with, BOB′. Even these events may require a full core sequence to facilitate *int* action. Since the secondary sites are numerous, it is now possible to obtain λ lysogens whose prophages are widely distributed on the chromosome. Such strains are best recovered by selecting lysogens from bacteria whose normal attachment site has been deleted. The lysogens are selected by challenging with a second phage (λcI) which is sensitive to the λ repressor but unable to make its own. Any non-lysogen infected with this phage is automatically killed. Lysogens on the other hand are already making repressor under the direction of their prophage. They will thus prevent growth of the λcI and survive. The method is powerful enough to select extremely rare phage integrations into specific genes. For example, λ insertion into the *bfe* gene (which codes for a membrane receptor for vit B_{12}, also the receptor site for phage BF23 and colicins E1, E2, E3) inactivates the gene and renders the strain resistant to the virulent phage BF23. This rare event was selected by a double challenge with λcI and BF23. The strain obtained has prophage, located at 88 minutes instead of 17. It has been used to create a transducing phage carrying the nearby *rpoBC* genes coding for RNA polymerase subunits. This derivative could be selected because the *rpoB* gene determines resistance or sensitivity to rifampicin. The λ*drif* transducing phage has since been used to characterize other genes in the region. It contains tRNA genes, ribosomal RNA and protein genes and a gene coding for the elongation factor EF-Tu. Most of these genes owe their discovery to studies on the λ*drif* transducing phage (Figure 15a).

The case of λ*drif* is only one of many where the new technique has proved invaluable. The method is useful

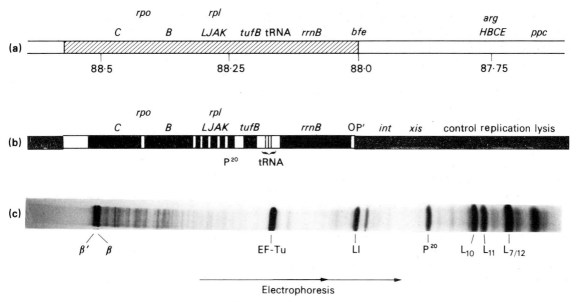

Figure 15. Detailed analysis of a short segment of the *Escherichia coli* chromosome.

(a) The region of the chromosome studied; the segment incorporated into the transducing phage λdrif[d]18 is hatched.

(b) The transducing phage λdrif[d]18, showing the contribution of bacterial genes (black) and phage genes (red). The detailed genetic structure of the bacterial DNA was deduced using a number of techniques including heteroduplex mapping against phage DNA and ribosomal RNA (to locate *rrnB*). In addition, specific fragments were cut from the bacterial segment by restriction endonucleases and used to direct protein synthesis *in vitro* to discover for which gene products they coded.

(c) A radioautograph of a polyacrylamide gel containing SDS and showing the gene products of λdrif[d]18 (kindly provided by Dr T. Linn). Bacteria were heavily irradiated with UV and infected with λdrif[d]18 in the presence of [55]S methionine. A lysate of the bacteria was electrophoresed on the gel. The bands (from largest molecular mass) are as follows: RNA polymerase subunits β' and β (*rpoC* and *rpoB*); the elongation factor of protein synthesis, EF-Tu (*tufB*); and proteins of the larger ribosomal subunit, L1 (*rplA*), L10 (*rplJ*), L11 (*rplK*), and L7/12 (*rplL*). The protein P[20] has not yet been assigned a gene. The product of the *rrnB* gene is a long RNA molecule which is subsequently processed to give 23S, 16S, and 5S ribosomal RNA's. It does not show on this picture. Very few phage-determined proteins appear on this gel. The bacteria used were λ lysogens and thus prevented phage gene expression.

in other respects. It allows diploid studies of the incorporated genes to be made. Gene dose effects can be studied by inducing a transductant to produce cells with many copies of incorporated genes. Finally more conventional mapping methods can be physically checked since the phage DNA can be visualized in the electron microscope and analysed and measured by heteroduplex techniques.

What we have discussed here amounts to a method for breaking up the chromosome into small self-replicating fragments (about 0·5–1·0% of the total). The technique has obvious value as a means of analysing genetic fine structure. It also enables us to detect genes which are not disclosed by the usual mutation techniques. Bacteria heavily irradiated with ultra-violet light make this approach possible. Such bacteria cease to make protein since their messenger RNA molecules decay and cannot be recopied from the extensively damaged DNA. However, undamaged

phage DNA infecting such cells can use the host's transcription and translation machinery to express its own genes and transducing phages can express their bacterial genes. The proteins synthesized in this way can be labelled with radioisotopes and fractionated by gel electrophoresis. Figure 15b shows the results of such an experiment. The phage is λdrif. The products of the *rpoB* and *rpoC* genes are accompanied by other proteins which have been shown by serological and other techniques to be EF-Tu, and the 50S ribosomal proteins L1, L7/L12, L10 and L11. The genes for most of these proteins were not previously known to map in the vicinity of *rpoBC*.

Transformation

Transformation, the first mode of bacterial gene transfer to be discovered, occupies an important position in the history of genetics since it provided the first direct evidence that the genetic material is nucleic acid. Transformation was described as early as 1928 by Griffith during his studies *in vivo* on virulence in pneumococcus. Later (1946) Avery, Macleod and McCarty showed that the agent for this process is chemically pure DNA. Until then it was thought that nucleic acids lacked the specificity necessary for the genetic material. This work and the studies of Hershey and Chase (1952), showing that phages inject essentially only their nucleic acid into their host, set the scene for the later structural studies on DNA of Watson and Crick (1953) which revolutionized the field.

Several bacterial species are now recognized to be able to take up exogenous DNA and incorporate it into their genome. Some of them, notably pneumococcus (*Diplococcus pneumoniae*) which causes pneumonia in Man, and *Haemophilus influenzae* which shares important morphological and pathogenic features with pneumococcus, are human pathogens which are able to undergo transformation in the course of normal growth. The transformable streptococci may also fall into this category. *Bacillus subtilis*, a saprophytic and much less fastidious bacterium, and some Actinomycetes have also proved to be readily transformable under normal physiological conditions. Other bacteria can now be rendered transformable by special treatments of a more or less drastic kind (see below).

Bacteria in the first category have provided most information about the mechanism of transformation. In these systems the recipient bacteria generally develop the ability to take up DNA under defined conditions. By appropriate growth regimes it is possible to obtain cultures in which the majority of bacteria are *competent*. In this state bacteria can take up DNA thus facilitating the first step of transformation.

Competence

Recipient cells of many transformable species vary in their ability to take up DNA. Those bacteria which can be efficiently transformed are said to be *competent*. The development of competence is not an inevitable feature of each round of the cell-cycle since bacteria in liquid culture may grow through many generations until, at about 10^8 cells per ml, the whole culture becomes competent. The change is relatively synchronous and is faster than the division cycle. These observations show that competence is a feature of the culture as a whole and suggests a degree of communication between the individual cells. In support of this idea, proteins inducing competence have been identified in culture supernatants of *Streptococcus* and *Pneumococcus*. So, at a certain stage in the growth of a culture, some cells appear to release or synthesize a signal protein which then causes other cells to become competent. The induction of competence is blocked by inhibitors of RNA and protein synthesis. Macromolecular synthesis may thus be required to bring about the changes associated with competence. Current evidence suggests that competent cells have in fact specific surface proteins which bind and transport the DNA into the interior.

The entry of DNA into competent cells has several detectable stages. There is an initial binding to the cells which later becomes irreversible and is associated with DNA degradation at the cell surface. Much of this breakdown may be irrelevant to the process of transformation itself. However, mutant studies in *Pneumococcus* indicate that an endonuclease is essential for transformation, nicking single strands of the transforming DNA into pieces of 2×10^6 u. A model for DNA uptake which incorporates these facts is illustrated in Figure 16. It shows the bound transforming molecule receiving a single-strand break at a site from which the broken single-strand is transported into the cell. There is evidence to suggest that DNA becomes single-stranded about the time of entry into the cell. In addition the uptake of DNA is commonly associated with oligonucleotide release at the cell surface.

Note that there are important differences between transformable species and that this model best accommodates the observations in the *Pneumococcus* system. For example there is evidence that *H. influenzae* and the bacteria discussed below all take up duplex DNA.

Once inside the cell, transforming DNA can recombine with the recipient chromosome. The incorporated fragment is typically single-stranded and about 2×10^6 u in size. Thus a single transformation event usually involves far shorter DNA molecules than a generalized transduction event (which may involve DNA fragments up to 200×10^6 u).

The range of transformable bacteria has recently been extended through the development by Mandel and Higa and others of a technique which renders the organisms able to take up DNA. Initially it was shown that pure phage DNA can infect treated cells and yield new intact phages. This process is called *transfection*. The technique was then extended to include transformation by plasmids and chromosomal DNA. The cells are made 'competent' by a regime of starvation in the presence of calcium ions at 0°C. Then DNA is added in the cold and the suspension is

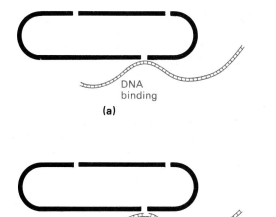

DNA
binding

(a)

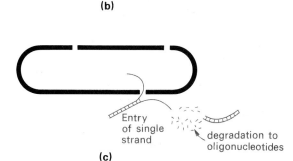

Cutting
gives single
strand break

(b)

Entry
of single
strand
degradation to
oligonucleotides

(c)

Figure 16. A model for DNA uptake during transformation.
(a) DNA binds to the cell surface.
(b) Single strand cuts (nicks) are made by an endonuclease at intervals of about 2×10^6 u.
(c) One strand of the DNA enters the cell, the remaining DNA being degraded outside the cell by nucleases.

heated briefly at 42°C before the mixture is plated for transformants or transfectants. By this method both Gram-negative (*Escherichia coli*, *Pseudomonas putida*, *Proteus vulgaris*, *Agrobacterium tumifaciens*) and Gram-positive (*Staphylococcus aureus*, *Micrococcus* spp) bacteria have been made amenable to transformation and transfection.

The techniques of transformation and transfection occupy an important position in present-day genetic studies since they bridge the gap between DNA manipulation *in vivo* and *in vitro*. Thus it is now possible to construct, *in vitro*, DNA from temperate phages and plasmids incorporating foreign DNA from any source. Since such elements are best replicated *in vivo* the manipulated, free, hybrid DNA must be reintroduced into host cells by transformation or transfection.

Plasmid-containing cells may be re-isolated usually by selection for a vector-determined drug resistance while phages may be recovered free, in lysates, or as prophages in lysogens. These techniques are described on p. 317 *et seq.*

GENETIC TRANSFER AND MAPPING

We are now going to consider how the techniques of conjugation, transduction and transformation are used to construct maps. Since genetic maps are linear, or at most circular, in principle we need only establish a single reference point from which we can measure marker distances to the 'left' or 'right'. In practice no single technique or reference point is perfect since it either generates errors in measurement or in orientation. A reliable genetic map is consequently the product of several approaches.

The broad features of a genetic map are most easily obtained in those bacteria in which Hfr conjugation occurs. The transfer of DNA is polarized so the method provides a means to measure distances between genes (time of entry). In addition, map positions can be cross-checked by comparisons between different Hfr strains. Finally, since the whole chromosome can be transferred, all markers can be mapped by conjugation. However, the resolution of transfer mapping is low because genes separated by say, two, minutes in time of entry may be separated by ~ 80 kb or 50–100 genes. We must therefore employ other techniques for *fine-structure mapping*.

In *B. subtilis* the general features of the map were established by a combination of transduction, using phage PBS1, and an interesting application of transformation to follow the sequence in which genes are replicated.

The general approach was devised by Sueoka and Yoshikawa who based their reasoning on two plausible conjectures about the mode of replication of the *B. subtilis* genome. The first was that replication always began at the same point on the (presumably 'circular') chromosome and proceeded along the chromosome until completed. The second was that in a stationary culture, in which growth had ceased because of the exhaustion of the medium, all the cells would have completed their last round of DNA replication. There was, in fact, some evidence that both these assumptions might well be true. It was argued that in an exponentially growing culture there would be twice as many DNA molecules just starting replication as there were just finishing replication, since every molecule finishing gives two ready to start. Thus

Table 4. Mapping of the *Bacillus subtilis* chromosome by sequential replication of markers (Yoshikawa and Sueoka's method).

Wild-type (ilv^+ leu^+ thr^+ ade^+ met^+) was used as the source of DNA for transforming, in different experiments, the triple mutants leu^- met^- his^-, leu^- met^- ilv^-, leu^- met^- ade^- and leu^- met^- thr^-. The ratios in each line were multiplied by a constant factor to give 1·00 in the last column.

Ratio of transformants	Phase of culture from which transforming DNA obtained			
	Exponential	End of exponential	Nearly stationary	Stationary
met^+/ilv^+	—	1·04	0·97	1·00
met^+/leu^+	1·37	1·24	1·05	1·00
met^+/his^+	1·55	1·28	—	1·00
met^+/thr^+	1·74	1·67	1·15	1·00
met^+/ade^+	1·96	1·92	1·22	1·00

Genetic map

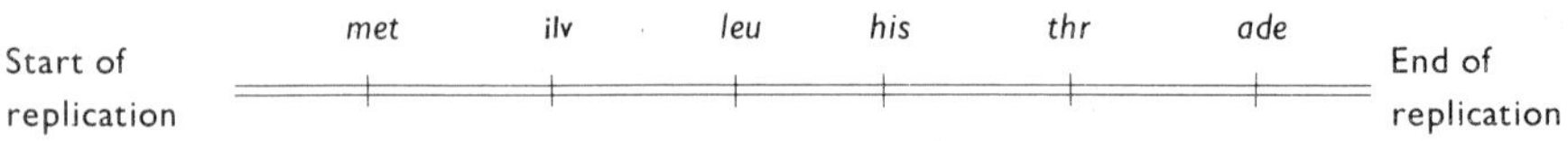

a genetic marker replicating near the starting point would have almost twice as much representation in the DNA as a marker replicated just before the finishing point. In a stationary phase population, on the other hand, all markers should be equally represented. Thus, following extraction of the DNA, the ratio of transforming efficiency of exponential phase DNA to that of stationary phase DNA should be almost twice as great for markers replicated early as for markers replicated late.*

Yoshikawa and Sueoka isolated DNA from samples of a *B. subtilis* culture at different stages of growth and used it to transform a multiply auxotrophic recipient strain. Some of their data are shown in Table 4. The ratios of transformation frequencies with respect to the different markers did indeed change with phase of growth in the way predicted. The strikingly consistent and orderly nature of the results permitted the placing of the markers in a linear order and vindicated the assumptions underlying the experiment. The markers *ilv* and *met* were very close together on the map obtained and they also show simultaneous transformation with significant frequency, confirming that they are contained within one relatively short piece of DNA.

* The ratio can exceed two in cultures growing very rapidly (due to the mode of chromosome replication).

A second ingenious method of mapping based on the idea of sequential replication is worth mentioning. It is technically more complicated than the first but gives rather more precise results. The method depends on the fact that incorporation of 5-bromodeoxyuridine (BDU) instead of thymine markedly increases the buoyant density of DNA and renders it easily separable from DNA of normal density of gradient centrifugation in caesium chloride (see pp. 204–5). DNA which is the product of one round of replication in BDU is 'half-heavy'—that is, it has one heavy and one light strand and forms a band in the ultracentrifuge intermediate in position between normal DNA and DNA with BDU in both strands—while after a second replication half of it becomes 'fully heavy'. In essence the procedure is to initiate a round of DNA replication by germinating spores of a thymine-requiring mutant in the presence of BDU, to extract and centrifuge the DNA from samples of the culture at different times thereafter, and to determine the distribution of different genes (assayed by transformation of multiple mutant recipient cells) between the different DNA bands. As each gene becomes replicated in turn it moves first from the normal to the half-heavy band, and, after a second replication, to the fully heavy band. The changes in the gene ratios in the different bands with time provide a sensitive indication of the replication sequence.

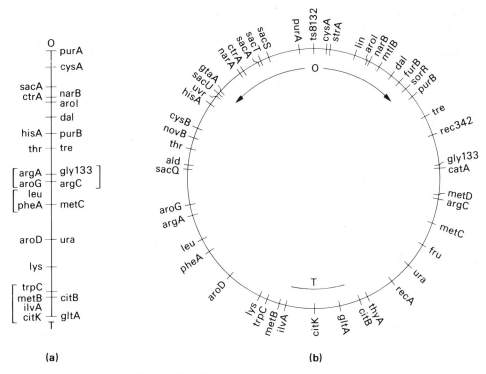

(a)   **(b)**

Figure 17. Revised genetic map of *Bacillus subtilis* 168.

(a) Order of time of replication for various markers, obtained by the density-transfer method. (The map is not to scale.) The order of markers enclosed within brackets cannot be determined with certainty from density transfer data.

(b) PBSI transduction map of the *Bacillus subtilis* chromosome. O is the origin of replication, T is the terminus, and the arrows indicate that replication is bidirectional. The map conforms approximately to distances derived from transduction data. From Kejzlarova-Lepesant, J., Harford, N., Lepesant, J.A. and Dedonder, R. (1975). In *Spores VI* (P. Gerhardt, R.N. Costilow and H.L. Sadoff, eds.) pp. 592–595. American Society for Microbiology, Washington, D.C.

When these experiments were first done it was thought that DNA replication was unidirectional. It is now known that replication in *B. subtilis*, and in *E. coli*, is bidirectional, and therefore this technique cannot be used to assign genes to the chromosome unambiguously. Nonetheless, an extensive comparison of density-transfer data with PBS1 transduction analyses has helped in the revision of the *B. subtilis* genetic map and has shown that DNA replication is fully bidirectional (see Figure 17). Note that the *his* gene in the map below Table 4 has been wrongly located because replication was assumed to be unidirectional.

Fine Structure Mapping

Higher resolution in conjugation mapping can be achieved by measuring linkage. As in classical genetics, the distance between genes on the bacterial chromo-some is proportional to the frequency of crossovers between them. The distance can therefore be derived by measuring the proportion of recombinant progeny bacteria formed in a cross. This principle is illustrated in Figure 18. In a cross between parent 1 $(A^-B^-C^-D^-)$ and parent 2 $(A^+B^+C^+D^+)$ suppose that progeny with crossovers in the interval between A/A^+ and B/B^+ amount to 2% of the total progeny. Likewise, those between B/B^+ and C/C^+ and between C/C^+ and D/D^+ represent 0·1% and 1·5%. We express the distance between the markers in *recombination units*. Each recombination unit represents 1% recombinant progeny. Of course a crossover in each interval can generate two types of recombinant progeny. In the first interval crossovers can give A^+B^- and A^-B^+. Thus the total crossovers is given by the *sum* of A^+B^- and A^-B^+. Generally speaking the number of A^+B^- equals the A^-B^+, so that recombinant frequency is often derived by doubling the frequency of one recombinant class.

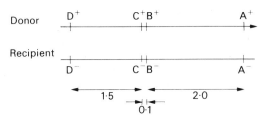

Figure 18. Measurement of map distances between genes following conjugation between two genetically marked strains of bacteria. The direction of transfer of the donor chromosome is shown by the arrow. The further two genes are apart, the more likely it is that crossover will occur between them to give recombinants of the type A^+B^-, B^+C^-, etc. Distances between markers are shown in *recombination units* which are based on the relative numbers of each type of recombinant progeny (see text).

In conjugation, as in all bacterial transfer systems, we have to take special precautions to standardize recombinant frequencies. The problem is that most of the zygotes made in conjugation receive only part of the donor DNA. Consequently, markers may be separated by *transfer* as well as by recombination. The solution is simple, exploiting the polarized nature of conjugative transfer. If we know that a given Hfr transfers markers A, B and C before D (easily established by preliminary time of entry studies) then we can study recombination for A, B and C by first selecting D^+ recombinants and scoring amongst them for the A, B and C properties. Because D^+ enters after A, B and C we can be sure that every D^+ recombinant come from a zygote that has received all the markers. The frequency of the different recombinant classes amongst the D^+ selected is taken to be a standardized measure of recombination frequencies between A, B and C. Estimates of genetic distance over short intervals (~ 10 base pairs) can be obtained in this way. More commonly, however, generalized transduction is used for fine structure mapping. For estimates of distance between markers a selective strategy is used. Transductants for one marker are selected and scored for the inheritance of a second marker. This gives the co-transduction frequency which was found by Wu to relate to map distance (in minutes) in *E. coli* according to the following useful empirical formula:

$$\text{frequency of cotransduction} = \left(1 - \frac{d}{L}\right)^3$$

where d is the distance between markers in minutes and L is the length of the effective transducing fragment in minutes. For P1 transduction of *E. coli*, L is taken to be 2 minutes (approximately 80 kilobases).

Close markers can be ordered by two main techniques. The first of these is the *three point cross* (Figure 19). It relies on the assumption that a recombinant made by four crossovers should occur much less frequently than one made by two crossovers. Figure 19 shows how we order A and B by two crosses knowing that marker C lies to the right of A and B. If the order is ABC we expect the wild type recombinants from cross I (donor $A^-B^+C^-$) to exceed those from cross II (donor $A^+B^-C^+$). If the order is BAC, on the other hand, we expect approximately equal numbers of wild type recombinants. The alternate method which, in general, has proved more reliable, exploits overlapping deletions (Figure 20). This technique was first used by Benzer in his intensive mapping of a gene (rIIB) of the phage T4. The figure shows how simple qualitative tests of this type can establish the order of the three genes.

Map distances and physical measurements between genes

There are several ways to relate map distance to physical distance. The lengths of chromosomal segments transposed to transducing phages or plasmids can be measured physically and compared to map distance. The phages and plasmids provide a concentrated source of intact, homogeneous DNA molecules which are suitable for analysis either in the electron microscope or by restriction endonucleases. By contrast, preparations of chromosomal DNA consist of a highly heterogeneous population of DNA molecules—physically indistinguishable fragments from all parts of the chromosome. Obviously, it would not be possible to use such preparations to compare map distance with physical distance between individual genes. The measurements on the transposed DNA are usually standardized to molecules of known size such as phage λ DNA and its restriction fragments (see below) to allow direct estimation of physical distance in kilobases or atomic mass units.

How do we know which part of the transducing phage or plasmid is chromosomal DNA—the segment we are interested in measuring? The solution to this problem is to compare the substituted and parent vectors, attributing the difference between them to the chromosomal DNA. We must now consider how these comparisons are made in practice.

Heteroduplex mapping

The DNA molecules in such preparations as we have discussed can be relatively easily displayed for viewing and measurement in the electron microscope. The size

Figure 19. Three point cross.

The object is to determine the relative positions of two genes, A and B which are both known to lie on the same side of gene C.

Two strains, containing between them the three auxotropic markers (A⁻, B⁻, C⁻) are crossed using each strain alternately as donor and recipient.

If the order is A B C, the first cross will require only two crossovers to produce prototrophs (A⁺ B⁺ C⁺). The reciprocal cross will require four crossovers so the number of prototrophic recombinants will be low. If the order is B A C, both the cross and the reciprocal cross will need only two crossovers to produce prototrophs.

In practice, the cross and reciprocal cross are made and the numbers of prototrophic recombinants are compared. If the numbers are roughly equal, the order B A C is deduced; if, however, one of the crosses yields a much larger number of prototrophs, then the order is A B C.

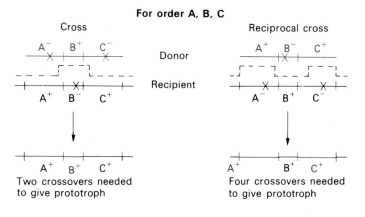

measurements are reproducible and, more important, the technique allows us to distinguish single- from double-stranded DNA. In this system, vector and derivative DNA molecules are compared by making heteroduplex molecules. The two DNA species are mixed and gently denatured. The separated single strands are then allowed to re-anneal. Complementary strands of the two species can thus form heteroduplex molecules. Those parts of the DNA which are identical will therefore appear as double-stranded DNA in the electron microscope. However, the DNA of chromosomal origin in the derivative vector remains unpaired in the heteroduplexes, appearing as a single-stranded loop in the microscope. The length of

Figure 20. Mapping by the method of overlapping deletions.

Three deletion mutants (1, 2 and 3) are used in which varying lengths of chromosome are deleted. They are defined by the fact that they do not back-mutate to wild-type and they cannot form wild-type recombinants with mutants containing point mutations in the segment concerned.

When the deletion mutants are crossed with the strains containing point mutations, wild-type recombinants are found as indicated in the table.

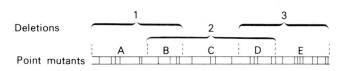

Whether or not recombinants formed in crosses

		Deletion mutants			Point mutants in region				
		1	2	3	A	B	C	D	E
Deletion mutants	1	−	−	+	−	−	+	+	+
	2	−	−	−	+	−	−	−	+
	3	+	−	−	+	+	+	−	−

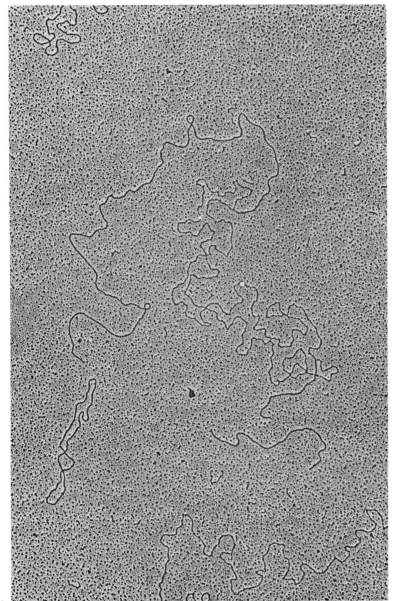

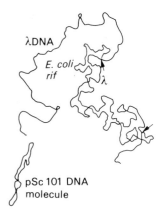

Figure 21. Heteroduplex formation between λ*drif*ᵈ18 and λ.

The interpretation of the electron micrograph shows the internal substitution (between the arrows) of λ*drif*ᵈ18, which does not anneal to λ, and the outer left and right arms which are homologous and thus able to form a duplex. Plasmid pSc101 DNA molecule was used as a reference for length measurements.

Reproduced with permission from J. Kirschbaum, J. Greenblatt, B. Allet, and J.-D. Rochaix (1976). In *RNA Polymerase* (eds. R. Losick and M. Chamberlin) Cold Spring Harbor Laboratory, New York.

the loop and its position in the heteroduplex molecule allows us to verify the origin and size of the chromosomal DNA it contains. An example of a heteroduplex molecule and its interpretation is illustrated in Figure 21.

Restriction Mapping

A second way of measuring map distances in physical terms makes use of restriction endonucleases. These enzymes have been purified from a number of bacteria

(e.g. *Eco*RI and *Eco*RII from *E. coli*; *Hin*III from *Haemophilus influenzae*) and each cuts double-stranded DNA only at a very specific nucleotide sequence, the *restriction site* (see p. 198). If a population of DNA molecules from a particular source (e.g. a phage vector) is treated with one of these endonucleases until complete digestion has occurred, then the DNA is cut into a distinct number of subpopulations. Within each sub-population, every fragment has the characteristic length determined by the distance between two consecutive restriction sites in the untreated DNA.

If this DNA, which has thus been completely digested with a restriction endonuclease, is subjected to electrophoresis on agarose gels, then the DNA fragments are separated according to their size. By calibrating the gel with appropriate standards the size of each fragment can be estimated accurately, and is usually expressed in kilobases (Kb). The single pattern of fragments obtained in this way does not give any idea of the order in which the fragments are arranged on the original DNA. Some idea of order can be obtained by *partially* digesting the original DNA with the same restriction enzyme, so that the products comprise longer fragments containing two or more of the fragments that are found on complete digestion. Further digestion of these partially-digested products and characterization by electrophoresis shows which of the fragments formed by complete digestion are adjacent to each other.

This analysis can be extended by combining similar results obtained using different restriction enzymes either singly, or combined. Different restriction enzymes cut at different sites, and ultimately a *restriction map* can be constructed showing where the DNA molecule is cut by a number of different enzymes. This mapping has been done for many plasmid and phage

DNAs, particularly where they are potential vectors for genetic manipulation, and a map of the promiscuous plasmid RP4 is given as Figure 22. Once such a map has been obtained for a vector, the 'restriction pattern' it gives after treatment with a restriction enzyme followed by agarose gel electrophoresis, can be compared directly with that from a derived vector containing an inserted DNA segment (for example a short region of the bacterial chromosome containing a number of genes of interest). The chromosomal DNA will change the size of the restriction fragment in which it is located, causing it to have a different position after electrophoresis on agarose gels. Moreover, new bands will appear if the inserted DNA contains its own restriction sites for the endonuclease. By summing the molecular masses of the new bands the physical size of the chromosomally-derived insertion can be estimated (see Figure 23), and by continuing the analysis with other restriction enzymes (or other inserted DNA-vector combinations) a restriction map of the inserted DNA sequence can be made.

Much information can now be obtained about the genes on individual restriction fragments. The fragments can be used as templates directing the synthesis of RNA and proteins in cell-free systems. Thus we can identify which DNA fragment carries a particular gene by showing that it codes for that gene's product.

Figure 22. Restriction map of plasmid RP4.

The RP4 genome is circular and is divided into 36 minutes. The location of genes for resistance to ampicillin (*amp*^r), tetracycline (*tet*^r), and kanamycin (*kan*^r) are shown, together with the sites at which various restriction endonucleases (*Eco*RI, *Bam*, *Sal*I, etc.) cleave the RP4 DNA. The outer ring shows in red the cleavage sites for all the enzymes shown and indicates the number and size of fragments which can be obtained by appropriate choice of nucleases.

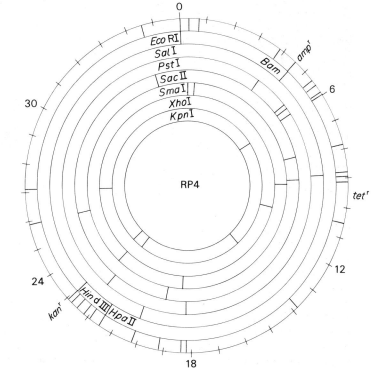

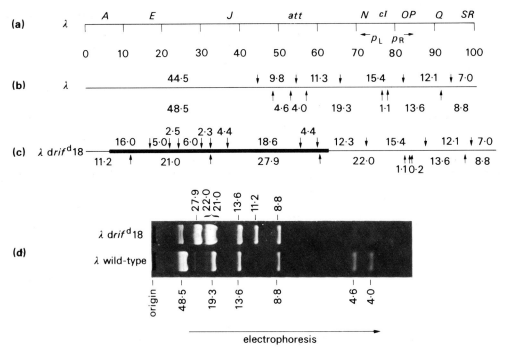

Figure 23. Restriction mapping of λ wild-type and $\lambda drif^d 18$.

(a) The genetic map of λ subdivided into 100 units. The positions of some of the genes are indicated.

(b) The restriction map of λ. The arrows above the line indicate sites of action of *Eco* RI; those below the line show those for *Hin*d III. The numbers represent the sizes of fragments produced, expressed in terms of Kb units.

(c) The corresponding restriction map for $\lambda drif^d 18$ (compare with Figure 15b, p. 305). The thick line represents DNA of bacterial origin.

(d) Products of digestion by *Hin*d III of λ wild-type and $\lambda drif^d 18$ DNA after electrophoresis on agarose gels. The wild-type fragments with mobilities corresponding to 48·5, 19·3, 4·6, and 4·0 units come from the region of the phage genome which is replaced by bacterial DNA in $\lambda drif^d 18$. Note that fragments shorter than 4·0 units are not shown in this gel, and that there is a small amount of cross-contamination of $\lambda drif^d 18$ with λ wild-type DNA.

This information can be combined with the size and order of sites to build up a physical map of the chromosomal fragment. The result of combined heteroduplex and restriction studies on a transducing phage, $\lambda drif$, is shown in Figure 15, p. 305.

A number of studies of this kind have allowed comparison of the genetic and physical maps for particular regions of the *E. coli* genome. The estimate of 41 kb per minute for the rate of transfer of DNA during conjugation in *E. coli* was obtained by measurement of the DNA of the F′ factor, F14. This estimate compares well with that found for the *nadA-gal* interval and the *aroE-rpsE* interval. However, discrepancies between genetic and physical measurements have been detected. For example physical measurements of the *bfe-argH* interval indicate a distance of 6 kB whilst the cotransduction frequency for these markers

(70%) converts to 9 kb by the Wu formula (p. 310). Resolving such differences may tell us more about the nature of transduction.

Complementation studies

The specialized transducing phages and F prime factors, which have figured prominently in an earlier section, have another use in genetic analysis—as a source of partial diploids.

Diploids are important in genetic analysis since they provide a means to study *complementation* between mutations. To illustrate this technique we shall take as an example mutations affecting lactose utilization. In *Escherichia coli* two contiguous genes are required to permit utilization of this sugar. One of

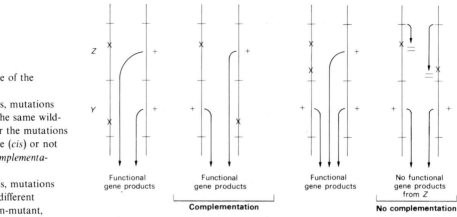

Mutations in *different* genes | Mutations in *same* gene

Figure 24. The principle of the complementation test.

(a) In partial diploids, mutations in different genes give the same wild-type phenotype whether the mutations are on the same genome (*cis*) or not (*trans*). In the latter, *complementation* is said to occur.

(b) In partial diploids, mutations in the same gene yield different phenotypes—*cis* are non-mutant, whereas *trans* are mutant since no complementation can occur.

them, *lacY*, codes for a permease which promotes the uptake of lactose; the other, *lacZ*, codes for β-galactosidase which hydrolyses lactose to glucose and galactose (Chapter 8, p. 331). If two different mutants unable to grow on lactose (Lac⁻ phenotype) are isolated, they might be altered in the same gene or in different genes. The question can be resolved if the two mutations are put into a single cell making it diploid. This can be done by constructing a partial diploid in which one of the mutations is carried on an F′ episome while the other is on the chromosome of the cell. This is the *trans* configuration. When the two mutations are in different genes, functional permease and function β-galactosidase—and therefore a Lac⁺ phenotype—will result, i.e. there will be *complementation*. Conversely, lack of complementation indicates that the mutations are in the same gene (Figure 24).

For a rigorous test, the same comparison should be made when the mutations are carried on the *same* genome (*cis* configuration; see Figure 24c). It will be apparent from the Figure that a Lac⁺ phenotype will be produced whether the two mutations are in the same gene or in different genes. In practice, it is often difficult to obtain both mutations on the same genome and the *cis* configuration is usually omitted. It does, nevertheless, guard against the rare case in which a loss mutation is dominant.

The term *cistron* was introduced by Benzer to describe a DNA sequence in which mutations do not complement each other. In general, *gene* and *cistron* are synonymous.

Gene transfer between species

Plasmid transfer

The first systematic studies on gene transfer between bacterial species began in Japan. Reviewed in English by Watanabe (1963), these studies showed that *Shigella* strains of clinical origin could simultaneously acquire resistance to several antibiotics. Mutants resistant to a single antibiotic are normally rare ($10^{-6} - 10^{-10}$), so the appearance of these multiply resistant strains could not be attributed to simultaneous mutations. It turned out that the resistance was caused by acquisition of plasmids called R-factors, which carry several antibiotic-resistance genes together with the determinants for conjugative transfer. The R-factors were found to be transmissible from *Shigella* to other members of the *Enterobacteriaceae* and indeed to genera as unrelated as *Serratia*, *Proteus* and *Vibrio*. These elements are very widespread amongst bacteria (both Gram-positive and -negative).

Plasmid classification

Many R factors have been isolated and it is becoming possible to classify them by criteria related to their phylogeny. The properties of some of them are summarized in Table 5.

In the long term, attempts to elucidate the taxonomy of individual plasmids may create some surprises. The reason for this is to be found in the mechanisms promoting exchange of blocks of genes

Table 5. Plasmids and other transposable elements.

Elements	$M_r \times 10^{-6}$ Average	Range	Essential properties	Ancillary properties
Plasmids	76	5–100	(a) Self replication (b) (Transfer)	Many (see Table 1)
Transposons	4	2–15	(a) Repeated terminal sequences (140–1400 base pairs) (b) Transposition	(a) Drug resistance (b) Tumour induction
IS elements	0.98	0.6–1.3	Genes for insertion	None

between these elements (see below). The world population of plasmids is probably always in flux, but nevertheless is stable enough to make plasmid classification clinically important. In broad terms we can consider R-factors as two entities combined. Segregation and transductional analysis shows that they often have a group of genes for their propagation and transfer, sometimes called RTF (for resistance transfer factor), and distinct segments with blocks of genes determining drug resistance. Current evidence points to considerable interchange between the latter. It is thus not surprising to find that more taxonomic emphasis is placed on propagation and transfer properties. The first of these is incompatibility.

It has long been known that two F-factors cannot stably cohabit in a bacterial cell. Their *incompatibility* causes rapid segregation of resident and newly infective plasmids. Incompatibility is a common feature of plasmids. It is assumed that the phenomenon is related to the low numbers of plasmids per nucleus. Perhaps the two elements must compete for a single replication site, the winner being able to replicate and segregate normally at division while the loser is left in the cytoplasm to compete in the next cell division cycle. Alternatively, a cell containing two incompatible plasmids per nucleus may accumulate an excess of a repressor blocking further plasmid replication until after they segregate to different daughters. Whatever its cause, it is assumed that incompatibility reflects an overlap in the replication specificity of two plasmids and that it is thus a fair measure of relatedness. It has been successfully used to separate plasmids into classes.

Other criteria may also be applied. First, plasmid-determined sex pili provide specific adsorption sites for certain bacteriophages. If the same bacteriophage adsorbs to pili determined by two plasmids then one or more of the pilus genes in each plasmid has a common origin. Secondly, techniques of plasmid iso-

lation are now so improved that it is possible to compare the purified DNA molecules. Molecular weight determinations have been used but, more effectively, related plasmid molecules are tested for their ability to form heteroduplexes. Short of direct sequence-determination, this technique is the most critical test of relatedness available; it demonstrates at least 80% homology between duplexed sequences. Thirdly, restriction endonucleases can be used to generate DNA fragments from each plasmid which are compared on gels. These fragments can be hybridized with labelled DNA from another plasmid to discover whether parts of two molecules are homologous.

Finally, characterization of individual plasmid gene products can provide a test of relatedness. For example, resistance to penicillin in Gram-negative bacteria is commonly due to a plasmid-determined β-lactamase. The enzyme from different sources can be shown to be the same by several criteria indicating that the plasmids from these strains have at least a common gene for penicillin resistance.

Transposons

An interesting feature of R factor structure has recently come to light. The genes for several drug resistances borne on these plasmids are contained in transposable elements termed *transposons*. These vary in size (2·5 − 20·5 kb) and other properties (see Table 5) but they have the ability to transpose to new locations on phages, plasmids or host chromosomes without using the conventional homologous (*rec*) recombination system. They have a characteristic nucleotide sequence at each end. For some transposons the repeat sequences are oriented in the same direction at both ends. These are called direct repeats. Other transposons have their repeats inverted with respect to one another (inverse repeats). The repeats are very interesting, since they are likely to form a part of the transposition machinery, and particularly since some transposons utilize repeat, or insertion, sequences (IS sequences) which are known to occur at sites on the bacterial chromosome itself. In other words, we now recognize that in bacteria reassortment of groups of genes can occur independently of the classical recombination process. How it does so is an interesting but unanswered question.

The autonomous mobility of transposons is undoubtedly important in plasmid evolution, particularly since the transposition of drug resistance between plasmids confers a selective advantage on their host bacteria faced with an antibiotic-loaded environment.

Transposon genetics gives a new dimension to the current repertoire of possible manipulations in bacteria.

Plasmid promiscuity

Pseudomonas aeruginosa is an opportunist pathogen of increasing importance, because of its high innate resistance to antibiotics. Nevertheless the organism remains susceptible to some drugs, notably carbenicillin, which was used successfully to control *Pseudomonas* infections of burns. However, in 1967, strains highly resistant to carbenicillin appeared and they rapidly became the predominant local form of Pseudomonas infection. Resistance was shown to be caused by an R factor, RP1, which could transfer to a wide range of Gram-negative organisms. In the laboratory it has been transferred, amongst others, to *Rhizobium*, *Myxococcus*, *Agrobacterium*, *Acinetobacter*, the Enterobacteriaceae, and several pseudomonads. Moreover new plasmids have been recovered in clinical isolates from widely different geographical locations and shown to belong to the same incompatibility group as RP1. Despite the fact that they were found in genera as different as *Bordetella*, *Klebsiella* and *Serratia*, these plasmids have many common features suggesting that they are related. Thus we are led to the conclusion that RP1 represents a class of plasmids which are common in nature and clinically important.

Chromosomal transfer

The promiscuous plasmids have been shown in the laboratory to be able to mobilize the host chromosome for transfer in *Escherichia coli*. In fact RP4, a variant of RP1, has been used successfully to initiate mapping studies in *Rhizobium meliloti* and *Acinetobacter calcoaceticus*, whilst Hfr derivatives of RP4 and R-prime factors occur in *Escherichia coli*. The plasmid R68.45 (in the same incompatibility group as RP1) can generate R-primes in *Rhizobium leguminosarum*. There is thus extensive laboratory evidence to suggest that chromosomal mobilisation can occur for transfer between species in nature. This may represent a significant source of variation in the microbial world.

GENETIC MANIPULATION

Conjugation, transduction and transformation provide the basic means of transferring genetic material between prokaryotic organisms, and we have seen how they can be used for straightforward genetic analysis, such as mapping and testing for dominance and complementation. Now, a range of techniques is available for more extensive manipulation of the genetic constitution of a bacterial cell. They are based on the three traditional transfer processes and on the recent development of a wide range of enzymic methods for specifically modifying DNA *in vitro*.

The new procedures are important because they have greatly improved our ability to locate genes, to determine their relative orientation on the chromosome, and to study the way their expression is controlled. Moreover, since they allow the handling of DNA molecules *in vivo* and *in vitro*, they provide a means of rapidly amplifying the DNA from a specific segment of a chromosome in amounts needed for chemical characterization. Because it is possible to splice different DNA molecules enzymically *in vitro*, and then transfer them into bacterial cells, these techniques can be used to study many aspects of the genetic organization of eukaryotes as well.

To illustrate, in part, the potential of genetic manipulation we will consider how it has simplified mapping in *E. coli*. The entire chromosome of *E. coli* has been subdivided into a set of known fragments (carried on natural vectors, mainly F' factors and transducing phages) of different size. These fragments provide a way of locating new genes, *sometimes in the absence of mutations* since genes carried on transducing phages can be expressed in hosts subjected to UV treatment to suppress host protein synthesis (see p. 305).

Also, since these gene fragments are attached to vectors, it is possible to study the orientation of genes, their promoters and regulatory sites, and their neighbours by exploiting gene expression under the control of the vector.

Thirdly, the DNA fragments can be purified in large enough quantities for their sequence to be determined, for restriction mapping, and for detection of particular mRNAs made *in vivo*. Ultimately, by exploiting suitable vectors *in vivo*, it may be possible to improve the yields of gene products of commerical significance, thereby facilitating their purification.

Chapter 7

Figure 25. Fusion mapping of operator/promoter region.

Two operons are shown. Deletions I and II leave the righthand operon still under the control of promoter and operator, P′ and O′. Deletion III, which extends into the P′/O′ region, results in the righthand operon coming under the control of P and O, i.e. the expression of Z and Y are subject to repressors of A, B, and C.

Basic techniques in genetic manipulation: deletion, transposition and incorporation of DNA fragments into vectors

Deletions

These were the first spontaneously occurring gross genetic events that were used to analyze gene expression. The removal of DNA intervening between two operons fuses them and may place the transcription of one under the control of another. Figure 25 illustrates how deletions can be used to locate the promoter of a gene. In bacteria this technique is limited, partly because large deletions are likely to remove vital genes. The problem may be solved by selecting for deletions in partial diploid bacteria, or more conveniently by fusing operons which are widely separated on the chromosome. In this case one of them must be moved closer to the other before fusion is attempted. This leads us to the other basic techniques.

Gene, excision, transposition and incorporation into vectors

The mechanisms which delete DNA from the chromosome do not automatically destroy the excised fragment. For instance, we have seen that excision of DNA-incorporating prophages or integrated plasmids can yield transducing phages and prime factors (e.g. F′) able to propagate as independent, self-replicating entities, and that these can be obtained from any part of the chromosome into which a prophage or suitable plasmid can integrate. There are, of course, many sites at which F normally integrates. Temperate phages such as λ, on the other hand, are normally much more attachment-specific. However, as we have seen (p. 295) many secondary attachment sites can be used by λ at a lower efficiency than usual. Such lysogens have given transducing phages for many parts of the chromosome.

The chromosomal fragments incorporated into a vector are said to have been *cloned*. They can be recovered by a variety of techniques (see, for example Figure 27). Thus DNA picked up by an F-prime factor can be transposed by reintegration into the chromosome along with its vector to form an Hfr strain. Since these events are relatively rare, probably using IS sequence interactions, it is important to be able to select the new Hfr strains. One effective technique employs a temperature-sensitive factor, $F_{ts}\,lac^+$, unable to replicate autonomously at 42°C. A bacterium with its chromosomal *lac* genes deleted but carrying $F_{ts}\,lac^+$ will grow on minimal lactose medium at 30°C since the $F_{ts}\,lac^+$ can replicate. At 42°C, however, bacterial growth will segregate Lac⁻ daughters unable to form colonies on minimal lactose. However, those rare cells in the culture which have reintegrated $F_{ts}\,lac^+$ into the chromosome will stay Lac⁺ and form colonies at 42°C. This technique can be combined with other procedures to allow direct selection of strains with $F_{ts}\,lac^+$ integrated into a specific gene. Two examples are shown in Figure 26. In the first, *lac* is inserted into a *gal* gene thus putting it next to the λ attachment site for subsequent transfer. In the other, *lac* is transposed next to the ϕ80 attachment site. Similar techniques can be used to transpose two operons together for subsequent fusion by deletion.

Directed integration of chromosomal genes into phage λ

A general method of gene integration into phage λ is illustrated in Figure 27. The crucial feature of this approach is to use a bacterial strain which has the λ attachment site, *att*λ, deleted and is thus unable to form normal lysogens. However, lysogens are formed at a low frequency. They each carry λ prophage integrated at one of many secondary sites. The pro-

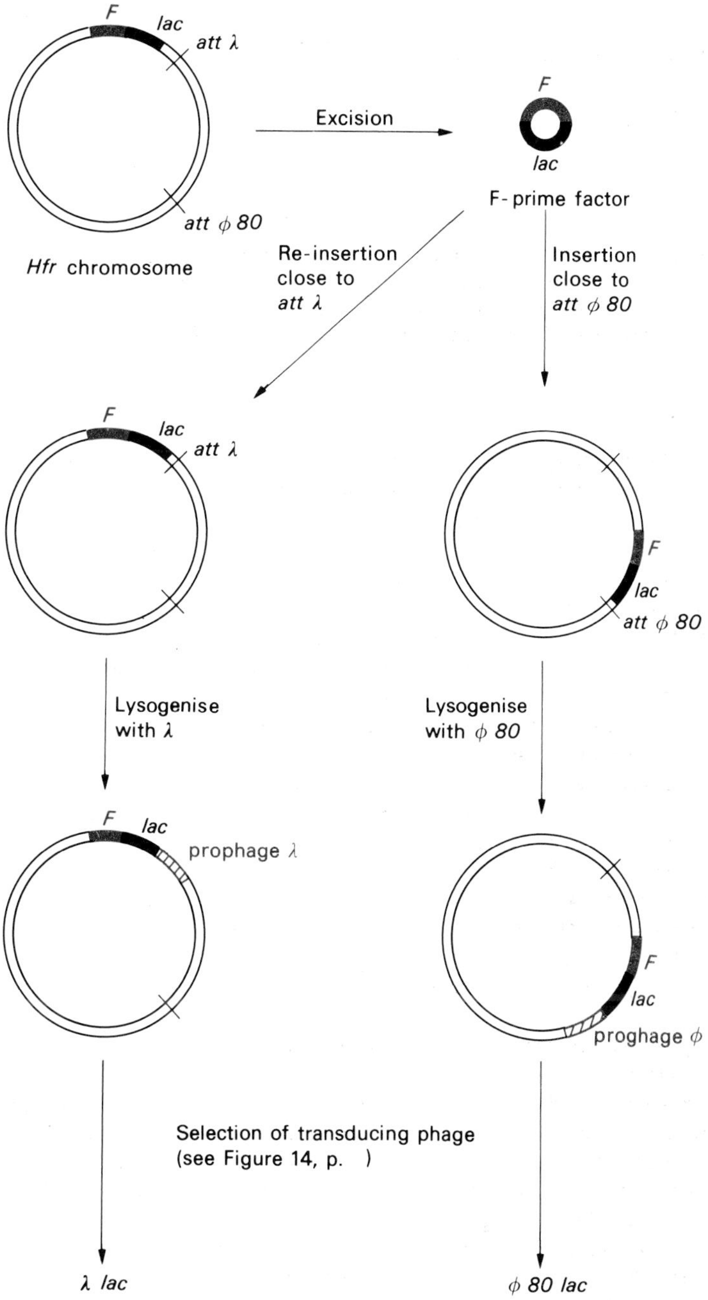

Figure 26. F-mediated transposition of *lac* genes.

An *Hfr* strain of *Escherichia coli* is used to select an *F'* factor carrying the *lac* genes. This *F'lac* can be re-inserted near to the attachment site for phage λ (*att*λ), or it can be inserted into the genome near the attachment site for phage φ80 (*att* φ80). Infection of these strains with the appropriate phage leads to the formation of a λ or a φ80 lysogen. These lysogens can then be used to yield λ*lac* or φ80 *lac* transducing phages.

phage is integrated through the normal POP' site and the event is *int* directed. It is possible to select these lysogens with a λcI⁻ which will kill the non-lysogenic survivors. Double selections can be used to 'direct' phage integration to a particular gene. Induction of these lysogens now yields rare transducing phages carrying DNA adjacent to the secondary site.

DNA manipulation *in vitro*

The methods so far described can be used to manipulate genes either within a single species (usually *E. coli*) or between those species which can act as common hosts for a promiscuous plasmid. They can be extended to genes (or DNA) from any source by using *in vitro* methods to construct hybrid molecules

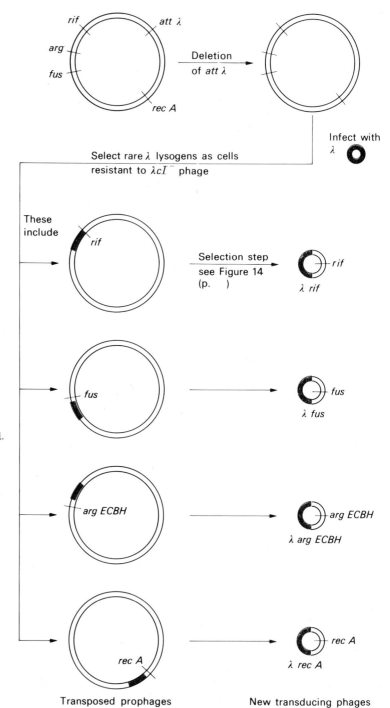

Figure 27. Transposition of λ prophage, and gene integration into phage λ.

Escherichia coli has its attachment
site for λ (*att*) deleted and is
then infected with λ. The rare
lysogens formed by integration of λ
at other sites are selected by their
resistance to phage λcI⁻ which lyses
its host unless that host contains a
normal λ prophage. In some of the
rare lysogens the phage DNA will be
integrated next to a selectable marker
e.g. *rif* or *fus* (conferring
resistance, respectively, to rifamycin
or fusidic acid) or to the operon
for arginine biosynthesis (*arg ECBH*)
or to the *recA* gene controlling recombination. Appropriate λ-transducing
phages can then be selected (for
example, λ*rif* can be isolated by
selecting for a transducing phage
conferring resistance to rifamycin).

containing DNA from the desired source, and from a suitable vector. The vector can be derived either from a plasmid, or from a temperate phage, and its role is to provide the information necessary for maintaining, and replicating, the recombinant DNA once it has been introduced into a suitable bacterial host.

Hybrid molecules can be formed in several ways. The most widely used method is to cut the vector molecule (which is circular) at a single site with a restriction endonuclease, thereby generating a linear DNA molecule with cohesive ends. For example, the enzyme *Eco*RI recognizes and cuts the following base pair sequence at the sites shown by the arrows:

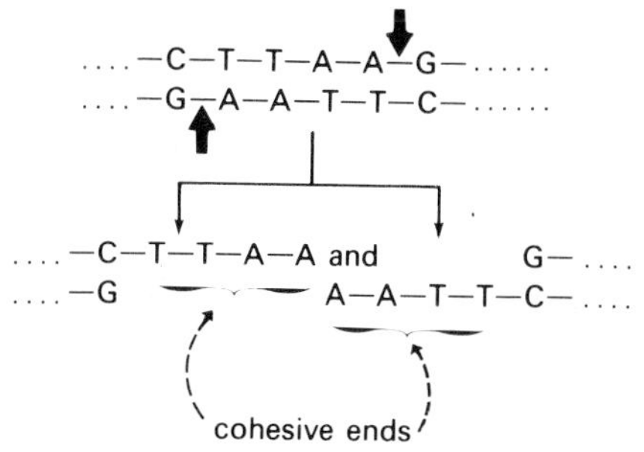

If the bacterial chromosome (or other DNA) is digested with *Eco*RI, fragments with the same cohesive ends are generated, and they can therefore form hybrid molecules by annealing with the *Eco*RI-treated vector. These hybrids are completed by sealing the DNA strands together using the enzyme *ligase*.

Alternatively, the enzyme *terminal transferase* can be used to attach a single-stranded 'tail' of adenine bases, poly(dA), at the 3'-OH termini of the vector. Randomly-sheared DNA fragments from any source can be tailed with a poly(dT) sequence using the same enzyme, and can then be combined into the vector by complementary base-pairing.

Transposition to plasmids

Transposition *in vitro* ideally requires a small plasmid which can be cut at a single position. The derivative pBR322, of the colicin factor Col E1, satisfies these requirements since it is cut at a single different restriction site by each of five different restriction endonucleases (*Eco*RI, *Pst*I, *Hin*dIII, *Bam*HI and *Sal*I). It has a molecular mass of 2.6×10^6 u, and it is now widely used in the study of DNA from bacterial and other sources. Hybrid molecules formed *in vitro* are introduced into *E. coli* by transformation, and bacteria containing the hybrid can be selected by using the resistance to tetracycline, or to ampicillin, conferred by the vector. The restriction site for *Pst*I is located in the ampicillin resistance gene of the plasmid, and the site for the enzyme *Hin*dIII is located in a gene conferring resistance to tetracycline. It is therefore possible to recognize bacteria containing a plasmid carrying recombinant DNA if either of these restriction enzymes was used in constructing the hybrid DNA, since a cut in either resistance gene will inactivate it.

Plasmids are useful for transferring DNA in the size range 0–30 kb, although there is a tendency for smaller sizes in that range to predominate. The application of this technique to the cloning of *Drosophila melanogaster* DNA on plasmid pSC 101 vector is summarized in Figure 28; it has already yielded much information about the genetic organization in *Drosophila* and other organisms.

Transposition to phages

Phage λ vectors containing a single site for a particular restriction enzyme have been developed to enable cloning of DNA fragments generated from any source by the appropriate restriction endonuclease; the method is outlined in Figure 29. Since the capacity of a phage for carrying additional DNA is limited by the amount of DNA which can be packaged into the phage head, the vectors constructed for cloning have been extensively modified to remove the non-essential phage genes, thereby maximizing their capacity for carrying other DNA. The hybrid DNA can be either transformed into *E. coli*, or it can be used in an *in vitro* assembly of infective λ particles from appropriate phage components.

Usually the DNA to be cloned is heterogeneous, for example it often originates from total chromosomal DNA, and therefore one needs to select for the desired recombinant clones. This can be done either by recovering a *lysogen* carrying the desired fragment, or by picking plaques containing the hybrid DNA. To recover lysogens, it is usually necessary to know that the DNA fragment desired from the donor population carries a selectable marker which is expressed in *E. coli* (this is rarely the case for eukaryotic genes).

Plaque-recovery techniques have been highly developed, and several methods can be used to recognize plaques resulting from a desired hybrid. If the pure RNA species (e.g. the mRNA for a single gene) transcribed from the desired gene is available it can be used as a hybridization probe in a process similar to replica plating. Nitrocellulose filters pressed on the surface of a plate carrying about 1000 plaques pick up some phage DNA from the region of the plaque. Alkali treatment of the filter produces single-stranded DNA, and the whole filter can be hybridized with the

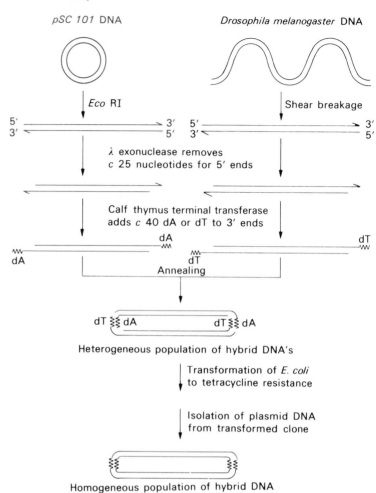

Figure 28. End-tailing method for generating recombinants *in vitro*.

Closed circular DNA of the tetra-cycline resistance plasmid, *pSC 101*, was cleaved with *Eco* RI endonuclease. *Drosophila melanogaster* DNA was sheared to about 8 Kb by stirring. About 25 nucleotides were removed from the 5′ terminus of each DNA by λ exonuclease and about 40 dA or dT residues were added to the 3′ ends of the DNA's with calf thymus terminal transferase. The mixture of DNA's was then annealed to yield only hybrid circles (the generation of homogenous circles is predicted by the end-tailing with dA or dT). These were then used to transform *Escherichia coli* to tetra-cycline resistance. Homogeneous hybrid DNA could then be isolated from the transformed clone. (See P.C. Wensink, D.J. Finnegan, J.E. Donelson, and D.S. Hogness (1974) *Cell*, **3**, 315–325)

radioactively-labelled RNA probe. Those plaques containing the appropriate DNA fragments are very clearly revealed by autoradiography. Another approach which is used to recognize plaques from phage carrying recombinant DNA uses specially-constructed phage vectors. Several of these contain the *E. coli lacZ* gene. In the appropriate Lac⁻ host, these phages will make plaques containing β-galactosidase. This enzyme can be detected by including a chemical indicator such as 5-bromo-4-chloro-3-indolyl-β-D-galactoside which is colourless, but which on hydrolysis by the enzyme yields a deep blue colour. Since there is an *Eco*RI restriction site within the *lacZ* gene (for β-

galactosidase), any vector in which this DNA has been replaced by foreign DNA will give a colourless plaque, rather than a blue one.

Different phage λ vectors have different capacities for carrying additional DNA. The maximum size if all non-essential genes were deleted is about 22 kb, although those commonly in use transfer rather less (0–11 kb for example).

This summary of some of the manipulative techniques now available illustrates how the combination of physical, chemical and biological techniques lends great power to current genetic analysis. For example, it is now a much less daunting undertaking to initiate genetic studies on a totally new bacterium,

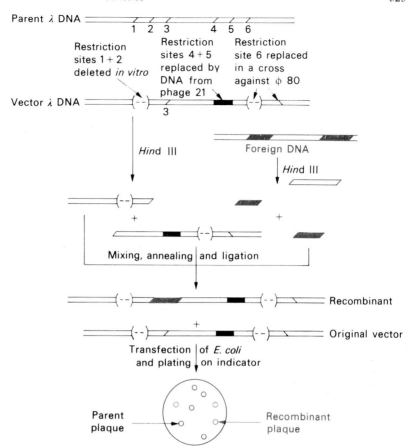

Figure 29. Phage λ as a vector for recombination *in vitro* using restriction enzyme *Hin*d III.

thanks to the discovery and exploitation of promiscuous plasmids. It is also easier to define the major aspects of regulation of a gene through deletion and transposition studies. Heterospecific gene expression can be investigated by transferring genes between species. Finally, DNA from eukaryotic sources can now be incorporated into vectors within the bacterial cell. The study of eukaryotic gene expression in such strains and their exploitation as a high-yielding source of cloned DNA fragments for hybridization *in situ* in the eukaryotic cell has opened wide a window on eukaryotic genetics at the molecular level which will be of great importance in the next decade.

FURTHER READING

1 Drake J.W. and Baltz R.H. (1976) The biochemistry of mutagenesis. *Ann. Rev. Biochem.* **45**, 11–37.
2 Achtman M. and Skurray R. (1977) A redefinition of the mating phenomenon in bacteria. In *Microbial Interactions* (ed. Reissig J.L.): *Receptors and Recognition*, Series B, Volume 3, 233–279. Chapman and Hall, London.
3 Hershey A.D. (ed.) (1971) *The Bacteriophage Lambda*. Cold Spring Harbor Laboratory, New York.
4 Bukhari A.I. Shapiro J.A. and Adhya S.L. (eds.) (1977) *DNA Insertion Elements, Plasmids and Episomes*. Cold Spring Harbor Laboratory, New York.
5 Lewin B. (1977) *Gene Expression 3*. Wiley-Interscience, New York.
6 Old R.W. and Primrose S.B. (1980) *Principles of Gene Manipulation*. Blackwell Scientific Publications, Oxford.
7 Glover S.W. and Hopwood D.A. (eds.) (1981) Genetics as a Tool in Microbiology. *Symp. Soc. Gen. Microbiol.* **31**.
8 Bachman B.J. and Low K.B. (1980) Linkage Map of *Escherichia coli* K-12. *Microbiol. Revs.* **44**, 1–56.
9 Henner D.J. and Hoch J.A. (1980) The *Bacillus subtilis* chromosome. *Microbiol. Revs.* **44**, 57–82.
10 Broda P. (1979) *Plasmids*. W.H. Freeman, Oxford.
11 *Moveable Genetic Elements* (1981) *Cold Spring Harbor Symp. Quant. Biol.* **45**.

Chapter 8
Co-ordination of Metabolism

INTRODUCTION

This chapter is concerned with the regulation of the flow of metabolites through the various degradative and biosynthetic pathways so that no compound is in excessive or inadequate supply. A bacterium might be expected to have acquired through natural selection two major features of metabolic control: the ability to use available nutrients as efficiently as possible, as regards both growth rate and yield, and the capacity to respond rapidly to environmental changes. In this chapter, the selective advantage of the control mechanisms found in bacteria in furthering one or other of these ends will often be self-evident; however, sometimes no plausible explanation will be apparent and we must then remember that an adaptive character may persist in a changed environment where it is selectively neutral.

Control mechanisms may be specific or non-specific, the former acting by stimulating or depressing the flow through a specific pathway, the latter involving a wide variety of pathways or an entire class of macro-molecules. We shall here pay more attention to specific controls, partly because more is known about them and partly because nonspecific controls often relate to growth or differentiation and so are discussed in Chapters 2 and 9. Most of this chapter will therefore discuss patterns and mechanisms of specific controls in Class I and Class II pathways, with a brief mention of Class III reactions.

Specific control mechanisms are of two major types: control (activation or inactivation) of enzyme *activity*, and control, by induction or repression, of enzyme *synthesis*. A minor type, controlled inactivation of a specific enzyme, will be mentioned but is less important in bacteria than in eukaryotic micro-organisms.

CONTROL OF ENZYME ACTIVITY

Here a metabolite though usually unrelated structurally to the substrate binds to the enzyme causing it to gain (*activation*) or lose (*inhibition*) catalytic activity. This is most easily demonstrated *in vitro*, e.g. with cell extracts or solutions of purified enzyme; addition of an *activator* metabolite causes the measured activity to increase, an *inhibitor* causes it to decrease. Activators and inhibitors are referred to as *effectors*, activation or inhibition sometimes as *modulation*. Modulation generally can increase or decrease the affinity of enzyme for substrate, so that a higher substrate concentration is needed in the presence of an inhibitor, and a lower concentration in the presence of an activator, to attain the reaction rate found in the absence of effectors.

The mechanism of control of enzyme activity

Enzymes in which activity is subject to modulation have certain characteristics which have prompted unifying schemes to explain their behaviour. Firstly, the effector binds to the enzyme at an *allosteric* site distinct from the active site. This follows from the fact that effector and substrate are often unrelated chemically or metabolically, and is consistent with the non-competitive kinetics of inhibition. It can sometimes be proven experimentally. For instance, the aspartate carbamoyltransferase of *Escherichia coli* which is *feedback inhibited* (i.e. inhibited by the end product of the reaction sequence involving it) by CTP (Chapter 4, p. 178) is an oligomeric* enzyme made up of two different types of polypeptide chain; on dissociation into monomers, the polypeptide chains of one kind bind substrate while those of the other kind bind effector.

How does the effector work if its binding site is distant from the active site? A possible answer is that binding of the effector results in a change in overall conformation of the enzyme, including the region of the active site; although, as we shall see below, one model suggests otherwise. This is consistent with the finding that exposure of an allosteric enzyme to conditions that destabilize protein folding (e.g. heat, high or low pH values, mercurials) often eliminates modulation while leaving catalytic function unimpaired. For instance, the first enzyme of histidine biosynthesis in

---

* An *oligomeric* protein (or *oligomer*—sometimes called *multimer*) is one whose molecule contains more than a single polypeptide chain (or *monomer*).

Salmonella typhimurium, if allowed to age, retains catalytic activity but loses sensitivity to inhibition by histidine, although histidine can still be shown to bind to it.

A second characteristic of allosteric enzymes is that plots of reaction velocity versus substrate concentration in the presence and in the absence of effector are usually *sigmoid* rather than hyperbolic as expected for classical Michaelis-Menten kinetics (see Figure 1(a)

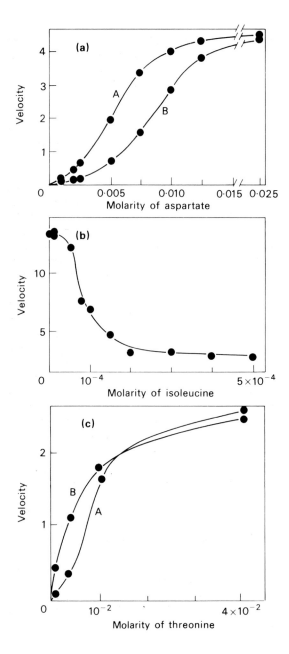

and (b)). This sigmoid appearance indicates a *cooperative effect*; that is, the affinity of enzyme for substrate is not constant but, at low substrate concentrations, increases with substrate concentration. Such a sigmoidal response may be physiologically advantageous in rendering reaction rate highly sensitive to small changes in substrate concentration; this simulates a threshold effect, so that at low substrate concentrations the reaction proceeds very slowly with consequent sparing of substrate. Sigmoidal curves are also obtained if reaction rate is plotted against activator or inhibitor concentration at constant substrate concentration (see Figure 1(b)).

Cooperative effects in which the molecules capable of binding to an enzyme (*ligands*) are identical are said to be *homotropic*, while those produced by different ligands are termed *heterotropic*. Homo- and heterotropic effects clearly have much in common, and the term 'allosteric' is often applied to both types, though where substrates are concerned their effects might in theory involve binding at the active site alone. We shall therefore call compounds capable of showing homo- or heterotropic effects *allosteric ligands*.

It appears that all allosteric proteins are built up of subunits, and aspartate carbamoyltransferase as already mentioned exemplifies this general characteristic of oligomeric structure. However, many allosteric enzymes differ from it in that they contain only one kind of monomer. The binding between subunits is often affected by the presence of allosteric ligands. For instance, threonine dehydratase of *Escherichia coli* is reversibly dissociated in the presence of 1·5 M urea, but the extent of dissociation is affected by threonine, valine or isoleucine, the first two favouring dissociation and the last association.

Figure 1. Oligomeric enzymes.

(a) Sigmoid response of reaction velocity to substrate concentration for aspartate carbamoyltransferase of *Escherichia coli*. Curve A, no inhibitor present; curve B, the inhibitor CTP present at 2×10^{-4} M. From J.C. Gerhart and A.B. Pardee, *Cold Spring Harbor Symposia on Quantitative Biology* **28** (1963) 491.

(b) Sigmoid response of reaction velocity to inhibitor concentration at constant substrate concentration shown by threonine dehydratase of *Escherichia coli*. The concentration of substrate (L-threonine) was 2×10^{-2} M. From J. Monod, J. Wyman and J.P. Changeux, *Journal of Molecular Biology* **12**, (1965) 88.

(c) Effect of the activator valine on threonine dehydratase of *Escherichia coli*. A, no valine added; B, valine present at 3×10^{-3} M. From M. Freundlich and H.E. Umbarger, *Cold Spring Harbor Symposia on Quantitative Biology* **28** (1963) 505.

Various schemes have been put forward to explain the interrelationships between these three characteristics. We shall now describe a particularly influential model, that of Monod, Wyman and Changeux. An interesting feature of this model is that it avoids the assumption that the binding of any molecule to the enzyme *causes* a conformational change; this is replaced by the concept of an equilibrium between molecules in different conformations. An allosteric enzyme is viewed as necessarily oligomeric, with a symmetrical arrangement of identical *protomers*; these may be monomers or some combination of identical or non-identical monomers. Four further postulates are: each protomer bears one binding site for each allosteric ligand; the conformation of a protomer is partly determined by its interaction with other protomers; the enzyme can exist in two or more interconvertible states which differ in the nature of the protomer–protomer interactions, so that the protomers adopt different conformations in the various states and—most important—the affinities of allosteric ligands for their binding sites may differ for the different states; and finally, the symmetry of the enzyme molecule, including the symmetry of interactions between protomers, is conserved in transitions between states. The significance of this last assumption is that it necessitates that all protomers in one enzyme molecule bear the same conformation.

The deductions that can be made from this model may be illustrated qualitatively as follows. Suppose we have an enzyme exhibiting a homotropic effect for a substrate S and heterotropic effects for an activator A and an inhibitor I. Suppose also that the enzyme can exist in two states, such that in one the protomers adopt a conformation with greater affinity for S than in the other (the states of the enzyme corresponding to greater and lesser affinity are conventionally termed 'R' and 'T', for *relaxed* and *tight*, respectively). In the absence of ligands, the R and T states will be in equilibrium. If a small amount of S is now added, there will be an increase in the proportion of enzyme molecules in the R state as opposed to the T state for purely thermodynamic reasons, because the affinity of S is greater for enzyme molecules in the former state than in the latter. This means that a low concentration of S causes the appearance of more available binding sites than were there in its absence (see Figure 2(a)); this constitutes the observed cooperative effect with resultant sigmoid dependence of reaction rate on substrate concentration.

The activator A may be taken to bind to its own allosteric site more strongly when the enzyme is in the R state than when it is in the T state. The inhibitor I shows just the opposite effect; the enzyme has greater affinity for it in the T state than in the R state. Hence A increases the proportion of enzyme molecules in the R state, with consequent enhancement of binding of S and therefore of reaction rate, while I favours the T state, leading to diminution of binding of S and of reaction rate (see Figure 2(b) and (c)). The sigmoid response of reaction rate to effector concentration can be explained in the same way as to substrate concentration.

This elegant model can be used as a basis for quantitative calculations of various enzyme parameters. Although the theoretical inferences can generally be made to agree with observation, there seem to be certain instances where the behaviour of an enzyme is not easily reconciled with the scheme in its strict form. Recently, modifications have been tried, especially with relaxation of the fourth postulate; that is, overall symmetry is not necessarily conserved. This allows for a range of intermediate states between the extreme R and T forms, providing for greater flexibility in fitting experimental results.

A rather different approach is that of Koshland. This replaces the Monod-Wyman-Changeux assumption of alternative conformational states in thermodynamic equilibrium by an *induced-fit* model. Enzyme molecules in the absence of ligands are supposed all to have the same conformation, but the binding of a ligand to a subunit induces a change in subunit conformation. When the ligand is a substrate the new conformation is one that facilitates the proper molecular alignment for the enzyme-catalyzed reaction. Heterotropic activation arises when an allosteric ligand induces the same conformation change as the substrate, heterotropic inhibition when an allosteric ligand holds the polypeptide chain in a conformation unfavourable for substrate binding. Koshland's model also allows for sequential changes in subunit conformation, i.e. the binding of a ligand to a subunit does not necessitate a change in conformation of the remaining subunits. Cooperative effects arise if associations of subunits in the same conformation are energetically more favourable than associations of subunits in different conformations.

Other models have been put forward, some of which involve further broadening of those already outlined, others assimilating parts of each, and yet others introducing different aspects of protein behaviour. Since each enzyme has presumably evolved its pattern of modulation independently, no unified theory that fails to allow for all potentialities of the protein molecule is likely to be able to explain in detail the properties of every enzyme.

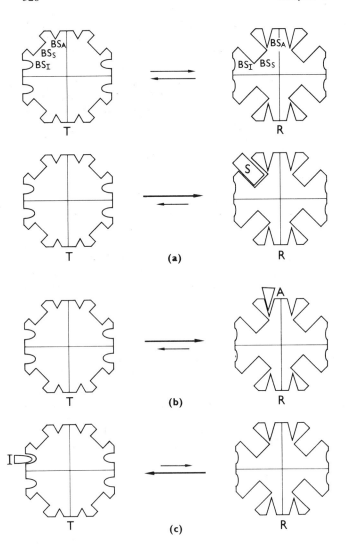

Figure 2. Monod–Wyman–Changeux model for control of enzyme activity.

A tetrameric enzyme is shown that can exist in 'tight' (T) and 'relaxed' (R) conformations. It has binding sites (BS_S) for substrate (S), and allosteric sites for an activator A (BS_A) and an inhibitor I (BS_I). The enzyme has greater affinity for S and A when in the R state, and for I when in the T state.

(a) In the absence of effectors, R and T are in equilibrium. If substrate is present this equilibrium will move towards a preponderance of R, creating more high-affinity binding sites for S so that a positive homotropic effect results.

(b) If activator is present, the $R \rightleftharpoons T$ equilibrium again shifts to favour R; more high-affinity binding sites for S appear and a positive heterotropic effect is observed.

(c) If inhibitor is present, the $R \rightleftharpoons T$ equilibrium alters, this time in the direction of T; the number of high-affinity binding sites for S diminishes, yielding a negative heterotropic effect.

A very different means of controlling enzyme activity involves chemical modification of the enzyme. An example is the glutamine synthetase of *E. coli* and other Gram-negative bacteria, which can be adenylated by the transfer of adenylyl groups from ATP. The *E. coli* enzyme molecule accepts up to 12 such groups (one per monomer), attached to tyrosine residues thus:

$$-\!\!\left\langle\!\!\bigcirc\!\!\right\rangle\!\!-O-\overset{\displaystyle O}{\underset{\displaystyle O}{\overset{\|}{P}}}-O-(5'\text{-C of adenosine}).$$

Adenylation is activated by glutamine and inhibited by α-oxoglutarate. The adenylated enzyme shows altered kinetic properties and affinities for divalent cations, and is more sensitive than the unadenylated enzyme to feedback inhibition by tryptophan, histidine, AMP and CTP, but less so by glycine and alanine. Removal of the adenylyl groups (which is not simply the reverse of the adenylation reaction, since the groups are liberated as ADP) is catalyzed by a separate enzyme activated by α-oxoglutarate and inhibited by glutamine.

CONTROL OF ENZYME SYNTHESIS

Rapid and specific alterations in the rates of synthesis of particular proteins consequent on changes in environment are a striking characteristic of enzymes catalyzing Class I and II reactions in most bacterial species. This type of regulation is entirely distinct from that exerted at the level of enzyme activity, and the description of its mechanism involves genetics as well as biochemistry.

Induction and repression

It is often found that the presence of a particular substance in the medium causes an increase (*induction*) or decrease (*repression*) in the cells' rate of synthesis of an enzyme relative to total protein synthesis. If the rate of synthesis is little affected by the concentration of any compound the enzyme is said to be *constitutive*. Because of the gap that often persists between our knowledge of events at the molecular as opposed to gross descriptive level, there arise problems of terminology. The compound whose addition to the medium causes induction is often initially called the *inducer*; later, it may be found that not this compound, but some derivative, is the active one in the molecular mechanism of induction, whereupon the latter may be characterized in some other way, e.g. as the 'true inducer', or the term 'inducer' dropped for the former. The word 'repressor' has a special meaning that precludes its use for a substance that causes repression (see below); *co-repressor* is (for reasons that will become clear) used for the compound active at the molecular level, and the substance whose addition to the medium results in repression may be referred to as, for instance, the 'repressing metabolite'.

The kinetics of induction and repression may be quantified as follows. The specific activity of an enzyme, in units of enzyme activity per unit of cell protein, may be determined for cultures grown under the appropriate different conditions. A better approach, which also allows one to look at the kinetics of induction or repression, uses the *differential plot* introduced by Monod. This is based on the fact that during balanced growth in a constant environment, for instance during early exponential growth in batch culture or in continuous culture, cell proteins are synthesized in fixed proportions. The *differential rate of synthesis* of an enzyme is defined as the fraction which it constitutes of total newly-synthesized protein. To obtain a differential plot, samples are taken at intervals during growth of a batch culture and enzyme activity and protein content determined for each. Units of enzyme activity per unit volume of culture are then plotted against mass of protein per unit volume. While culture conditions remain constant, a straight line is usually obtained with a slope that measures the differential rate of synthesis. Towards and following the end

of exponential growth the plot may deviate from linearity as metabolites (which may themselves induce or repress) accumulate in the medium. If a compound that induces or represses the enzyme is added during growth of the culture, the plot will show two linear regions with slopes that give the different rates of synthesis in the absence and presence of the controlling compound.

Figure 3(a) shows a differential plot for the inducible *E. coli* enzyme β-galactosidase, which hydrolyses lactose to glucose plus galactose and is therefore necessary for lactose catabolism. The enzyme is *induced* in the presence of lactose or certain structurally related compounds. Some of these inducers are not metabolized, in which case they are said to be *gratuitous*; examples are various synthetic alkyl thio-β-D-galactosides, such as the *iso*propyl derivative, 'IPTG'. In Figure 3(a), IPTG has been added at a point during growth, leading to induction of the enzyme; the dashed line indicates that if it is removed a short time later, the effect of the inducer rapidly ceases. Figure 3(b) shows a similar plot for the arginine-repressible *E. coli* K12 enzyme ornithine carbamoyltransferase, which is involved in arginine biosynthesis (Chapter 4, p. 169). Following addition of the amino acid to a growing culture, the differential rate of synthesis quickly changes to the repressed level; the dashed line shows that if the arginine is removed, release from repression (or *derepression*) again rapidly ensues.

The ratio of the fully induced to non-induced rates of enzyme synthesis is termed the *induction ratio*, and of the fully derepressed to fully repressed rates of synthesis the *repression ratio*.

Co-ordinate control
The enzymes of a pathway are often induced or repressed in parallel. When β-galactosidase is induced, two other proteins behave likewise; galactoside permease, which transports lactose and many related substances (including thio-β-D-galactosides) into the cell, and β-galactoside transacetylase, whose physiological role remains obscure. Similarly, in *E. coli* K12, arginine represses the other seven enzymes of arginine biosynthesis in addition to ornithine carbamoyltransferase. A group of enzymes that are induced or repressed in parallel, though not necessarily co-ordinately, is sometimes called a *regulon*. The proteins of lactose catabolism show *co-ordinate control* in that their rates of synthesis are in a constant ratio under all growth conditions; however, of the arginine biosynthetic enzymes, some show co-ordinate control but not all. In particular, ornithine carbamoyltransferase is derepressed by a factor of 100

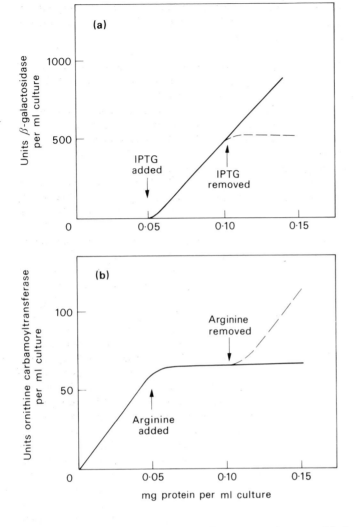

Figure 3. Kinetics of induction and repression.

(a) Differential plot of β-galactosidase activity against protein concentration for a culture of *Escherichia coli* growing exponentially in minimal salts-maltose medium. The gratuitous inducer IPTG was added to give a final concentration of 10^{-4} M. The specific activity of β-galactosidase (given by the slope of the linear plot) in the presence of inducer is about 10^4 units/mg protein; in the absence of inducer it is about 10 units/mg, which is not discernible on this scale. The dashed line indicates the activity when a sample of the culture is filtered and the cells grown further in the absence of IPTG.

(b) Differential plot of ornithine carbamoyltransferase activity against protein concentration for a culture of *Escherichia coli* growing exponentially in minimal salts-glucose medium. Arginine was added to repress enzyme synthesis. The specific activity in the absence of arginine is about 1200 units/mg protein; in the presence of arginine it is about 1·5 units/mg, which is not discernible on this scale. The dashed line indicates the activity in a sample when the culture is filtered and the cells grown further in the absence of arginine.

when the cells are transferred from medium containing arginine to medium lacking it, while none of the other enzymes are derepressed to the same degree (most of them by a factor of about 10). A possible explanation is that ornithine carbamoyltransferase competes for one of its substrates, carbamoyl phosphate, with the enzyme aspartate carbamoyltransferase, which catalyses the first step in pyrimidine synthesis (see Chapter 4, p. 180), so that under conditions of low intracellular arginine it is necessary to have a high level of ornithine carbamoyltransferase to ensure that sufficient carbamoyl phosphate is channelled into the arginine pathway. Of course, the reverse should be true also, and in fact aspartate carbamoyltransferase is derepressed in the absence of uracil or cytosine far more than the other pyrimidine biosynthetic enzymes.

Mechanisms of specific control of protein synthesis

The elucidation of several mechanisms of control of gene expression has (beginning with the *operon model* of Jacob and Monod) been one of the triumphs of molecular biology. The operon model was put forward to describe the lactose catabolism system of *E. coli* already described. Later work, on this system and others, shows that there exist a number of possible molecular mechanisms for regulation of bacterial gene expression. Most systems show some combination of these. The operon model in its original form describes one of them, the *repression control* mechanism. Other common mechanisms may be termed the *activation control* mechanism and the *termination/anti-termination* mechanism. We shall consider here in detail the *E. coli* K12 lactose and tryptophan biosynthetic systems, which are probably the most completely understood and also exemplify all three mechanisms, discussing briefly other systems to show the variety of ways in which these mechanisms may be combined.

The *E. coli* lactose system

We begin by describing the physiological and genetic characteristics of this system. To recapitulate: there are three proteins involved, β-galactosidase, galactoside permease and β-galactoside transacetylase. These proteins are induced co-ordinately by a variety of compounds structurally related to lactose, some of which (e.g. IPTG) are 'gratuitous' inducers. It has recently been found that lactose itself is not a 'true' inducer; it causes induction through being transformed intracellularly by β-galactosidase (which is essentially a β-D-galactosyl-transferase rather than a hydrolase) into the 'true' inducer, *allo*lactose, in which the glucose and galactose units are attached 1 to 4 rather than 1 to 6. The system when 'induced' by lactose therefore shows *product induction*: i.e. the true inducer is the product of the first enzyme acting on a substrate, rather than the substrate itself. Its occurrence requires an appreciable *basal* (i.e. non-induced) level of the first enzyme.

In addition to induction, a second physiological control acts on the rate of synthesis of the lactose proteins, namely *catabolite repression*. It is found that the maximum induced rate of synthesis of the lactose proteins depends on the growth rate permitted by the medium in which the cells are growing: the faster the cells grow, the lower the maximum induced level of synthesis. Induction and catabolite repression act independently of one another. As an illustration we can compare the activities of β-galactosidase in cells which have been grown rapidly with glucose, which is a good carbon source, and those which have been grown slowly with maltose, a poor carbon source. The basal activity of uninduced cells in glucose medium may be 0·1 unit/mg protein. If IPTG is added and the culture is grown further the maximum activity will be about 100 units/mg protein. In the medium containing maltose, there will be very little catabolite repression and the corresponding values will be about 10 times greater, i.e. about 1 unit/mg for the basal activity and 1000 units/mg after induction.

The term 'catabolite repression' was suggested by Magasanik in view of its non-specific nature. He pointed out that rather than the carbon source acting directly, it was more likely that the relative availability of metabolites produced from it by catabolism caused the effect.

The structural genes for the three lactose proteins (*lacZ* for β-galactosidase, *lacY* for the permease, and *lacA* for the transacetylase) are *clustered*, i.e. they form a contiguous block on the chromosome, the order being *Z-Y-A* moving anticlockwise with the conventional orientation of the genetic map of *E. coli* (see Chapter 7, p. 293). It can be shown that the genes are transcribed to a single RNA molecule, i.e. a *polycistronic* mRNA. The promoter region of the DNA (*lacP*) at which the RNA polymerase holoenzyme binds preparatory to transcription (see Chapter 6, pp. 270–272), lies to the *lacZ* side of the cluster, so that the genes are transcribed in the order *Z-Y-A*. Part of the evidence for this will be mentioned later in this chapter under 'polarity' (p. 338).

A very important question we can now consider is: at what stage—transcription, translation, or assembly of preformed peptides—does specific control of protein synthesis occur? That the last-named step is not involved was shown for β-galactosidase by adding inducer to a culture together with radioactively labelled sulphate. The enzyme showed maximum specific radioactivity as soon as it appeared, demonstrating that it is synthesized very rapidly *de novo* from amino acids rather than assembled from peptide precursors. Later experiments measured levels of *lac* mRNA in cells grown with various degrees of induction and/or catabolite repression. The *lac* mRNA was estimated as amount of pulse-labelled RNA capable of hybridizing (Chapter 5, p. 203) with denatured (i.e. single-stranded) DNA—usually immobilized on a nitrocellulose membrane—containing the *lac* cluster. For adequate sensitivity the DNA 'probe' must be rich in *lac* sequence; suitable DNA molecules are F' *lac* and λ*lac* genomes (see Chapter 7, pp. 318–319). It was found that the level of *lac* mRNA determines the rate of synthesis of the *lac* proteins, indicating that the regulation of synthesis is exerted at the stage of transcription rather than translation. The susceptibility of mRNA to degradation explains the rapidity of the regulatory response, in particular during de-induction or repression (see Figure 3).

It is known that the rate of movement of RNA polymerase along the DNA, like that of ribosomes along mRNA, is constant for a given set of growth conditions. Control at the level of transcription, therefore, must reflect specific control of the effective *frequency of initiation* of transcription of the *lac* genes. We will now describe how this is achieved through two of the three control mechanisms mentioned.

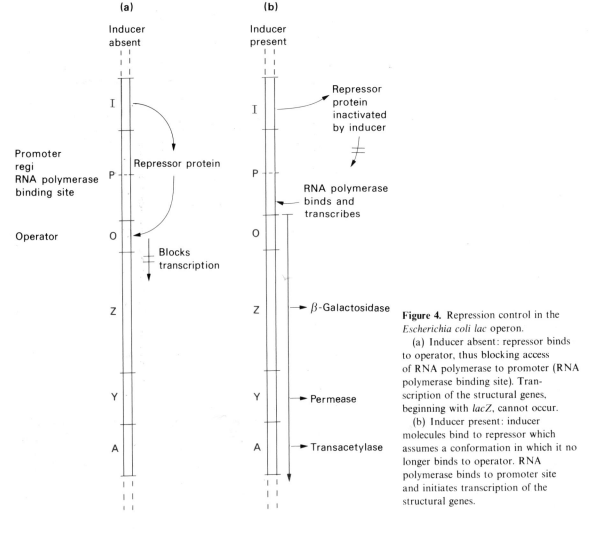

(a)
Inducer absent

(b)
Inducer present

I

Promoter regi
RNA polymerase binding site   P

Operator   O

Repressor protein

Blocks transcription

Z

Y

A

Repressor protein inactivated by inducer

RNA polymerase binds and transcribes

β-Galactosidase

Permease

Transacetylase

Figure 4. Repression control in the *Escherichia coli lac* operon.

(a) Inducer absent: repressor binds to operator, thus blocking access of RNA polymerase to promoter (RNA polymerase binding site). Transcription of the structural genes, beginning with *lacZ*, cannot occur.

(b) Inducer present: inducer molecules bind to repressor which assumes a conformation in which it no longer binds to operator. RNA polymerase binds to promoter site and initiates transcription of the structural genes.

Repression control in the *lac* system

We must here introduce two further DNA sequences involved in regulation of expression of the *lac* genes, namely *lacI*, the *regulator gene* ('I' standing for inducibility), and *lacO*, the *operator* region (see Figure 4). *lacI*, which maps to the left of *lacP* (the cluster customarily being written *lacPZYA*), is the structural gene for a polypeptide which aggregates to give a homotetrameric protein, the lactose *repressor*. *lacO*, which maps within, and to the right-hand side of, *lacP*, is the lactose *operator*. The essential feature of this control mechanism is that the repressor, which binds fairly strongly in a non-specific manner to DNA is capable of binding extremely tightly to the operator, and that *lacO* does not code for a protein but is a binding region. When repressor is bound the promoter is inac-

cessible to RNA polymerase and so transcription is blocked. The repressor, however, is an allosteric protein existing in two forms, only one of which binds the operator. The repressor can also bind small molecule effectors which stabilize one conformation or the other. True inducers, such as *allo*lactose or IPTG, favour the operator-non-binding conformation; in the absence of effectors, or in the presence of others such as *o*-nitrophenyl-β-D-fucoside (ONPF), the operator-binding conformation is favoured. Hence in the absence of inducer, transcription of the *lac* genes occurs at a very low rate which is increased by addition of inducer. ONPF acts as a competitive antagonist of induction. This picture of repression control is depicted in Figure 4.

The 'operator', a new concept when proposed by

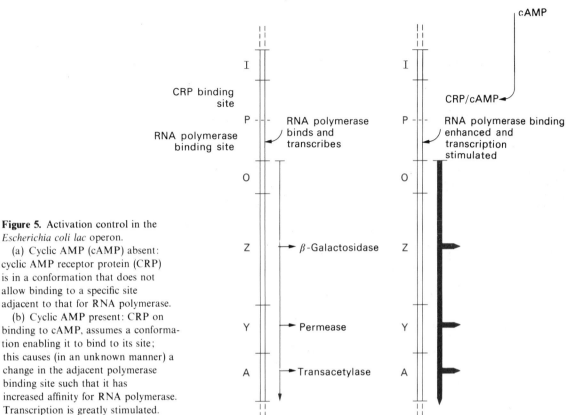

Figure 5. Activation control in the *Escherichia coli lac* operon.

(a) Cyclic AMP (cAMP) absent: cyclic AMP receptor protein (CRP) is in a conformation that does not allow binding to a specific site adjacent to that for RNA polymerase.

(b) Cyclic AMP present: CRP on binding to cAMP, assumes a conformation enabling it to bind to its site; this causes (in an unknown manner) a change in the adjacent polymerase binding site such that it has increased affinity for RNA polymerase. Transcription is greatly stimulated.

Jacob and Monod in 1961, gave its name to the *operon*, a gene or gene cluster transcribed as a single unit under the control of an operator. The *lac* cluster is thus frequently referred to as the *lac* 'operon'. We will comment on this term later.

Activation control in the *lac* system
Paradoxically, catabolite repression turns out to involve a system for stimulation of transcription that is often referred to as 'activation'. The relevant genetic elements are: the *crp* gene (unlinked to the *lac* cluster), which is the structural gene for the *cyclic 3′,5′-AMP receptor protein* (CRP); and an as yet unnamed DNA region (frequently taken to be part of the promoter), lying between *lacI* and *lacP*, that constitutes a binding site for CRP. The CRP again may assume two conformations, in which it respectively does or does not recognize its DNA binding site. In the presence of the effector, cyclic 3′,5′-AMP (cAMP), the DNA-binding form is favoured; in its absence the non-binding form is favoured. Binding of the cAMP-CRP complex to the DNA binding site in some way enhances the affinity of the promoter for RNA polymerase. Also, for reasons that are still unclear, the intracellular level of cAMP correlates inversely with

growth rate. The result is that when *E. coli* is growing rapidly with glucose as carbon source, the intracellular cAMP pool is low, most of the CRP does not have cAMP bound and is therefore unable to bind to the CRP binding site, and hence the promoter shows its basal level of affinity for RNA polymerase; whereas when the cells are growing more slowly, e.g. with maltose as carbon source, the intracellular cAMP is elevated, most of the CRP can bind cAMP and therefore also the CRP binding site, and the promoter consequently shows an enhanced affinity for RNA polymerase. This mechanism is illustrated in Figure 5. Addition of cAMP to a catabolite-repressed culture will tend to reverse the catabolite repression.

Evidence for the control mechanisms of the *lac* operon

Regulatory gene mutations of the *lac* operon
All the controlling elements mentioned—promoter, repressor, operator, CRP, CRP binding site—may be altered by mutation. Properties of some of the most informative mutants are detailed below.

Repressor gene mutations

I^- mutants produce an inactive repressor that can no longer bind to the operator. Transcription occurs at the level corresponding to full induction, even in the absence of inducer. These mutants are said to be *constitutive*.

Partial diploid *E. coli* strains can be constructed that possess two copies of the *lac* operon, one on the chromosome and the other on, for instance, an F' (F-prime: Chapter 7, p. 298) carrying this region of the chromosome. If each copy carries a different *lacI* allele, the dominance relationship of the alleles can be determined. It is found that most I^- alleles are recessive to I^+. This was historically an important indication that the *lacI* product is a *negatively-acting* repressor—that is, the protein product prevents functioning of the *lac* operon. In the partial diploid, containing the non-functional I^- product and the active I^+-specified repressor, the latter would be effective unless inducer were present (Figure 6(a)). The alterna-tive, that the *lacI* product is a *positively-acting* activator (or, in early discussions, an enzyme involved in synthesis of an internal inducer) would lead to the opposite prediction that I^- would be dominant to I^+ (Figure 6(b)).

Another type of I mutant is the I^S. These mutants are *super-repressed*, in that they are uninducible (though low levels of induction can sometimes be produced by high inducer concentrations). The mutant I^S product is able to bind to the operator but unable to bind inducer. As expected, I^S is dominant to I^+ or I^-.

Operator mutations

In O^c mutants, the operator is altered so that it binds repressor more weakly than in the wild-type O^+. The mutants are therefore *semi-constitutive*, expressing the *lac* operon in the absence of inducer at levels inter-mediate between the induced and non-induced wild-type.

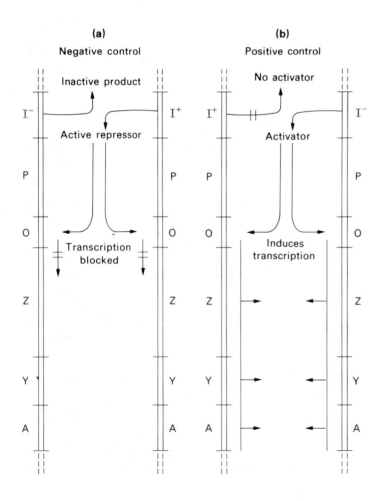

Figure 6. Consequences of negative and positive control models.

(a) *Negative control.* The I^+/I^- diploid contains functional repressor product of I^+ and inactive I^- product; in the absence of inducer the former is able to prevent transcription irrespective of the presence of the I^- product, i.e. inducibility is dominant.

(b) *Positive control.* In the I^+/I^- diploid in the absence of exogenous inducer, I^- determines the production of an activator and I^+ does not. Hence transcription occurs even without exogenous inducers, i.e. constitutivity is dominant.

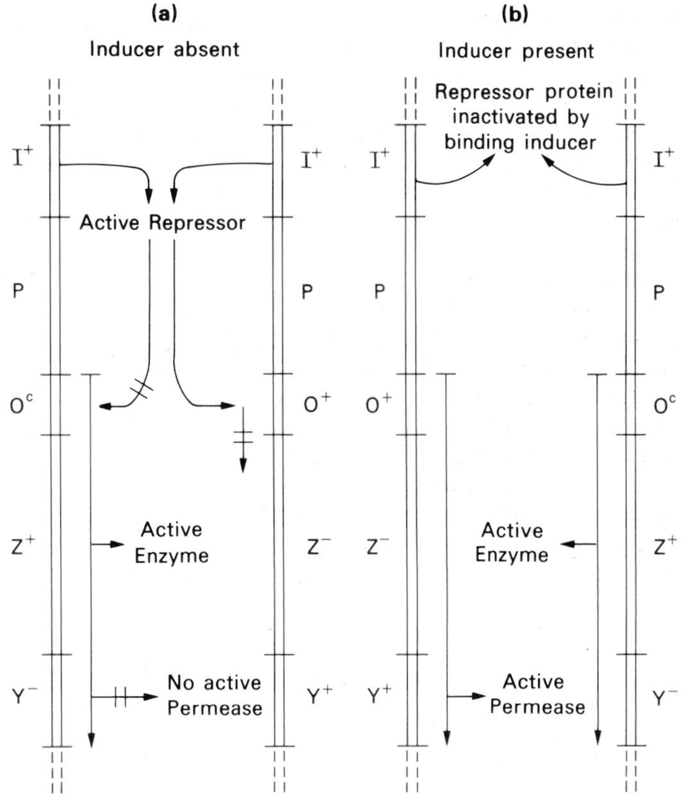

Figure 7. *Cis*-dominance of O^c mutation in partial diploid $I^+O^cZ^+Y^-/I^+O^+Z^-Y^+$.

(a) In the absence of inducer, repressor acts at O^+ but cannot bind to O^c. *Cis*-dominance of O^c means that it affects only those cistrons on the same genome as itself, i.e. Z^+ and Y^- are expressed but Z^- and Y^+ are not. Hence active β-galactosidase is made but permease is not.

(b) Inducer binds to repressor and abolishes repression at O^+, allowing expression of Y^+ as well as Z^+. Thus permease is inducible whereas β-galactosidase is constitutive.

A characteristic of O alleles, unique when they were first characterized, is *cis*-dominance: that is, in partial diploids each O allele exerts its effect on the expression of the *lac* operon adjacent to it irrespective of any other O allele in the cell. For instance, in a partial diploid O^c Z^+ Y^-/F' O^+ Z^- Y^+, β-galactosidase activity is constitutive but permease activity is inducible (Figure 7). *Cis*-dominance is to be expected of control regions in DNA.

Mutations in *crp*

These mutants make inactive CRP and so appear to be 'catabolite repressed' even at slow growth rates. There are also mutants deficient in adenyl cyclase, the product of the *cya* gene, which cannot produce cAMP and so show a similar phenotype to *crp* mutants. The difference between them is that addition of cAMP to the medium will remove the effect of a *cya* mutation but not of a *crp* mutation.

CRP-binding site mutations

These are still often described as promoter mutations. Ability to bind CRP ($+$cAMP) is partially or completely lost, so that again the mutant appears permanently catabolite repressed.

Promoter mutations

These can be of three types: (i) the affinity for RNA polymerase may be diminished ('down-promoter' mutation), so that the level of expression of the *lac* operon is reduced by a constant factor under all growth conditions; (ii) the affinity for RNA polymerase may actually be enhanced ('up-promoter' mutation) and the level of expression is likewise increased by a constant factor; or (iii) the promoter may mutate so as to allow efficient RNA polymerase binding without CRP ($+$cAMP) having bound at the CRP-binding site. The mutant appears then to be partially or completely resistant to catabolite repression. Promoter and CRP-binding site mutations are *cis*-dominant.

In vitro transcription of the *lac* operon

Systems for *in vitro* transcription of the *lac* operon are of two types. Both types require a suitable *lac*-containing DNA template, such as a λlac (see Chapter 7, p. 319), RNA polymerase, and ribonucleoside triphosphates. In the first type, these compounds suffice, and incorporation can be measured if one of the ribonucleoside triphosphates is radioactive and the *lac* mRNA produced is assayed by hybridization with a suitable denatured DNA. In the second, the *lac* mRNA produced is also translated: a complex mixture of ingredients is thus required, including ribosomes, amino acyl tRNA's, and protein factors for translation. Transcription of the *lac* operon is then detected by assay of the β-galactosidase produced.

In both cases, *lac* transcription decreases on addition of repressor, and this decrease is prevented by inducer. Production of *lac* mRNA increases, on the other hand, if CRP and cAMP are added together, though neither is effective alone. When mutationally altered components are used, the results are as predicted from the *in vivo* properties of the mutants. The picture derived from *in vivo* results has thus been confirmed to a remarkable degree by *in vitro* studies.

Sequence of the *lac* DNA control region

Over the past few years rapid progress has been made in methods for determining DNA sequences and also for isolating small purified homogenous DNA fragments suitable for sequencing. Techniques for the former are mentioned in Chapter 5, p. 195. Techniques of isolation include: (i) 'heteroduplexing' of DNA chains complementary only in the region of interest (this is the method used in the first isolation of part of the *lac* operon: Chapter 5, pp. 205–208): (ii) protection of a DNA region by binding a specific protein to it, then digestion of all other DNA with DNAase—this was initially employed to isolate the *lac* operator obtained as the *lac* repressor-protected region; and (iii) isolation of restriction endonuclease-cleaved fragments (Chapter 5, p. 195).

The sequence of the *lac* control region was shown in Chapter 6. Some interesting regularities are to be seen. Two of the protein-binding regions, the CRP-binding site and the operator, contain what is often termed a *palindromic* sequence, but is more properly referred to as a sequence possessing a 2-fold rotational axis of symmetry. The function of this type of symmetrical sequence (found in many DNA binding sites for specific proteins) is obscure. It theoretically permits base-pairing within each DNA chain, producing a cruciform structure (illustrated in Figure 8 for the *lac* operator), but the significance of this is disputed. It could reflect some necessary symmetry property of the binding protein molecules. Or it might just represent the form adopted by a sequence that is recognized by protein molecules moving along the

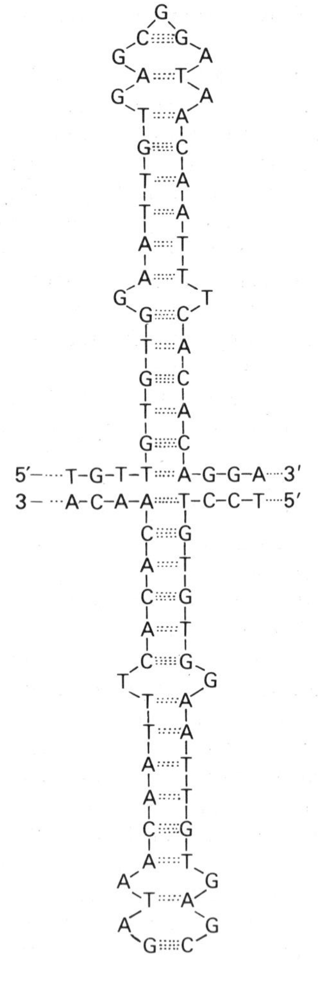

Figure 8. Hypothetical looped-out or 'cruciform' conformation of the *Escherichia coli lac* operator. (See also Figure 24, p. 272.)

DNA double helix in either direction. The promoter, on the other hand, does not show such a symmetry: instead, it displays an A-T rich region flanked by two G-C rich regions. The lack of a 2-fold rotational symmetry may reflect the fact that RNA polymerase, after attaching at the promoter, transcribes one strand in one direction only.

The detailed molecular interactions involved in recognition of these DNA regions by their binding proteins are not understood. This represents one of the major problems still to be solved in the *E. coli lac* control system.

The *E. coli* tryptophan biosynthetic system

This pathway is depicted in Chapter 4, Figure 1, and its regulatory characteristics in *E. coli* are described on pp. 160–163. The genes form a cluster, *trpEDCBA*, with a promoter at the *E* end from which the genes are transcribed on to a single polycistronic mRNA molecule. The system is repressible, in that expression of the cluster appears to be determined solely by the intracellular concentration of tryptophan; there appears to be no equivalent to the catabolite repression of the *lac* operon. Tryptophan represses the enzymes in cells growing in minimal medium; in the absence of exogenous tryptophan, the enzyme levels correspond to a certain extent of repression exercised by the endogenous tryptophan pool. If the latter decreases, a further degree of derepression ensues. The amount of tryptophan in the pool can be affected in various ways: for instance by a 'leaky' mutation in one of the tryptophan biosynthetic enzymes, which leads to restricted flow through one step of the pathway; or by addition of indole 3-propionate, a tryptophan analogue which probably binds competitively with tryptophan to apo-repressor (see below) to give a complex inactive in repression.

Tryptophan exerts its effect *via* two distinct mechanisms. The first is a repression-type control similar to that in the *lac* system. The second is totally different in that control is exerted, not on the frequency of initiation of transcription of the cluster by RNA polymerase, but on the frequency with which the RNA polymerase crosses a 'barrier' that lies between the promoter and the first structural gene.

Repression control in the *trp* system

Here again, as in the *lac* system, there is a regulator gene, *trpR*, and an operator, *trpO*. *trpR* is unlinked to the *trpEDCBA* cluster. This arrangement differs from that in the *lac* system (though the linkage of *lacI* to the *lacZYA* cluster is not functionally significant). Its product is not an active repressor in the absence of tryptophan; on binding tryptophan, it adopts the conformation of active repressor, and will then bind to the tryptophan operator. In its inactive state (i.e. uncomplexed with tryptophan), the *trpR* product is termed the *apo-repressor*; tryptophan is said to be the *co-repressor*. As in the *lac* system, *trpO* lies partly within and partly to the right of the promoter, *trpP*. The means by which the repressor (i.e. apo-repressor + tryptophan complex) blocks transcription is the same as for the *lac* operon; namely, bound repressor excludes RNA polymerase from the promoter, which therefore presumably overlaps the

operator. The presence of an operator enables one to use the term 'operon' for the *trp* cluster without ambiguity. One may here discuss the use of the term 'operon'. From our discussion of the *lac* and *trp* systems, it is clearly possible to have a gene cluster that is transcribed on to a single polycistronic mRNA but which is not regulated by repression control, i.e. does not possess an operator. Should such a cluster be called an operon? Although strictly the answer should be 'no', the term now seems to be used to indicate a gene cluster transcribed as a unit, and it is in this more general sense that we shall use it.

Constitutive *trpR* mutants are known, corresponding to the *lacI⁻* mutants of the *lac* operon; *trpOᶜ* mutations, whose mutant operator fails to bind repressor, are *cis*-dominant like *lacOᶜ*.

Termination/anti-termination control in the *trp* operon

A further control is exerted on the frequency with which transcription events, initiated at the promoter, in fact reach the beginning of the first structural gene. The *trp* transcript starts with 162 ribonucleotides before the AUG corresponding to the beginning of the *trpE* transcript: this region is called the *leader sequence*. However, under most conditions 80–90% of transcripts initiated terminate after about 140 ribonucleotides. The DNA sequence which can apparently act as a signal for termination of transcription is called the *attenuator*.

A component for transcription termination at the attenuator may be the protein called *rho factor*. In some mutants with lowered rho factor activity, virtually all transcription continues past the attenuator into the structural genes. The overall picture of transcription termination is shown in Figure 9. The role of rho in termination at the attenuator is unclear but rho is certainly necessary for transcription termination at the end of the *trp* operon.

Under conditions where the endogenous trytophan pool is very low, transcription does not stop at the attenuator even in the presence of rho. The mechanism for this sensitivity of termination to the level of trytophan is discussed in more detail at the end of this Chapter (pp. 352–3). However, an indication is provided by the observation that mutants with a lower activity of tryptophan tRNA synthase also show reduced attenuation. It seems likely that either charged tryptophan tRNA is required for efficient termination, or uncharged tryptophan tRNA antagonizes termination.

The *trp* leader sequence has been determined and is shown in Figure 9. Interesting features are the G-C and A-U rich region (derived from DNA with dyad

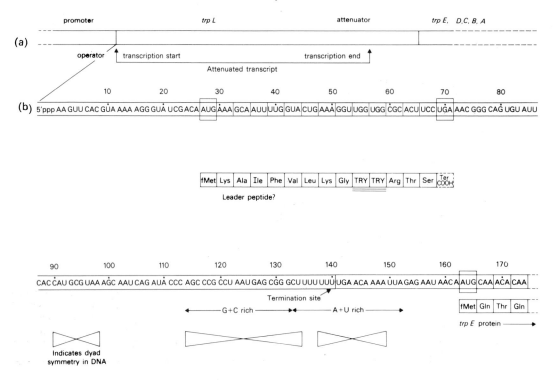

Figure 9. Attenuation control in the *trp* operon of *Escherichia coli*.

(a) The control region of the operon.

(b) Transcription from the leader region (*trpL*) together with postulated translation to give a 'leader peptide' of 14 amino acid residues. At high concentrations of tryptophan, 80–90% of the transcripts end at the termination site defined by the attenuator; at low concentrations there is complete readthrough into the structural genes *trpE, D, C, B*, and *A*, and the polycistronic mRNA is translated into the enzyme proteins for tryptophan synthesis.

Note the two codons for tryptophan (54–56 and 57–59) in the attenuated transcript. It is believed that they have a role in attenuation (see end of chapter, p. 352).

symmetry) and the sequence of eight U's with which the attenuated transcript ends. That the last corresponds to the attenuator itself is suggested by the properties of deletion mutants with deletions penetrating various distances into the leader sequence from the *trpE* site. Deletions that do not reach the U_8 sequence result in wild-type attenuation; those which remove the entire sequence result in no attenuation; while one in which just four of the U's are removed results in an intermediate degree of attenuation.

Polarity

A feature of operons in *E. coli* and *Salmonella* spp. is *polarity*. In this phenomenon a mutation (such as a 'nonsense' mutation, or a frameshift that gives rise to a nonsense triplet in the altered reading frame: see Chapter 7, p. 281) produces premature polypeptide

chain termination. This allows normal expression of genes between the promoter and the mutation, whereas genes on the promoter-distal side are poorly expressed. The effect is useful (in *E. coli* and *S. typhimurium*, at least, the only organisms in which it is known to occur) in establishing the operon nature of a gene cluster and the direction of its transcription. The effect presumably results in the first instance from detachment of ribosomes from the nascent mRNA molecule as they reach the nonsense triplet, but exactly how this produces polarity is still not clear. The most puzzling features are: the widely varying extent of the polar effects of different chain-terminating mutations (which tend to correlate with position within the gene, those closest to the beginning of a gene having the strongest polar effect); and the observed involvement of transcription as well as translation—the level of mRNA corresponding to the

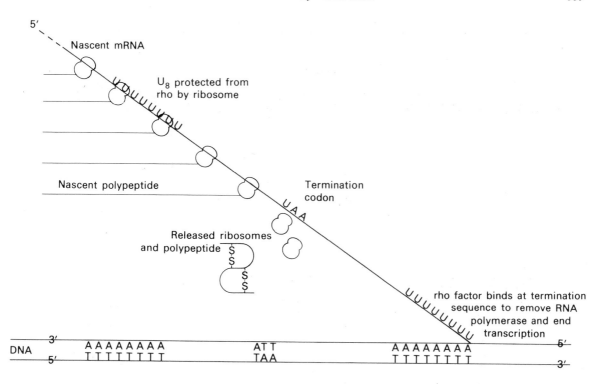

Figure 10. A possible mechanism for termination of transcription. Ribosomes dissociate from the nascent mRNA they are translating when they reach a termination codon (UAA, UAG or UGA) at the end of a cistron in a wild-type or a 'nonsense codon' (same three triplets) resulting from mutation. The following region of mRNA is being transcribed by RNA polymerase but is unprotected by ribosomes. Once a transcription termination sequence (shown here as 8 U's) has been synthesized, it is presumed that rho factor can bind to it and, in an unknown way, remove the polymerase and terminate transcription.

A similar U_8 sequence shown preceding the UAA codon would be inaccessible to rho as it would be protected by ribosomes.

The same mechanism may explain termination of transcription in the leader sequence (attenuation) of the *trp* operon except that here there may be no protected region of mRNA.

distal part of the cluster decreases in the same proportion as the levels of polypeptide products, although sometimes hyperlabile mRNA just distal to the site of chain termination can be detected.

The following explanation of polarity seems likely to turn out correct, although it is not yet conclusively proven. Biochemical studies with purified rho factor suggest that in causing termination of transcription it interacts with RNA rather than DNA; for instance, rho factor acts enzymically as an RNA-dependent ATPase. The explanation takes into account that in bacteria, transcription and translation are coupled (see also Chapter 5, p. 231), so that the leading ribosomes are moving along a nascent RNA molecule only just behind the RNA polymerase. It is suggested that rho factor can bind to RNA at an appropriate site—for instance, a run of U's as in the *trp* leader sequence—if the RNA is unprotected by ribosomes.

On doing so, it rapidly moves along the RNA to interact with RNA polymerase in such a way as to terminate transcription. The RNA unprotected by ribosomes may arise through premature polypeptide chain termination, caused by a nonsense mutation; or through normal termination of the final polypeptide of an operon; or, as in the *trp* operon where there is an attenuator in a leader sequence. This mechanism is depicted in Figure 10.

Polarity can be explained in terms of this scheme. The polar effect of a chain-terminating triplet could be assumed to depend entirely on its potential to cause premature termination of transcription following premature termination of translation. The likelihood of a nonsense mutation within an operon exerting a polar effect would depend on the presence of a rho factor recognition site between it and the ribosome binding site (see Chapter 6, p. 262) at the start of the next gene. On a random basis, the chance of this occurring would be proportional to the distance to the ribosome-binding site, so that the strength of the polar effect could be correlated with the position of the mutation within the gene as described above. Of course, there might be a wide variety of somewhat different rho factor binding sites with different probabilities of terminating transcription, so that wide variations could be expected and in fact occur.

MOLECULAR CONTROL MECHANISMS IN OTHER METABOLIC SYSTEMS

A wide variety of other systems have been studied in *E. coli* and *Salmonella typhimurium*. Although in most of these the detailed information available is less than in the *E. coli lac* and *trp* systems, it seems likely that no additional molecular control mechanisms are involved. However it is also clear that varying combinations of the modes of control are possible. A few examples from *E. coli* and *Salmonella typhimurium* will be mentioned briefly.

In the *ara* operon, whose genes encode enzymes of L-arabinose catabolism, the product of the regulator gene *araC* is a repressor in the absence of inducer but is converted by the binding of inducer into an activator without which 'induction' cannot take place. In the *mal* system involved in maltose catabolism, the genes form not one but two clusters, both under the control of the same regulator gene product—this latter functions both as a repressor and as an activator, as in the *ara* system above; one of the clusters has its control region not at one end but in the middle, and transcription takes place outwards in both directions (this is called *divergent transcription*). It may be noted, incidentally, that in the *ara* and *mal* systems, and in some others, catabolite repression takes the same form as in *lac* and is mediated by the same CRP. In the *hut* (histidine utilization) system of *S. typhimurium*, the gene cluster comprises two contiguous operons transcribed in the same direction; one of these contains the regulator gene itself, which thus controls it own expression (a situation termed *autogenous regulation*).

Biosynthetic pathways also show idiosyncracies. In the *ilv* (isoleucine-valine) and *his* (histidine) systems, there seems to be no repression control and only termination/anti-termination control obtains; in both, the mechanism that renders this control sensitive to endogenous amino acid level seems to involve, as well as the corresponding tRNA, the first enzyme of

the pathway, demonstrating autogenous regulation again. In the *bio* (biotin) and *arg* (arginine) systems, divergent transcription is again found; in the former, in one gene cluster that contains all the genes and in the latter, in a cluster containing four of the nine genes, the other five being scattered singly.

No molecular control mechanism has been dissected in detail in any species other than *E. coli* and *S. typhimurium*, although in some cases, as we shall see below, some genetic information has been obtained as part of studies on the physiology of control.

PATTERNS OF CONTROL IN CLASS I PATHWAYS

We shall now describe the physiological patterns of control found in some Class I systems. Among these we will distinguish between the 'peripheral' sequence (in which the carbon source is converted into an intermediate of one of the glucose-degrading pathways or of the tricarboxylic acid (TCA) cycle) and the sequences constituting the central metabolic pathways (Figure 11).

Not all these sequences, of course, serve a purely catabolic function. The Embden-Meyerhof pathway, for instance, works in reverse in cells grown on a TCA cycle intermediate to produce sugars and derivatives for synthesis of cell wall components. Pathways which can act either degradatively or biosynthetically are said to be *amphibolic*; clearly they might be expected to show peculiarities as regards control. Another class of reactions with special function comprises those which replenish the TCA cycle when the carbon source feeds into the glycolytic sequence at or before pyruvate; they are termed *anaplerotic* (see Chapter 3, p. 151).

Control of the central pathways

The controls of activity of enzymes in the Embden-Meyerhof pathway and TCA cycle, and of the anaplerotic enzymes pyruvate carboxylase and phospho*enol*pyruvate (PEP) carboxylase, are known in some detail. These pathways provide, for many bacteria, the main route for the formation of carbon skeletons for biosynthesis by degradation of carbohydrates and related compounds and for the production of energy and reducing power needed in biosynthetic reactions. Effectors of modulation tend, as might be expected, to

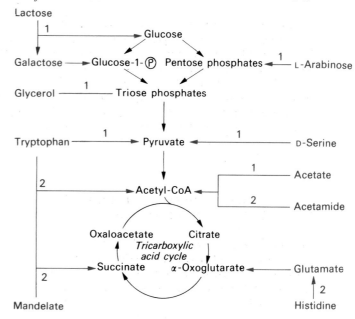

Figure 11. Patterns of catabolic pathways for carbon/energy sources. The 'central' pathways of glucose breakdown via the Embden–Meyerhof and pentose phosphate routes and the TCA cycle are shown in black, while the 'peripheral' sequences leading into these pathways from some less common carbon sources are shown: 1 for *Escherichia coli* and 2 for *Pseudomonas spp.*

be either compounds that reflect the availability of metabolites at various points in the pathway, or compounds that reflect the energy state or level of reducing power in the cell. Often an enzyme is inhibited by a compound formed in a reaction many steps later in the pathway, or by a compound immediately derived from it. Examples are the inhibition of phosphofructokinase by PEP in *E. coli*; of PEP carboxylase by aspartate (formed by transamination of α-oxoglutarate), malate or fumarate in *E. coli* and *S. typhimurium*; and of citrate synthase by α-oxoglutarate in these two species. This is a form of *end-product* or *feedback control*, which reduces the flow through the early part of a sequence in which products of later steps are accumulating so that wasteful excessive flux through the pathway is avoided.

Another common phenomenon (termed *precursor activation*) is the activation of an enzyme late in a sequence by a metabolite coming some way before it. For instance, fructose 1,6-bisphosphate activates one of the two pyruvate kinase isoenzymes of *E. coli* and the PEP carboxylase of *S. typhimurium*. Presumably this ensures that a key enzyme is mobilized to deal with a flow of substrate shortly to reach it from breakdown of the activating 'precursor'. The PEP carboxylase is also activated by acetyl coenzyme A, perhaps to facilitate a continued supply of oxaloacetate for condensation with the acetyl CoA to form citrate. Activation of the two effectors fructose 1,6-bisphosphate and acetyl CoA is cooperative (Chapter 4, p. 159).

Energy state

The energy state of the cell is principally indicated by the concentration of ATP and other nucleoside triphosphates. The nucleoside mono- and di-phosphates may, however, be equally useful in a regulatory role because of the ready interconversion of mono-, di- and tri-phosphates within the cell: a drop in ATP concentration, for example, is equivalent to a rise in that of AMP or a somewhat smaller rise in that of ADP (see the discussion of *energy charge* in Chapter 3, p. 127). For example, in *E. coli* phosphofructokinase is activated by ADP and GDP, and the pyruvate kinase *not* activated by fructose 1,6-bisphosphate (see above) *is* activated by AMP. Both of these activations promote flow through energy-yielding sequences in response to accumulation of compounds signalling an energy dearth.

NADH inhibits several enzymes, most of which can be regarded as coming early in sequences that result in its formation. It can therefore be regarded as a feedback inhibitor, although its function may be to indicate the level of reducing power within the cell. NADH inhibits citrate synthase (whose functioning produces it) in Gram-negative organisms but not Gram-positives. Furthermore, in obligately aerobic Gram-negative species such as *Pseudomonas*, *Acetobacter* and *Chromobacterium* this inhibition is counteracted by AMP, whereas in facultative anaerobes such as the enterobacteria it is not. This striking difference may lie in a physiological peculiarity of the enterobacteria and probably other facultative anaerobes. These organisms generate energy primarily by anaerobic glycolysis when growing on glucose even with good aeration; the TCA cycle functions here primarily biosynthetically, as suggested by the fact that *E. coli* mutants lacking succinate dehydrogenase or certain co-factors essential for oxidative phosphorylation grow aerobically on glucose in the same way as the wild-type. However, organisms requiring the operation of the TCA cycle for energy production either cannot allow its functioning to be inhibited merely by accumulation of a by-product, as in the Gram-positives, or need at least to ensure that this inhibition can be reversed by a signal of low ATP level, as in the obligately aerobic Gram-negatives. In the facultatively anaerobic Gram-negatives, it is sufficient for end-product inhibition of citrate synthase to be counteracted (as in *E. coli*) by its substrates, oxaloacetate and acetyl CoA antagonizing the effects of NADH and α-oxoglutarate respectively.

The complexity of controls for these pathways in bacteria in contrast to those in eukaryotes may reflect the isolation of energy-generating enzyme systems of the latter in mitochondria; this facilitates control of flow through different systems by keeping them physically apart. It has been suggested that the diminished extent of regulation of citrate synthase in Gram-positive as opposed to Gram-negative bacteria reflects the organization of the TCA cycle enzymes into a membrane-bound system (possibly the mesosome: Chapter 1, p. 89) in the former. However, the complexity may in part result from the amphibolic nature of many reactions. The enzymes concerned belong not only to Class I in making available biosynthetic intermediates and energy during growth on carbohydrate, but also to Class II in biosynthesis of cell wall components, ribose and deoxyribose, lipids and storage carbohydrates when the bacteria grow on a TCA cycle intermediate or related compound as sole carbon source.

Modulation of enzymes
It is worth while considering here what reactions turn out to be catalyzed by enzymes subject to modulation. They usually (i) are thermodynamically irreversible and (ii) come immediately after a branch point of two (or more) metabolic routes. This is true of phosphofructokinase, pyruvate kinase, PEP carboxylase and citrate synthase. Fructose 6-phosphate can be phosphorylated *via* phosphofructokinase to yield fructose 1,6-bisphosphate, or it can react with glutamine to give glucosamine 6-phosphate in the first step towards the biosynthesis of cell wall constituents *N*-acetylgluco-

samine and *N*-acetylmuramic acid; PEP can yield pyruvate *via* pyruvate kinase or oxaloacetate *via* PEP carboxylase; acetyl CoA can condense with oxaloacetate *via* citrate synthase, forming citrate, or with CO_2 *via* acetyl CoA carboxylase, forming malonyl CoA in the first step of fatty acid biosynthesis.

These characteristics are probably physiologically advantageous for the following reasons. Firstly, the extent of a reversible reaction will be to some extent dictated by the mass-action equilibrium between reactants and products; changes, caused by modulation, in the enzyme's affinity for the reactants will therefore be only partly effective in altering the rate of their conversion to products. But where the reaction is irreversible, the flow through it is entirely determined by the catalytic activity of the enzyme, so that regulation here is more effective. Secondly, suppose for illustration that the production of E in the sequence

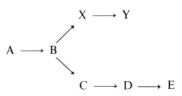

is to be regulated by inhibition of one of the enzymes by E. It would clearly be undesirable to inhibit the enzyme catalyzing A → B, for then flow through the branch B → X → Y would be curtailed unnecessarily and perhaps deleteriously. It would also be inefficient to modulate one of the enzymes after C, for then C or D would accumulate: not only would this waste B, which could otherwise pass into the B → Y branch, but also—since the inhibition can be partly overcome by increase in substrate concentration—once sufficient C had accumulated, the reaction B → C would again proceed. The most effective point of control is therefore B → C, immediately after the branch point; this leaves the pathway from B → Y undisturbed, and any B that accumulates can be diverted into this branch.

Control of synthesis of the enzymes of glycolysis and the TCA cycle and the associated anaplerotic enzymes seems to be of less importance than control of enzyme activity. Most of the glycolytic pathway enzymes, in particular, seem to be constitutive in the enterobacteria, although the AMP-activated pyruvate kinase is induced during growth on succinate, and the components of the pyruvate dehydrogenase complex (see below, p. 348) are induced during growth on pyruvate. The TCA cycle enzymes, in this group and some *Bacillus* species, show some induction and repression under certain conditions; for instance, the enterobacterial enzymes are induced during growth on nutrient broth, when the cycle is essential for energy production, and also following a shift from anaerobic to aerobic conditions.

As regards controls in other pathways of glucose catabolism, in *E. coli* glucose 6-phosphate dehydrogenase is inhibited by NADH, which perhaps accords with the idea that the main purpose of the pentose phosphate cycle in this organism is to generate NADPH for biosynthesis. In those organisms possessing them, enzymes of the Entner-Doudoroff and phosphoketolase pathways are highly inducible; in addition, the glucose 6-phosphate dehydrogenase of *Pseudomonas aeruginosa* differs from that of *E. coli* in being inhibited by ATP.

Control of the peripheral pathways

Control of the enzymes of a peripheral sequence is usually at the level of enzyme synthesis, although occasionally (especially in amino acid-degrading pathways) the first enzyme in the sequence may be inhibited by a compound that indicates the energy state of the cell. For instance, the histidase of *Pseudomonas aeruginosa* (the first enzyme of histidine catabolism) is inhibited by pyrophosphate, accumulation of which indicates a high intracellular energy supply, and this inhibition is counteracted by AMP and GDP. Typically, these enzymes are inducible, by the carbon source itself or by the first intermediate in the sequence (product induction: see above, p. 331), and are subject to catabolite repression. In *E. coli*, catabolite repression is particularly strongly exerted by glucose, glucose 6-phosphate or gluconate; however, in other organisms catabolite repression of enzymes of glucose dissimilation by other carbon sources can occur; for instance, citrate represses the enzymes of the Entner-Doudoroff pathway in *Pseudomonas aeruginosa*.

The utilization of acetate by *E. coli* does not fit into our generalization that the enzymes induced for utilizing a new carbon source are those of a peripheral sequence leading to the central metabolic pathways. In this instance the enzymes induced are those of the glyoxylate cycle (see Chapter 3, p. 151), which serve an anaplerotic rather than a purely catabolic function. They replenish the TCA cycle as intermediates are drawn off for biosynthesis. The apparent inducing effect of acetate may depend on antagonism of repression by PEP or some metabolite immediately derived from it.

Inducibility of synthesis of enzymes specific for the degradation of carbon sources that are only occasionally or rarely present in the cell's environment allows the cell to produce the enzymes only when they can be useful. Catabolite repression also ensures that even if such a carbon source is present, it will not be metabolized if a more efficiently utilizable alternative is available. The advantages to the cell of these forms of control are clear.

Although induction and catabolite repression are widespread among bacteria, it is not clear to what extent their mechanisms resemble those in *E. coli* discussed above. There are indications, however, that similar mechanisms occur in some cases. One can often isolate constitutive mutants of inducible systems that resemble *lacI⁻* mutants. One such system is the inducible amidase of *Pseudomonas aeruginosa* (whose normal substrate appears to be acetamide); in this case, the catabolite repression exerted by succinate or propionate is counteracted by addition of cAMP to the medium, suggesting also a similar mechanism of catabolite repression. A mutant has also been isolated that has properties expected of a catabolite-insensitive promoter mutant.

Regulation in the catabolically versatile pseudomonads: sequential induction, convergent and divergent pathways

Several bacterial genera contain species with great versatility in utilizing unusual compounds as sources of carbon or nitrogen. Many strains of *Pseudomonas*, for instance, can utilize a wide variety of aromatic compounds, by use of especially long peripheral sequences. The control of the latter has certain unique features.

The pathways for degradation of D-mandelate and *p*-hydroxy-D-mandelate are shown in Figure 12. Down to benzoate and *p*-hydroxybenzoate, the same broad-specificity enzymes catalyse the homologous reactions in both branches of the pathway. In *Ps. putida*, the first four enzymes are induced together by D-mandelate, or by benzoylformate, or by their *p*-hydroxylated derivatives. (It is interesting that in the related *Ps. aeruginosa*, L-mandelate induces L-mandelate dehydrogenase and the resulting benzoylformate induces benzoylformate decarboxylase and benzaldehyde dehydrogenase.) Next, benzoate oxidase and *p*-hydroxybenzoate hydroxylase are induced by benzoate and *p*-hydroxybenzoate respectively. A difference now appears between the branches in that whereas protocatechuate induces protocatechuate

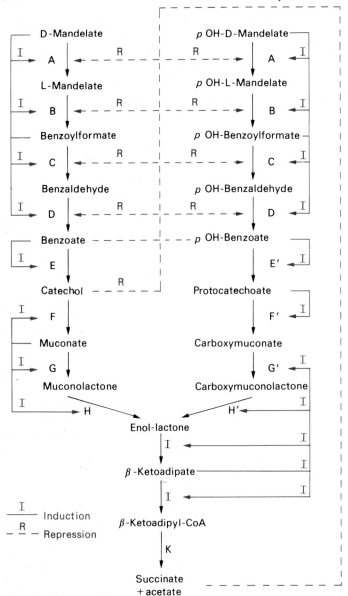

I ———— Induction

– R – Repression

Figure 12. Inducers and repressors of the enzymes of the mandelate and *p*-hydroxymandelate pathways in *Pseudomonas putida*. Enzymes A, B, C and D are common to both pathways.

oxygenase, catechol oxygenase shows product induction by muconate. The latter also induces muconate lactonizing enzyme and muconolactone isomerase, although the latter two can be shown to comprise a different regulatory unit from catechol oxygenase. Finally, β-ketoadipate, at which the branches converge, exerts product induction on enol-lactone hydrolase, carboxymuconate lactonizing enzyme, and carboxymuconolactone decarboxylase, as well as inducing β-ketoadipate succinyl-CoA transferase. We see, then, that the sequences break down for regulatory purposes into blocks, each block being induced by the substrate of the block or by a product.

Many of the enzymes also show *metabolite repression* by pathway intermediates coming later in the sequence. For instance, benzoate, *p*-hydroxybenzoate, catechol and succinate repress the first four enzymes of the pathway, and benzoate oxidase is repressed by catechol. Repression of the former block of enzymes by benzoate and *p*-hydroxybenzoate is so strong that in the presence of mandelate and one or other of these substances, utilization of mandelate is negligible and the growth occurs principally on the later intermediate (it may be that antagonism of mandelate uptake is also involved). This effect, being specific, is probably mechanistically distinct from general catabolite repression, which is exerted on these five enzymes by, for example, succinate.

Another remarkable feature of this system is that catechol can be catabolized by another route different from that shown in Figure 12. There, catechol is split by 1,2-oxygenation 'ortho' cleavage). In the alternative reaction, oxygenative fission takes place at the 2,3 position ('meta' cleavage), yielding 2-hydroxymuconic semialdehyde which is degraded further and eventually yields acetaldehyde and pyruvate. In *Ps. putida*, both types of cleavage may be found in the same organism. The specificity of regulation, however, is such that in general only one or other of the cleavage routes will be induced. Catechol 1,2-oxygenase is induced by muconate, and induction therefore requires a considerable accumulation of catechol and muconate. The 'meta' pathway enzymes are usually substrate induced. For instance, in one strain of *Ps. putida* which utilizes phenol, this substance induces the 'meta' system, the catechol derived from it never reaching a sufficiently high endogenous concentration to allow induction of the 'ortho' oxygenase.

In many cases, *Pseudomonas* catabolic enzymes controlled as a block are encoded by clustered genes, which appear to be regulated as a unit in that mutations can often be found which lead to all being simultaneously rendered constitutive (as in *lacI⁻* mutants, p. 332). A remarkable feature is that catabolic clusters seem to form 'super-clusters'; in *Ps. putida*, for instance, it has been shown by transduction that genes or clusters for catabolism of benzoate, histidine, *p*-hydroxybenzoate, mandelate, nicotinate, phenylacetate, phenylalanine, quinate, shikimate and tyrosine map within 10–15% of the chromosome. Catabolic gene clusters are also sometimes located on plasmids (often self-transmissible): examples include those for camphor, naphthalene, *n*-octane and toluene. These observations probably reflect a tendency for genetic information valuable only under rather specialized conditions not to be maintained in all cells of a species but to be kept in a relatively easily mobilizable state in the population. Under appropriate selective circumstances, the useful potentialities can then spread rapidly through the population.

Control of nitrogen assimilation

Just as glucose is a preferred carbon source for *E. coli*, in the sense of permitting faster growth than other sugars such as lactose, a particular nitrogen source may be preferred over others by a given bacterium. One might therefore expect to find a phenomenon analogous to catabolite repression exerted by a favoured nitrogen source on the production of enzymes involved in the catabolism of less efficient nitrogen sources. A striking instance of this is the repression of nitrogen fixation enzymes of *Azotobacter vinelandii* and *Clostridium pasteurianum* by NH_4^+. In the former, nitrate (which the cells can reduce to NH_4^+) inhibits, but does not repress, the nitrogenase. In the latter ADP is the inhibitor, which is perhaps understandable in view of the large ATP requirement of the process.

Another example of 'nitrogen catabolite repression' is seen in *S. typhimurium* and *Klebsiella aerogenes*. In these bacteria, as in many others, nitrogen can be obtained by assimilation of NH_4^+ or through catabolism of various amino acids, the former apparently being preferred. NH_4^+ can be assimilated in two ways: by the action of glutamate dehydrogenase (Chapter 4, p. 167) or by the glutamine synthase-catalysed amidation of glutamate to yield glutamine, followed by the reaction of glutamine with α-oxoglutarate to yield two molecules of glutamate; the enzyme involved in these reactions is sometimes termed glutamate synthase. Glutamate dehydrogenase operates at high (greater than 1 mM) endogenous NH_4^+ concentrations, the alternative system at low concentrations. Several amino acids can be catabolized to nitrogen-containing components from which the nitrogen can be incorporated into other cell components. Of these, the best studied is histidine, whose conversion to glutamate is mediated by four enzymes (the *hut* system).

It is found that control is exercised through glutamine synthase, which acts both as an enzyme and as a regulatory protein: it regulates its own synthesis (autogenous regulation again: see above, p. 340), represses glutamate dehydrogenase synthesis and activates the *hut* system. The dual role of glutamine synthase can be explained by the fact that it exists in two forms, adenylated and non-adenylated (see p. 328). The relative amounts of the two forms reflect the nitrogen balance in the cell; adenylation is favoured by a high ratio of glutamine to α-oxoglutarate, corresponding to conditions of NH_4^+ excess, while de-adenylation is favoured by a high ratio of α-oxoglutarate to glutamine, corresponding to NH_4^+ deficiency. Adenylated glutamine synthase represses its own synthesis, so that at high external NH_4^+ concentrations most nitrogen is assimilated *via* glutamate dehydrogenase. Under low external NH_4^+, the de-adenylated form accumulates, allowing depression of its own synthesis; it is this form also which represses glutamate dehydrogenase synthesis and activates the *hut* system. The *hut* system is therefore subject to three forms of control: repression, cAMP-mediated carbon catabolite repression, and the nitrogen catabolite repression described here.

In *Bacillus subtilis*, NH_4^+ seems not to be the preferred nitrogen source—for instance, glutamine, but not NH_4^+ represses induction of the arginine-degrading enzyme arginase. This may be correlated to the tendency of *B. subtilis* to associate in the soil with particles of organic matter and with fungal hyphae.

Anaerobic respiration

If we take Class I reactions to include those yielding energy only as well as those yielding both energy and carbon skeletons for biosynthesis we should mention how cells control the use of inorganic ions such as nitrate or thiosulphate instead of O_2 as terminal electron acceptors for respiration. This process is called *anaerobic respiration*. Control is again primarily at the level of enzyme synthesis. For instance, the nitrate reductases of *E. coli* and *Proteus mirabilis* are induced by nitrate and repressed by O_2, while the thiosulphate reductase of *Proteus mirabilis* is induced by thiosulphate and repressed by O_2 or nitrate. The latter enzyme is interesting because if an induced culture is exposed to O_2 or nitrate, not only is further synthesis of thiosulphate reductase repressed, but the enzyme present in the cells is rapidly inactivated, possibly by a protease which has to be newly synthesized since the inactivation is blocked by chloramphenicol. This *inactivation repression* is common in eukaryotic microorganisms but less so among bacteria. It is not certain (in either case) whether the inactivation involves inhibition of the enzyme or its selective degradation. If the latter, the effect exemplifies regulation by control of breakdown of specific proteins, which as pointed out earlier differs fundamentally from controls at the levels of enzyme activity or synthesis.

THE CONTROL OF CLASS II REACTIONS

Control of synthesis of amino acids and purine and pyrimidine nucleotides is almost always by feedback inhibition and end-product repression, either separately or together. The controls have presumably evolved to prevent wasteful flow through a pathway and the unnecessary synthesis of enzymes when the end-product is in good supply. Feedback inhibition generally operates on enzymes immediately following metabolic branch points (for reasons already mentioned), and repression usually affects enzymes leading from the branch point to the end-product. Straightforward examples where the pathways are unbranched (e.g. arginine and histidine biosynthesis) will be found in Chapter 4.

Branched pathways

In many Class II systems, complications arise through branching of pathways. Consider a pathway with a single branch (Figure 13). One finds typically (at least in the enterobacteria) that end-products E and G each partially inhibit and/or repress enzyme 1: that each *may* also partially repress enzyme 2; and that in addition, E inhibits enzyme 3 and represses enzymes 3 and 4, while G may inhibit enzyme 5 and may repress enzymes 5 and 6. In other words, the end-products *in combination* control the shared part of the pathways, while they *separately* control their individual branches. Control of enzyme 1 can be *concerted, cooperative* or *cumulative*. Alternatively, the organism may have two isoenzymes catalysing the A → B reaction, one of which is controlled by E, the other by G. Repression of enzyme 2, where it occurs is usually concerted or cooperative. Examples of all these types have been given in Chapter 4. Another type of regulation, *sequential* feedback inhibition (Chapter 4, p. 159) was first described for the aromatic amino acid system in *Bacillus subtilis*. In the scheme shown in Figure 13, it would involve E inhibiting enzyme 3, and G inhibit-

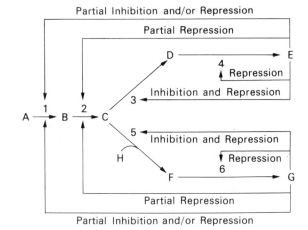

Figure 13. Branched pathways. Schematic diagram of branched pathways, and commonly found inhibition and repression patterns of control (see text).

ing enzyme 5, thereby accumulating C which would inhibit enzyme 1.

Finally, *compensatory* antagonism of feedback inhibition can be exemplified by the control of carbamoyl phosphate synthase in *E. coli*. Its product, carbamoyl phosphate, is utilized in both arginine and pyrimidine synthesis (see Chapter 4, Figures 6b, 16). The synthase is inhibited by UMP, this inhibition being antagonized by ornithine (which combines with carbamoyl phosphate to yield citrulline). In Figure 13, this would represent the inhibition of enzyme 1 by E and the antagonism of this inhibition by H.

Such controls in the shared part of a branched pathway clearly avoid its complete closure when only one end-product is present, which would lead to starvation for the other end-product, but nevertheless provide some economy in its operation.

Biosynthesis of aromatic amino acids

The different systems in various bacterial species for control of the aromatic amino acid biosynthetic pathways demonstrate particularly clearly how diverse are the regulatory patterns that have evolved. Some of these (especially in the enterobacteria) have been described in Chapter 4, pp. 160–163. Figure 14(a) shows controls in the pathway up to chorismate in two related species of *Bacillus*, *B. subtilis* and *B. licheniformis*. The sequential feedback inhibition of the first enzyme has been mentioned above. *B. subtilis* possesses two distinct shikimate kinases (E) and two chorismate mutases (H), while *B. licheniformis*, although it also possesses two H enzymes, has only one E enzyme. One of the E isoenzymes of *B. subtilis*

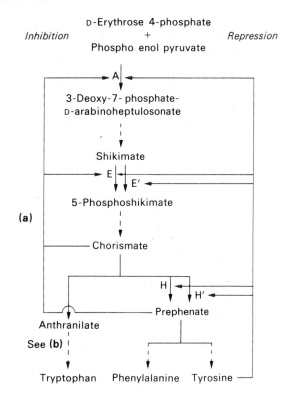

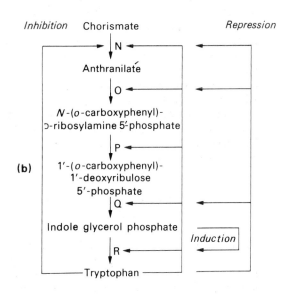

Figure 14. Control of the aromatic acid biosynthetic pathway.

(a) Tyrosine and phenylalanine pathway in *Bacillus subtilis*. Control in *Bacillus licheniformis* is similar but although it also possesses two H enzymes, it has only one E enzyme which is constitutive and is inhibited only by chorismate.

A	Phospho-2-keto-3-deoxyheptonate aldolase
E and E′	Shikimate kinase
H and H′	Chorismate mutase

(b) Tryptophan terminal pathway in *Bacillus subtilis*, *Pseudomonas putida* and *Pseudomonas aeruginosa*.

N	Anthranilate synthase
O	Anthranilate phosphoribosyltransferase
P	Phosphoribosyl-anthranilate isomerase
Q	Indoleglycerol phosphate synthase
R	Tryptophan synthase

is inhibited by chorismate or prephenate, like enzyme A; however, the single enzyme of *B. licheniformis* is inhibited only by chorismate. In *B. subtilis*, A, E and H activities are repressed by tyrosine, while in *B. licheniformis*, tyrosine represses A and H activities but E is constitutive. Enzyme aggregation patterns also differ between the species (see p. 349).

The pattern of A activity is highly distinctive for the various bacterial groups. The sequential feedback inhibition found in *Bacillus* species is also observed in the staphylococci. In *Streptomyces*, a single amino acid, tryptophan, appears to be the sole inhibitor. In *Pseudomonas* species, there is also one major inhibitor, tyrosine, but it appears that maximum inhibition requires the combined presence of tyrosine and phenylpyruvate (the immediate precursor of phenylalanine), while partial inhibition is produced by tryptophan. Inhibition by tyrosine and tryptophan is competitive with (i.e. can be overcome by) phosphoenolpyruvate, and that by phenylpyruvate is competitive with D-erythrose-4-phosphate. In other words, the economy affected by end-product inhibition is particularly marked when the starting materials of the pathway are themselves in short supply.

Regulation in the terminal pathway for tryptophan is shown for *Bacillus subtilis*, *Pseudomonas aeruginosa* and *Ps. putida* in Figure 14(b). In *B. subtilis*, the pattern is like that seen in enterobacteria, in that the first enzyme (anthranilate synthase, enzyme N) is inhibited, and it and the other enzymes are repressed, by tryptophan. In the pseudomonads, N is again inhibited by tryptophan, but the regulation of enzyme synthesis is different: enzymes N, O, and Q are repressed by tryptophan, P is constitutive, while R (the tryptophan synthase complex) is induced by its substrate, indoleglycerol phosphate. These differences in control of enzyme synthesis are reflected in the location of the respective genes. In *B. subtilis*, these form a single cluster, as in the enterobacteria, while in *Ps. aeruginosa* and *Ps. putida* the genes for enzymes N, O and Q constitute one cluster and those for two components of tryptophan synthase another, the gene for enzyme P being isolated. Again, we shall discuss gene aggregation patterns later, on p. 350.

Finally, it may be mentioned that apparent differences in physiology of control may hide resemblances at the genetic level. An example is afforded by the arginine biosynthetic enzymes of *E. coli* strains K-12 and B, which are repressed by arginine in the former but are constitutive or slightly induced by arginine in the latter. The differences in behaviour are governed entirely by differences at a single genetic locus *argR* that codes for the apo-represser.

Control of vitamin and cofactor biosynthetic pathways

We know less about the control of these pathways than of amino acid pathways. It appears that like other Class II pathways they are usually regulated by end-product inhibition and repression. For instance, enzymes of biotin synthesis in *E. coli* are repressed by biotin; while the first enzyme of haem synthesis in *Staphylococcus aureus*, δ-aminolaevulinate synthase, is both inhibited and repressed by haemin.

Spatial control of Class I and Class II reactions: enzyme aggregates

We have so far discussed enzymes as if these are randomly distributed among the 'soluble' components of the cell. However, even apart from the association of certain enzymes with the cell membrane (Chapter 1, p. 86), enzymes of the same pathway are often found to *aggregate* to form complexes held together by non-covalent forces. These often cohere in cell extracts and may do so through many stages of purification. We shall discuss a Class I instance, the pyruvate dehydrogenase complex of *E. coli*, and several Class II examples, namely aggregates combining activities in aromatic amino acid pathways in various organisms, and the *E. coli* aspartokinase-homoserine dehydrogenase complex.

The pyruvate dehydrogenase complex of *E. coli* can be isolated intact from cell extracts following ultracentrifugation or protamine sulphate fractionation, and has a relative molecular mass of about four million. It catalyses the reactions shown in Figure 15. Chromatographic separation in the presence of 4M urea or at pH 9·5 (neither of which treatments breaks peptide bonds) causes complete dissociation into components with the individual activities E1 (decarboxylase), E2 (transacetylase) and E3 (flavoprotein). Each of these components is in turn composed of subunits. The components can be reconstituted to give the active complex.

The complex and the component enzymes can be seen in electron micrographs as regular arrays of subunits. This information and the biochemical data suggest that a single molecule of the complex contains 24 transacetylase polypeptide chains, 24 decarboxylase chains and 12 flavoproteins, the relative molecular masses of the chains being 40 000, 90 000 and 55 000 respectively. The transacetylase forms eight aggregates, each of three chains, located at the vertices of a cube, with 12 decarboxylase dimers along the edges and 6 flavoprotein dimers in the faces.

(1)   Pyruvate + TPP-E$_1$ $\longrightarrow$ Hydroxyethyl-TPP-E$_1$ + CO$_2$

(2)   Hydroxyethyl-TPP-E$_1$ + Lipoyl-E$_2$ $\longrightarrow$ S-Acetyl dihydrolipoyl-E$_2$ + TPP-E$_1$

(3)   S-Acetyl dihydrolipoyl-E$_2$ + CoA-SH $\longrightarrow$ Acetyl $\sim$ S-CoA + Dihydrolipoyl-E$_2$

(4)   Dihydrolipoyl-E$_2$ + FAD-E$_3$ $\longrightarrow$ Lipoyl-E$_2$ + FADH$_2$-E$_3$

(5)   FADH$_2$-E$_3$ + NAD$^+$ $\longrightarrow$ FAD-E$_3$ + NADH + H$^+$

Pyruvate + CoA-SH + NAD$^+$ $\longrightarrow$ Acetyl $\sim$ SCoA + NADH + H$^+$ + CO$_2$

Figure 15. Reactions catalysed by the pyruvate dehydrogenase complex.

E1	Decarboxylase
E2	Dihydrolipoyl transacetylase
E3	Dihydrolipoyl (flavoprotein) dehydrogenase
TPP	Thiamine pyrophosphate

We can surmise that the speed and efficiency of reactions 2 and 4 (Figure 15), which involve pairs of enzymes and enzyme-bound intermediates, will be much greater if the enzymes are juxtaposed. It is interesting that the decarboxylase and transacetylase are synthesized coordinately in an equimolecular ratio, the proportions in which they appear in the complex. It would clearly be wasteful for one component to be made in excess and to remain uncomplexed.

The aromatic amino acid pathways we have described display interesting differences in patterns of enzyme aggregation. The aggregation in enterobacteria of chorismate mutases with prephenate dehydrogenase or prephenate dehydratase, of anthranilate phosphoribosyltransferase with the *trpE* product in anthranilate synthase, and of the A and B components of tryptophan synthase, have already been described (Chapter 4, p. 163). In these examples, the aggregated enzymes catalyse successive steps in a pathway and the advantage presumably lies again in the ease with which a molecule produced in the first step becomes available as substrate for the second stage, never leaving the enzyme aggregate (though proof of this is not often available). It may be noted that the catalysis of reactions P and Q (Figure 3, p. 164) by a single enzyme, although formally the opposite of enzyme aggregation, serves essentially the same purpose, namely the more efficient use of a reaction product by allowing it to move to the substrate-binding site for the ensuing reaction without diffusing away. We shall now consider systems in which the existence of aggregates is harder to understand in that the aggregated enzymes do not catalyse successive steps in the pathway.

In *Bacillus subtilis* the aldolase, one of the shikimate kinase isoenzymes, and one of the chorismate mutase isoenzymes form an aggregate; however, in *B. licheniformis* there is no evidence for aggregation of any of these enzymes. In the chorismate to tryptophan

sequence in *B. subtilis*, *Pseudomonas aeruginosa* and *Ps. putida*, the complex found in enterobacteria with both anthranilate synthase and anthranilate phosphoribosyltransferase activities does not occur, and steps P and Q are catalysed by separate enzymes. Aggregation of the two components of tryptophan synthase, however, seems to be general in bacteria.

Finally, the aspartokinase and homoserine dehydrogenase isoenzymes of *E. coli* resemble the enterobacterial enzyme that catalyses steps P and Q in the tryptophan pathway in comprising a single polypeptide chain with both activities. Here, however, the enzymes are oligomers. In *E. coli* K-12, the threonine-inhibited aspartokinase I and homoserine dehydrogenase I activities are associated in a single protein consisting of six identical chains, while the methionine-repressed aspartokinase II and homoserine dehydrogenase II activities are found together in another protein consisting of four identical chains.

In such cases it is harder to point to advantages to the cell of having a single protein with more than one function, whether this takes the form of an enzyme aggregate or a single multifunctional species. However, in the *B. subtilis* case an interesting suggestion is that the aggregation of chorismate mutase with the aldolase and shikimate kinase may serve to associate the latter two enzymes with another that has binding sites for chorismate (its substrate) and prephenate (its product). This may have facilitated the evolution of their presumably advantageous sensitivity to feedback inhibition by chorismate or prephenate. The unexpected susceptibility of the shikimate kinase to inhibition may be of selective value in that *Bacillus subtilis* is permeable to shikimate; shikimate kinase is the first enzyme in the further metabolism of this compound and hence might be expected to need a mechanism for modulation. *B. licheniformis* is impermeable to shikimate; this may explain why its shikimate kinase is not subject to control.

General comments on the control of Class I and Class II reactions

In the last few sections, we have given some idea of the wide variety of ways in which bacteria regulate the flow through catabolic and biosynthetic pathways by ringing the changes on simple mechanisms for altering enzyme activity or synthesis. We have attempted to explain the existence of these varied control mechanisms in terms of their selective advantage in permitting the most economical exploitation of nutrients in the environment and the most rapid and efficient adaptation to changes therein. While there are often puzzling variations in patterns of control between closely related species (or even different strains of the same species), some aspects at least of the means of regulation of a given pathway are common to most or all members of a given group. Regulatory characteristics have therefore taken their place among other bacterial properties used for taxonomic purposes.

The value of control mechanisms to the organism has not often been conclusively proven experimentally. However, a mutant of *E. coli* in which the proline pathway was no longer inhibited by proline was shown to be gradually overgrown by the wild-type; and a mutant of *Bacillus subtilis* in which the tryptophan enzymes were not repressible by tryptophan was found to be at a selective disadvantage when grown in mixed culture with the parent strain.

Possible value of gene clusters

What is the significance of the clustering of functionally related genes? Although this is a marked bacterial characteristic, it seems to occur to different extents in different groups, being for instance much less frequent in amino acid biosynthetic sequences in *Pseudomonas*. We may ask whether this clustering is advantageous for reasons connected with the control of gene expression, is advantageous for some other reason, or is an evolutionary relic; and we would like to know why, whichever of these hypotheses is correct, its frequency varies between the different groups.

As usual in such cases, we can only make informed guesses. Although clustering was originally regarded as an integral part of the operon system, facilitating coordination and economizing on promoter and operator regions (and hence on the amount of DNA to be replicated), this is less convincing in the light of the variety of systems now studied. In particular, the existence even in *E. coli* of many regulons (p. 329) made up partially or completely of scattered genes, sometimes showing coordinate control, indicates that the correlations between clustering, coordinate control, and economy in use of DNA, are not complete.

What of other possible explanations? Two speculations may be mentioned. Firstly, it has been proposed that some degree of clustering is advantageous to organisms among which, as is true for bacteria, genetic material is transferred only in small pieces. It may not be useful for an organism that has lost several catalytic functions of a pathway to acquire just one functional gene, whereas it may be greatly advantageous for it to pick up all the functional genes as part of a cluster. Second, the evolution of metabolic pathways has been postulated to proceed *via* tandem duplication, giving rise to two contiguous copies of the gene determining the last step, followed by independent evolution of one of these copies to become the gene determining the next-to-last step, and so on. This process clearly generates clusters of functionally related genes, which could then separate fortuitously.

THE CONTROL OF CLASS III REACTIONS

Here, in contrast to the detailed schemes described in previous pages, we are mainly concerned with the synthesis or breakdown of whole classes of macromolecules, and in most cases these processes reflect the growth state of the cell. Controls which affect the relative affinities of RNA polymerase for different classes of promoter site have been discussed in Chapter 5, pp. 234–236. There remains one topic, however, which is conveniently treated here.

BREAKDOWN OF PROTEIN AND RNA

We shall enlarge here on a point arising from experiments on nutritional 'shift-down' (Chapter 2, p. 118). During the lag before protein and RNA synthesis resume, considerable breakdown of protein and RNA may occur. What is the function of this protein and RNA turnover?

Protein breakdown can be investigated by adding a radioactively labelled amino acid to the cells under study, and then transferring them to a medium containing a high concentration of the unlabelled amino acid as a 'trap'. It is assumed that when labelled amino acid incorporated into protein is liberated by proteolysis it will be so diluted in the 'trap' as to have a negligible chance of being re-incorporated during protein synthesis. Hence the amount of labelled amino acid released in a given time is a measure of the rate of protein breakdown.

By such means it has been shown that some protein breakdown accompanies normal exponential growth, although its accurate determination is difficult; for *E. coli*, rates of 0·5% to 2·5% of the protein per hr have been suggested. However, under 'shift-down' conditions or when growth is stopped by an inhibitor or by the exhaustion of a nutrient, the rate of protein breakdown increases immediately, to about 5% per hr in *E. coli* and to nearly 8% in *Bacillus cereus*. It seems likely that during exponential growth, a limited class of proteins are broken down relatively rapidly, while under less favourable conditions a wider range of proteins is degraded. On 'shift-down' or during starvation, the amino acids freed are rapidly re-utilized, so that there is an efficient turnover process in which the cell protein content remains more or less constant.

In parallel with the increased breakdown of protein attendant on 'shift-down' or growth inhibition, there is an enhanced rate of degradation of RNA (mostly ribosomal), which in *E. coli* also reaches about 5% per hour. Again, most of the released nucleotides are rebuilt into RNA, but under certain conditions a fraction of them may be further broken down and the ribose used as an energy source.

These degradative processes accelerate during temporary or permanent cessation of growth but subside rapidly when growth resumes. How they are controlled is as yet unclear. However, their usefulness in enabling adaptation to occur under non-growing conditions is patent. Consider, for instance, the diauxic growth of *E. coli* on a mixture of glucose and lactose (p. 110). When the glucose is exhausted growth temporarily stops; the lactose cannot yet be utilized because induction of the *lac* operon proteins has been prevented by catabolite repression on the part of the glucose. There ensues a lag during which protein and RNA are degraded. As conditions now favour induction of the *lac* operon proteins, resynthesis using nucleotides and amino acids liberated during this degradation results in the formation of *lac* mRNA and the *lac* operon proteins, so that gradually lactose begins to be utilized and the second stage of growth commences.

Similar events follow 'shift-down' from nutrient broth to minimal medium, but here the enzymes of many pathways, not merely one, need to be synthesized. As a result, the lag may last many hours.

SUMMARY

Complex controls of bacterial metabolism are to be expected in view of the strong selective pressures on bacteria to utilize available nutrients as efficiently as possible, and to achieve optimal growth rates while doing so. These controls may be classed as *specific*, where flow through a particular pathway is regulated, or *general*, where formation or breakdown of a whole class of macromolecules is involved.

Specific controls act either on enzyme activity or enzyme synthesis. In the former, an enzyme is *modulated*, being either *activated* or *inhibited*; the activator or inhibitor (*effector*) molecule is usually unrelated structurally to the substrates or products, and binds to an *allosteric* site distinct from the active site. Enzymes subject to allosteric modulation have certain peculiar properties, notably a subunit structure and sigmoidal, rather than hyperbolic (Michaelis-Menten) dependence of reaction velocity on substrate concentration in the presence (and often absence) of effector. This latter property creates a physiologically advantageous approximation to a threshold effect whereby the change from very low to near-maximal velocity takes place over a narrow range of substrate concentrations.

Specific controls of enzyme synthesis comprise *induction* and *repression*. A number of enzymes may be induced or repressed together; if their rates of synthesis are always proportionate, they are said to be *co-ordinately controlled*. These controls act at the level of frequency of transcription of the structural groups for the regulated enzymes into messenger RNA; a cluster of contiguous genes, rather than a single gene, may be transcribed from a given *promoter*, to form a polycistronic mRNA molecule. Three molecular mechanisms have been recognized so far. In the first, *repression control*, an *operator* region of DNA overlaps the promoter; a *repressor* can bind to the operator, preventing transcription. The repressor usually has allosteric properties in that its adoption of either an active (DNA-binding) or an inactive (non-DNA-binding) conformation is enhanced by its binding a small molecule effector. The second control mechanism is *activation control*; an activator protein binds to a site adjacent to the promoter and this stimulates the promoter's affinity for RNA polymerase. Again, the activator protein may adopt conformations of different activity consequent on binding a small molecule effector. The third mechanism, *termination/anti-termination control*, differs from the previous two in that the level of regulation is not initiation of transcription from the promoter; instead, there is a site (*attenuator* or *termination site*), in a region before translation starts, at which transcription may terminate so that the structural genes fail to be transcribed. Efficiency of termination may be subject to control.

These mechanisms are exemplified in the *E. coli lac* (lactose catabolism) and *trp* (tryptophan biosynthesis) systems. The *lac* operon proteins are induced by certain analogues of lactose and are repressed (*catabolite repression*) when the cells are growing rapidly, e.g. on a sugar such as glucose which permits better growth than lactose. Induction is mediated by a repression control mechanism; the inducers bind to repressor which then adopts the inactive conformation, permitting transcription to occur. Catabolite repression is mediated by an activation control mechanism; a protein termed CRP, when it has cyclic AMP bound, can bind to a DNA region adjacent to the promoter and stimulate *lac* transcription. Cyclic AMP levels are inversely correlated to growth rate, which explains the catabolite repression effect. The tryptophan biosythetic enzymes are repressed by tryptophan. This is achieved both by repression control and termination/anti-termination control mechanisms. Tryptophan binds to an inactive *apo-repressor* which thereupon adopts an active repressor conformation and blocks transcription. In conditions of tryptophan sufficiency, transcription frequently terminates at an attenuator; in tryptophan starvation, this premature termination does not occur.

Class I sequences can be divided into *central*, comprising the main routes of glucose breakdown and the TCA cycle and associated anaplerotic reactions, and *peripheral*, comprising the steps leading from other carbon sources to the central pathways. The central reactions are controlled principally at the level of enzyme activity. Controls are of two types, those indicating the flux through parts of the sequence coming before or after the step at which control is exerted, and those indicating the availability of energy or reducing power within the cell. The former are most often *feedback inhibitors*, in which a metabolite occurring late in a pathway inhibits an enzyme catalysing an earlier reaction. Sometimes they are *precursor activations*, in which a metabolite occurring early activates an enzyme catalysing a later step. In the latter type, energy-yielding reactions may be inhibited by nucleoside triphosphates or activated by nucleoside mono- and/or diphosphates. NADH may inhibit certain reactions involved in sequences that generate the reduced coenzymes.

The peripheral sequences are mainly controlled at the level of enzyme synthesis, by induction and catabolite repression (often mediated by cyclic AMP). Long peripheral pathways often show *sequential induction*; the enzymes are induced in blocks, the product of one block usually inducing the next. Nitrogen assimilation may also occur through peripheral pathways, in which case a 'nitrogen catabolite repression' (elicited by, for instance, NH_4^+) may exist in addition to the carbon catabolite repression.

The most characteristic controls in Class II sequences are feedback inhibition, usually by the end-product of a pathway, and end-product repression of enzyme synthesis. However, varied patterns exist, in part reflecting the problems created by branched pathways.

Multi-enzyme aggregates are often found in Class I and Class II sequences. These often, though not always, comprise enzymes mediating successive steps, in which case they probably promote efficient utilization of intermediates by eliminating the need for their diffusion throughout the cell.

General controls involving Class III reactions are often determined by the growth conditions. An important control that enables the organism to adapt to a changed environment is the sharp increase in rates of degradation of many cell components, especially RNA and protein, under conditions unfavourable for growth; this provides a supply of nucleotides and amino acids which facilitates the synthesis of new proteins.

Further comments on attenuation

Attenuation as a means of control of expression of the *trp* operon of *Escherichia coli* was described by Yanofsky in 1973 and has since been observed for other amino acid biosynthetic operons in this and other organisms. The RNA transcripts of the leader sequences which precede the structural genes are thought to contain segments which are translated. If this is so, the short 'leader peptides' contain a number of residues of the amino acid(s) in question. Figure 16 shows the amino acid sequences predicted from the nucleotide sequences of the leader transcripts in several such operons.

The results are indeed striking and indicate a very much higher relative abundance of the amino acid(s) than in overall cell protein. Based on these and many other observations, Yanofsky has formulated a theory

Operon	Predicted amino acid sequence of leader peptide	Relative abundance (moles of amino acid(s))	
		Peptide	General protein
trp	fMet-Lys-Ala-Ile-Phe-Val-Leu-Lys-Gly- TRP-TRP -Arg-Thr-Ser	2/14	1%
pheA	fMet-Lys-His-Ile-Pro- PHE-PHE-PHE -Ala- PHE-PHE-PHE -Thr- PHE -Pro	7/15	3%
his	fMet-Thr-Arg-Val-Gln-Phe-Lys- HIS-HIS-HIS-HIS-HIS-HIS-HIS -Pro-Asp	7/16	1%
leu	fMet-Ser-His-Ile-Val-Arg-Phe-Thr-Gly- LEU-LEU-LEU-LEU -Asn-Ala-Phe-Ile-Val-Arg-Gly-Arg-Pro-Val-Gly-Gly-Ile-Gln-His	4/28	8%
thr	fMet-Lys-Arg- ILE Ser- THR-THR-ILE-THR-THR-THR-ILE -THR-ILE-THR-THR -Gly-Asn-Gly-Ala-Gly	12/21	9%
ilv	fMet-Thr-Ala- LEU-LEU -Arg- VAL-ILE -Ser- LEU-VAL-VAL-ILE -Ser-VAL-VAL-VAL-ILE-ILE-ILE -Pro-Pro-Cys-Gly-Ala-Ala- LEU -Gly-Arg-Gly-Lys-Ala	15/32	18%

Figure 16. Leader peptides in operons of *Escherichia coli* and *Salmonella typhimurium*.

The predicted amino acid sequences for the leader peptides of the *trp*, *pheA*, *his*, *leu*, *thr*, and *ilv* operons. Amino acids which regulate the respective operons are in red. Their relative abundances in the leader peptides and in general cell protein are also given.

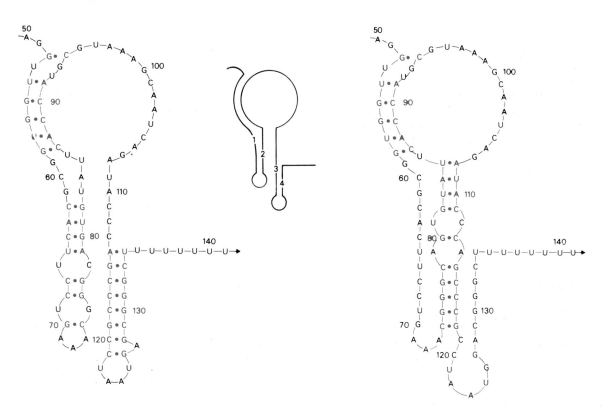

Figure 17. Alternative secondary structures of the *trp* leader transcript of *Eschesichia coli*. Nucleotide residues 50–141 showing (a) intra-chain H-bonding between strands 1 and 2 and between 3 and 4; and (b) intra-chain bonding between strands 2 and 3. The numbering of the strands is indicated in (c).

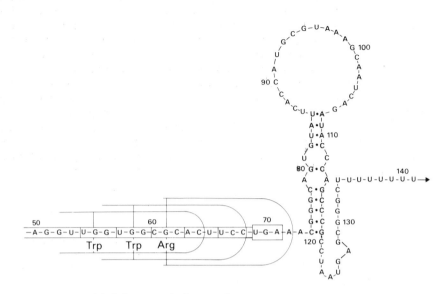

(a) Cells starved of tryptophan

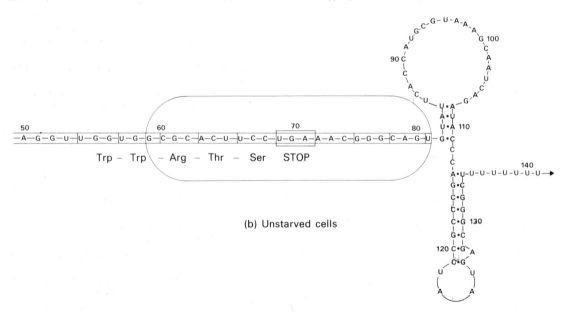

(b) Unstarved cells

Figure 18. Representation of ribosomes translating part of the *trp* leader transcript of *Escherichia coli.* (After Yanofsky, 1981)
   It is proposed that a ribosome translating the transcript sterically blocks about 10 nucleotides downstream and thus determines which of the two alternative secondary structures may form during transcription. If it is the 3:4 form, then the RNA polymerase terminates transcription at the attenuator.
   (a) Cells starved of tryptophan: a ribosome 'stalled' at a tryptophan codon (UGG) for lack of Trp-tRNA^{trp} is thought to allow formation of the 2:3 secondary structure and prevent pairing of 3 with 4 while the polymerase is synthesising strand 4. This permits read-through into the structural genes and the enzymes for tryptophan biosynthesis can be made.
   (b) Non-starved cells: the leader peptide is translated to the termination codon UGA (69–71) and the 2:3 pairing does not occur. However, 3:4 pairing does, and this causes termination of transcription at the attenuator and the structural genes are not read.

which is compellingly attractive. For the *trp* operon, the leader transcript of *trpL* (see Figure 9, p. 338) can adopt the alternative secondary structures shown in Figure 17a and b. If the parts of the sequence are designated 1, 2, 3, and 4 (see Figure 17c), intra-chain H-bonding occurs between 1 and 2 and between 3 and 4 in Figure 17a but between 2 and 3 in Figure 17b. The theory proposes that soon after a ribosome-binding site has been transcribed, a ribosome begins translating the leader region from the codon AUG/(27–29) and approaches the advancing RNA polymerase ahead of it. If tryptophan is deficient, the ribosome stalls when it reaches one of the Trp codons (UGG, 54–56 or 57–59); if tryptophan is plentiful, translation proceeds to the leader peptide stop codon UGA (69–71). Stalling allows formation of the 2:3 secondary structure and permits the polymerase to transcribe the structural genes *trpE, D, C, B,* and *A*. Otherwise, the structure 3:4 forms and causes the polymerase to end transcription at the attenuator (see Figure 18).

Investigation of other amino acid biosynthetic operons (see Figure 16) has broadly supported the theory that changes in the secondary structure of transcripts are responsible for the regulatory responses (see Yanofsky, 1981).

FURTHER READING

1 Atkinson D.E. (1970) Enzymes as control elements in metabolic regulation. In *The Enzymes*, volume 1 (ed. Boyer P.D.), 3rd edition, p. 461.

2 Baumberg S. (1981) The evolution of metabolic regulation. In *Molecular and Cellular Aspects of Microbial Evolution, 32nd Symposium of the Society for General Microbiology* (eds. Moseley B.E.B., Collins J.F. and Carlile M.J.), in press.

3 Chamberlin M. and Losick R. (eds.) (1976) *RNA Polymerase*. Cold Spring Harbor Laboratory, Cold Spring Harbor, N.Y.

4 Clarke P.H. and Ornston L.N. (1975) Metabolic pathways and regulation. In *Genetics and Biochemistry of Pseudomonas* (eds. Clarke P.H. and Richmond M.H.), Chapters 6 and 7. Wiley, London.

5 Cole J.A. (1976) Microbiol gas metabolism. *Adv. Micro. Physiol.* **14**, 1 (refers to nitrogenase and anaerobic respiration).

6 Crawford I.P. and Stauffer G.V. (1980) Regulation of tryptophan biosynthesis. *Ann. Rev. Biochem.* **49**, 163.

7 Englesberg E. and Wilcox G. (1974) Regulation: positive control. *Ann. Rev. Genet.* **8**, 219.

8 Goldberger R.F. (ed.) (1979) *Biological Regulation and Development, vol. 1: Gene Expression*. Plenum Press, New York and London.

9 Jacob F. and Monod J. (1961) Genetic regulatory mechanisms in the synthesis of proteins. *J. Mol. Biol.* **3**, 318.

10 Jacob F. and Monod J. (1961) On the regulation of gene activity. *Cold Spring Harbor Symp. Quant. Biol.* **26**, 193 (two classic papers).

11 Koshland D.E. Jr. (1970) The molecular basis for enzyme regulation. In *The Enzymes*, volume 1 (ed. Boyer P.D.), 3rd edition, p. 342.

12 Lewin B. (1974) *Gene Expression – 1: Bacterial Genomes*. Wiley, London (in particular pp. 272–344).

13 Magasanik B. (1976) Classical and postclassical modes of regulation of the synthesis of degradative bacterial enzymes. *Progr. Nucl. Acid. Res. Mol. Biol.* **17**, 99.

14 Magasanik B., Prival M.J., Brenchley J.E., Tyler B.M., DeLeo A.B., Streicher S.L., Bender R.A. and Paris C.G. (1974) Glutamine synthetase as a regulator of enzyme synthesis. *Current Topics in Cellular Regulation* **8**, 119.

15 Miller J.H. and Reznikoff W.S. (eds.) (1978) *The Operon*, 2nd edn. Cold Spring Harbor Laboratory, Cold Spring Harbor, N.Y.

16 Monod J., Changeux J.-P. and Jacob F. (1963) Allosteric proteins and cellular control systems. *J. Mol. Biol.* **6**, 306.

17 Monod J., Wyman J. and Changeux J.-P. (1965) On the nature of allosteric transitions: a plausible model. *J. Mol. Biol.* **12**, 88 (two influential papers).

18 Ornston L.N. and Parke D. (1977). The evolution of induction mechanisms in bacteria; insights derived from the study of the β-ketoadipate pathway. *Current Topics in Cellular Regulation*, **12**, 209.

19 Sanwal B.D. (1970) Allosteric controls of amphibolic pathways in bacteria. *Bact. Rev.* **34**, 20.

20 Switzer R.L. (1977) The inactivation of microbial enzymes *in vivo*. *Ann. Rev. Microbiol.* **31**, 135.

21 Tyler B. (1978) Regulation of the assimilation of nitrogen compounds. *Ann. Rev. Biochem.* **47**, 1127.

22 Umbarger H.E. (1978) Amino acid biosynthesis and its regulation. *Ann. Rev. Biochem.* **47**, 533.

23 Yanofsky C. (1981) Attenuation in the control of expression of bacterial operons. *Nature*, **289**, 751.

Chapter 9
Morphogenesis and Differentiation

INTRODUCTION

Previous chapters have been concerned with the production of the building blocks for cell growth, the generation of energy, the synthesis of biosynthetic intermediates and the way these are made into the polymers needed for growth and reproduction. There is however more to growth than the mere accumulation of essential polymers. Cells are composed of organelles, and each of these has a characteristic form and shape. Cells themselves have distinctive shapes depending on the species, and cell division requires not only the development of two cells resembling the parent but also the accurate distribution of newly-replicated chromosomes into the daughter cells. We are therefore concerned in this chapter with how structures are assembled from their component macromolecules, initially into simpler organelles, but ultimately into the cell itself. This is *morphogenesis*, which is defined as the development of form.

In addition to the question of how the structures found in cells are assembled, there is another level at which morphogenesis is important. In the life cycle of many organisms (particularly in eukaryotes, but also in a number of bacteria) there are several different types of cell, each with a distinctive shape, structure and often function. To take a simple example, when cells of *Bacillus subtilis* are faced with starvation for certain substrates, an actively dividing cell is gradually transformed into a dormant, highly resistant structure called an *endospore* (Figure 1). This type of morphogenetic transformation, involving cells or aggregates of cells in changes of form and function, is an example of *differentiation*. Sporulation in a single cell is the simplest type of differentiation; at the other end of the scale of complexity is the series of changes that converts the fertilized mammalian egg into the adult organism with its many different tissues and specialized cell types.

There is a fascinating range of differentiation processes to be found in bacterial systems. In most cases these, like sporulation, involve single cells, and with the exception of the streptomycetes and myxobacteria, there are few examples of co-operation between groups of cells.

Examples of differentiation in bacterial systems

The following are some of the more extensively studied examples of development in micro-organisms. These provide an insight into fundamental aspects of morphogenesis and they illustrate some of the basic developmental mechanisms which have arisen in the course of evolution.

Bacteriophage development
The simplest morphogenetic systems are those in which lytic bacteriophages undergo development in their hosts, and the molecular events that take place during the development of some are known in detail. Figure 2 illustrates the infection of *E. coli* by phage T4 and the subsequent development of the phage. For each infected cell under optimal conditions about one hundred phage particles are produced within about 25

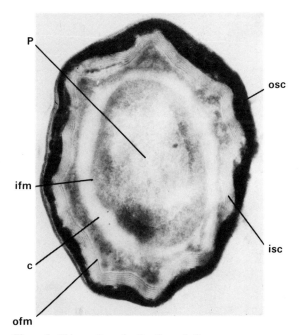

Figure 1. Thin section of a *Bacillus subtilis* spore.
The structures are: **osc**, outer spore coat; **isc**, inner spore coat; **c**, cortex; **ofm**, outer forespore membrane; **ifm**, inner forespore membrane; P, protoplast.

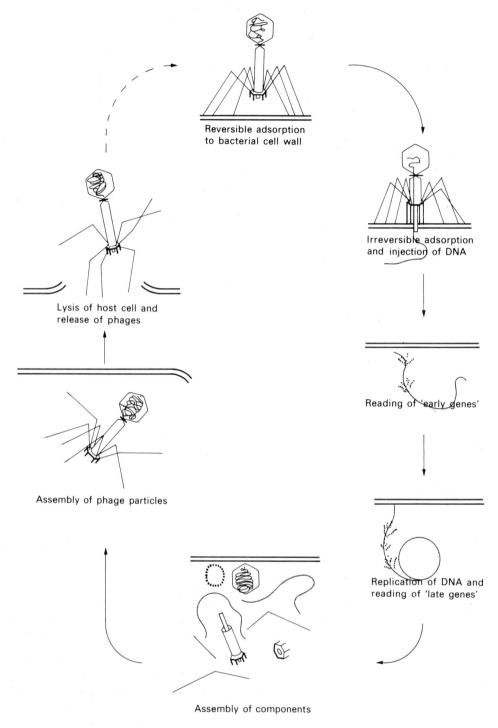

Figure 2. The lytic cycle of bacteriophage T4.

    The first step in infection is recognition of the host cell surface by the tail fibres of the phage. The tail-spikes on the baseplate of the phage then interact with the cell wall causing contraction of the sheath and injection of the DNA. 'Early' phage genes are read, some leading to switching off of host protein synthesis, others to switching on of the later genes. Phage DNA replication begins and follows a 'rolling circle' mechanism. Eventually, the phages are assembled from proteins synthesized late after infection, DNA is cleaved to the right size for packaging, and the host lyses to release the T4 progeny.

minutes. This process has several interesting features, beginning with the *recognition* of the host by the phage. This recognition is highly specific since most phages can infect only a narrow range of hosts, usually limited to a single bacterial species. Here then is a simple instance of interaction between organisms. After adsorption to the host the phage particle injects its DNA, and there follows an *ordered programme of gene expression*: 'early' phage genes are expressed and they have two functions. They switch off the host's RNA and protein synthesis and switch on reading of 'later' phage genes. These are concerned with replication of the phage DNA and with synthesis of the components needed for assembly of complete phage particles and lysis of the host cell.

For temperate phages (Chapter 7, p. 302) such as *E. coli* phage λ, there is an alternative outcome of infection—integration of the phage DNA into the host by a specific recombination process. In this state the information for lytic development is repressed and the phage DNA is transmitted to daughter cells of the host in the normal replication of the host chromosome. Phage λ has, therefore, often been used as a model to illustrate how genetic information used only by the cell during a particular stage of morphogenesis remains unexpressed at other stages of the life cycle of the organism.

The cell division cycle

Cell growth and division is seldom cited as an example of a morphogenetic process since there is no specialization of the cell and the initial and final states are similar. Nevertheless, the final state (two cells) is quite different from that initially (one cell) and several 'marker' events can be discerned including the various steps in DNA replication and septum formation. The bacterial cell division cycle is thus the simplest example of morphogenesis at the cellular level.

The cell cycle is relevant to the study of morphogenesis since it provides a simple system for examining biological *timing control*, the *growth and assembly of structures* such as cell surfaces, and the *regulation of cell size*. It will also be seen that bacterial sporulation is an example of a morphogenetic process that is related to cell division, and that it can only be induced in cells at a particular stage of the cell division cycle.

Asymmetric cell division: A prelude to bacterial differentiation

Some prosthecate (stalked) bacteria, notably *Caulobacter*, have a curiously modified cell cycle. Cell division occurs asymmetrically leading to daughters of

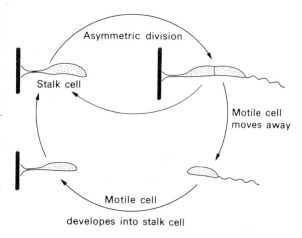

Figure 3. The life cycle in *Caulobacter*.

A sessile stalk cell attached to a surface undergoes asymmetric division to form a *motile cell* and a *stalk cell*. The latter can resume cell division immediately, while the motile cell swims away. After some time it sheds its flagellum, forms a stalk, attaches itself to a new surface, and begins cell division.

different cell type (Figure 3). Stalk cells divide by fission to produce a stalk cell and a motile cell. The stalk cell can continue growing and dividing, whereas the motile cell swims around for some time before losing its flagellum, developing a stalk and undergoing cell division. Asymmetric cell division plays an important part in the early development of higher organisms, where the products of initial cell divisions differ and eventually give rise to different cell types, and *Caulobacter* provides a simple microbial system in which the control of this asymmetry may be investigated.

Sporulation and spore germination

Endospore formation is the most extensively studied example of morphogenesis in bacteria. In the Gram-positive genera *Bacillus*, *Clostridium* and *Sporosarcina*, sporulation occurs in cells facing starvation for either carbon or nitrogen substrates. Through a highly ordered sequence of morphological and biochemical steps a vegetative cell gives rise to a dormant spore which differs markedly from the vegetative cell in structure (see Figure 1) and chemical composition as well as being very resistant to heat, organic solvents and ionizing or ultraviolet radiation.

The morphological changes occurring during sporulation in *Bacillus* and *Clostridium* species have been extensively characterized by electron-microscopy and

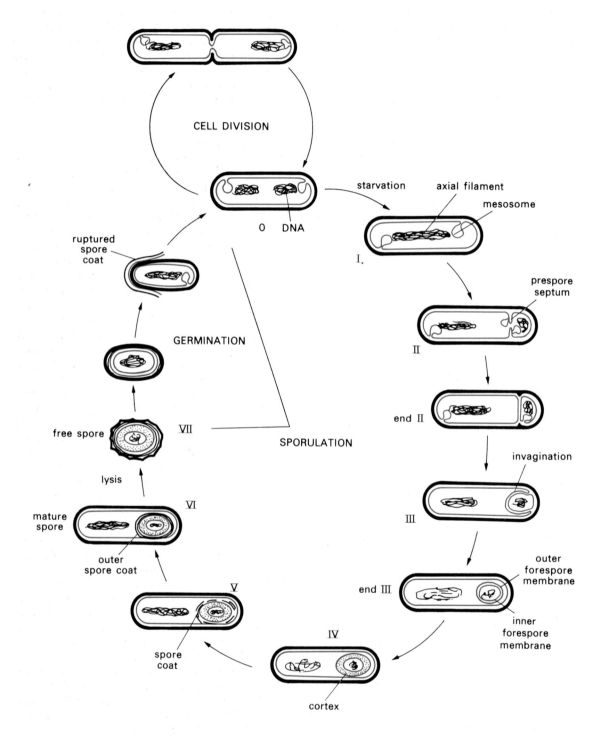

Figure 4. The life cycle of *Bacillus subtilis*, including sporulation.

The seven stages of sporulation are indicated by Roman numerals. The cell membrane is shown in red to indicate the processes of septation and invagination of the prespore.

are outlined in Figure 4. Sporulation has been arbitrarily divided into seven stages to provide a convenient system for correlating the biochemical, morphological and genetic events taking place.

Stage O represents the vegetative cell. The first recognizable event is the formation of an axial filament of chromatin (Stage I) followed in Stage II by an invagination of the cell membrane near one of the poles of the cell. This gives rise to a septum separating the forespore and the mother cell, each containing a complete chromosome. Note that these changes resemble an asymmetric cell division, although cell wall material is not deposited in the developing septum. During Stage III, growth of the mother cell membrane leads to gradual engulfment of the forespore to form a prespore which is now enclosed within the mother cell membrane. The prespore at the end of Stage III is completely enclosed by two membranes which, if one examines their mode of formation, can be seen to have *opposite polarity*. In other words if both prespore and mother cell protoplasts were stimulated to synthesize peptidoglycan, it would be formed by both in the space between the two membranes of the developing spore. In fact, during Stage IV a poorly staining material accumulates between these membranes, this is the *cortex* and it is composed of a modified peptidoglycan. Once this cortex is almost complete a densely-staining layer composed of a protein (containing relatively large amounts of cystine and lysine) assembles in lamellae around the outside of the spore (Stage V) to form the *spore coat*. During this stage the spore becomes refractile and bright when viewed in the phase-contrast microscope. In Stage VI the spore matures and is released by lysis of the mother cell (Stage VII).

The sporulation process involves a sequence of events occupying about 8 hours which is long in relation to the cell division cycle (about 1 hour), and presents an important system in which to study the control of initiation, ordering and timing of events during development. Sporulation can be made to occur synchronously in a population of cells if they are rapidly harvested and resuspended in a starvation medium. Most studies of the control of spore formation have been done using *Bacillus subtilis*, since, for this species, techniques have been developed which allow extensive genetic analysis by transduction and transformation.

Bacterial spores are dormant with no detectable metabolic activity and they can remain viable for decades, possibly, centuries. They germinate on being placed in a suitable solution, which need contain only a low concentration of a single specific compound

(e.g. L-alanine). Germination involves a sudden contraction of the cortex, the loss of Ca^{2+} and dipicolinic acid (DPA) from the spore core, and a swelling of the spore protoplast as water is taken up. If sufficient nutrients are available in the medium, the germinated spore will undergo *out-growth*. This is a sequence of changes ending in restoration of the vegetative state.

Cyst and exospore formation

Cysts are thick-walled, dormant structures derived from an entire vegetative cell by the synthesis of the thick outer layers external to the protoplast membrane. In some genera, including the nitrogen-fixing *Azotobacter* and some of the methane-oxidizing bacteria encystment occurs in individual cells. In contrast, the Myxobacteria produce cysts in multicellular fruiting structures. These often have a complicated architecture (see Figure 5) and their assembly requires a considerable degree of co-operation and co-ordination of individual cells within the mass. Indeed, the life cycle of the Myxobacteria involves a variety of highly organized cell–cell interactions which are not found in other prokaryotes but which bear a striking resemblance to those seen in the eukaryotic slime moulds such as *Dictyostelium discoideum*.

The life cycle of *Myxococcus xanthus*, which forms less complicated spherical fruiting structures, is illustrated in Figure 6. Myxospores germinate in the presence of any of a number of nutrients and grow out to form long slender rod-shaped cells. These move by *gliding motility*, a process not requiring flagella but involving the secretion of a slime layer on the surface. On solid surfaces the myxobacteria migrate as a group by a phenomenon known as *swarming*. Individual cells at the edges of the swarm can move away but rapidly return to the main body of the swarm. There is obviously considerable communication between cells even in the vegetative state and this can be seen very clearly in '*signal waves*' (i.e. waves of cell movement) propagated through swarms. Fruiting is induced by a reduction in the concentration of specific amino acids in the medium. At points within the swarm, centres of aggregation form and the myxobacterial cells stream to these centres in response to a chemotactic stimulus. They then form fruiting bodies which, depending on the genus concerned, may be simple mounds of cells or complicated stalked structures. Within these, the vegetative cells gradually undergo a shape transition to form rounded, resistant, optically refractile and dormant cells called myxospores. In some myxobacteria the 'spores' are contained in specialized structures termed cysts (not to be

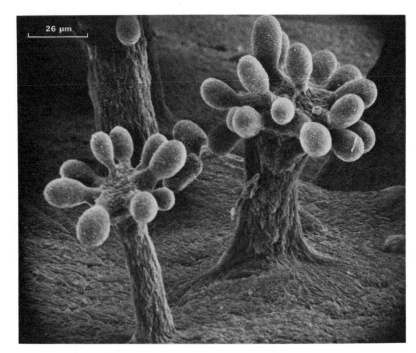

Figure 5. Mature fruiting body of *Chondromyces crocatus* as seen by scanning electron microscopy.

This myxobacterial species produces a relatively complicated fruiting body containing myxospores within the branch-like myxocysts. The bar represents 10 μm. Reproduced with permission of P. Grilione and J. Pangborn. *J. Bacteriol.* **124** (1975) 1558–1565.

confused with the *Azotobacter* cysts which resemble the myxospores).

The myxobacteria therefore represent an ideal *prokaryotic* system in which to study intercellular interactions such as communication between cells.

Two other groups of prokaryotic organisms provide further interesting examples of morphogenesis. In the filamentous actinomycetes there are two types of 'spores' formed. In *Thermoactinomyces vulgaris* endospores are formed in a very similar way to those in *Bacillus*, while in *Streptomyces coelicolor* aerial mycelia form septa synchronously at regular intervals and a thick spore wall forms around each 'cell' ultimately giving chains of spores held together within the remnants of an old cell wall. The filamentous form and sporulation of *Streptomyces coelicolor* are shown in Figure 7.

In the filamentous blue-green algae, two differentiated cell types occur in addition to the vegetative cell. One stage is an exospore form termed an *akinete*; the other, the *heterocyst*, is a granular cell which has a much thicker wall than the vegetative cell. These heterocysts are formed in the absence of readily assimilable nitrogen substrates and appear to be functionally specialized as sites at which the fixation of nitrogen is localized. Blue-green algae produce oxygen during photosynthesis, and heterocysts may provide the highly reducing conditions needed for nitrogen

fixation in a filamentous organism. It is interesting to note that the number of vegetative cells between heterocysts is a direct function of the concentration of ammonium ion in the medium: possibly an NH_4^+ gradient is set up through the cells between the adjacent heterocysts, and new heterocysts develop between previous ones where the NH_4^+ concentration was minimal.

The main features of differentiating systems

We have already stressed that differentiation involves the *assembly of cellular structures*. New structures are formed by an ordered assembly of subunits (usually macromolecules) and the aim of morphogenetic research is to determine how molecules aggregate, what controls operate to ensure precision and order in assembly, and what determines the shapes, not only of cells, but also of the various structures found within cells. This also requires knowing how molecules move from their point of synthesis in the cell to the site at which they are incorporated into the final structure.

A second feature of differentiating organisms is their capacity to control the expression of genetic information concerned with each particular development process. For example, populations of

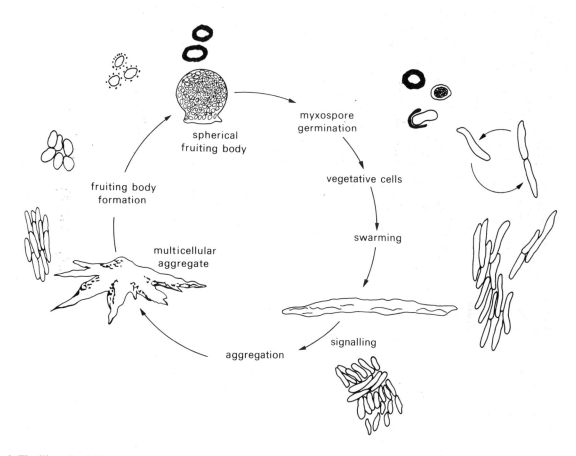

Figure 6. The life cycle of *Myxococcus xanthus.*
Myxospores germinate to form rod-shaped, vegetative cells which swarm over the surface of the substrate. On starvation, the cells begin aggregating with considerable signalling between cells (seen as wave-like patterns in the cell mass). Eventually, the multicellular aggregate forms into a relatively simple fruiting body. During formation of this, the cells change from slender rods to spherical, thick-walled myxospores. The appearance of cells at different stages of the life cycle is shown in red.

sporulating bacteria can exist indefinitely in a vegetative, dividing state. Then, on receiving the appropriate stimulus (starvation) their metabolism is diverted into a series of changes leading ultimately to formation of spores. There are two main points here: first, cells are switched from one phase of the life cycle to another in response to an environmental stimulus. This process is often termed *initiation* or *induction*. Secondly, once initiated the cell undergoes a sequence of changes that are strictly controlled both in their *timing* and their *order*.

Finally, in the differentiation of multicellular organisms there is often a great deal of co-ordinated activity involving many types of intercellular interactions.

These assume more importance in eukaryotic organisms but there are a few instances in bacteria in which interactions between cells in a population are significant. For instance, during the adsorption of a bacteriophage to its host and during bacterial conjugation (Chapter 7, p. 294) specific *recognition* events take place at the cell surface. At many stages in the morphogenetic development of multicellular organisms, and particularly during cellular aggregation, cells need to *communicate* with each other over a distance. After aggregation of cells there follows one of the most intriguing and least understood processes in biology: the *formation of patterns*. These result from the interaction of cells with each other to produce, for

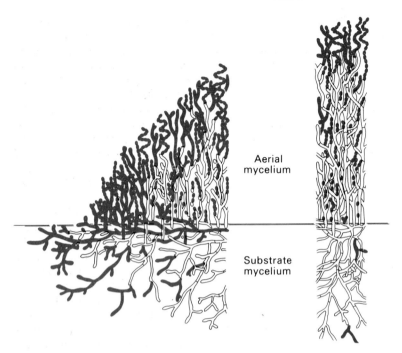

Figure 7. Sporulation in *Streptomyces coelicolor*.

Spores form in chains within aerial mycelia. The diagram shows a vertical section through the centre of a colony on agar. The white areas represent zones at which lysis is occurring. Reproduced with permission of H. Wildermuth. *J. gen. Microbiol.* (1970) **60**, 43–50.

example, the delicate structures seen in some myxobacterial fruit bodies, or the regular array of heterocysts in blue-green algae.

Even the relatively simple bacterial systems undergoing differentiation are biochemically very complicated and there are therefore many gaps in the present knowledge of even the most extensively studied bacterial system, sporulation. It is probably best therefore to begin at a simpler level of morphogenesis, and gradually build up a picture of what is known in more complicated systems.

ASSEMBLY AND SHAPE DETERMINATION

In order to understand how assembly occurs it is necessary not only to separate a structure into its basic subunits, but also to *reconstitute* it if possible. Reconstitution experiments using different combinations of subunits, or even modified subunits (for example those derived from mutants), reveal not only the *order* and possible *mechanisms* of their assembly but also the particular role of each component in the structure. When the final structure is composed of many species of subunit as is the ribosome, some components may have a catalytic role, others may be required to maintain the shape, while still others may play little part in

the active function of the structure but are required during the assembly process to determine the final shape.

There is a long way to go before the assembly, or reconstitution, of anything as complex as a bacterial cell can be attempted. Considerable progress has however been made with isolated smaller structures such as pili, flagella, ribosomes and even with membranes and cell walls, although probably the best understood systems are those involved in the assembly of bacteriophage, notably phage T4 of *E. coli*. To begin, we can consider the organization of components in cellular structures, and see how nearly all biological structures are built up by the *repetition* of smaller subunits.

Subunits in assembly

The assembly of a large structure by the repetitive association of a few types of subunits is informationally economical. This economy in coding is seen most clearly when one considers the small RNA bacteriophages, such as R17 and Qβ. These phages contain an RNA molecule enclosed within a nearly spherical coat (or *capsid*). The coat is composed of 180 molecules of a specific polypeptide and therefore, only one gene is needed to code for the whole phage coat structure. This is just as well since the RNA of these phages can code for only three proteins.

In addition to conferring economy in coding, assembly processes making use of identical subunits have a reduced likelihood that errors will accumulate in the final structure, assuming that defective subunits are rejected before assembly takes place. To take an example which is somewhat oversimplified, but nonetheless illustrative, Caspar calculated that there would be a 90% chance of making a correct virus shell if it were assembled from 1000 identical proteins, each containing 100 amino acid residues, and if the chance of an error at each step in protein synthesis were 1 in 10^4. With the same probability of error at each step in synthesizing a single protein molecule of comparable size (10^5 amino acid units), the chances of building a correct structure would be 0.005%.

There are numerous examples in bacteria of structures composed of subunits. At a very simple level, many enzymes are oligomeric; this property has been discussed in Chapter 8, p. 326. Flagella and pili when viewed by negative-straining methods (see p. 78, Chapter 1) are seen to consist of subunits arranged regularly around a hollow core. In addition to these tubular forms there are sheet-like structures. These may be composed of a regular array of subunits as seen in the bandage-like spore coats or the sheet-like exosporium of *Bacillus* spp. (Figure 8), or of the fluid, semicrystalline array of non-identical subunits seen in membranes. Even the peptidoglycan and some other polymers of the bacterial cell wall are composed of repeating units, although these differ from those in other structures since they are covalently linked during assembly. Ribosomes illustrate a rather different type of assembly since they are composed of a large number of non-identical subunits (in *E. coli* there are 3 RNA species and 55 different polypeptides, p. 243) which are nonetheless assembled into an extremely ordered structure.

In most of these examples the subunits are proteins and one can ask how protein molecules can recognize each other, whether any energy must be provided by the organism for assembly to occur, and how subunits can aggregate to form correct shapes. To answer these questions in detail we can begin by examining one of the simplest examples of interaction between polypeptides—that seen in the formation of oligomeric proteins.

Aggregation of subunits: Self assembly of multimeric proteins

A polypeptide chain contains sufficient information in its amino acid sequence for it to fold spontaneously to its native secondary and tertiary structure, and no input of energy is required for this process. This has been shown with a number of enzymes which can be denatured *in vitro* to a random configuration with no enzymic activity; activity and the native configuration may be regained after careful removal of the denaturing agent. Even the association of several polypeptides to form oligomeric enzymes (i.e. with quaternary structure) occurs spontaneously, but the bonding is not usually covalent and the degree of association of subunit polypeptides and their precise conformation can be influenced markedly by low molecular mass ligands.

At first sight, self-assembly appears to be thermodynamically anomalous since there is an apparent decrease in entropy (i.e. an increase in order) as the components associate. However, this has been resolved for a number of interactions, including those involved in the dimerization of insulin, in the formation of the trypsin–trypsin inhibitor complex and in the association of subunits in haemoglobin. These interactions occur in solution and the thermodynamic system includes not only the polypeptide chains which are associating, but also the solvent molecules in contact with the subunits and with the assembled complex. On association of subunits there is a favourable contribution to the overall free energy change (see p. 125) as a result of a decrease in surface area of the polypeptides accessible to water. Some water molecules displaced from the interface of the two associating proteins gain entropy as they pass from the more ordered state at the surface of the protein to the less ordered one of molecules in solution. This is part of the phenomenon usually referred to as *hydrophobic bonding*, and it involves mainly the non-polar side chains of the amino acids. The more polar side chains can also contribute to the binding. In the case of the interaction of trypsin-inhibitor with trypsin some polar side chains are 'buried' during complex formation, and most of them (25 of 29) form hydrogen bonds. These contribute to the stability of the complex, but much less than do the hydrophobic bonds.

Assembly of subunits is a highly specific process since mixing experiments indicate that association only occurs between subunits of a particular complex and not between those from different complexes. It is believed that the specificity is conferred by the complementary nature of the two surfaces which from the interface between the adjacent polypeptides. This view is supported by X-ray diffraction studies which indicate 'close-packing' of the amino-acyl residues at the interface of associated polypeptides.

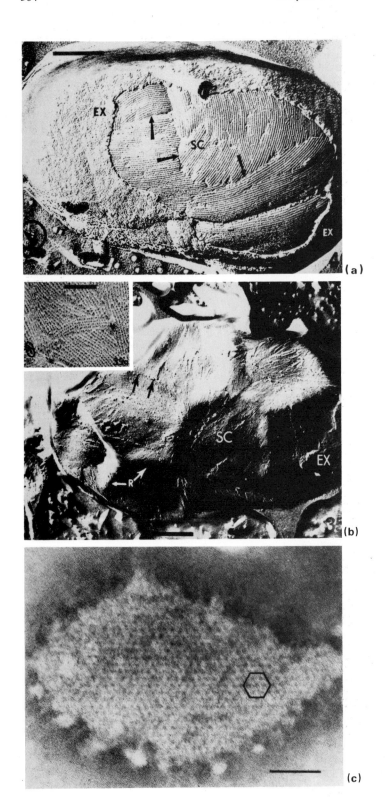

Figure 8. Subunits in the assembly of spore structures.

Freeze-etch electron micrographs showing the structure of the exosporium (ex) and spore coats (sc) of *Bacillus licheniformis* (a) and of *Bacillus fastidiosus* (b) with an inset at higher magnification to show that the spore coat is composed of particles of about 0·5 nm in diameter. In (a) and (b) the bars represent 0·25 µm, in the inset to (b) 0·1 µm. Reproduced with permission of S.C. Holt and E.R. Leadbetter. *Bacteriol. Rev.* **33** (1969) 346–378.

(c) Hexagonal array of subunits seen in a sheet reconstituted from the exosporium of *Bacillus cereus*; the bar represents 0·05 µm. Reproduced with permission of T.C. Beaman, H.S. Pankrantz and P. Gerhardt. *J. Bacteriol.* **107** (1971) 320–324.

Self assembly of larger structures

In addition to the oligomers that we have discussed spontaneous assembly of subunits accounts for the formation of filamentous structures which are much larger. These include: filamentous viruses (such as tobacco mosaic virus, TMV); tail components of many bacteriophages (including phage T4); flagella and pili of bacteria; and, microtubules and actin filaments found in eukaryotes, all of these are cylindrical structures formed by *helical* association of subunits. A helix is formed when identical subunits associate by the repetition of a single two-site bonding pattern (Figure 9). In such a structure all subunits, except those that are terminal, occupy equivalent positions with respect to their neighbours.

Reconstitution studies on filamentous structures have been very successful, particularly with TMV and with bacterial flagella. Probably more is known about assembly of the virus, but for the present purpose we will discuss flagellar formation since it concerns a cellular structure and is also not complicated by the interaction with the viral nucleic acid (RNA) seen in TMV assembly.

Flagella of many bacterial genera, including *Proteus*, *Bacillus* and *Salmonella* have been studied extensively. These long filamentous structures contain many thousands of subunits which are probably arranged in a three-fold helical array. They can be detached from cells by gentle shear and when treated with heat, acid, alkali or denaturing agents, disaggregate into single molecules of a polypeptide, *flagellin*, of relative

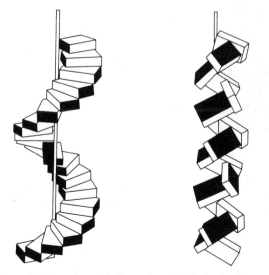

Figure 9. Formation of helices when identical subunits are joined with repetition of the same bonding pattern (see D.J. Kushner (1969) *Bacteriol. Rev.* **33**, 302–345.

molecular mass 20 000 to 40 000, depending on the species. Flagellins have been re-aggregated by a number of techniques which do not require the input of energy. For example, gradually increasing the pH of acid-dissociated flagella of *Bacillus pumilis* leads to the formation of straight and wavy (flagella-like) filaments. Straight filaments are less stable and can be converted to the wavy state seen in native flagella. Studies with non-motile mutants have shown that the shape of flagella is determined by the primary sequence of amino acids in the flagellin molecule. Thus, a non-motile mutant of *Bacillus subtilis*, having a single amino acid alteration in the flagellin molecule, produces straight instead of wavy flagella, and flagellin isolated from this mutant re-aggregates to give straight flagella. In general, flagellins contain relatively high proportions of non-polar amino acyl residues, and they can polymerize under conditions of high ionic strength in which intermolecular hydrophobic bonds are more important than ionic ones.

So far we have discussed structures composed of identical subunits. In considering those composed of different subunits we need to know whether assembly is still spontaneous, whether the subunits assemble in any particular order, and if so, what dictates this order.

The bacterial ribosome affords a good example of a complicated structure which is built up from many non-identical subunits, but a fair amount is known about this very complex assembly process.

Assembly of the bacterial ribosome

The structure of the ribosome has been dealt with in Chapter 6. To summarize, the 70 S ribosome is composed of two subunits. The term 'subunit' is used in a special sense since both of the ribosomal subunits are themselves built up from many components. The 30 S subunit contains one RNA species (16 S RNA) and 21 polypeptides and the 50 S contains two RNA species (5 S and 23 S) and 34 polypeptides. The RNA species are transcribed as a single large precursor RNA. This is then cleaved at certain sites in a number of steps involving ribonuclease III and several other RNAses to produce eventually the 16 S and 23 S species (see Chapter 5, Figure 38, p. 227). The assembly in the cell of ribosomal proteins on the RNA probably starts before this *processing* is complete; also some of the RNA bases are methylated after transcription. Such modifications to the RNA may be of importance in the assembly of the final structure, or in stabilizing it, or even in its activation. (In yeasts which are eukaryotic and have rather larger ribosomes of different composition, blocking methylation of the RNA does interfere with the ribosome assembly process.) In addition to these changes to the RNA, there are post-translational alterations in ribosomal proteins. Some are phosphorylated by a protein kinase catalysing the transfer of phosphoryl groups from ATP to serine residues in the proteins. Such changes may occur during assembly to assist the overall process—or they may occur at a later stage to modify or regulate the activity of the ribosome in protein synthesis.

Despite these changes, Nomura and his colleagues have managed to reconstitute 30 S particles *in an active state* from the components isolated from the mature 30 S subunit, as well as from isolated polypeptides and high molecular weight precursor RNA. *This indicates that most, if not all, of the information for assembly of the 30 S subunit is intrinsic to the 16 S RNA and the 21 polypeptides, and that the process of assembly can occur spontaneously.* It has been shown that the reconstituted subunits are fully functional in protein synthesis.

The success of these studies depended on the painstaking search for optimal conditions such as the concentration of magnesium ions, ionic strength, and the temperature for the reconstitution. These conditions would be expected to affect reconstitution if ionic bonding contributed substantially to assembly and stability. Similar work has been done with the 50 S ribosomal subunit. Reconstitution was first achieved with the larger subunit from *Bacillus stearothermophilus* and then with that from *E. coli.*

Order of assembly

The question of whether or not the components of the ribosome have to be assembled in a specified order has been answered by *in vitro* reconstitution experiments in which each component of the 30 S subunit is omitted in turn from the reconstitution mixtures, and by examining the ability of isolated proteins to bind to the RNA. Early studies indicated that only some of the proteins (S4, S7, S8, S15, S17, S20) bound directly to 16 S RNA and that subsequent binding of other proteins depended on co-operative interactions between them and some of the proteins already bound, and also the RNA. On the basis of these results complex assembly maps have been constructed (Figure 10). Recently, however, it has been shown that the protein-binding properties of the 16 S RNA are altered if the RNA is differently prepared. Protein binding sites can therefore be either obscured, or exposed, according to the conformation of the isolated RNA.

From this it is clear that there is some order in the assembly of the ribosomal proteins, although *in vivo* it may differ from that shown in Figure 10. Presumably the binding of the first proteins assists further assembly not only by promoting protein–protein interactions, but also by assisting in the appropriate folding of the RNA species to facilitate RNA-protein interactions as well. These reconstitution experiments are valuable not only in studies of assembly, but, taken in conjunction with biochemical studies of appropriate mutants, they have given insight into the role of each component in protein synthesis.

So far it would seem that assembly is a spontaneous process. The components of the structure are formed, and when they reach a certain concentration begin to aggregate without any further intervention by the cell. This is not, however, the complete picture since there are many examples in which proteins not appearing in the final structure intervene to assist or direct the course of assembly. This is seen most clearly in the assembly of the more complicated bacteriophages.

Bacteriophage assembly

Before considering the mechanisms of assembly of complex bacteriophages, it is useful to know how the subunits are arranged in the simpler spherical viruses. These are not strictly spherical but are composed of subunits arranged in icosahedral patterns. These represent the lowest free-energy arrangement in which subunits can form closed shells. Icosahedra have 12 vertices, sides and faces. Each vertex is pentameric, that is it has five faces meeting at the vertex. If subunits aggregate with two or more types of bond (cf. helical arrangements), they can form sheet-like structures. When hexagonally packed, the subunits in such sheets are equivalent. If these same subunits are however arranged into pentamers they have similar bonding angles and spatial relationships to each other, but a *vertex* is formed. Twelve such vertices are needed to form an enclosed structure. The capsids of some simpler viruses are in fact composed entirely of pentameric arrangements of subunits. For example, the 60 subunits in the phage ϕX174 are arranged in equivalent positions to form the twelve pentamers. In larger, symmetrical, viruses hexamers are interspersed among the twelve pentamers in a regular way. In such a structure the subunits differ slightly depending on whether they occur in hexamers or pentamers; despite this the structures have great stability.

So far, several of the simpler viruses have been reconstituted with some degree of success, although in few cases has there been a high recovery of viability. In using the RNA and protein extracted for re-aggregation studies, structures resembling the native particles are formed spontaneously but have very low

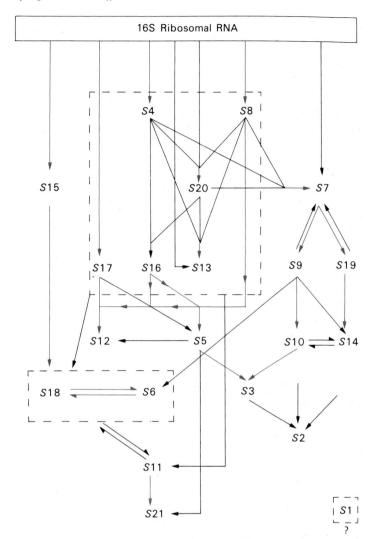

Figure 10. Assembly map for reconstitution of 30 S ribosomal subunits of *Escherichia coli.*

Arrows between proteins (S) indicate facilitation of one protein on the binding of another—red arrows indicate a major effect. See M. Nomura and W.A. Held in *Ribosomes* (eds. M. Nomura, A. Tissières and P. Lengyel), p. 200 (1975). Cold Spring Harbor Laboratory Press.

infectivity. Either the correct conditions of pH, ionic strength, and ionic composition have yet to be found or there is some process occurring during 'natural' assembly which leads to an irreversible change. A change of this type occurs in the assembly of the head of the much more complicated T4 bacteriophage. The genome of phage T4 codes for at least 80 genes, and at least 25 gene products appear in the intact phage particle. The particle is composed of several distinct structures—a head and a tail (consisting of an outer contractile sheath and an inner hollow core) terminating in a hexagonal baseplate to which are attached short tail pins and long tail fibres (Figure 2). Conditionally-lethal mutants of phage T4 have been characterized and at least 40 genes have been found to be involved in the formation or assembly of the various

phage structures, since mutations in these genes give rise to incomplete or abnormal phage particles. Biochemical studies of these mutants have contributed significantly to this research. The proteins produced by phage T4 after infection of the host cell can be separated by polyacrylamide electrophoresis (see page 372), and in many cases the phage genes coding for them have been identified. It is the accepted convention to describe each protein by the arbitrary number assigned to the gene coding for it (thus P24 is the protein produced from gene 24).

The assembly of the phage T4 head typifies that of larger structures in that there is more 'direction' to the process than there is in the straightforward self-assembly of the subunits. At least 17 phage gene products and two host proteins are involved, although

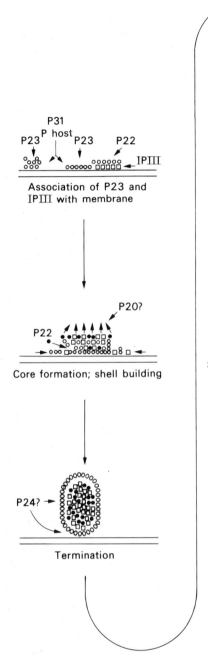

Figure 11. Proposed pathway for the early steps in assembly of phage T4 head.

Phage gene products are designated P23 for the product of gene 23, etc. (The gene for IPIII has not been identified.) An asterisk indicates gene products modified by proteolytic cleavage during assembly. See M.K. Showe and E. Kellenberger (1975). *Symp. Soc. gen. Microbiol.* **25**, 407–438.

the head is composed mainly of a single phage protein, P23.

A simplified pathway for T4 head assembly is illustrated in Figure 11. It should be noted that the proteins P31 and P22 are essential for normal assembly of the phage head but neither appears in the final phage particle. Here then is a clear example of an assisted assembly process. The preheads are released from the membrane by the action of protein P24 and are morphologically distinct from the final head, being smaller and ovoid. *Subsequently a proteolytic enzyme (the product of gene 21) cleaves P23, P24 and*

IP III. This brings about a conformational change transforming the thick-shelled prehead into the thin-shelled final structure. Such proteolytic cleavage of P23, P24 and IP III (to give P23*, P24* and IP III*) takes place only when these proteins have formed the prehead structure; they are insensitive to such proteolysis in solution.

Specific proteolytic cleavage is a common feature in the development of other bacteriophages, and appears to play a similar part in cellular development (e.g. bacterial sporulation, see page 379). Several possible functions have been ascribed to proteolysis. It may be needed to cause conformational changes and thereby extend the range of subunit interactions to accommodate a sequence of changes. It may also introduce irreversibility into an assembly process at a stage at which it might otherwise be susceptible to disarray or be energetically unfavourable.

Assembly lines

Bacteriophage T4 provides an excellent example of how larger, more complex, structures are built up from substructures. Wood and Edgar used *in vitro* complementation tests involving many combinations of phage T4 mutants to show that heads, tails and tail fibres are assembled in separate sequences and only later joined to form the complete virus. In their experiments, extracts of *E. coli* infected with individual phage mutants were mixed in pairwise combinations and the mixtures tested for the presence of viable phage particles. In general, phage mutants unable to produce normal heads could produce normal tails which accumulated in the host. Similarly, tail-less mutants accumulated normal heads. Mixing the extracts from a tail-less mutant with that from a headless one led to the production of infective particles.

Three parallel assembly lines operating in the development of phage T4 have been established using reconstitution experiments and assembly mutants. In each pathway there is stepwise addition of gene products according to the scheme set out in Figure 12. In summary, many biological structures are formed from subunits which aggregate spontaneously by means of interactions which are stabilized largely by hydrophobic bonding. The specificity with which these subunits (usually proteins) recognize each other is determined by the matching of the conformations at their sites of contact. Helical structures are formed by the assembly of identical subunits associating with each other by a single pattern of bonding; sheet and icosahedral shapes are formed by subunits interacting with each other in at least two different bond patterns. Regular sheet structures have identical subunits in

equivalent positions and, therefore, in the most stable array when hexagonally packed. More complicated structures are built up by an ordered assembly of non-identical subunits (or substructures) and for these the interaction of subunits with each other may be directed or assisted by proteins not appearing in the final structure. One of the ways in which this is accomplished is by specific proteolytic cleavage of the subunits.

ORDER AND TIMING IN DEVELOPMENT

So far we have assumed that the subunits for assembly have been present in the right amount when required. During cellular differentiation there is, however, an additional dimension, that of time, since the various morphological changes in the differentiating cell occur in a sequence over a period which is often long relative to the time taken for a cell to divide. We will now consider the timing and ordering of morphological change, partly because a lot more is known about sequential changes during cellular differentiation than about assembly and shape determination of the bacterial cell.

The main interest in this aspect of differentiation is not in just cataloguing the changes in terms of their time of appearance, but in determining how the cell controls the biochemical events, and how these bring about the morphological changes. In order to understand how a clock works it is not enough to pull it to pieces; one must observe the way in which each component interacts with, or controls the activity of, the other parts. Thus, in differentiating systems one can ask: Is there a need for new genes to be expressed during differentiation? At what level (transcription or translation) is the control of this exerted? How are the genes concerned with differentiation read to produce the necessary proteins so that they are active in the right amount at the right time? Is any control exerted after proteins have been made, by processes which modify their activity?

To try to answer these questions we will begin by examining the timing and regulation of gene expression during phage development following infection of its host and then seeing what progress has been made in the most extensively studied cellular system—bacterial sporulation. For many reasons progress has been more rapid with phage than with sporulation. This is due partly to the great thrust forward in *E. coli* genetics and the interest this engendered in research on phage, and partly to the relative simplicity of the

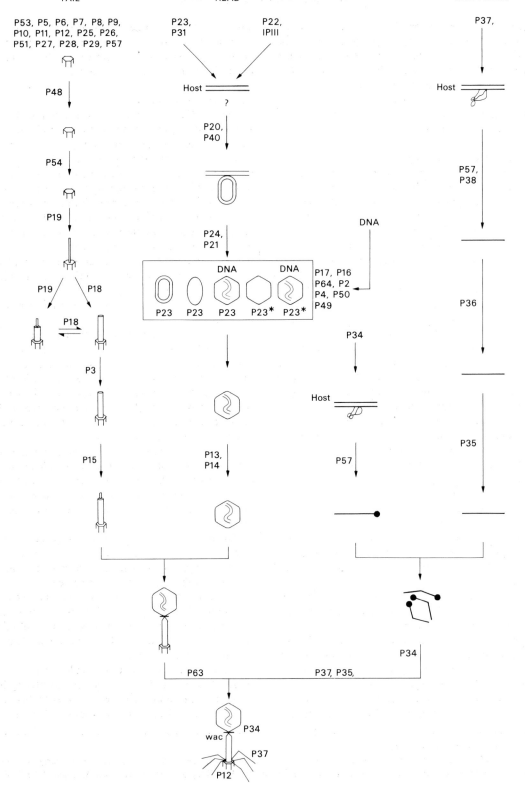

TAIL HEAD TAIL-FIBRES

phage genome. Furthermore, it is much easier to separate phage gene expression, and phage gene products from those of the host cell, than it is to separate genes and gene products concerned with sporulation in *Bacillus subtilis* from those for the vegetative functions with which they are intermingled. (Vegetative functions are those concerned with cell growth and division rather than with sporulation.)

Lytic bacteriophage development

The typical lytic cycle of phage development in its bacterial host is outlined in Figure 2. Following infection, the phage genes are read, not all at once, but according to a definite programme. Some are read only briefly, others for longer periods. This timing of gene expression clearly leads to efficient development since it ensures that the appropriate proteins for particular steps in morphogenesis are formed at the right time and in adequate amounts. As will be seen, the 'early' proteins produced are those which subvert the machinery of the host cell to the needs of the phage, and those which exclude competing host cell functions. The proteins concerned with replication of the phage nucleic acid are synthesized next, followed by synthesis of the subunits needed for assembly of the phage coats and other structures, as well as the lytic enzymes required to burst the host cell.

The sequence of gene expression can be demonstrated in a number of ways. For example, assays of individual phage gene products (where their activity is known and can be assayed) show that there is a definite order in the timing of their appearance. The sequence can be demonstrated by polyacrylamide gel electrophoresis of phage-coded RNA and proteins which have been pulse-labelled with radioactive precursors at different stages in development of the phage. Figure 13 shows the pattern of protein synthesis seen at different times after infection of *E. coli* by phage T4.

In the development of the DNA-containing bacteriophages we will see that much of the control over gene expression is exerted at the level of gene *transcription*, since phage-coded mRNA species are produced sequentially after infection, and inhibitors of transcription in bacteria (rifamycins or actinomycin D) added at different stages of phage development in the host shut off the appearance of proteins produced at later stages. However, control of transcription in

this way is not possible in RNA phages but they still show some regulation of gene expression and provide interesting models of how protein synthesis may be controlled at the level of translation.

RNA phage development

The small RNA bacteriophages (including R17, MS2 and Qβ with *E. coli* as host) have a single-stranded RNA genome coding for only three proteins: a maturation protein (concerned with the packing of RNA into the head and its release during infection); a coat protein; and a replicase (concerned in replication of the RNA). They are produced with different kinetics, since electrophoretic studies have shown that the maturation protein is produced at a linear rate, the coat protein is synthesized at an exponentially increasing rate throughout the lytic cycle, while the replicase synthesis is exponential early in infection, but is later shut off.

Several control mechanisms account for these differences:

1 Secondary structure (intramolecular base pairing) in the RNA is very extensive (see p. 270 for the base sequence and proposed secondary structure of phage MS2 RNA). The initiation site for synthesis of the replicase is buried in one such region and is inaccessible for ribosome binding until the coat protein has been translated. This therefore introduces an element of timing in the gene expression.

2 The coat protein may act as a translational repressor of the synthesis of the replicase by binding to the initiation site of its gene. In fact replicase synthesis is inhibited *in vitro* when purified coat protein is added to a protein-synthesizing system using phage RNA as messenger.

3 Differences in the nucleotide sequences which precede each gene, and which are the sites of ribosome binding, may lead to differences in the rate of initiation of the synthesis of each protein.

Such controls could ensure not only differences in the timing of gene expression but also in the relative amounts of the proteins synthesized. In R17 infections coat protein, replicase, and maturation protein are formed in the ratio 20:5:1.

These controls are relatively simple compared with those found in the DNA phages or sporulating bacteria and their importance may be rather to determine the amounts of the gene products rather than the time

---

Figure 12. The pathways for assembly of phage T4.

Numbers refer to the gene involved at each step in assembly. The pathways have been inferred from a large number of studies of structures seen in defective mutants. See M.K. Showe and E. Kellenberger (1975). *Symp. Soc. gen. Microbiol.* **25**, 407–438.

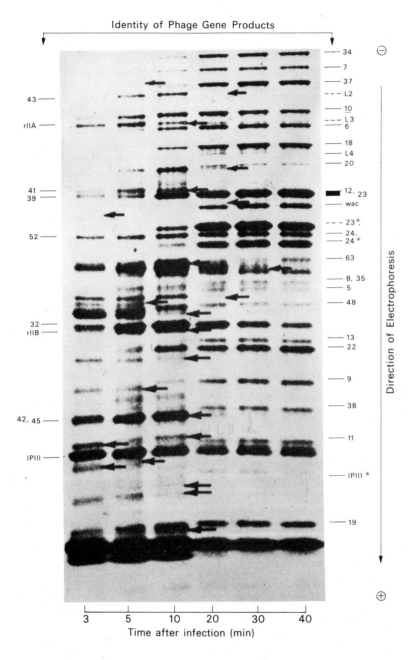

Figure 13. The changing pattern of phage-coded protein synthesis during development of phage T4 in *Escherichia coli*.

Proteins synthesized in infected cells were labelled with ^{14}C amino acids for 5 minute periods, each beginning at the time after infection indicated. Samples were electrophoresed in polyacrylamide gels to separate the polypeptides mainly according to molecular mass. Cessation of synthesis of early proteins and the beginning of synthesis of later proteins can be seen. The numbers correspond to the gene numbering for phage T4 and indicate the positions of particular gene products in the gel. Reproduced with permission of R. Wu, E.P. Guideschek, and A. Cascino. *J. molec. Biol.* **96** (1975), 539–562.

of their appearance. Several double-stranded DNA phages have been studied in sufficient detail for the mechanisms controlling gene expression to be understood and we will briefly consider three of them. (Several comprehensive articles and books are cited in the references.)

Bacteriophage T7

T7 has a simpler icosahedral structure than the T-even phages and has a smaller genome (2.5×10^7 u) which codes for about 25 polypeptides. The genes coding for these polypeptides are arranged on the genome as shown in Figure 14, on which the gene

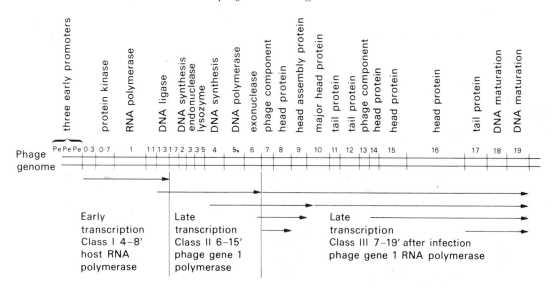

Figure 14. Genetic map of bacteriophage T7.

The genes read soon after infection are clustered at the left, and are under the control of three adjacent promoters each of which is recognized by the RNA polymerase of *Escherichia coli*. The late genes are also clustered into operons each of which is read from the late promoters which are recognized by the phage-coded RNA polymerase. Red arrows indicate the lengths of polycistronic mRNA's; the positions of the late promoters are uncertain but some transcripts overlap, for instance DNA ligase is under the control of an early and a late promoter. The numbers refer to genes.

products are indicated where known. Note the clustering of genes for related functions and that the order of these clusters is related to the sequence in which protein synthesis occurs.

The early genes (those coding for the first polypeptides to be formed after infection) code mainly for enzymes concerned with switching off transcription of *E. coli* genes and switching on the reading of the late phage genes. The late genes are concerned with replication of the phage DNA, formation of the phage structures and release by lysis of the host. The early genes form a single operon under the control of three adjacent 'early' promoters which are recognized by the *host cell RNA polymerase*. One of the products of this early transcription is a new (T7) RNA polymerase. Another is a protein kinase which catalyses phosphorylation of the host RNA polymerase and thereby prevents its transcribing either the early phage genes or the host genes. Therefore the primary switching mechanism which introduces some of the timing into protein synthesis is concerned with a change in the RNA polymerase: the host cell enzyme is inactivated and replaced by a newly synthesized enzyme.

The new polymerase transcribes the late genes only and the sizes of the RNA transcripts produced indicate that there are probably several promoters located at different points throughout the right-hand three-

quarters of the genome. Differences in the times of reading of late genes are due in part to the arrangement of these promoters on the genome, and in part to the order of genes in each cluster under the control of a promoter.

Bacteriophage T4

The lytic T-even phage (T2, T4, T6) have *Escherichia coli* as host and form a closely related group with the complicated structure whose assembly has already been discussed. About 25 proteins are found in the intact phage particle (about 8 in the head; about 15 in the tail and 3 in the tail fibres) while the T4 phage genome codes for at least 80 polypeptides. T4 is therefore much more complex than T7 and this is reflected in the regulation of its development. As with T7 the phage-coded proteins appear in an ordered way after infection (see Figure 13), moreover the corresponding mRNA species also appear sequentially, so that much (but not all) of the control is exercised at the level of transcription. Unlike phage T7 the T4 genes are not rigidly clustered according to gene function, but are grouped into quite a large number of relatively short operons. The mRNA's were once classified into 'early' and 'late' classes. However each of these classes is often subdivided again: early into *immediate early* and *delayed early*; synthesis of the latter depends on the synthesis of a phage-coded protein and is prevented by the addition of chloramphenicol. Late messenger RNA's can be classified into *middle* (or quasi-late) and *late* (true-late) since the latter are not transcribed until phage DNA replication has taken place.

Immediate early genes read before $1\frac{1}{4}$ minutes) are transcribed from promoters which are recognized by the host RNA polymerase, and the *E. coli* σ factor protein determines the specificity of this transcription (see page 231). The size of early mRNA transcripts and *in vitro* experiments using the ρ termination protein (see page 236) suggest that the switch to reading delayed early genes is due to synthesis from an immediate early phage gene of an anti-termination (anti-ρ) protein factor (see Figure 15).

Another product of the early genes is an enzyme which can modify the host cell RNA polymerase by adenylylation (addition of an AMP-moiety) of its α subunits. This, and possibly other changes (which may include the synthesis of a new phage-coded σ factor) lead to inhibition of early transcription and also of host cell transcription within 10 minutes from infection. The middle mRNA's are then produced until the completion of viral assembly, from a new set of middle promoters. Since many 'early' genes continue to be transcribed they are presumed to be under the dual control of early and middle promoters.

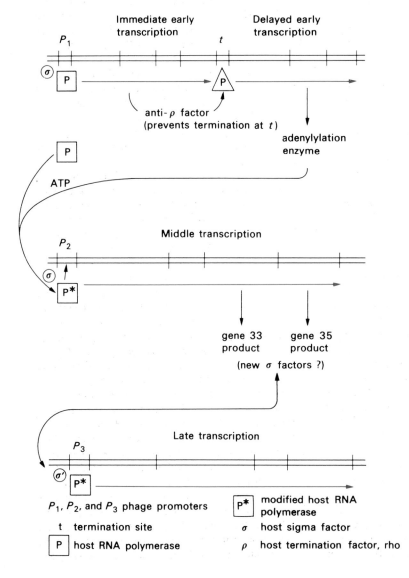

Figure 15. Representation of timing control during development of bacteriophage T4.

For simplicity, three types of operon are shown, including an early one containing a σ factor-dependent termination signal which separates genes into those which are transcribed immediately after infection from those read a little later. One product of the immediate early genes is an anti-termination factor which allows transcription to proceed into the delayed early genes. At least initially, all early genes are transcribed by the host RNA polymerase, but this is modified by an enzyme produced from one of the delayed early genes. This modification shuts off most of the early gene transcription, and the modified polymerase then recognizes a different set of phage promoters to read the middle operons. However, some early genes continue to be read—probably from new promoters. The products of the middle genes include two proteins (from genes 33 and 35) which may act as new σ factors altering the specificity of transcription again to enable reading of the late genes to occur.

The late genes, including those for structural proteins and those concerned with assembly of the complete T4 particle, are read from late promoters. The precise nature of the switching mechanism is unknown but it is dependent on the products of the early genes 33 and 55. These code for polypeptides (12 000 u and 22 000 u respectively) which bind to the modified RNA polymerase present in the late stages of infection. These may be new phage σ factors. Late *transcription* is also dependent on replication of phage DNA (or at least on the generation of single strand 'nicks' in the DNA). The molecular basis of this control is not known, however it has significance for the overall regulation of morphogenesis. The late gene products include those needed in very large quantities to package the phage DNA (i.e. the head and tail proteins). Once DNA replication has begun, many more copies of the phage DNA are available to act as templates for these proteins; this represents a form of *gene amplification*.

So far we have been discussing lytic phage development, which provides useful models for the control of a developmental process *once it has been initiated* but does not give much insight into the initiation or induction processes seen in most differentiating systems. There is, nevertheless, one group of bacteriophages which do provide an excellent model since under certain circumstances the phage genome is integrated into that of its bacterial host, and the genes for lytic development are held unexpressed. Phage λ is the best known example.

Maintenance of lysogeny in bacteriophage λ

λ is a temperate phage; after it infects *E. coli* its genome can be integrated into the host chromosome at a specific attachment site (p. 302) to form a *lysogen* (an *E. coli* containing integrated phage λ DNA). In this lysogenic state the host is immune to further infection by λ and several other phages. Most of the λ genes are not expressed, the genome is replicated as part of the host chromosome and is transmitted to all of the host's progeny. The lysogenic state is therefore in many respects analogous to the vegetative growth phase in cellular systems in which genes for a particular development (for example sporulation) are present but are not expressed.

The alternative to lysogeny is *lytic* development. In some of the host cells infected with λ DNA the process of phage development begins before lysogenization can occur and culminates in the production of several hundred phage particles released by lysis of the host cell. Even in lysogenized cells lytic development can be induced by treating them with ultraviolet light or a number of chemicals. These treatments cause the integrated λ DNA to be excised from the host chromosome; it is converted to the circular form and its genes begin to be expressed.

In order to understand how the lysogenic state is established after infection and maintained through many generations of the host, it is helpful to examine the location on the phage genetic map of the genes controlling lysogeny and the early stages of lytic development. This region, shown in Figure 16, represents about 5% of the total phage λ genome.

In the lysogenic state only one gene, *c*I, is transcribed and translated to give the *repressor* protein, see pp. 303–305. Mutations in *c*I lead to *lytic* phages which are unable either to establish or to maintain lysogeny. When *E. coli* is multiply infected with a *c*I mutant and a wild-type λ phage, the host cell becomes *lysogenic*, and so the wild-type allele of the *c*I gene is dominant. Since both amber-suppressible and temperature-sensitive mutations in *c*I have been isolated, the normal *c*I gene codes for a protein which acts to switch off all other λ genes including those of the prophage and those of any other phage which is 'super-infecting' a lysogenic cell.

Rare *virulent* mutants of λ exist which are unable to lysogenize and which differ from *c*I mutants since they can super-infect and lyse host cells which are already lysogenic for λ. Virulent mutants are therefore not affected by the presence of normal repressor protein. Virulence results from the simultaneous presence of three mutations v_1, v_2 and v_3 which map in two regions, v_1 and v_3 very close to each other and immediately to the right of *c*I, v_2 to the left of *c*I. These mutations are *cis*-dominant and therefore behave as operator-constitutive mutations which are unable to recognize the *c*I repressor protein (analogous mutations in *lac* operon are discussed on pp. 333–335).

The maintenance of lysogeny therefore depends on the inhibition of transcription of phage genes by a phage-coded repressor protein which acts (negatively) at two operator sites (O_L and O_R) to prevent transcription leftward from the promoter P_L and rightward from the promoter P_R. This has been confirmed in biochemical studies. The repressor protein has been isolated in pure form and its specific binding to λ DNA demonstrated *in vitro*; repressor binding to DNA isolated from λv_2 and λv_3 avirulent mutants is less, and it hardly binds at all to DNA from the double mutant $\lambda v_1 v_3$. Both leftward and rightward operator/promoter sites have been sequenced and the binding sites for repressor protein and RNA poly-

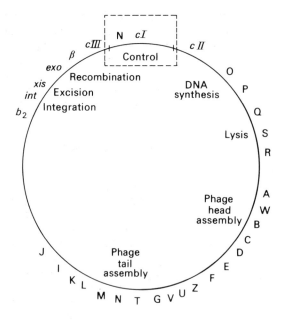

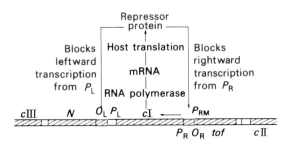

cI	gene for repressor protein
cII, cIII	genes for proteins affecting the establishment of lysogeny
P	promoter gene
O	operator gene
N and tof	early control genes
⟶	direction of transcription

Figure 16. Control region for bacteriophage λ during maintenance of lysogeny. (The inset shows the whole genome.)

merase have been identified in terms of the base sequences these proteins protect from nuclease digestion.

Induction of lysis in a λ lysogen therefore occurs when the cI repressor protein in inactivated; in wild-type strains this protein is very sensitive to UV radiation. For laboratory convenience, induction is brought about by a short heat treatment of a λcI mutant with a thermolabile repressor protein. Once the λ repressor is inactivated, transcription begins and the early RNA

transcripts produced *in vivo* and *in vitro* will hybridize with DNA from the regions immediately adjacent to the $O_L P_L$ and $O_R P_R$ sites.

The various mechanisms controlling gene expression in bacteriophage development can be summarized. They include repression or induction, and also those which alter the activity of RNA polymerase in its ability to initiate transcription at particular promoter sites, or else to terminate it. Some timing is introduced by the arrangement of genes into operons. Control is also exerted at translation, but this may be concerned mainly with regulating the amount of each protein produced.

While they provide models for differentiation in cells, even complicated bacteriophage complete their development in a short time (about 20 min for T4) relative to that taken by cells undergoing morphogenetic development (8 h for sporulation in *Bacillus*). Cellular differentiation therefore probably involves a more complicated set of changes than those seen in phage T4 but at the moment none of the mechanisms controlling bacterial sporulation, or any other cellular differentiation process, has been worked out in detail. Progress has been made, however, in analysing the changes which occur during bacterial sporulation, in determining how many genes are involved, and in describing some aspects of the timing of gene expression. We will therefore turn to sporulation in bacteria as the most extensively studied example of cellular differentiation.

BACTERIAL SPORULATION

Initiation of sporulation in *Bacillus*

Sporulation is induced in most *Bacillus* species when either the carbon (and therefore energy) source or the nitrogen source is exhausted, or to a lesser extent when phosphate ions are unavailable. It is not an all-or-none response of every cell in a population to the complete depletion of a particular nutrient since cultures can be obtained in which a proportion of the cells are initiated to form spores, while the remainder continue to divide and so contribute to growth of the culture. This situation arises when relatively poor carbon or nitrogen sources are used to support growth.

Experimentally, sporulation can be induced fairly synchronously by transferring cells grown in a rich nutrient medium to one containing a poor source of carbon. Even under these conditions sporulation is not completely synchronous since the cells in an exponential culture are asynchronous with respect to the cell division cycle, and cells can only begin to sporulate when they reach a certain stage of their cell cycle, or more specifically, of their DNA replication cycle. For instance, sporulation in sister cells, which obviously begin the cell cycle at the same time, is synchronous. Furthermore, if cells of a mutant which is temperature sensitive in the *initiation* of chromosome replication are allowed to complete a round of replication at the restrictive temperature, they 'align' their chromosomes in the completed stage. On shifting such a culture to the permissive temperature there follows synchronized DNA replication in the population. In such a culture there is a peak period about 15 minutes after the beginning of chromosome replication when the cells are maximally sensitive to induction of sporulation.

It is clear that in cellular systems a morphogenetic process such as sporulation should not be viewed in isolation since some events may be connected with the previous stage of the life cycle which provided the foundation for subsequent development. This is particularly true for those processes concerned with *assembly* of developing structures.

The mechanism of initiation

The way in which vegetative cells of *Bacillus* species switch from the vegetative state to begin sporulation is not understood. One hypothesis is based on the similarity between conditions which cause an alleviation of the catabolite repression of enzyme synthesis (see p. 331) and those leading to initiation of sporulation. There may be present within the cell a repressor or activator which prevents or promotes the reading of genes concerned with sporulation by binding to control sites on the bacterial genome. This binding is presumed to be sensitive to a molecule or molecules which monitor the cell's ability to continue further growth. Such a model must of course be modified to take account of the fact that the reading of the early sporulation genes can occur only at a certain point during DNA replication. There is no evidence that cyclic AMP, the effector modulating catabolite repression in *E. coli*, is involved in sporulation although there are several other species of molecules which could act in a repression system. These include the highly phosphorylated nucleotides which are produced in ribosome-idling reactions. These reactions are involved in the stringency mechanism by which cells monitor their capacity for protein synthesis. Idling ribosomes of vegetative cells of *Bacillus subtilis* synthesize such compounds as guanosine-3′,5′-bis-diphosphate (ppGpp) and guanosine 3′-diphosphate, 5′-triphosphate (pppGpp). However, cells in sporulation medium rapidly begin to synthesize the adenosine nucleotide (pppAppp) instead in reactions catalysed by membrane-bound enzymes. Subsequently, pppApp and ppApp are synthesized by ribosomes of sporulating cells. Each of these nucleotides as well as a number of other compounds have at one time or other been proposed as being involved in the initiation mechanism.

Sequential changes during sporulation

Once begun on a programme of development, a cell undergoes a series of morphological changes. These changes involve the building of new structures on old foundations, and for this a series of components must be provided. Some of these are unique to the differentiated state, others (including for example many membrane components) are present in vegetative cells and are also needed during development. In addition to subunits for assembly there are some cellular components, such as metabolic enzymes concerned with the generation of energy or provision of monomers for polymer synthesis, which must be produced or altered during development. These include elements needed to control the overall processes. The synthesis of some 'vegetative' enzymes may also need to be switched off. In this section we are concerned not only with the nature of these changes, but also with how they are ordered and co-ordinated.

The morphological changes occurring during sporulation in *Bacillus subtilis* and their arbitrary classification into seven stages has been outlined (Figure 4). Concurrent with these structural changes is a series of biochemical events, some of which are shown in Figure 17. There are two points concerning this list. First, it probably represents only a small proportion of the changes occurring in sporulating cells. Secondly, the relevance, if any, of some of the known changes is not clear. Sporulation is usually induced by a change in conditions equivalent to a shift-down, and this would lead to a considerable alteration in the pattern of metabolism in non-sporulating organisms. The question of relevance will be discussed later, but there are some changes which are directly concerned with the forming of a spore, and which are therefore specific to the process.

Specific sporulation events

The spore coat is formed by the aggregation during stage V of many molecules of a polypeptide ($M_r = 13\,500$) which is synthesized in a soluble form much earlier during stage II. This polypeptide is rich in hydrophobic amino acids and cysteine and can be

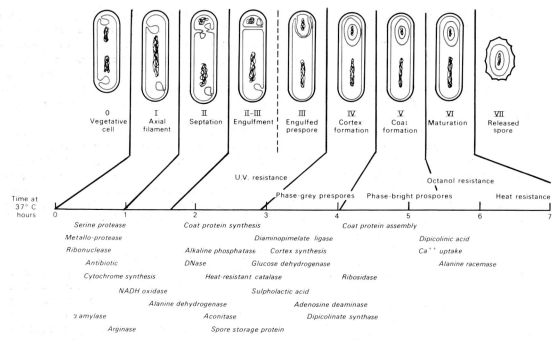

Figure 17. Morphological and biochemical changes during sporulation in *Bacillus* spp. The approximate timing of the morphological changes is indicated.

extracted from mature spores only by using denaturing agents in the presence of compounds rupturing disulphide bonds. It is immunologically distinct from the proteins found in vegetative cells, and is therefore unique to the spore.

In spores of *Bacillus subtilis* and *B. sphaericus*, the cortex is composed mainly of Type I peptidoglycan (see p. 54), but there are significant differences between it and the vegetative cell polymer (Type I for *B. subtilis*; Type II for *B. sphaericus*) and these differences necessitate the synthesis during sporulation of several 'new' enzymes. The cortical peptidoglycan is much less cross-linked (6% per disaccharide compared with 60% for vegetative wall) and contains the lactam of muramic acid, which is not found in vegetative peptidoglycan.

The cortical polymer is a loose anionic compound, and therefore its volume increases markedly as divalent cations are removed from it. The cortex expands during maturation of the spore, probably as divalent cations (mainly Ca^{2+}) are removed into the spore core. This expansion within a strong protein coat, together with the osmotic properties of this polymer, are considered to be important in maintaining the spore core relatively free of water, thereby maintaining the dormant, heat-resistant state.

Another substance which is found only in spores or sporulating cells is dipicolinic acid (DPA). It is synthesised probably in the mother cell during stage IV, and is concentrated in the developing spore core mainly as the chelate with Ca^{2+}; it can account for up to 10% of the dry mass of the spore. Initially it was thought that the Ca-DPA chelate was responsible for the heat resistance of spores; now however it is considered to participate in germination instead since DPA-less mutants of several *Bacillus* species retain heat resistance, but germinate poorly.

Other biochemical events occurring during sporulation include the synthesis of enzymes. Some, including enzymes of the TCA cycle, may or may not be present in vegetative cells, but are essential for sporulation. Others found in the spore are never found in vegetative cells (for example, glucose dehydrogenase), yet appear when the appropriate stage of sporulation is reached. The catalogue of sporulation events is currently far from complete. One approach to studying sporulation is to find as many changes as possible in the pattern of protein synthesis using the technique of two-dimensional polyacrylamide gel electrophoresis. This technique is capable of resolving the polypeptides of *B. subtilis* into about a thousand 'spots' and its application to sporulating cells has indicated that a significant number of new polypeptides appear during sporulation. This could indicate that genes silent in the vegetative phase are activated during sporulation. However, interpreting these data requires caution since during sporulation very active intracellular proteases appear and these cleave some vegetative proteins in a specific way to produce smaller polypeptides. Some of the new proteins may have been evolved by the step-down conditions.

Roles for sporulation proteases

Early in sporulation of *Bacillus subtilis* several proteases are produced, including one or more serine proteases (inhibitable by di-isopropylfluorophosphate, which inactivates enzymes with an active serine at the catalytic site), a Zn^{2+}-activated metalloprotease and an enzyme with esterase activity. Of these a serine protease may be the most important since mutations affecting this enzyme apparently affect the ability to sporulate, whereas mutants producing a defective metalloprotease are able to sporulate normally. There are a number of possible roles for these proteases during sporulation and these include:

1 The breakdown by extracellular enzymes of proteins in the culture, to provide substrates for further growth of the cells.

2 The hydrolysis of unwanted intracellular proteins to provide amino acids for further protein synthesis. In many bacteria, including non-sporulating species, *turnover* of proteins occurs at the onset of starvation; during sporulation this turnover appears to be essential, at least in the aerobic spore-formers.

3 The selective cleavage of enzymes to alter their activity or stability. The extreme heat resistance of bacterial spores has prompted study of the ways the enzymes in spores are stabilized against heat injury. In several instances, including *aldolase*, the heat-stable enzyme isolated from spores is related immunologically to the vegetative form of the enzyme, but has lower molecular mass. It results from cleavage by a sporulation protease of the heat-sensitive vegetative form of the enzyme to give a smaller, and thermodynamically more stable spore form. This modification can be reproduced *in vitro* using purified aldolase from vegetative cells, and protease from sporulating cells.

The post-translational modification of enzymes or structural proteins also has wider significance in morphogenesis. It can lead to changes in the control of gene expression (for example by modification of the transcriptional machinery, see p. 383) or to essential alterations in the configuration of structural proteins during their assembly (see p. 369).

Sequence of biochemical events

The biochemical changes set out in Figure 17 do not all occur immediately after initiation of sporulation; instead they appear in an ordered sequence. This is illustrated in the figure; it can be seen that biochemical changes continue to take place right up to the late stages of spore development. The appearance of each biochemical event can be correlated with a particular morphological stage of sporulation and this has been indicated.

Since many of these events involve the *de novo* synthesis of proteins, *the existence of an ordered sequence of biochemical events can be taken to imply that a high degree of control exists over the expression of those genes concerned with sporulation.*

Gene expression during sporulation

We have indicated that genes silent during vegetative growth are activated during sporulation. How many 'sporulation-specific' genes are involved? So far there is no precise answer, and this is mainly due to the lack of adequate techniques for doing complementation tests in *Bacillus subtilis*. It is however easy to map mutations preventing spore formation (*asporogenous*). Those presently known map in about 30 clusters which are distributed throughout the *Bacillus subtilis* genome and it is estimated that at least 50 clusters, presumed to be operons, are probably involved in sporulation.

Study of asporogenous mutants has provided insight into the nature of sporulation. In these the mutations are usually highly *pleiotropic* (that is, many characters are affected by the presence of a single mutation), since cells progress through sporulation until they reach a block and none of the subsequent morphological or biochemical changes seen in normal

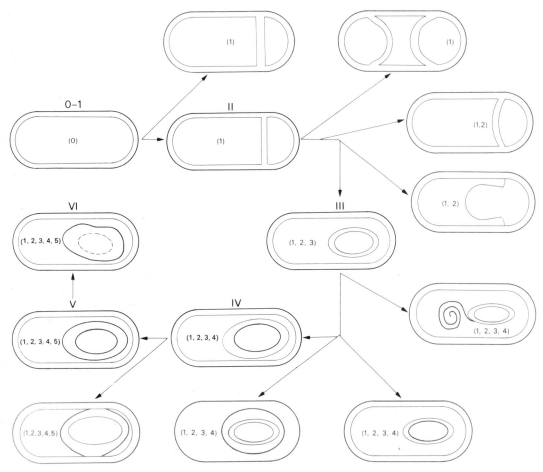

Figure 18. Diagram illustrating some of the aberrant phenotypes seen in asporogenous mutants.

The probable order of expression of some sporulation genes represented by the mutants is also given. The scheme was devised by considering phenotypes of double mutants and the biochemical properties of each strain (see text). Numbers in brackets refer to the presence of biochemical or morphological characters, as follows: (1) protease, (2) alkaline phosphatase, (3) glucose dehydrogenase, (4) refractile spores formed, (5) dipicolinic acid, (0) indicates the absence of all of these. See Coote J.G. and Mandelstam J. (1973) *J. Bacteriol.* **114**, 1254–1263.

cells can be detected after this 'arrest'. Sporulation mutants can be found blocked at any stage (except I) and every mutant can therefore be classified according to the stage at which it is blocked (e.g. *spoII*). Sometimes the mutants continue to develop aberrantly from the point of block. The terminal phenotypes seen in a number of asporogenous mutants are shown in Figure 18.

On the basis of this 'unidirectional' pleiotropy it has been inferred that sporulation proceeds by a series of events, each of which is dependent on the successful completion of previous events. It is also possible that there are several sets of interdependent sequences,

since after stage II there are two chromosomes, one in the mother cell, the other in the prespore, and each is bounded by a different set of membranes. Experiments with forespores isolated after disruption of the mother cell indicate that proteins synthesized in the forespore differ from those appearing in the mother cell. The generalization about dependent sequences does not apply strictly to later stages of sporulation and there are mutants in which further development continues despite the absence of a particular step. The best example is a mutant of *Bacillus megaterium* in which no cortex (normally appearing at stage IV) can be seen but a normal spore coat (stage V) is

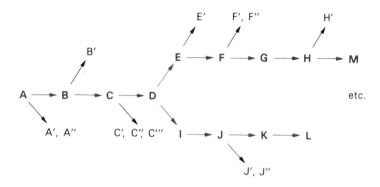

Figure 19. Hypothetical scheme for dependent biochemical sequences during later stages of sporulation.

The main sequences of indispensible events are shown in red; those in black are apparently regulated in the same way as are essential events, but are not necessary for sporulation.

formed. These events might involve a complicated set of interrelated sequences as indicated in Figure 19. Concomitant with these essential events is a series of changes which are coupled into the sporulation sequence, but which are not necessary for sporulation to continue (for example, the formation of dipicolinic acid).

The point has already been made that some biochemical changes found in sporulating cells may be irrelevant to sporulation, merely resulting from changes in culture conditions needed to initiate the process. Asporogeneous mutants provide one way of checking this since any event normally connected with sporulation would fail to appear in all mutants blocked after the stage at which it would normally occur. Ultimately the relevance of each biochemical event can be determined by isolating mutants which are affected specifically in the function, and testing their ability to sporulate. DPA is not essential in the formation of a spore, since DPA-less mutants form heat-resistant spores. These, however, germinate poorly, showing that some events that occur during sporulation are not essential for assembly of the spore but may be needed for maintenance of dormancy or for subsequent germination and outgrowth.

Order and timing of gene action

With some asporogenous mutants one can determine when the protein coded by a sporulation gene is required to act in the sporulating cell, and in what order these gene products must act.

The *order* in which the products of sporulation genes are required to act is usually inferred from the stage at which mutations in that gene block the sporulation sequence. When two mutations act at about the same time however, the order of 'gene action' can be determined by using the double mutant, *provided that each mutation produces a distinctive phenotype.* When

put into sporulation conditions the double mutant will display the phenotype of the earlier of the two events defined by the mutations.

Interactions of this type between two mutations present in the same cell are known as epistasis relationships. They have been used extensively in determining the order of gene function in *Bacillus* sporulation, particularly during stages II and IV. In some cases with late sporulation mutations the phenotype of the double mutant does not resemble that of either of the single mutants. This has been taken as indicating that there are branched sequences of dependent events.

Timing (and therefore to some extent order) can be studied by using temperature-sensitive sporulation mutants, and with these it is often possible to determine when a gene product must first begin to act, and for how long it must remain active. Figure 20 illustrates the method. Sporulation in a temperature-sensitive (*ts*) *spo* mutant is initiated at the high (restrictive) temperature and at intervals thereafter samples of this culture are shifted to the low (permissive) temperature. Spores will only form in those cultures transferred to the permissive temperature *before* the *ts* gene product is required to act. These experiments therefore define the beginning of the period when the gene product is needed. From a similar approach it can be seen that the end of the period of essential activity can be defined by the first time at which a transfer to the restrictive temperature leads to normal sporulation. This approach has so far been used for only a few mutants of *Bacillus subtilis*; of those studied however, one was affected in a function which is needed to act for only 15 minutes during stage II, another in a process required throughout stages III and IV, whereas yet another was blocked at the restrictive temperature at stage O, and the normal gene product was needed at all subsequent stages of sporulation.

(a) Shift-up experiment

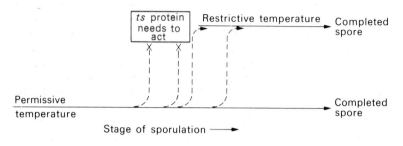

(b) Shift-down experiment

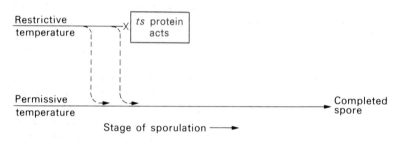

Figure 20. Use of temperature-sensitive mutants to determine the period during which an essential sporulation gene must function.

(a) A temperature shift-up experiment defines the *end* of the period during which the temperature-sensitive gene product (ts protein) needs to function. This is given by the first time at which cells shifted from the permissive to the restrictive temperature complete sporulation at the usual time.

(b) The shift-down experiment defines the *beginning* of the period during which the temperature-sensitive gene product would normally need to act.

Control of gene expression during sporulation

When inhibitors of RNA or protein synthesis are added at any time to sporulating cells of *Bacillus*, spore formation is prevented. Moreover, the appearance of those enzyme activities associated with sporulation is sensitive to the addition of chloramphenicol up to the time at which the activities appear. Therefore, much of the control of timing, and possibly also of amount of gene expression during sporulation results from ordered *de novo* synthesis, rather than from the activation or modification of proteins already extant. There are certain exceptions. Most of the spore coat proteins are synthesized during stages II–IV but are assembled on the spore surface during stage V, and a number of vegetative cell enzymes are modified to give more heat-resistant forms during sporulation.

The mechanisms which control the sequential expression of sporulation genes are not understood in detail. Some of this control is exerted at the level of gene *transcription* since inhibitors of RNA synthesis prevent spore formation, even if added at relatively late stages. Similar results have been obtained using a mutant with a thermolabile RNA polymerase. Moreover DNA/RNA hybridization experiments indicate that some of the mRNA species formed at any one stage of sporulation differ from those present at other stages.

So far two general models have been proposed to account for the timing of gene expression. One is based on changes in the specificity of transcription due to alterations in RNA polymerase; the other follows from an analogy drawn between sporulation and the process of *sequential induction* seen in the metabolism of aromatic compounds by pseudomonads (see p. 343).

Changes in specificity of RNA polymerase

This model was first proposed in view of changes which were observed in the activity of RNA polymerase during sporulation in *Bacillus subtilis*. When the phage ϕe infects vegetative cells of *B. subtilis* normal lytic development ensues. If, however, it infects a sporulating cell, the injected viral genome is not immediately expressed, but can be packaged into the developing spore and expressed after germination and outgrowth. One explanation for this is that during sporulation the transcriptional machinery of the host cell is modified so that 'sporulation' genes can be read while genes used in the vegetative state (including some of those in ϕe DNA) are switched off. That this alteration involves a change in the specificity of the enzyme responsible for transcription, DNA-dependent RNA polymerase, was suggested from several findings. First, some mutations leading to rifamycin resistance were found to cause a simultaneous loss of sporulation ability. Rifamycins are known to bind to the β-subunit of RNA polymerase, and it has been shown that the resistance mutations affecting sporulation are located in the gene coding for this β subunit. Secondly the RNA polymerases isolated from vegetative and sporulating cells have been found to differ in terms of their subunit composition and their capacity to bind the vegetative cell σ factor which is responsible for correct initiation of transcription at *promoter* sites (see p. 231). At the moment, the precise nature of these changes is uncertain since the proteases produced very early during sporulation are very active and can cause modification of the RNA polymerase during extraction of the enzyme. While modifications to RNA polymerase may play some part in regulating the sporulation sequence, as they do in phage development, it seems unlikely that there could be enough of them to account for all the steps in sporulation. On the basis of this model it is also difficult to account for the dependency of the sporulation sequence.

Sequential induction

This model is illustrated in Figure 21. In its simplest form, the initiation mechanism leads to the reading of the first operon concerned with sporulation. The product of one of the genes in this operon acts to activate the next operon, and so on.

What these models have in common is that both involve modulation of the frequency of reading of sporulation operons. They differ in that one requires the existence of promoters specific to sporulation, the other of specific operators. Changes of both types may play a part.

Translational control of sporulation

There are a number of indications that translational controls are important during bacterial sporulation. For example, in *Bacillus subtilis*, studies using low concentrations of actinomycin D as an inhibitor of transcription have shown that there may be a time lag between transcription of sporulation genes and their translation. Inhibitor studies of this nature need to be interpreted with caution, but there is other evidence that some form of translational control of sporulation exists. Some mutations to streptomycin resistance cause an alteration in one ribosomal protein and also inhibit sporulation completely. Moreover, other mutations which suppress *spoOA* mutants blocked at the very start of sporulation map in the region of ribosomal protein genes, and lead to altered or missing ribosomal proteins. Since such mutants grow normally, it is clear that during sporulation there is a need for the ribosome to function in a way not required during growth.

It is clear that much remains to be done before the control of gene expression during sporulation can be understood completely. Rapid progress is however being made, particularly in implementing genetic engineering techniques. Already sporulation genes have been cloned on phage vectors, and this approach will enable not only more detailed study of the transcription of specific spore genes, but will also permit tests of dominance and of complementation of sporulation mutations to be done. It should also lead to the development of *in vitro* protein synthesizing systems that will enable detailed analysis of control mechanisms.

While knowledge of the biochemical events that occur during sporulation and the way they are regulated is far from complete, even less is known about the biochemistry of the structural aspects of sporulation, or of any other differentiation process.

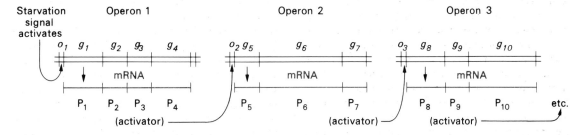

Figure 21. Sequential induction model for control of sporulation.
It is assumed that sporulation involves the successive activation of a series of operons, each under the control of a different operator. In each operon there is a gene (g_3, g_7, etc.) coding for a protein which is responsible for activating the next operon in the sequence.

Thus, it is not yet known what determines the shape and size of cells, the position of septum formation, the segregation of replicated chromosomes, and the assembly of large structures such as spores. So far, most progress has been made towards an understanding of the processes involved in the growth and division of cells, and these have been covered already in detail in Chapters 1 and 2.

INTERCELLULAR INTERACTIONS

Previous sections have dealt with morphological changes in individual cells. In the development of eukaryotic organisms there are concurrent with these intracellular changes many instances of co-operation between cells to produce multicellular differentiated structures. This co-operation frequently depends on extensive communication between cells, as well as highly specific surface contact recognition processes. Although there are few examples of multicellular development in prokaryotes, there are several which illustrate very clearly cellular recognition and communication phenomena.

Cellular recognition

The adsorption of a bacteriophage to its host, and the conjugation of bacterial cells, are two examples of very specific recognition processes. Most bacteriophages can infect only a narrow range of bacterial

can act as receptors include lipopolysaccharide of Gram-negative bacteria (*E. coli* and *Salmonella* phages ϕX174, T3, T4, T7, P22, S13, G4, G6) or exopolysaccharides of capsules or slime layers (phage 29 of *E. coli* K29) or proteins in the cell envelope (*E. coli* phage λ) or, in Gram-positive bacteria, the teichoic acids (e.g. *Bacillus subtilis* phages ϕ29, SPO2, SP3, etc.).

This binding resembles that occurring between an enzyme and its substrate and is illustrated by the group of *Salmonella* phages with lipopolysaccharide as their receptor. The structure of *Salmonella typhimurium* lipopolysaccharide is known, not only for the *smooth* wild type, but also for a series of *rough* and *semi-rough* mutants which are affected in their ability to synthesize the complete lipopolysaccharide (the structure of this molecule, in the wild type and mutants is outlined in Chapter 1, p. 68). The ability of a number of phages to infect this range of host mutants is given in Table 1. A phage may be unable to bind either because the receptor site is not accessible to it (for example ϕX174 is unable to lyse smooth *Salmonella* strains whose lipopolysaccharide contains many O-antigen side chains) or because the receptor site is absent (P22 cannot adsorb to any strain which lacks the terminal O-antigen side chains) and it is possible from a consideration of the data in Table 1 to infer which sugar residues are probably involved in the recognition of each phage. For example, an octasaccharide of structure:

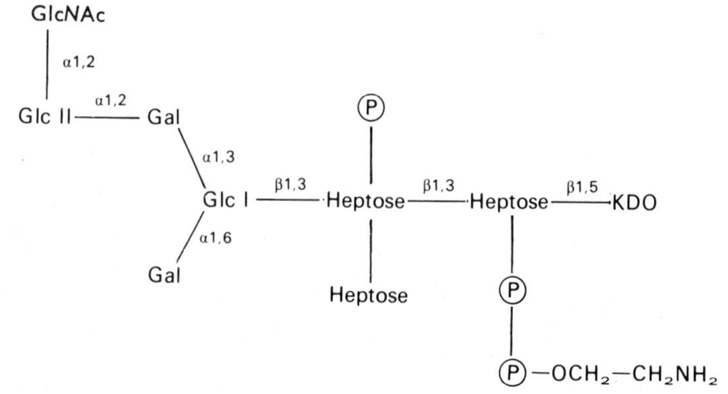

GlcNAc
| α1,2
Glc II —— α1,2 —— Gal
                        α1,3
                        Glc I —— β1,3 —— Heptose —— β1,3 —— Heptose —— β1,5 —— KDO
                        α1,6
Gal   Heptose

can be isolated from an appropriate mutant. This binds specifically to ϕX174.

For some phages attacking the capsular exopolysaccharide, the phage protein/host receptor interaction is, in fact, that of an enzyme and its substrate. Protein H (one of eight) in the tail-spikes of phage 29 of *E. coli* K29 recognizes and hydrolyses the host's exopolysaccharide. This gradually opens a path

species, and each phage has a particular site on the host to which it attaches. This may be the flagellum, the pilus or one of the surface polymers. Many such phage–host interactions have been studied, and in all of them the specificity of recognition is due to the binding of a component of the phage surface (usually a single protein) and another *receptor* molecule on the cell surface. Molecules on the host cell envelope which

through the capsule to the cell wall where a second recognition process occurs. This phage will not adsorb to non-capsulated mutants.

Binding reactions of this type also occur during conjugation in *E. coli* between the pili of plasmid-carrying strains and the surface of susceptible cells and are very common in morphogenetic processes of cell agglutination and mating in eukaryotes.

Communication between cells

When microbial cells interact in liquid media they can make contact with each other by a process of diffusion and random collision. This occurs in the phenomena already discussed. When, however, the morphogenetic process involves aggregation of cells on the surface of a solid substrate, the cells must be motile and have a mechanism for communicating with each other over a distance which is large in comparison with the length of the cell.

This communication or 'signalling' over a distance is seen very clearly in the myxobacteria, and in the much more extensively studied cellular slime mould, *Dictyostelium discoideum*. The signalling between swarming or aggregating cells of these organisms is very intense and is best seen by time lapse photography. Figure 22 is one frame from a film and shows an interference pattern of signal waves passing through an aggregating swarm of the myxobacterium *Myxococcus fulvus*.

Many studies have been devoted to the nature of these signals, how they are propagated, and how they lead to aggregation of the cells. So far most progress has been made with the eukaryotic slime mould, *D. discoideum*, although a number of similar observations have been made on myxobacteria.

Signalling in slime moulds begins soon after the onset of starvation. A number of amoebae in the population begin to secrete cyclic adenosine-3′,5′-monophosphate (cAMP) in regular bursts or pulses. This compound is a powerful attractant for slime mould amoebae, and in the presence of a gradient of this chemotactic agent they will migrate towards the higher concentration. During aggregation, cells receive pulses of cAMP (they have cAMP-receptor proteins located in their surface membrane). Each pulse stimulates a cell to move briefly towards

Figure 22. Oscillating waves in the myxobacterium *Myxococcus fulvus*.

The myxobacteria are growing as a swarm on water agar with a streak of living *Micrococcus luteus* as food source. The edge of the colony is clearly recognizable. Part of the colony has a smooth surface, but in the central parts of the swarm can be seen the oscillating waves. Magnification × 21. Courtesy of Dr. Hans Reichenbach.

Chapter 9

the source of the pulse and also to secrete a pulse of cAMP itself. This leads to waves of inward movement which gradually change into radial streams of amoebae moving towards centres of aggregation. An important part of the transmission of the signal as a wave is the ability of the amoebae to degrade the cAMP after receiving a pulse, and all cells have a very active phosphodiesterase which hydrolyses cAMP. The reception of a pulse of cAMP causes the cells to change their shape from an amoeboid form to an

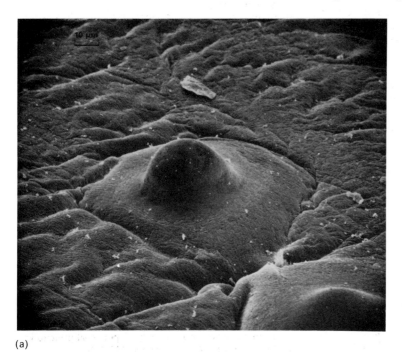

(a)

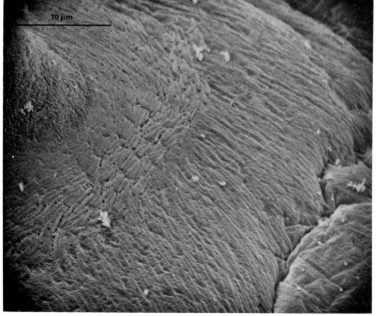

(b)

Figure 23. Scanning electron micrographs of the initial stages of fruiting-body formation in *Stigmatella aurantica.*

(a) Development of an aggregation centre at the 'fried egg' stage.

(b) Higher magnification of cells surrounding the 'yolk' region.

Reproduced with permission of P.L. Grilione and J. Pangborn. *J. Bacteriol.* **124**, 1558–1565.

elongated form, and it also alters their ability to adhere to each other. The mechanisms underlying these changes are not yet understood.

Pattern formation

The intricate myxobacterial fruiting bodies can be seen by scanning electron microscopy to result from the ordered arrangement of cells with the structure. This arrangement of cells can be seen at early stages of aggregation (Figure 23). Such patterns presumably result from unknown and complicated interplay of cell communication and recognition events. The unravelling of these mechanisms remains one of the challenging areas of developmental biology.

FURTHER READING

1 Mandelstam J. (1976) Bacterial sporulation: a problem in the biochemistry and genetics of a primitive developmental system. *Proc. Roy. Soc., Lond., B,* **193**, 89–106.

2 Piggot P.J. and Coote J.G. (1976) Genetic aspects of bacterial endospore formation. *Bact. Rev.* **40**, 908–962.

3 Ashworth J.M. and Smith J.E. (eds.) (1973) *Microbial Differentiation. Symp. Soc. gen. Microbiol.,* Volume 23.

4 Halvorson H.O., Hanson R.S. and Campbell L.L. (eds.) (1972) *Spores V.* American Soc. Microbiol., Washington, D.C.

5 Gerhardt P., Costilow R.N. and Sadoff H.L. (eds.) (1975) *Spores VI.* American Soc. Microbiol., Washington, D.C.

6 Chambliss G. and Vary J.C. (eds.) (1978) *Spores VII.* American Soc. Microbiol., Washington, D.C.

7 Lewin B. (1974 and 1977) *Gene Expression,* Volumes 1 and 3. Wiley-Interscience, New York, London.

8 Kushner D.J. (1969) Self assembly of biological structures. *Bact. Rev.* **33**, 302–345.

9 Casjens S. and King J. (1975) Virus assembly. *Ann. Rev. Biochem.* **44**, 555–611.

10 Chothia C. and Jaim J. (1975) Principles of protein-protein recognition. *Nature,* **256**, 705–708.

11 Showe M.K. and Kellenberger E. (1975) Control mechanisms in virus assembly. *Symp. Soc. gen. Microbiol.* **25**, 407–438.

12 Hershko A. and Fry M. (1975) Post-translational cleavage of polypeptide chains: role in assembly. *Ann. Rev. Biochem.* **44**, 775–797.

13 Brimacombe R. (1978) The structure of the bacterial ribosome. In *Symp. Soc. gen. Microbiol.* **28**, 1–26.

14 Linberg A.A. (1977) Bacterial surface carbohydrates and bacteriophage adsorption. In *Surface Carbohydrates of the Prokaryotic Cell* (ed. I.W. Sutherland), pp. 289–356. Academic Press, London & New York.

15 Reichenbach H. (1965) Schwarmentwicklung und Morphogenese bei Myxobakterien—*Archangium, Myxococcus, Chondrococcus, Chondromyces.* Film C-893, Institut für den Wissenschaftlichen Film, Göttingen, West Germany.

16 Campos J.M. and Zusman D.R. (1975) Regulation of development in *Myxococcus xanthus:* effect of 3′:5′-cyclic AMP, ADP, and nutrition. *Proc. natn. Acad. Sci., USA* **72**, 518–522.

17 Jenkinson H., Sawyer W.D. and Mandelstam J. (1981) Synthesis and order of assembly of spore coat proteins in *Bacillus subtilis. J. gen. Microbiol.* **123**, 1–16.

Appendixes

Appendix A
Bacterial Classification

There are two questions to answer before bacterial classification can be considered in detail: what are bacteria and how are they related to the other major groups of micro-organisms? Traditionally micro-organisms have been divided into protozoa, algae, fungi and bacteria. This division is quite arbitrary and unrealistic, but although Haeckel suggested in 1866 that all micro-organisms should be grouped together in the 'Protista' it took some considerable time for the idea to be accepted that this group should include all unicellular organisms and those multicellular organisms in which the cells are the same and there is no tissue differentiation.

Prokaryotic and eukaryotic cells

Micro-organisms are still divided into protozoa, fungi, algae and bacteria but cytological and biochemical studies have demonstrated a more fundamental division within this group. Two distinct cell types* have been observed and these have been called *prokaryotic* and *eukaryotic* cells.

All living organisms have a basic pattern of metabolic processes which enables them to grow and divide. In the eukaryotic cell many of these complex processes are associated with definite intracellular structures. Two examples of these are the mitochondria which contain enzymes concerned with production of energy and the chloroplasts which are associated with the utilization of light energy for biosynthetic purposes. Mitochondria and chloroplasts have apparently descended from a cell line quite distinct from that which led to the rest of the eukaryotic cell and may be the end result of an essentially symbiotic relationship between two ancestral types.*

These membrane-bounded organelles are absent from prokaryotic cells and the corresponding metabolic processes are associated with cell membranes. There are fundamental differences in both the structure and division of nuclear (genetic) material. In prokaryotic cells the nuclear material consists only of DNA which usually exists as a single circular double strand attached to the plasma membrane, but not separated from the remainder of the cell contents (cytoplasm) by a membrane. Eukaryotic cells are different in that the nuclear DNA is associated with basic proteins (histones) and occurs in the form of definite structures (chromosomes) surrounded by a membrane. During division there is a complex sequence of events to ensure that the genetic material is correctly distributed between the two daughter cells. In eukaryotic but not in prokaryotic cells the cytoplasmic contents often stream in an organized manner. Both types of cells may possess flagella: in prokaryotic cells these are relatively simple structures but in eukaryotic cells the flagellum is a complex arrangement of microtubules surrounded by a membrane. The detailed structure of the prokaryotic bacterial cell is described in Chapter 1 and the major structural differences between the two types of cells are summarized in Table 1 as are also a number of metabolic differences. It is very important to remember that the major differences between the two cell types are organizational and not functional.

Micro-organisms showing the typical prokaryotic cellular organization may be further subdivided on the basis of mechanism of movement and general cellular metabolism into Blue-green algae and Myxobacteria which have no obvious organelle associated with movement, but move by 'gliding'; Spirochaetes, where undulations of the cells are caused by axial filaments lying between the membrane and the cell wall; and bacteria which, if motile, have characteristic flagella. The discussion in this appendix will be

---

* The names *prokaryon* and *eukaryon* were derived by Dougherty in 1957 from the Greek πρό (before) and κάρυόν (kernel) and ευ (well) and κάρυόν respectively to distinguish between the nuclear structures of the two cell types.

It is of interest to note that the existence of a third ancestral type of cell—the archaebacteria—has recently been demonstrated. The archaebacteria are represented by the methanogenic bacteria, the extreme halophiles, and a thermoacidophile. Evidence for their separation from prokaryotic

organisms comes largely from analysis of the base sequence of 16 S ribosomal RNA, the absence of peptidoglycan from their cell walls, and the possession of unusual lipids. For reference see Fox *et al.* (1980).

Table 1. A comparison of Prokaryotic and Eukaryotic cells.

	Prokaryotic cells	Eukaryotic cells
Cellular Structures		
Cell wall composition	peptidoglycan present	peptidoglycan absent
Membrane composition	sterols absent from bacterial membranes	sterols abundant in mammalian membranes
Mitochondria	absent	present
Chloroplasts	absent	present in photosynthetic organisms
Nuclei		
Nuclear membrane	absent	present
Chromosome number	one	more than one
Mitotic apparatus	absent	present
Nucleolus	absent	present
Histones	absent	present
Golgi apparatus	absent	present
Mesosomes	present	absent
Microtubules	absent	present
Ribosomes (sedimentation coefficient)	70 S	cytoplasmic ribosomes 80 S chloroplast and mitochondrial ribosomes similar to 70 S
Movement		
Cytoplasmic streaming	does not occur	may occur
Amoeboid movement	does not occur	may occur
Flagella	if present are simple structures	if present are of the '9 + 2' type
Metabolism		
Oxidative phosphorylation associated with	membranes	mitochondria
Photosynthesis associated with	membranes	chloroplasts
Reduced inorganic compounds as energy source	may be used	cannot be used
Non-glycolytic mechanisms for anaerobic energy generation	may occur	do not occur
Poly β-hydroxybutyrate as a reserve storage material	may occur	does not occur
Nitrogen fixation	may occur	does not occur
Peptidoglycan synthesis	occurs	does not occur
Exo- and endo-cytosis	do not occur	may occur

limited to the bacteria, but the general principles of biological classification are applicable to all organisms.

Some general properties of bacteria

When we examine the bacteria we are at once impressed by their ubiquity and diversity. The range of environmental conditions in which bacteria will grow is wide: growth temperatures can range from less than 0° to more than 60°; salt concentrations from that of the Dead Sea to that of distilled water; and pH values from 0 to more than 10. The ability of bacteria to grow in a wide variety of environmental conditions is a reflection of their metabolic versatility. All bacteria have the same fundamental requirements for growth—a source of energy, of reducing power, of carbon, nitrogen and other elements—but there is considerable variation in the sources that can be utilized for these purposes. Energy for bacterial growth may be obtained from light, from oxidation-reduction reactions involving inorganic compounds for example H_2S, Fe^{2+} and H_2, or from oxidation-reduction reactions involving organic compounds. The sources of carbon may be carbon dioxide or any one of a large number of organic compounds. The hydrogen donors can be either inorganic or organic compounds. However, these metabolic differences are in fact modifications and extensions of a basic metabolism common to all bacteria.

Structurally there is again great diversity and bacteria can vary in shape from a sphere to a cork-screw shaped rod, and in size from $0.1~\mu m^3$ to $500~\mu m^3$. Internally there are varying degrees of cellular organization often associated with membranous bodies (see Chapter 1). All these differences, although superficially large are variations on the basic 'prokaryotic theme'. Nevertheless the bacteria comprise a diverse and complex group of organisms and the necessity for bringing order into this complicated situation requires that we should attempt to classify them.

The function of classification

The function of a classification is to group together bacteria having similar properties and to separate those with dissimilar properties. This can obviously be an arbitrary process and the form it takes may be decided by the interests of the person making the classification. For example a medical bacteriologist might choose to start by dividing bacteria into saprophytic (free-living) and parasitic organisms and then to divide the latter into pathogens and non-pathogens. The number and size of groups in the subdivisions will depend on the number of common properties that bacteria must have before we decide to include them in any particular group. If a large number of properties is required the groups will consist of a small number of bacterial types, whereas if there are only a few properties in common, large numbers of bacterial types will be included in that group. This difference in group size is the basis for a hierarchical classification. A classification has the advantage over a catalogue in that it will show relationships between bacteria and between groups of bacteria. Once a bacterium has been assigned to a group it should also be possible to predict other characters that it might possess. The aim of most systematists is that the classification should not be arbitrary but 'natural', i.e. it should show the relationship be-

tween various groups of bacteria. It is implicit in this definition that relationship means relationship by ancestry, i.e. phylogenetic, and if we assume that all bacteria existing today are derived from a common ancestral type then the degree of similarity between various groups will depend on the time at which their evolutionary pathways diverged. We can only speculate on the ancestry of bacteria and it is preferable to define relationship as 'overall similarity' and to adopt Gilmour's definition that a natural classification is a general arrangement intended for general use by all scientists.

The basic unit for most biological classifications is the individual organism. Similar organisms are grouped together to form a species, related species are grouped into a genus, related genera are grouped into a family, related families into an order, and so on. This hierarchical system is satisfactory for plants and animals but are we justified in applying it directly to bacteria and, if we do, will the taxonomic groups (taxa) have the same significance for microbiologists as they have for other systematists? It is unrealistic to use the individual bacterium as the basic unit, so the isolate, a bacterial culture derived from a single cell, is the fundamental unit in bacterial classifications. Isolates having similar properties are grouped together to form the next taxonomic unit, the species. The species is a concept that exists only in the mind of the scientist making the classification and hence is a unit without parallel in science. In *The New Systematics* (1940) a species is defined as a self-perpetuating unit the members of which have a definite geographical distribution area, are morphologically distinguishable from other groups, and do not normally interbreed with members of related groups. This definition cannot be applied to bacteria and there is no agreement among bacterial systematists as to a suitable definition of a 'bacterial species'. The difficulty in defining a bacterial species arises from the failure of bacteriologists to use similar criteria when classifying different types of bacteria. Ravin (1963) has distinguished three types of 'bacterial species'; these are: *taxospecies* in which the bacteria are phenotypically similar (this is the normal sense in which the term species is used); *genospecies* in which the bacteria can exchange genetic information; and finally *nomenspecies* where we have a group of bacteria bearing a particular binomial name irrespective of any other claim it may have to be a species. Later in this appendix we will consider precise alternatives for 'species' and 'genus', but until then these terms will be considered only as names for two convenient taxonomic groups.

Properties used in bacterial classification

Before we draw up a classification, we have to consider the various characteristics that are used to classify bacteria. A hundred years ago knowledge of bacteria was limited to their size and shape, and early classifications were based entirely on these properties. Later, the classifications were modified to include physiological properties. Today there is an extremely wide range of properties used in classifying bacteria.

Morphological characters

These include the size, shape, arrangement of the individual bacterial cells and the occurrence of extracellular appendages. For example, the occurrence of fimbriae and flagella and their arrangement on the bacteria cell are characters used extensively in classifications. Detailed knowledge of bacterial structure has increased rapidly during the past few years as a result of improvements in the techniques of electron microscopy. This has extended and in some ways supplanted earlier methods using specific staining reactions for various cellular constituents and components.

Cultural characteristics

The majority of bacteria are able to grow under a wide variety of cultural conditions in the laboratory and in certain cases the appearance of the culture under particular conditions of growth is characteristic of the organism and hence of value in classification.

Resistance tests

Some bacteria are renowned for their ability to resist adverse conditions by spore formation (see Chapter 9). Other bacteria can withstand high temperatures or the presence of antimicrobial agents. In a number of bacteria resistance to certain antimicrobial agents is mediated by plasmids (Chapter 7) and as a result this particular property may be of only limited use in classification. Resistance tests reflect metabolic or structural characteristics of the different bacteria and are useful in bacterial classification.

Metabolism and nutrition

The range of compounds that can be used for growth is characteristic of any given organism as are the products of metabolism which are often excreted into the growth medium. Although many bacteria can utilize relatively simple compounds as sources of carbon or nitrogen they may also require minute amounts of specific growth factors which they cannot synthesize. As for oxygen requirements, bacteria show a complete range from the aerobic bacteria which will grow only in the presence of oxygen, through the facultatively anaerobic bacteria which will grow in the presence and absence of oxygen, to the obligate anaerobes which are inhibited even by traces of oxygen.

Biochemical reactions

There are many specific and characteristic enzymic reactions which can be easily demonstrated.

Molecular structure

In this category we can include DNA base composition and hybridization data, comparisons of ribosomal proteins and RNA composition.

Genetic relationship

Genetic information may be transferred from one strain of bacteria to another by conjugation, transformation or transduction (Chapter 7) and, in general, gene transfer is much more frequent between phenotypically similar strains than between strains showing considerable differences in their phenotype.

Conventional methods of classification

Bergey's system

These then are the types of character by which bacteria may be classified, and at this point we must outline a scheme in which bacteria are grouped satisfactorily and the relationships between the different groups are demonstrated. The most widely-used system of classification and identification is that described in *Bergey's Manual of Determinative Bacteriology*. In earlier editions of this manual a complete hierarchical classification was given with the object of demonstrating the phylogenetic relationships between the different groups of bacteria. However this approach has been dropped from the current (8th) edition as the editors are of the opinion that a complete and meaningful hierarchy is impossible since for most groups of bacteria, genera and species are the only categories that can be recognized and defined with reasonable precision. Where possible genera have been grouped into families, but there are a number of genera that cannot be placed into any of the accepted families or grouped together into new families on the basis of the information available. The manual is meant to assist in the identification of bacteria and since it describes all genera and species of bacteria is an excellent reference book. For most purposes it is sufficient to have a simple scheme which will enable us to separate the major groups of bacteria and to outline their properties.

A number of such schemes have been suggested and the one presented in this appendix is based on the information given in the 8th edition of *Bergey's Manual of Determinative Bacteriology*. Its function is to outline major groups of bacteria and the properties used to distinguish them, not the phylogenetic relationships between the different groups.

The first major division we can make is into the Gram-positive and the Gram-negative bacteria. The Gram-positive bacteria are those from which the dye, crystal violet, cannot be removed by washing with ethanol whereas from Gram-negative bacteria it can. Within the Gram-positive group we can distinguish a number of morphological types: the cocci, the non-flagellate bacilli and the flagellate bacilli. The morphological characters, together with pathways for sugar metabolism, enable us to divide Gram-positive bacteria into the major families. The families are then further subdivided into genera on the basis of arrangement of cells after cell division, ability to produce resistant spores, and oxygen requirements (Table 2a).

The classification of Gram-negative bacteria is more complicated. These organisms do not produce resistant spores but as with Gram-positive bacteria many distinguishing features are associated with the shape of cells and the number and position of flagella. Among the Gram-negative bacteria we find the Pseudomonadaceae and Spirillaceae, two large groups in which either a single flagellum or a tuft of flagella is found at either one or both ends of the cell and the flagella are not distributed over the whole bacterial surface. The ability to grow in the presence or absence of oxygen together with the mechanism of sugar metabolism is an important factor in distinguishing various groups. The Gram-negative bacteria show a great versatility in their ability to utilize various substrates and this is used to distinguish the Nitrobacteriaceae which oxidize inorganic nitrogenous compounds, and the Azotobacteriaceae and Rhizobiaceae which fix gaseous nitrogen. Members of the Azotobacteriaceae are free-living organisms whereas the Rhizobiaceae live in a symbiotic association with plants (Table 2b). This scheme is intended only as a guide to the genera and families of bacteria, not as a definitive 'natural classification'. The list of genera in Tables 2a and 2b is not comprehensive but includes most of the commonly-occurring groups of bacteria. It is important to remember that although the majority of members of a particular genus or family will have the properties given in the table there will be exceptions. For example there are motile strains of

Lactobacilli and non-motile strains of Salmonellae but as these strains have so many other properties in common with members of the Lactobacilli and Salmonellae they are included in these genera.

Adansonian principles: numerical taxonomy

In the scheme we have described, and in many others proposed in the past, undue importance has been given to certain features. It is this indiscriminate weighting of characters that has led to so much confusion and controversy among bacterial systematists. The alternative is to assume that all observable characters are of equal importance and to assign them equal weight in the classification scheme. This principle is not new and was first proposed by Adanson in 1757. Adanson's views on the construction of taxonomic groups remained in relative obscurity until their revival by Sneath and other systematists as the bases for a 'Numerical Taxonomy' (see Sokal and Sneath, 1963).

The Adansonian principles have been clearly defined by Sneath (1958) and Sokal and Sneath (1963). They are as follows:

(1) All characters are of equal importance in creating natural groups.

(2) These groups should be based on as many features as possible.

(3) The relationship between the groups is a function of the similarities of the characters which are being compared.

It follows from these principles that phylogenetic considerations are not taken into account in constructing taxonomic groups and the groups are constructed in an empirical manner. If the Adansonian principles are accepted, classification becomes a mathematical exercise in handling data from studies of bacterial properties. This approach to classification is now called 'Numerical Taxonomy' which is defined as 'Numerical evaluation of the affinity or similarity between taxonomic units and the ordering of these units into taxa on the basis of their affinities' (Sokal and Sneath, 1963). The affinity (or similarity) between strains is defined as the ratio of the number of characters in common to the total number of characters compared. This value is usually expressed as a percentage. Characters which are negative in both strains of bacteria may or may not be included. Before an attempt can be made to classify bacteria within a given group the affinities between all strains must be

Table 2a. Gram-positive bacteria

Shape and arrangement	Spores	Growth	End products of glucose fermentation	Miscellaneous properties	Genus (species)	Family
(cocci in packets/clusters)	–	Anaerobic	CO_2, H_2, acetic acid, ethanol or butyric acid		Sarcina	Peptococcaceae
	–	Aerobic			Micrococcus (*M. luteus*)	Micrococcaceae
	–	Aerobic and facultatively anaerobic	Lactic acid and acetoin	Produce extracellular enzymes and toxins	Staphylococcus (*S. aureus*)	
	–	Aerobic and facultatively anaerobic	Lactic acid	Minimal growth requirements complex	Streptococcus (*S. pyogenes*)	Streptococcaceae
	–	Aerobic and facultatively anaerobic	Lactic acid, CO_2 ethanol and/or acetic acid	Minimal growth requirements complex	Leuconostoc (*L. mesenteroides*)	
(rods)	–	Anaerobic but oxygen tolerant	Lactic acid	Minimal growth requirements complex	Lactobacillus (*L. delbrueckii*)	Lactobacillaceae
	–	Aerobic and facultatively anaerobic		Important animal and plant pathogens	Corynebacterium (*C. diphtheriae*)	
(often pleomorphic)	–	Anaerobic to aerotolerant	Propionic and acetic acids		Propionibacterium	Propionibacteriaceae
Ranges from rods to branched filaments	–	Aerobic and facultatively anaerobic	Mixed acids but not propionic acid	Most species form a filamentous microcolony	Actinomyces	Actinomycetaceae

396

Table 2a (*continued*)

Shape and arrangement	Spores	Growth	End-products of glucose fermentation	Miscellaneous properties	Genus (species)	Family
Slender hyphae produce a branched mycelium	+	Aerobic		Produce a variety of antibiotics	Streptomyces	Streptomycetaceae
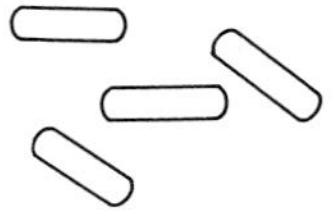 (rod)	+	Aerobic			Bacillus (*B. subtilis*)	Bacillaceae
(rod)	+	Anaerobic	Include acetic and butyric acids		Clostridium (*C. butyricum*)	
 (cocci)	+	Aerobic		Minimal growth requirements complex	Sporosarcina	

397

Table 2b. Gram-negative bacteria

Shape and arrangement	Growth	End-products of glucose fermentation	Miscellaneous properties	Genus (species)	Family
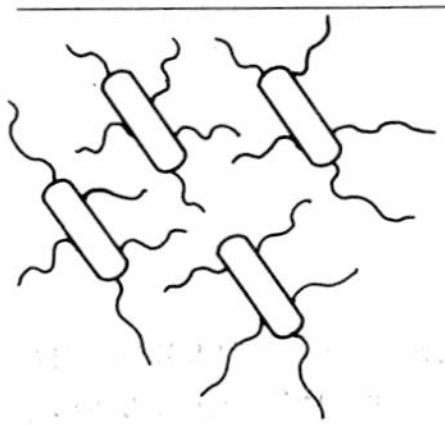	Aerobic		Form thick walled cysts Fix atmospheric nitrogen	Azotobacter	Azotobacteriaceae
	Aerobic		Fix atmospheric nitrogen when in symbiotic association with leguminous plants	Rhizobium	Rhizobiaceae
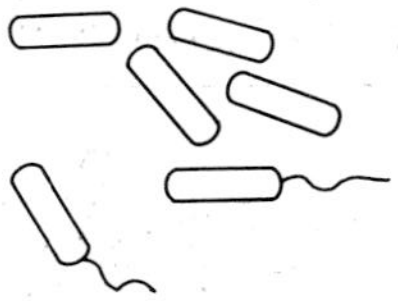	Aerobic		Oxidise nitrite to nitrate	Nitrobacter	Nitrobacteriaceae
	Aerobic		Oxidise ammonia to nitrite	Nitrosomonas	
	Aerobic		Energy derived from the oxidation of reduced sulphur compounds	Thiobacillus	
	Aerobic			Pseudomonas (P. aeruginosa) Xanthomonas	Pseudomonadaceae
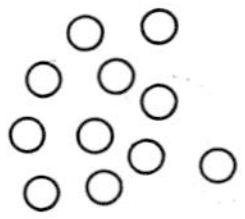	Aerobic		Parasites of mammalian mucous membranes	Neisseria	Neisseriaceae
	Anaerobic	Glucose not metabolised	Parasitic to man and other animals	Veillonella	Veillonellaceae
Highly pleomorphic	Aerobic and facultatively anaerobic		Lack a true cell wall (i) require sterols for growth (ii) do not require sterols for growth	Mycoplasma Acholeplasma	Mycoplasmataceae Acholeplasmataceae
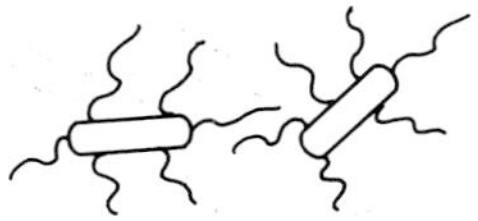	Aerobic and facultatively anaerobic	Lactic, acetic and formic acids, CO_2 and H_2	Grow readily on simple nutrient media	Escherichia (E. coli)	Enterobacteriaceae
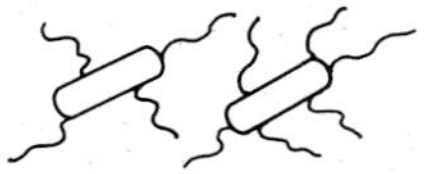	Aerobic and facultatively anaerobic		May be pathogenic to man and other animals	Salmonella (S. typhi)	
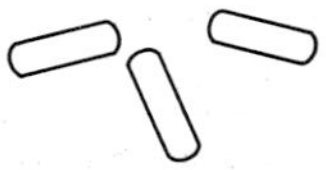	Aerobic and facultatively anaerobic		Non-motile	Shigella	

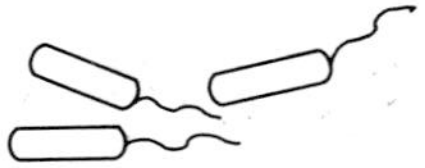

Table 2b. Gram-negative bacteria (*continued*)

Shape and arrangement	Growth	End-products of glucose fermentation	Miscellaneous properties	Genus (species)	Family
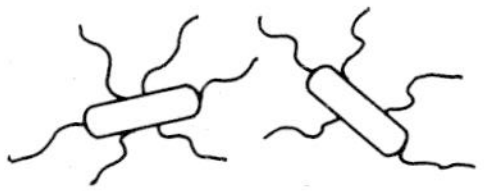	Aerobic and facultatively anaerobic			Proteus (*P. vulgaris*)	
	Aerobic and facultatively anaerobic	2,3-butanediol	Non-motile but often fimbriate	Klebsiella	
	Aerobic and facultatively anaerobic	Mixed products but no CO_2 or H_2		Vibrio (*V. cholerae*)	Vibrionaceae
	Aerobic and facultatively anaerobic	Mixed products which may include CO_2 and H_2		Aeromonas	
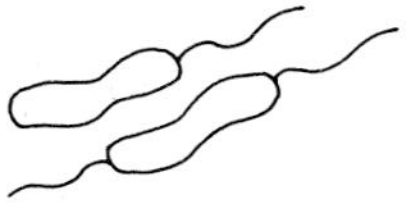	Aerobic or microaerophilic		Polyhydroxybutyrate granules usually present in the cytoplasm	Spirillum (*S. volutans*)	Spirillaceae
	Aerobic or facultatively anaerobic		Strict parasites requiring growth factors present in blood	Haemophilus (*H. influenzae*)	
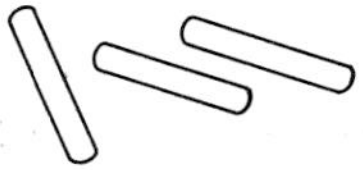	Aerobic		Motile by gliding motility Produce fruiting bodies containing refractile myxospores	Myxococcus (*M. fulvus*)	Myxococcaceae

N.B. Motility and the distribution of bacterial flagella are shown diagrammatically in the first column and a representative species is shown in parenthesis beneath the genus.
The classification given in Tables 2a and 2b is derived from the 8th edition of *Bergey's Manual of Determinative Bacteriology*.

determined and then strains grouped together on the basis of these affinities. Presentation of the data in such a way that 'natural groupings' are immediately obvious may be difficult, particularly when large numbers of bacterial types are involved and the procedure usually adopted is best understood by considering a simple example. If 8 bacterial strains are examined and affinities between them determined we can tabulate the results as in Table 3a. From an examination of the table it is difficult to find subdivisions within this group. The information can however be rearranged so that strains with high affinities are grouped together and separated from strains which have lower affinities. The resulting similarity matrix (Table 3b) shows two distinct groups; strains 1, 7, 5 and 6 are in one group and strains 3, 2, 4 and 8 are in the other. The similarities of the bacteria within the two groups all exceed 70%, but with bacteria from the different groups the values do not exceed 40%.

Similarity matrices

One of the first examples of a computer-based classification was that of the chromobacteria carried out by Sneath (1957). He examined 45 strains for properties ranging from the size and shape of the bacteria to the presence or absence of certain enzymes. He then used the data to calculate the affinities between the various strains and, after arrangement by computer so that strains with highest affinities were grouped together, obtained the full similarity matrix shown in Figure 1. Here instead of tabulating percentage similarity as in Table 3, the values are represented by different degrees of shading and the 'natural groups' stand out very clearly. This technique has been widely applied to the classification of bacteria. Similarity matrices of this type are but one of many ways in which these data may be presented, and readers are referred to *The Principles of*

Table 3a. The affinities (characters in common/characters tested × 100) of strains 1 to 8 are tabulated randomly and there are no obvious groupings among the strains. In Table 3b the similarity matrix has been re-arranged so that related strains are grouped together. There are two groups A and B; the affinities of the strains within the groups exceed 70% and between the groups does not exceed 40%.

Strain Number	1	2	3	4	5	6	7	8
1	100							
2	5	100						
3	10	95	100					
4	0	90	95	100				
5	80	15	35	15	100			
6	70	25	40	10	80	100		
7	95	10	20	10	90	75	100	
8	5	90	75	80	15	25	5	100

Table 3b.

Strain Number		A	A	A	A	B	B	B	B
		1	7	5	6	3	2	4	8
A	1	100							
A	7	95	100						
A	5	80	90	100					
A	6	70	75	80	100				
B	3	10	20	35	40	100			
B	2	5	10	15	25	95	100		
B	4	0	10	15	10	95	90	100	
B	8	5	5	15	25	75	90	80	100

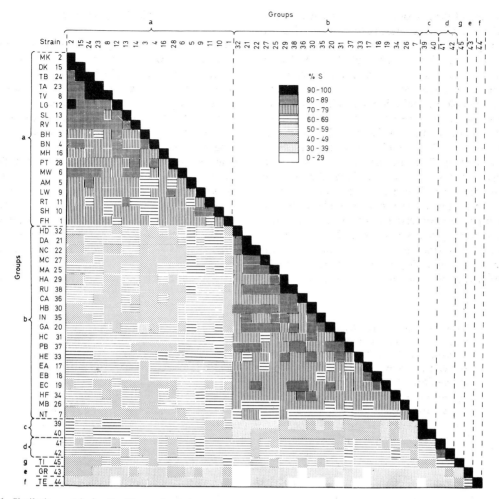

Figure 1. Similarity matrix for the Chromobacteria.

The percent similarity for the 45 strains of Chromobacteria studied was based on 105 features and the similarity matrix is shaded to represent the degree of overall similarity.

Numerical Taxonomy by Sokal and Sneath (1963) for description of other methods.

Taxonomic groups can be constructed from similarity matrices by defining limits within which organisms must fall if they are to be included in any particular group. Organisms with the highest affinities will be in the smallest group and a hierarchical system can be built up depending on the limits imposed. The smallest groups of bacteria would have a very high degree of similarity and the larger groups would have less. It would be possible to continue to use the names 'species', 'genus', 'family' but to avoid confusion it is preferable to adopt the nomenclature suggested by Sneath and Sokal (1962) where groups based on cell phenotype are called 'phenons'. There is no need to coin other words in building up a hierarchical system of classification as the various taxonomic groups can be clearly defined by prefacing the word 'phenon' with a number giving the level of similarity. For example the 90-phenon would be a small groups of organisms having affinity values of 90% or more, whereas the 70-phenon would include a large number of bacterial types (Figure 2).

Other possible modes of classification

Will systematists decided eventually that cell phenotype is unsatisfactory and that genotype must be used for microbial classification? In a number of species of

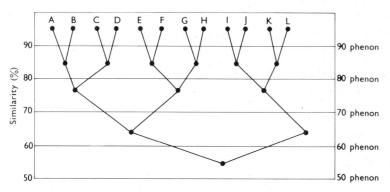

Figure 2. A hierarchical system of classification based on the percentage similarities.

A to H represent groups of organisms in which the affinities between individual bacterial strains exceeds 90%, i.e. these are '90-phenons'. The groups are related to each other in that the affinities between A and B, C and D etc. are between 80–90% and these 90-phenons form part of an 80-phenon. Similarly the affinities of all bacteria in groups A to H exceeds 60% and all the strains are included in the 60-phenon.

bacteria the basic information necessary for survival and growth is in one large replicon and additional genetic information which may be required to enable the organism to fill a specific ecological niche is found in smaller replicons or plasmids (see Chapter 7). These plasmids may contain 1–2% of the total DNA of the cell. Any attempt to classify bacteria on the basis of nucleotide sequences of the cellular DNA must take this genetic flexibility into account. It may be some time before it becomes possible to determine a complete nucleotide sequence for a replicon and even then the amount of work necessary to determine sequences for the replicons of all bacteria would be excessive.

Analytical techniques are currently available which apparently give some measure of the genetic relatedness between different micro-organisms. The ratio of the bases guanine + cytosine/adenine + thymine is not constant for all bacteria and varies from 1:3 to 3:1. Examination of the data for a number of genera, e.g. *Micrococcus* and *Vibrio* shows a considerable variation in the DNA base ratios of organisms within these groups, whereas all the Enterobacteria have similar DNA compositions (Figure 3). Results of this type emphasise the difficulties associated with bacterial identification and classification, as it is to be expected that closely related bacteria (i.e. those within the same genus)

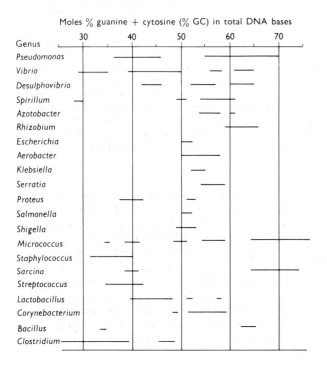

Figure 3. Variation of DNA base ratios within selected genera of bacteria. The horizontal lines represent the range of % GC values reported for each genus. Only genera for which at least two species or strains have been investigated for DNA base composition are included (Data from Hill 1966).

would have similar base compositions. Comparison of the structure of DNA molecules from different bacteria by hybridization methods (Chapter 5) gives some measure of the similarities of base sequences. Studies of this type have obvious potential uses. A system of bacterial classification could be devised where taxonomic groups are arbitrarily defined by the degree of hybridization between heterologous DNA's. The information might be used as a check on existing bacterial classifications since, where it has been possible to compare hybridization data with phenetic similarities, correlations have been found (Figure 4).

A correlation exists between the degree of hybridization and the time that has elapsed since the evolutionary paths for various animals have diverged (Figure 5) and hybridization data might provide an insight into possible phylogenetic relationships between bacteria.

Finally some reference must be made to nomenclature and identification. The correct identification and naming of bacteria are necessary so that scientists can

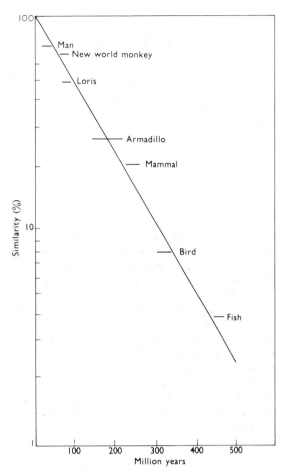

Figure 5. Correlation of hybridization data with time of divergence of evolutionary pathways.

The similarity of DNA's from various sources to DNA from the Rhesus monkey was determined from their ability to form hybrid double stranded DNA. The logarithm of this value was then plotted against the time that has elapsed since the evolutionary pathways diverged from that of the Rhesus monkey, and a linear relationship was found between these values. (Taken from R.J. Britten, Ref. 3.)

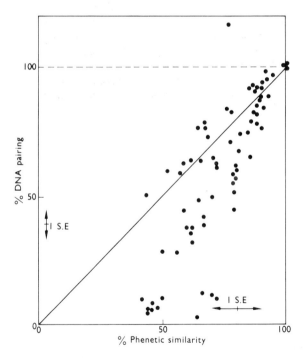

Figure 4. Congruence between DNA pairing and phenetic similarity for strains of the genera Pseudomonas, Xanthomonas, Agrobacterium, Rhizobium, Vibrio and Chromobacterium. The arrows show standard errors and it can be seen that the scatter is of the same order of magnitude as twice the standard errors (approximating to 95% confidence limits). (Taken from Jones and Sneath, Ref. 10.)

compare their observations, and as a result there are international regulations for the proposal and selection of names for bacteria. These are embodied in *The International Code of Nomenclature of Bacteria* (1966). The object of this code is to see that bacterial types are precisely defined by Latin binomials which provide characteristic internationally recognizable labels. A binomial system has one advantage in that the relationship between units is indicated by the first or generic name but apart from this it gives little information of the properties of bacteria included in

Table 4. A descriptive code for bacterial nomenclature. (Cowan, 1965.)

	Character	
Number	Before dash	After dash
1	Gram-positive	Catalase-positive
2	Gram-negative	Catalase-negative
3	Sphere	Oxidase-positive
4	Rod	Oxidase-negative
5	Acid-fast	Glucose not attacked
6	Not acid-fast	Glucose attacked by fermentation
7	Spore-forming	Glucose attacked by oxidation
8	Not spore-forming	Gas produced from glucose
9	Motile	Gas not produced from glucose
0	Non-motile	Does not grow in air

The number after the colon is an arbitrary number allotted to a species.
Examples: 13680–1469:1 *Staphylococcus aureus*
13680–2469:1 *Streptococcus pyogenes*
14670–1479:7 *Bacillus anthracis*
24689–1469:1 *Salmonella typhimurium*
14679–24590:1 *Clostridium tetani*

the group. Cowan (1965) suggested that a more elaborate descriptive code might be adopted for labelling bacteria (Table 4). In this system certain 'important' characters are given numbers which then form a sequential code for a particular bacterium. This system is somewhat analogous to the recommendations of the 'Enzyme Commission' for enzyme nomenclature. One important objection to this system is that although it is often difficult to remember Latin binomials it is almost impossible to remember a sequence of ten numbers.

The identification of an unknown bacterium depends on comparison of its properties with those of known strains in an existing classification. In general it is not difficult to place an unknown isolate into a genus although allocation to a species may prove more difficult and require the help of a specialist reference laboratory. Cowan (1965) describes three methods by which bacteria can be identified: the blunderbuss approach in which many tests are made before attempting to compare the unknown with known bacteria: the intuitive approach where we think we know the answer; and the progressive approach using dichotomous keys. Identification, unlike classification depends largely upon the constancy of a limited number of characters. For example,

most clinically important bacteria can be correctly identified on the results of not more than ten tests. If more than a limited number of tests are to be used in identifying unknown bacteria then dichotomous keys and diagnostic tables become complicated and mechanical aids are required to handle the information. These can be simple systems in which the characters are recorded on punched cards which are then compared with cards containing the characters of known organisms, or better still an electronic computer can be programmed to do this and so bring the 'blunderbuss' approach up-to-date.

RECENT OBSERVATIONS

Recent advances in molecular sequencing methods of nucleic acids and proteins have not only demonstrated the existence of a third ancestral type of organism, the archaebacteria, but also made it possible to measure directly the genealogical relationships between different bacteria and to construct a phylogenetic family tree (Fox *et al*, 1980). In many instances the groupings obtained from analysis of the base sequence of the 16 S ribosomal RNA are consistent with the traditional classifications, but there are exceptions where the traditional groupings are not phylogenetically related. These differences reflect the undue weight that has in the past been placed on characters such as shape, mode of cell division and absence of a cell wall. These studies have resulted in the revision of our ideas on early cellular evolution (Ford Doolittle, 1980) and made it possible to draw up a 'Natural classification' of bacteria which reflects their phylogenetic relationships. It is clear that we are entering a new phase in the study of cell evolution.

FURTHER READING

1 Ainsworth G.C. and Sneath P.H.A. (eds.) (1962) Microbial Classification. *Twelfth Symposium of the Society for General Microbiology.* Cambridge University Press.

2 Britten R.J. (1963) In *Carnegie Institution of Washington Year Book,* **62**, 303.

3 Britten R.J. (1964) In *Carnegie Institution of Washington Year Book,* **63**, 366.

4 Buchanan R.E. & Gibbons N.E. (eds.) (1974) *Bergey's Manual of Determinative Bacteriology,* 8th edition. The Williams & Wilkins Company, Baltimore.

5 Charles H.P. and Knight B.C.G. (eds.) (1970) Organisation and Control in Prokaryotic and Eukaryotic cells. *Twentieth Symposium of the Society for General Microbiology.* Cambridge University Press.

6 Cowan S.T. (1965) Principles and practice of bacterial taxonomy, a forward look. *J. Gen. Microbiol.* **39**, 143.

7 Ford Doolittle W. (1980) Revolutionary concepts in evolutionary cell biology. *Trends in Biochemical Science,* **5**, 146.

8 Fox G.E. *et al* (1980) The phylogeny of Prokaryotes. *Science,* **209**, 457.

9 Hill L.R. (1966) An index to deoxyribonucleic acid base compositions of bacterial species. *J. Gen. Microbiol.* **44**, 419.

10 Huxley J.S. (ed.) (1940) *The New Systematics,* Clarendon Press, Oxford.

11 International Code of Nomenclature of Bacteria (1966) In *Int. J. Syst. Bacteriol.* **16** (4), 459–490. Edited by the Editorial Board of the Judicial Commission of the International Committee on Nomenclature of Bacteria.

12 Jones D. and Sneath P.H.A. (1970) Genetic transfer and Bacterial taxonomy. *Bacteriol. Revs.* **34**, 40.

13 McCarthy B.J. and Bolton E.T. (1963) An approach to the measurements of genetic relatedness among organisms. *Proc. Natl. Acad. Sci., U.S.* **50**, 156.

14 Mandel M. (1969) New approaches to Bacterial Taxonomy: Perspective and prospects. *Ann. Rev. Microbiology* **23**, 239.

15 Marmur J., Falkow S., and Mandel M. (1963) New approaches to bacterial taxonomy. *Ann. Rev. Microbiol.* **17**, 329.

16 Sneath P.H.A. (1957) The application of computers to taxonomy. *J. Gen. Microbiol.* **17**, 201.

17 Sneath P.H.A. (1958) Some aspects of Adansonian classification and of the taxonomic theory of correlated features. *Ann. Microbiol. Enzimol.* **8**, 261.

18 Sokal R.R. and Sneath P.H.A. (1963) *Principles of Numerical Taxonomy.* W.H. Freeman & Co., San Francisco.

19 Stanier R.Y. (1964) Towards a definition of the bacteria. In *The Bacteria,* Vol. 5: Heredity (eds. Gunsalus I.C. and Stainer R.Y.), p. 445. Academic Press, New York.

20 Stanier R.Y., Adelberg E.A., and Ingraham J.M. (1977) *General Microbiology,* 4th edition. Macmillan & Co., London.

21 Steel K.J. (1965) Microbial identification. *J. Gen. Microbiol.* **40**, 143.

Appendix B
Enzyme Mechanisms: Functions of Vitamins and Co-enzymes

CLASSES OF ENZYMES

A living organism depends on its catalytic proteins for its identity and continued existence. At first sight it appears enormously complicated but on analysis it is found that biological reactions are of six relatively simple kinds—such as oxidation by removal of a pair of hydrogen atoms. Complex changes and formation of intricate molecules are usually brought about in a step-wise manner. Many proteins (*apo*-enzymes) require the co-operation of a non-protein *co-enzyme* or *prosthetic group* and many of these are derivatives of the animal vitamins known as the B group. The same mechanisms involving the same coenzymes occur in all kinds of organisms including bacteria.

Although this cannot be a treatise on enzymology, we can perhaps illustrate the mechanisms of the six kinds of reaction and give examples of each together with the involvement of the coenzymes where appropriate. Many coenzymes and prosthetic groups are dinucleotides containing adenosine-5'-monophate (AMP) as one half. This will be represented thus AM℗ in what follows:

The six classes of enzyme are:

(1) Oxido-reductases.   (4) Lyases.
(2) Transferases.   (5) Isomerases.
(3) Hydrolases.   (6) Ligases.

(1) Oxido-reductases

$$XH_2 + Y \rightleftharpoons X + YH_2$$

Dehydrogenases: the immediate acceptor, Y, is not directly oxygen except in oxidases. Usually the pair of H atoms (or electrons) passes through a series of carriers of increasing oxidation-reducton potential ultimately reaching O_2. Carriers include nicotinamide adenine dinucleotide (NAD^+) and the corresponding phosphorylated compound ($NADP^+$), flavine mononucleotide (FMN) and flavine adenine dinucleotide (FAD), and the cytochromes which are proteins with haem prosthetic groups in which the iron can be reversibly oxidized and reduced.

Vitamin	**Coenzyme**	**Reaction**

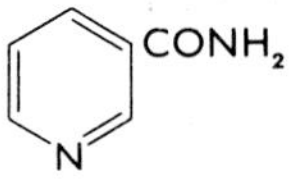

Nicotinamide

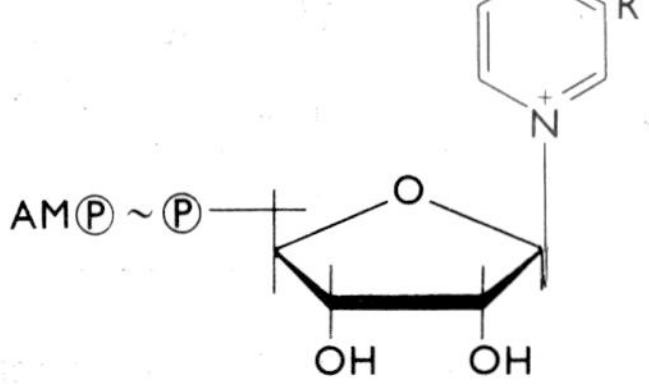

Nicotinamide adenine dinucleotide (NAD^+) ($NADP^+$ has additional phosphate on AM(P))

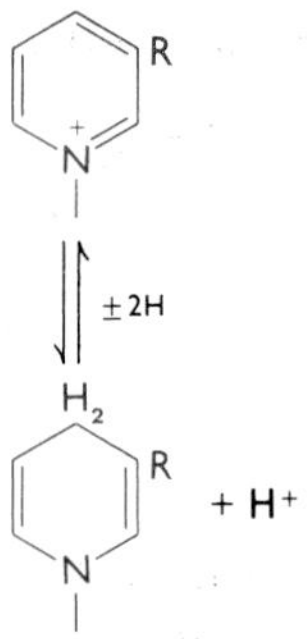

$$NAD^+ \xrightleftharpoons[\pm 2H]{} NADH + H^+$$

$$NADP^+ \xrightleftharpoons[\pm 2H]{} NADPH + H^+$$

Vitamin	**Prosthetic group**	**Reaction**

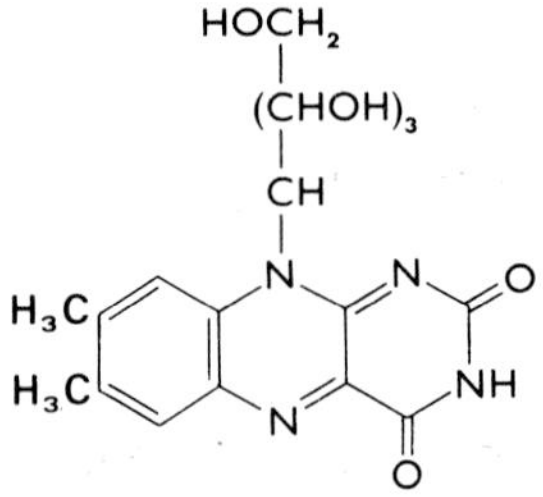

Riboflavin

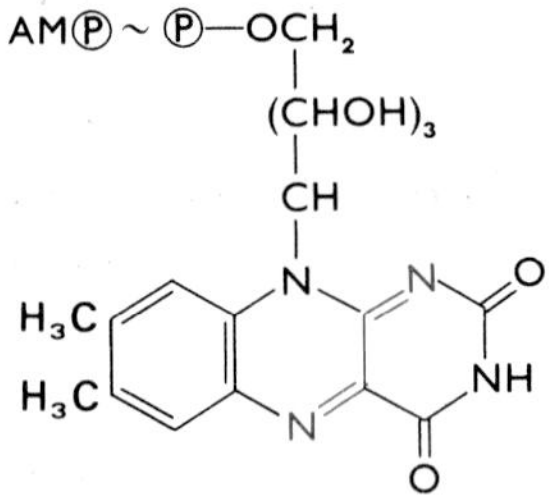

Flavin adenine dinucleotide (FAD) (Flavin mononucleotide (FMN) lacks the AM(P))

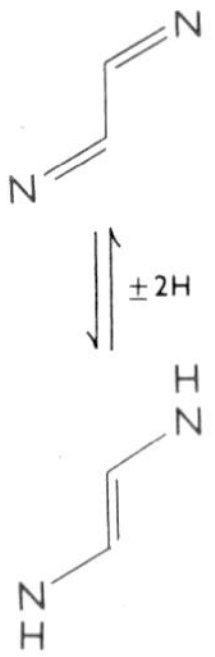

$$FAD \xrightleftharpoons[\pm 2H]{} FADH_2$$

$$FMN \xrightleftharpoons[\pm 2H]{} FMNH_2$$

Carrier	**Reaction**

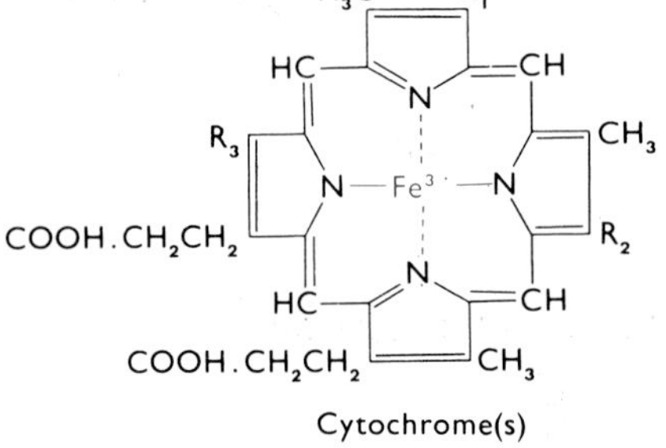

Cytochrome(s)

$$Fe^{3+} \xrightleftharpoons[\pm e^-]{} Fe^{2+}$$

A common sequence is for the pairs of H's to pass from substrate to NAD^+, to FAD, through several cytochromes (as electron transport) and finally to reduce oxygen to water:

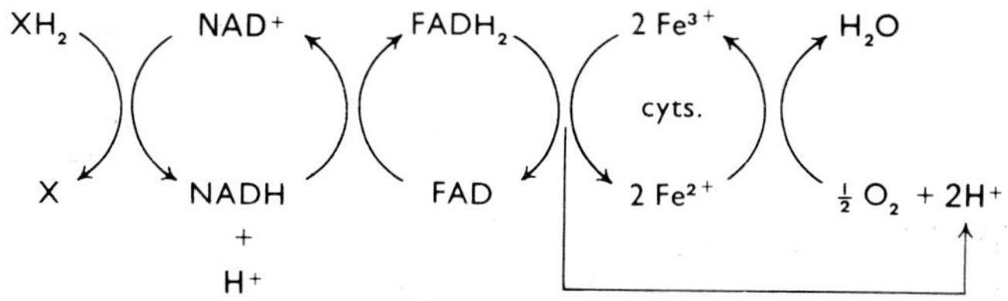

$$
\begin{array}{ccccc}
XH_2 & NAD^+ & FADH_2 & 2\,Fe^{3+} & H_2O \\
& & & \text{cyts.} & \\
X & NADH & FAD & 2\,Fe^{2+} & \tfrac{1}{2}\,O_2 + 2H^+ \\
& + & & & \\
& H^+ & & &
\end{array}
$$

The overall energy of the oxidation $XH_2 + \tfrac{1}{2}O_2 \rightarrow X + H_2O$ is thus released in several stages and coupled reactions can make use of it to form ATP. In the step-wise oxidation of NADH by oxygen 3 moles of ATP may be made:

$$H^+ + NADH + \tfrac{1}{2}O_2 + 3\,ADP + 3\,P_i \longrightarrow NAD^+ + H_2O + 3\,ATP$$

This is referred to as *oxidative phosphorylation* and is the principal way in which free energy of chemical reactions is converted into a form utilizable for biological purposes such as chemical synthesis, mechanical work (movement), light production, transport against a concentration gradient, etc.

Examples of dehydrogenases:

Lactic

$$CH_3CHOH.COOH + NAD^+ \rightleftharpoons CH_3CO.COOH + NADH + H^+$$
Lactic acid   Pyruvic acid

Alcohol

$$CH_3CH_2OH + NAD^+ \rightleftharpoons CH_3CHO + NADH + H^+$$
Ethanol   Acetaldehyde

Succinic

$$COOH.CH_2CH_2COOH + FAD \rightleftharpoons COOH.CH{=}CH.COOH + FADH_2$$
Succinic acid   Fumaric acid

N.B. NAD^+ was earlier called DPN or CoI and $NADP^+$ was TPN or CoII.

Lipoic acid

Lipoic acid (thioctic acid) may also act as a H-carrier (and as an acyl carrier):

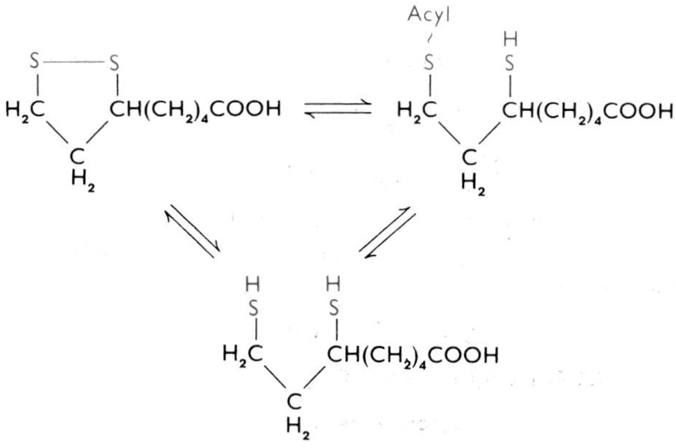

e.g. Pyruvic dehydrogenase (and α-oxoglutaric dehydrogenase):

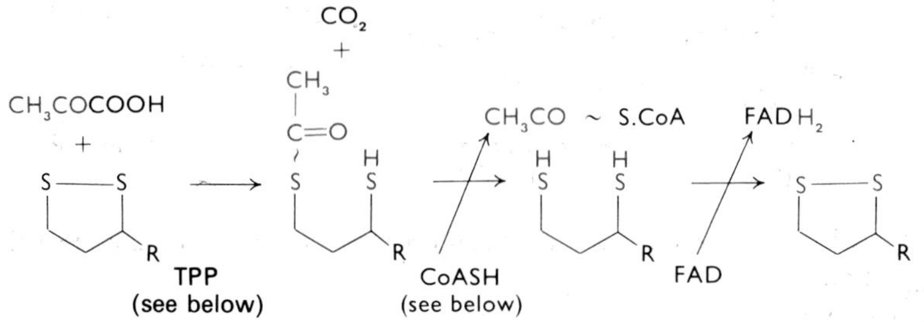

(2) Transferases

$$XR + Y \rightleftharpoons YR + X$$

where R is acyl, amino, formyl, hydroxymethyl, methyl, glycosyl, or phosphate.

Transacetylases, transaminases, transmethylases, transglycosidases and kinases come under this heading. The group transferred is frequently carried by an appropriate co-enzyme.

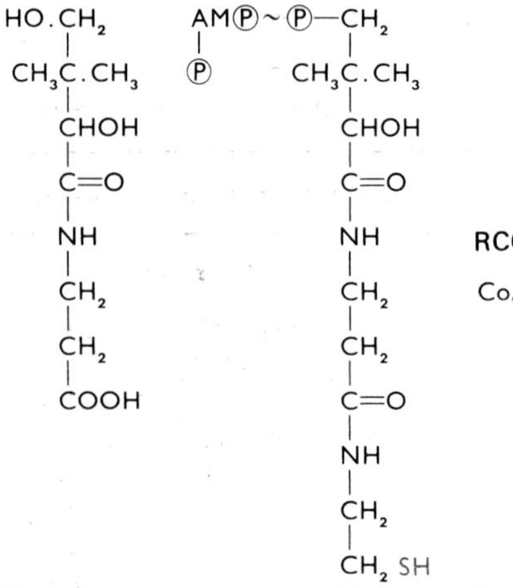

Examples of transferases:
Phosphotransacetylase

$$CH_3CO \sim S.CoA + Pi \rightleftharpoons CH_3CO \sim \text{℗} + CoA.SH$$

Acetyl CoA   Acetyl phosphate

Transketolase (needs thiamine pyrophosphate as coenzyme; see below, p. 413).

$$R.CHOH.CO.CH_2OH + R'.CHO \rightleftharpoons R.CHO + R'.CHOH.CO.CH_2OH$$

Ketose   Aldose   Aldose   Ketose

Vitamin   **Coenzyme**   **Reaction**

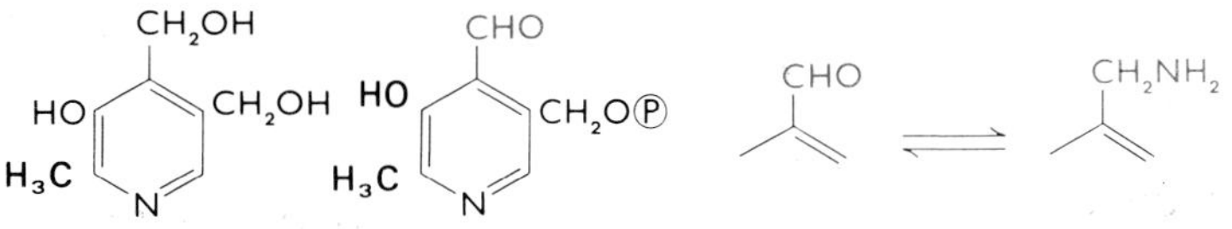

Pyridoxin   Pyridoxal phosphate   P-al P P-amine P
(P-in)   (P-al P)

Pyridoxal phosphate is the coenzyme in a variety of other reactions of amino acids (see below) as well as those involving transamination.

Example of transamination:

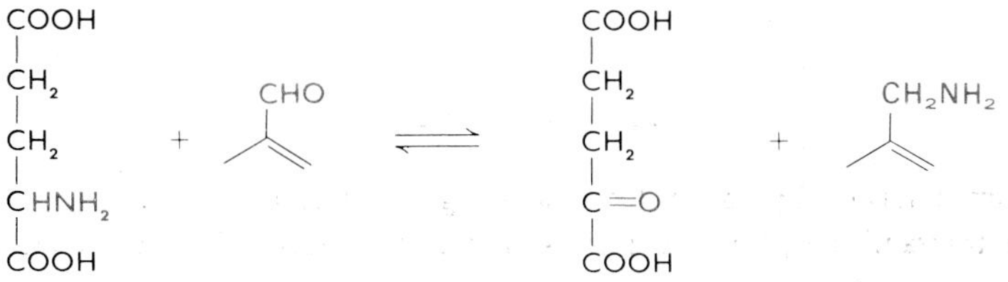

Glutamic acid   P-al P α-Oxoglutaric acid   P-amine P

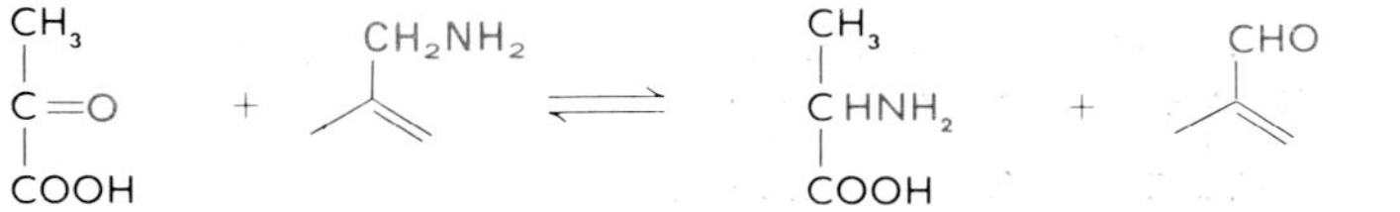

Pyruvic acid   P-amine P Alanine   P-al P

Glutamic acid + Pyruvic acid $\rightleftharpoons$ α-Oxoglutaric acid + Alanine

Vitamin   **Coenzyme**

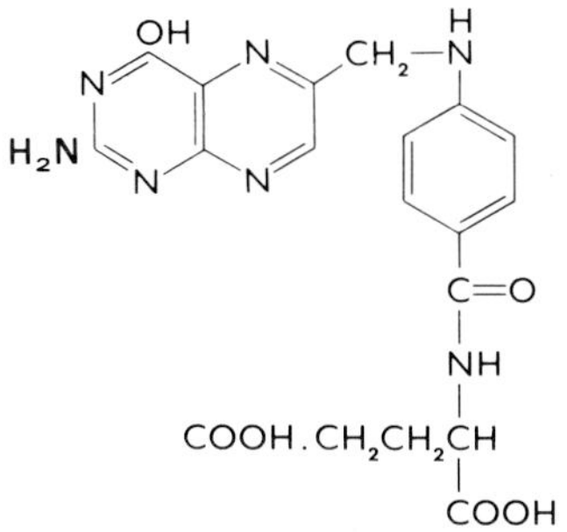

Folic acid 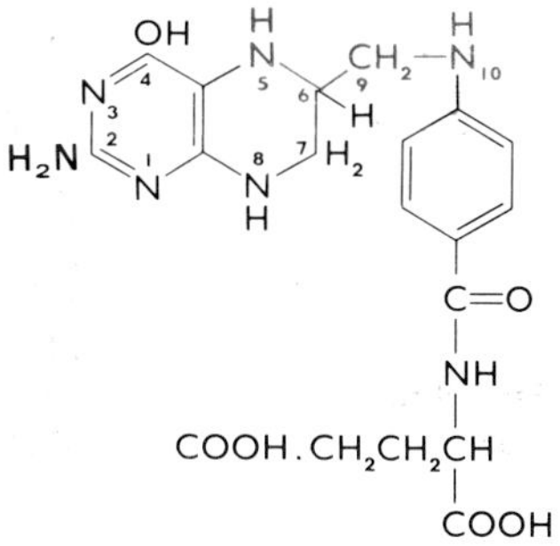   5,6,7,8 tetrahydro-folic acid (FH₄)
Coenzyme F (CoF)

Reaction

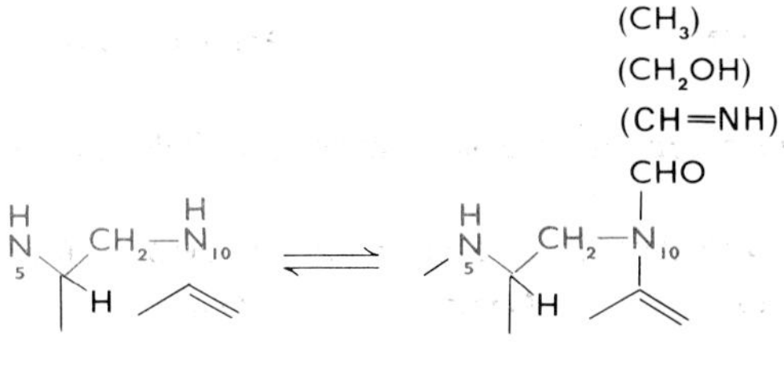

(CH₃)
(CH₂OH)
(CH=NH)
CHO

Coenzyme $F(FH_4)$ carries 1-C fragments—formyl, formimino, hydroxymethyl and methyl. Frequently this is on the N^{10} position but it can also be on N^5 or, after dehydration, as a ring between N^5 and N^{10}.

Examples of 1-C transfer:

$$H_2O + CH_2NH_2COOH + N^5,N^{10}\text{-methylene-}FH_4 \rightleftharpoons FH_4 + CH_2OH.CHNH_2COOH$$

Glycine   Serine

$$\begin{cases} \text{Vitamin } B_{12} + N^5\text{-methyl-}FH_4 \longrightarrow \text{Methyl-}B_{12} + FH_4 \\ \text{Methyl-}B_{12} + \text{Homocysteine} \longrightarrow \text{Methionine} + B_{12} \end{cases}$$

$$5'\text{-Phosphoribosyl-glycinamide} + N^5, N^{10}\text{-methenyl-}FH_4 \longrightarrow$$
$$5'\text{-phosphoribosyl-}N'\text{-formyl-glycinamide} + FH_4$$

Carrier	Reaction
ATP	$ATP + Y \longrightarrow ADP + Y\,\textcircled{P}$

Kinases transfer phosphate from a nucleoside di- or tri-phosphate to an acceptor.

Example:
Hexokinase

$$\text{Glucose} + ATP \longrightarrow \text{Glucose-6-phosphate} + ADP$$

(3) Hydrolases

$$X{-}Y + HOH \longrightarrow X{-}H + Y{-}OH$$

Peptidases, glycosidases, esterases, phosphatases, cause hydrolysis of the corresponding compounds. Coenzymes are *not* needed.

Examples of hydrolases:
Dipeptidase

$$H_2N.CH.CONH.CH.COOH + HOH \longrightarrow H_2N.CH.COOH + H\,NH.CH.COOH$$
$$\quad\; R_1 \qquad\quad R_2 \qquad\qquad\qquad\qquad R_1 \qquad\qquad\quad R_2$$

β-Galactosidase

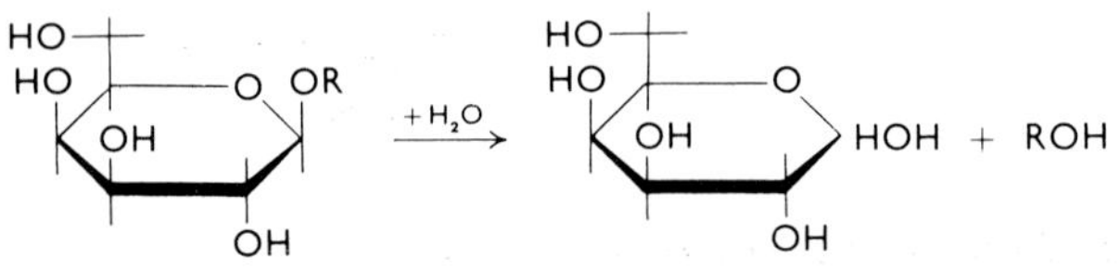

β-Galactoside   Galactose   Alcohol

Lipase

$$\begin{array}{ll} CH_2OCO.R_1 & CH_2OH + R_1COOH \\ | & \\ CHOCO.R_2 \xrightarrow{+3\,H_2O} & CHOH + R_2COOH \\ | & \\ CH_2OCO.R_3 & CH_2OH + R_3COOH \end{array}$$

Glyceryl tri-ester   Glycerol   Fatty acids

Glucose-6-phosphate phosphatase

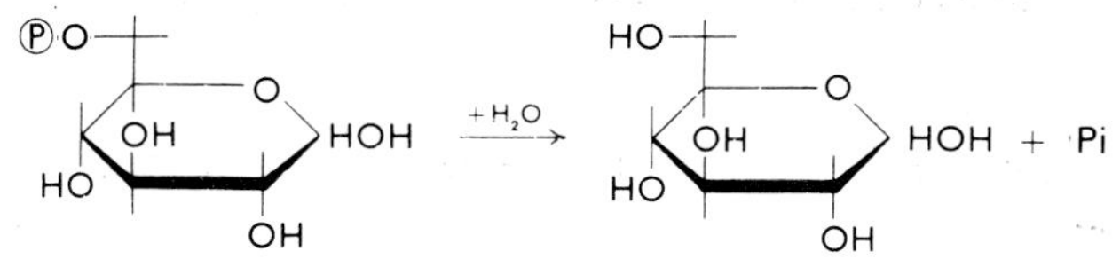

Glucose-6-phosphate   Glucose   Inorganic phosphate

(4) Lyases

$$X-Y \longrightarrow X + Y$$

Some decarboxylases, deaminases, aldolase

Vitamin   **Coenzyme**

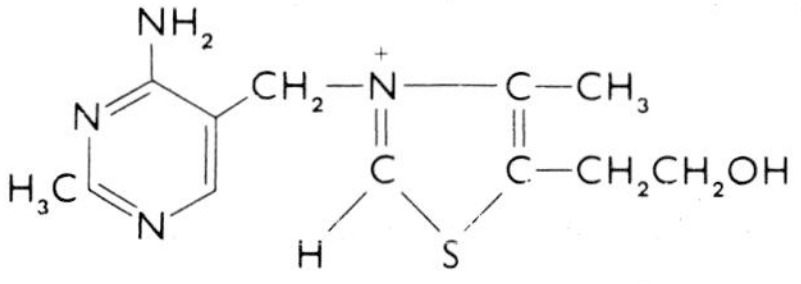

Thiamine   Thiamine pyrophosphate (TPP)

Reaction

$$R.CO.COOH + TPP$$

$$\downarrow$$

$$(R.CHO.TPP) +$$

Some decarboxylases use thiamine pyrophosphate as coenzyme
   Examples:
Pyruvate decarboxylase ('carboxylase')

$$CH_3CO.COOH + TPP \longrightarrow (CH_3CHO.TPP) + CO_2$$

    Pyruvic acid

$$\downarrow$$

$$CH_3CHO + TPP$$

    Acetaldehyde

Some deaminases use pyridoxal phosphate as coenzyme (see p. 411)
   Examples:
Serine dehydrase

$$CH_2OH.CHNH_2COOH \longrightarrow CH_2{=}C(NH_2)COOH + H_2O$$

    Serine

$$\downarrow$$

$$CH_3CO.COOH + NH_3 \xleftarrow{+H_2O} CH_3C({=}NH)COOH$$

    Pyruvic acid

Cysteine desulphurase

$$CH_2SH.C\ NH_2COOH \longrightarrow CH_2{=}C(NH_2)COOH + H_2S$$

Cysteine

$$CH_3CO.COOH + NH_3 \xleftarrow{+H_2O} CH_3C({=}NH_2)COOH$$

Pyruvic acid

Aldolase does not need a coenzyme, nor does citrate synthase

Examples:

Aldolase

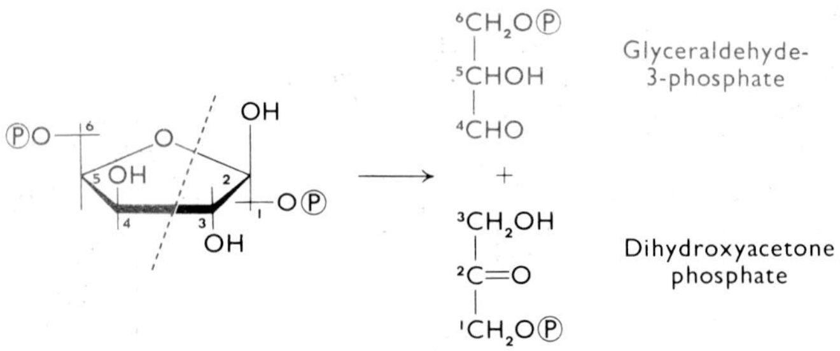

Glyceraldehyde-
3-phosphate

$^6CH_2O\text{\textcircled{P}}$
|
^5CHOH
|
^4CHO

+

$^3CH_2OH$
|
$^2C{=}O$ Dihydroxyacetone
| phosphate
$^1CH_2O\text{\textcircled{P}}$

Citrate synthase
(Condensing enzyme)

$$CH_3CO \sim S.CoA + \underset{\substack{|\\ CO.COOH}}{CH_2COOH} \rightleftharpoons \underset{\substack{|\\ HO.C.COOH\\ |\\ CH_2COOH}}{CH_2COOH} + CoA.SH$$

$+ HOH$

Acetyl CoA   Oxaloacetic   Citric acid
                         acid

(5) Isomerases

$$X.R{-}Y.S \rightleftharpoons X.S{-}Y.R$$

Examples:

Triosephosphate isomerase

$$\underset{\substack{|\\ CHOH\\ |\\ CHO}}{CH_2O\text{\textcircled{P}}} \rightleftharpoons \underset{\substack{|\\ C{=}O\\ |\\ CH_2OH}}{CH_2O\text{\textcircled{P}}}$$

Glyceraldehyde   Dihydroxyacetone
3-phosphate   phosphate

D, L-Alanine racemase

$$\text{D-Alanine} \underset{\xleftarrow{\hspace{2cm}}}{\overset{\text{Pyridoxal phosphate}}{\rightleftharpoons}} \text{L-Alanine}$$

Methylaspartate and methylmalonyl-CoA mutases require a vitamine B_{12} derivative as coenzyme.

Vitamin

CH₂OH

O
P

CH₃—CH

CH₂

NH

CO

CH₂

CH₂

CH₃

C

CH₃

H₃C

CH₃

N

Co

N

R

CH₂CH₂CONH₂

CONH₂

CH₃

CH₃

CONH₂

CH₂

CH₂

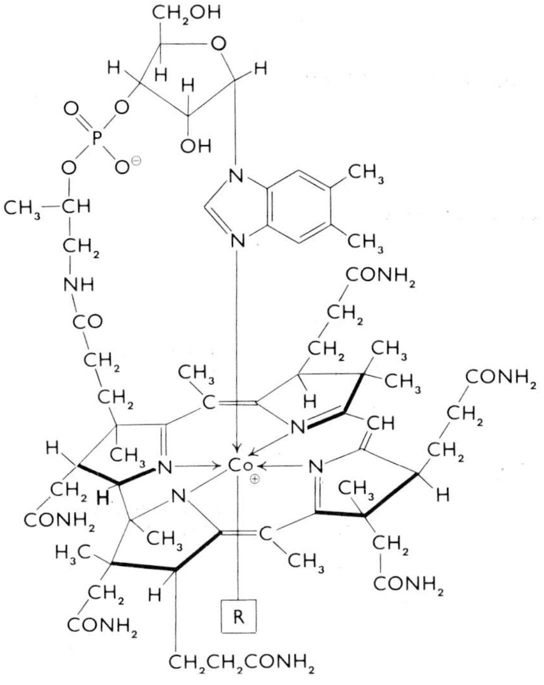

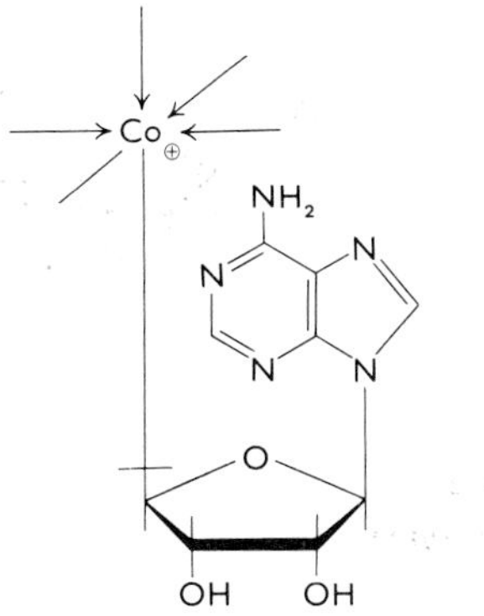

CH₃

CH₂

CONH₂

In vitamin B_{12} as isolated R = CN but
this may be an artefact.

Coenzyme

Co

NH₂

N

O

Adenosine

OH   OH

In coenzyme, R = adenosine but
other derivatives are active.

Examples of isomerases using B_{12} coenzymes:

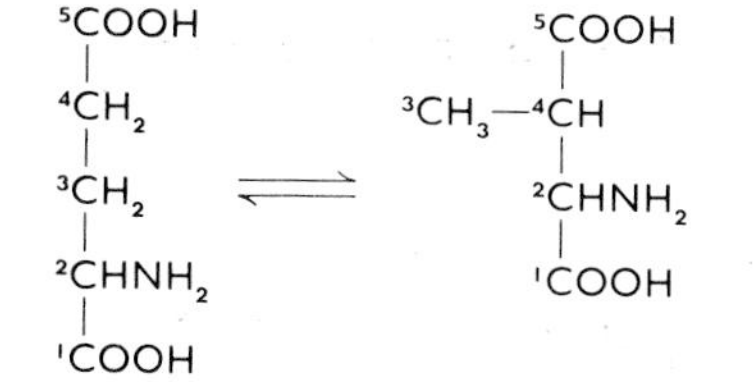

$$
\begin{array}{ccc}
^5\text{COOH} & & ^5\text{COOH} \\
| & & | \\
^4\text{CH}_2 & & ^3\text{CH}_3\text{—}^4\text{CH} \\
| & \rightleftharpoons & | \\
^3\text{CH}_2 & & ^2\text{CHNH}_2 \\
| & & | \\
^2\text{CHNH}_2 & & ^1\text{COOH} \\
| & & \\
^1\text{COOH} & &
\end{array}
$$

Glutamic acid   β-Methyl aspartic acid

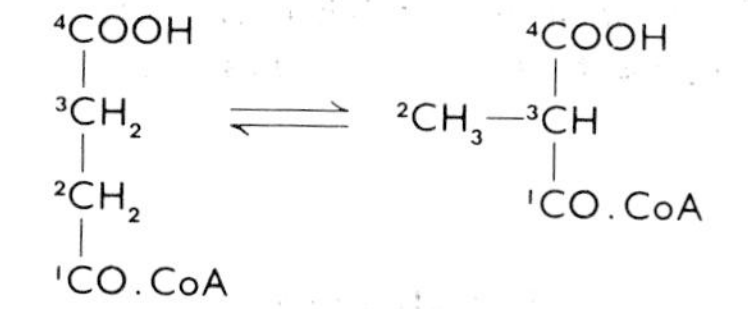

$$
\begin{array}{ccc}
^4\text{COOH} & & ^4\text{COOH} \\
| & & | \\
^3\text{CH}_2 & \rightleftharpoons & ^2\text{CH}_3\text{—}^3\text{CH} \\
| & & | \\
^2\text{CH}_2 & & ^1\text{CO.CoA} \\
| & & \\
^1\text{CO.CoA} & &
\end{array}
$$

Succinyl-CoA   Methylmalonyl-CoA

(6) Ligases (synthases)

$$X + Y + ATP \longrightarrow X—Y + \begin{array}{c} AMP + PPi \\ \textit{or} \\ ADP + Pi \end{array}$$

The mechanism may be as follows:

$$X + ATP \longrightarrow X.AMP + PPi$$

$$X.AMP + Y \longrightarrow X—Y + AMP$$

Frequently the inorganic pyrophosphate is rapidly hydrolysed thus establishing a reaction with the equilibrium position to the right.

Example:

Amino acid activating enzymes

$$\underset{\substack{| \\ R \\ \text{Amino acid}}}{H_2N.CH.COOH} + ATP \longrightarrow \underset{\substack{| \\ R \\ \text{Amino acyl-AMP}}}{H_2N.CH.CO} \sim AMP + PPi$$

$$\underset{\substack{| \\ R \\ \text{Amino acyl-AMP}}}{H_2N.CH.CO} \sim AMP + tRNA \longrightarrow \underset{\substack{| \\ R \\ \text{Amino acyl-tRNA}}}{H_2N.CH.CO}—tRNA + AMP$$

$$\underset{\substack{| \\ R}}{H_2N.CH.COOH} + tRNA + ATP \longrightarrow \underset{\substack{| \\ R}}{H_2N.CH.CO}—tRNA + AMP + PPi$$

Some carboxylations occur by means of a ligase with biotin acting as carrier.

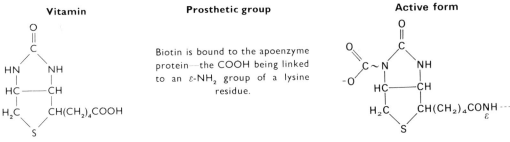

Vitamin	**Prosthetic group**	**Active form**
	Biotin is bound to the apoenzyme protein—the COOH being linked to an ε-NH$_2$ group of a lysine residue.	

Example:

Pyruvate Carboxylase

$$CO_2 + ATP + Biotin—Enz \rightleftharpoons Carboxybiotin—Enz + ADP + Pi$$

$$Carboxybiotin—Enz + \underset{\text{Pyruvic acid}}{CH_3CO.COOH} \rightleftharpoons \underset{\text{Oxaloacetic acid}}{COOH.CH_2.CO.COOH} + Biotin—Enz$$

$$CO_2 + Pyruvic acid + ATP \rightleftharpoons Oxaloacetic acid + ADP + Pi$$

BACTERIAL GROWTH FACTORS

The following members of the B group of vitamins play roles in enzymic reactions (see above) and if a bacterium is unable to synthesize any of them, they must be provided in the growth medium unless all possible products of the enzymic reactions are supplied.

> Thiamine
> Nicotinamide
> Riboflavin
> Pyridoxin
> Pathothenic acid
> Folic acid
> Biotin
> Vitamin B_{12}
> Lipoic acid

Some bacteria are unable to synthesize certain amino acids or purines, etc. and may require them. These also must be supplied as *growth factors*.

FURTHER READING

Dixon M., Webb E. C., Thorne C.J.R. and Tipton K.F. (1979) *Enzymes*, 3rd edition. Longmans, London.

Appendix C
Glossary

Abortive transduction. Transduction in which the donor DNA is not integrated with the recipient chromosome (as in *complete* transduction) but persists as a non-replicating fragment.

Allele. A form of a gene. Two genes which are derived from the same gene by mutation, and which are alternative occupants of the same chromosomal locus, are described as *allelic* with respect to each other.

Allosteric effect. A change produced in the properties of an enzyme by the specific action of a small molecule acting at a site other than the active site.

Antibiotic. Substance produced by an organism which is toxic to one or more other types of organism.

Antibody. A protein produced in an animal when a substance normally foreign to its tissues gains access to them: the antibody combines chemically with the foreign substance (antigen).

Anti-codon. A sequence of three nucleotides (in an amino acid transfer-RNA) complementary to the codon triplet in a messenger-RNA.

Antigen. A substance capable of stimulating production of an antibody. Each antigen, when injected into a suitable vertebrate evokes the synthesis of a specific antibody in the animal.

Auxotroph (auxotrophic mutant). A mutant strain differing from the normal or wild type of the organism in having an additional nutritional requirement.

Bacteriophage. A virus whose host is a bacterium.

Capsule. A surface component of the bacterial cell lying outside the cell wall.

Catabolite repression. A form of general repression, produced by glucose or other 'rich' substrates, of the synthesis of enzymes associated with the breakdown of alternative carbon and energy substrates.

Chromatophore. A structural component of photosynthetic bacteria surrounded by a unit membrane and containing chlorophyll.

Chromosome. A structure carrying genes in an ordered array. In bacteria and bacteriophage, a single large circular molecule of DNA without evident joins.

Cis. The arrangement of the alleles of two genes in a diploid (or partial diploid) in which the recessive alleles of the two genes are on the same chromosome or chromosome fragment from one parent, and two corresponding dominant alleles are present together on that from the other parent (cf. *trans*).

Cistron. A segment of chromosome which determines a single phenotypic function in a *cis-trans* complementation test. Thus, most pairs of mutations within the cistron give a mutant phenotype when present in a diploid in the *trans* configuration but a normal phenotype in the *cis* configuration (cf. *cis* and *trans*). In some instances certain pairs of mutations within a cistron can complelent each other (not necessarily completely) in a phenomenon known as allelic complementation.

Clone. Genetically identical cells derived by successive divisions from a single common ancestor.

Codon. A sequence of three nucleotides (in a nucleic acid) which codes for an amino acid or the end of a polypeptide chain.

Complementation. Complementary action within a diploid, or partial diploid cell, of different mutant genomes or homologous fragments of genomes. This is the general rule for recessive mutations in different genes, and failure of two recessive mutations to complement each other indicates that they occur in the same gene. In some instances alleles of the same gene can complement each other, but usually not strongly.

Conjugation. Attachment of bacterial cells, permitting the transfer of genetic material from one to another.

Co-ordinate regulation. The induction or repression of a group of enzymes in such a way that each enzyme is synthesized at a rate proportional to that of every other enzyme in the group.

Cross. An encounter between organisms which permits the formation of genetic recombinants. In bacteria, crosses can be performed by conjugation, transduction or transformation.

Deletion. Loss of a chromosome segment.

Diploid. Containing two homologous sets of chromosomes. In bacteria diploidy is usually partial, and restricted to a certain region of the chromosome.

Dominant (allele, mutation). One allele of a gene is dominant to another if its phenotype is expressed in a diploid heterozygous for both alleles.

419

Doubling time (mass or cell number). Time interval during which a population doubles in either cell mass, or cell number.

Enzyme. Biological catalyst of high specificity; protein with or without additional non-protein prosthetic group or coenzyme (see Appendix B).

Episome. Dispensable piece of genetic material endowed with the capacity of independent replication; able to multiply independently of the chromosome but capable of reversible integration into the chromosome.

Exchange (in the genetic sense). The exchange between chromosomes of homologous segments of DNA.

Exponential phase (log phase). State of growth in which population doubles regularly each mean doubling time (q.v.) i.e. increases 1, 2, 4, 8, 16, ... in equal time intervals.

Feedback inhibition. Inhibition of the activity of an enzyme specifically caused by a small molecule which is usually the end-product of a biosynthetic pathway.

Fimbria. See *pilus*.

Flagellum. The organelle of locomotion. In bacteria, thread-like structures composed of protein subunits of flagellin. The diameter varies between 12 nm and 30 nm.

Gene. Originally defined as the genetic unit which determines the development of a particular character; now usually considered as the length of DNA transcribable into a length of RNA of discrete function, i.e. a specific transfer-RNA or ribosomal RNA or a messenger specifying a specific polypeptide chain. This definition does not include control regions on DNA such as operators or promoters, which some authors describe as genes.

Genetic code. The relationship between the sequence of nucleotides in DNA or RNA which specifies a sequence of amino acids. The code is 'triplet' in the sense that a run of three nucleotides codes for each amino acid.

Genome. A complete single set of genetic material—i.e. a complete set of genes including the non-transcribed regions between them.

Genotype. The genetic constitution, or assembly of genes.

Haploid. Possessing a single genome.

Heterozygote. A cell carrying two different alleles of the same gene (cf. *homozygote*).

Homozygote. A cell carrying two identical copies of a gene (cf. *heterozygote*).

Induction. An increase in the rate of synthesis of an enzyme, specifically caused by a small molecule

which is generally the substrate or a compound closely related to it.

L-form. Some protoplasmic elements without defined morphology which no longer possess a rigid bacterial form.

Linkage. The tendency of genetic markers which are relatively close together on the chromosome to be transmitted together through crosses and not separated by recombination.

Log phase. See *exponential phase*.

Lysogeny. The harbouring by a (lysogenic) bacterial strain of a temperate bacteriophage, capable of uncontrolled multiplication and causing lysis in response to an inducing treatment or, occasionally, spontaneously.

Marker. A genetic mutation with a distinctive observable effect on the organism, serving to *mark* the chromosome locus at which it occurs, thus enabling the transmission of the locus through cell divisions and genetic crosses to be followed.

Mean generation time. The average of the generation times of the viable cells in a population. Usually a close approximation to, and often confused with, the doubling time (q.v.).

Merozygote. A zygote in which there is one complete genome and a fragment of a second genome.

Mesosome. A membranous involution of the protoplast membrane.

Messenger-RNA (mRNA). The RNA which specifies the amino acid sequence for a particular polypeptide chain.

Minimal medium. Medium containing the minimal nutrients required by the wild-type of an organism—incapable of supporting growth of auxotrophs.

Mutagen. A substance capable of inducing mutation.

Mutant. A cell or clone of cells carrying a mutation.

Mutation. A change in nucleic acid producing a stably inherited change in the properties of the organism.

Operator. A region in the chromosome capable of interacting with a specific repressor (or activator) to control the transcription of the operon contiguous with it.

Operon. A group of adjacent genes whose transcription is regulated from a single operator region.

Phenotype. The observable characteristics of an organism. The phenotype does not necessarily reflect the complete genetic constitution or genotype, since the expression of some genes may be masked for one reason or another.

Phosphatide. A fat consisting of glycerol and one or two fatty acid residues linked to a nitrogenous base through a phosphoric acid residue. (A lipid containing phosphoric acid and a nitrogenous base).

Plasmid. Chromosomal element replicating independently of the bacterial chromosome, and transmitted to the progeny of the cell.

Pilus (pili). Surface appendage of certain Gram-negative bacteria, composed of protein subunits and approximately 7 nm in diameter. Pili are shorter and thinner than flagella.

Polycistronic messenger. A length of messenger-RNA containing the information for the synthesis of more than one polypeptide chain.

Polynucleotide. Polymer consisting of nucleotides linked by phosphodiester links between positions 3′ and 5′ of successive sugar units:

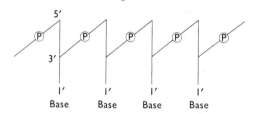

Polypeptide. Polymer consisting of amino acids linked by condensation of the carboxyl group of one with the amino group of the next:

$$NH_2CH.COOH + NH_2CH.COOH + etc.$$
$$\quad\; |R_a \qquad\qquad\quad |R_b$$

$$— \to NH_2CH.CONH.CH.CO\ldots etc.$$
$$\qquad\qquad |R_a \qquad\quad |R_b$$

Polyribosome (polysome, ergosome). Several ribosomes (q.v.) associated with the same strand of messenger-RNA.

Promoter. A region in the chromosome at which RNA polymerase binds to initiate transcription of a gene or an operon.

Prophage. Temperate bacteriophage in its latent state in the host cell.

Protein. Large molecule formed from one or more polypeptide chains and having 'native' secondary, tertiary and sometimes, quarternary structure. Many have non-protein prosthetic groups.

Protoplast. A structure derived from a vegetative cell by removal of the entire cell wall.

Prototroph. A strain of bacteria having the minimal nutritional requirements of the wild-type organism (cf. *auxotroph*).

Recessive (allele, mutation). An allele of a gene is recessive to another if the phenotype of the other is the one that is expressed in a heterozygous diploid.

Recombinant. A cell or clone of cells resulting from genetic recombination.

Recombination. The formation of a new genotype by reassortment of genes following a genetic cross (see also *exchange*).

Replica plating. Replication of a pattern of colonies from one plate to another: a disc of sterile material (often velveteen) is pressed on the surface of the first plate, and the adhering bacteria are printed on the second.

Repression. A decrease in the rate of synthesis of an enzyme specifically caused by a small molecule which is usually the end-product of a biosynthetic pathway (e.g. an amino acid, nucleotide, etc.).

Restriction. A process whereby a bacterial cell or a lysogen (q.v.) prevents the entry or development of an infecting bacteriophage or the infection of the cell by foreign nucleic acids.

Restriction endonuclease. An enzyme produced by a bacterial cell and capable of recognizing a specific sequence of base pairs in double-stranded DNA. Both strands of the DNA are cleaved at these sequences to generate specific fragments of defined length.

Reversion (genetic). A mutation in an already mutant gene which restores in part or in full the original function of that gene.

Ribosomal-RNA (rRNA). The RNA of the ribosome particles as distinct from mRNA and tRNA. 70 *S* ribosomes yield one molecule each of 5 *S* rRNA, 16 *S* rRNA and 23 *S* rRNA coming from the two sub-units, 30 *S* and 50 *S*.

Ribosome. Particle composed of protein and RNA to which messenger-RNA and amino acyl-tRNA are attached during synthesis of polypeptide chains. Usually designated by sedimentation coefficient, e.g. 70 *S*. These functional ribosomes can dissociate 70 *S* ⇌ 30 *S* + 50 *S*.

Segregation. The separation into daughter cells of two distinct forms (or alleles) of the same gene, originally present in the same cell.

Shift-down. The transfer of a bacterial culture to a poorer medium, i.e. one supporting a lower growth rate.

Shift-up. The transfer of a bacterial culture to a richer medium, i.e. one supporting a higher rate of growth.

Spheroplast. A form of a bacterium in which the cell wall structure has been modified but in which typical cell wall components are still present.

Suppressor mutation. A mutation which masks the effect of another mutation elsewhere in the genome.

Temperate bacteriophage. A bacteriophage whose genome can be integrated into the host cell genome.

In this prophage (q.v.) state it is replicated with the host chromosome and is transmitted to all of the progeny of the host without necessarily causing lysis.

Trans. The arrangement in a diploid (or partially diploid) cell of two mutations on different chromosomes, or chromosome fragments, with each one linked to the non-mutant homologue of the other (c.f. *cis*).

Transcription. The synthesis of RNA on a DNA template such that the bases in the product are complementary to those in the template.

Transduction. Transfer of a fragment of the genome from a donor to a recipient strain of bacteria by infecting the recipient with bacteriophage particles grown on the donor strain.

Transfer-RNA (tRNA). Molecule of RNA able to combine with a specific amino acid which becomes esterified to the terminal adenosine. Each species of tRNA has a specific trinucleotide sequence which interacts with a complementary sequence in mRNA. Also called soluble-RNA (sRNA) and acceptor-RNA.

Transformation. Transfer of a fragment of the genome from a donor to a recipient strain of bacteria by treating recipient cells with DNA isolated from the donor.

Transition. A mutation consisting of a change in one base pair of the DNA, a different purine being substituted for the purine and a different pyrimidine for the pyrimidine.

Translation. The synthesis of polypeptide whose amino acid sequence is specified by successive triplets of nucleotides in messenger-RNA.

Translocation. A rearrangement of the order of genes on a chromosome involving the moving of one segment of the chromosome to a new location.

Transposon. A region of DNA bounded at both ends by a specific sequence of nucleotides known as repeat sequences. This element of DNA can be transposed to new locations on phage, plasmid or host chromosomes, becoming integrated at repeat (or insertion) sequences. This transposition is independent of the normal recombination mechanism.

Transversion. A mutation consisting of a change in one base pair of the DNA, a purine being substituted for the pyrimidine and a pyrimidine for the purine.

Zygote. The cell which is the immediate product of a cross—at least partly diploid.

Index

Note: a page number may refer to the text, to a figure, or to both.